Scramble for the Skies

Scramble for the Skies

The Great Power Competition to Control the Resources of Outer Space

Namrata Goswami and Peter A. Garretson

LEXINGTON BOOKS
Lanham • Boulder • New York • London

Published by Lexington Books
An imprint of The Rowman & Littlefield Publishing Group, Inc.
4501 Forbes Boulevard, Suite 200, Lanham, Maryland 20706
www.rowman.com

6 Tinworth Street, London SE11 5AL, United Kingdom

British Library Cataloguing in Publication Information Available

Library of Congress Control Number: 2020943851

ISBN: 978-1-4985-8311-4 (cloth)
ISBN: 978-1-4985-8312-1 (electronic)

This book is dedicated to our parents, who taught us to shoot for the stars.

To Tarun Chandra Goswami, who always reminded me to gaze at the night sky against the backdrop of the majestic beauty of the Barail mountain ranges in Northeast India. To Era Goswami, who taught me to be curious and kind, and the spirit of Vasudhaiva Kutumbakam *(the world is one family).*

To Geraldine Garretson, who read me books on mysterious black holes and astronomy, sparking a lifetime fascination for the Universe and our moral purpose within it. To Peter Walter Garretson, on whose lap I watched Star Wars, Star Trek, and who first taught me how cool space could be while launching model rockets.

Contents

List of Tables and Figures

TABLES

FIGURES

Abbreviations

AAS	Austrian Academy of Sciences
AEC	Atomic Energy Commission (India)
AFRL	Air Force Research Lab (United States)
AIAA	American Institute of Aeronautics and Astronautics
ARM	Asteroid Redirect Mission (NASA)
ARPA	Advanced Research Projects Agency (United States)
ASAT	Anti-satellite
ASD	Alliance for Space Development
ASEAN	Association of South East Asian Nations
ATM	Automated Teller Machine
BCE	Before Common Era
BLSS	Bioregenerative Life Support System
BRI	Belt and Road Initiative
CAE	Chinese Academy of Engineering
CALT	China Academy of Launch Vehicle Technology
CAPS	Centre for Airpower Studies (India)
CAS	Chinese Academy of Sciences
CASC	China Aerospace Science and Technology Corporation
CCAMLR	Convention for the Conservation of Antarctic Marine Living Resources
CCDEV	Commercial Crew Development (NASA)
CCIRICMI	Chongqing Collaborative Innovation Research Institute for Civil-Military Integration (China)
CEIP	Carnegie Endowment for International Peace
CEO	Chief Executive Officer
CLEP	China's Lunar Exploration Program
CLOCS	Celestial Lines of Communication

CLPS	Commercial Lunar Payload Service (United States)
CLTC	China Satellite Launch and Tracking Control General
CMC	Central Military Commission (China)
CNAS	Center for New American Security
CNSA	China National Space Administration
CO2	Carbon dioxide
CoC	Code of Conduct
COMM	Communications
COO	Chief Operating Officer
COPUOS	Committee on the Peaceful Uses of Outer Space (United Nations)
COTS	Commercial Orbital Transportation System
CPC	Communist Party of China
CSLCA	U.S. Commercial Space Launch Competitiveness Act
CSR	Committee for Space Research (India)
CSSAR	Centre for Space Science and Applied Research (India)
DAE	Department of Atomic Energy (India)
DIPAC	Defence Imagery Processing and Analysis Centre (India)
DIU	Defense Innovation Unit (United States)
DN-3	*Dong Neng*-3 (China)
DOD	Department of Defense (United States)
DOE	Department of Energy (United States)
DRDO	Defense Research and Development Organisation (India)
DSA	Defence Space Agency (India)
DSI	Deep Space Industries
ERIN	Environmental Research and Innovation
ECS	East China Sea
EEC	European Economic Community
EIU	Economist Intelligence Unit
EU	European Union
FAA	Federal Aviation Administration (United States)
FCC	Federal Communication Commission (United States)
FFRDC	Federally Funded Research and Development Center (United States)
GAD	General Armament Division (China)
GDP	Gross Domestic Product
GEO	Geosynchronous Orbit
GLAE	Groupement Luxembourgeois de l'Aéronautique et de l'Espace (Luxembourg)
GPS	Global Positioning System (United States)
GSLV	Geosynchronous Satellite Launch Vehicle (India)
HEOMD	Human Exploration Mission Directorate (NASA)

HDI	Human Development Index
H.R.	House Resolution (United States)
HSRWG	Hague Space Resources Working Group
IAC	International Astronautical Congress
IAF	Indian Air Force
ICBM	Inter Continental Ballistic Missiles
IDIQ	Indefinite Delivery Indefinite Quantity
IDSA	Institute for Defence Studies and Analyses (India)
IGO	International Governmental Organizations
IISL	International Institute of Space Law
IISS	International Institute of Strategic Studies
IIST	Indian Institute of Space Technology
IMF	International Monetary Fund
INCOSPAR	Indian National Committee for Space Research
ISDC	International Space Development Conference
ISECG	International Space Exploration Coordination Group
ISRO	Indian Space Research Organisation
ISRU	In-Situ Resource Utilization
ISU	International Space University
ITS	Interplanetary Transportation System (SpaceX)
JPL	Jet Propulsion Laboratory (United States)
KE-ASAT	Kinetic Energy Anti-Satellite
L1	Lagrange 1
L4	Lagrange 4
L5	Lagrange 5
LIST	Luxembourg Institute of Science and Technology
LEO	Low Earth Orbit
LSE	Luxembourg Stock Exchange
MBRSC	Mohammad Bin Rashid Space Center (UAE)
MEA	Ministry of External Affairs (India)
MER	Market Exchange Rates
MIIT	Ministry of Industry and Information Technology (China)
MoU	Memorandum of Understanding
MSF	Military Space Force
MTOE	Million Tons of Oil Equivalent
NASA	National Aeronautics and Space Administration (United States)
NASIC	National Air & Space Intelligence Center (United States)
NEAs	Near-Earth Asteroids
NEOs	Near-Earth Objects
NEOSM	Near-Earth Object Surveillance Mission (NASA)
NGO	Non-Governmental Organization
NPT	Nonproliferation Treaty

NRC	National Research Council (United States)
NRO	National Reconnaissance Office (United States)
NSC	National Security Council (United States)
NSpC	National Space Council (United States)
NSI	National Space Institute
NSS	National Space Society
NSSO	National Security Space Office (United States)
NSSC	National Space Science Center (China)
NTRO	National Technical Research Organization (India)
OECD	Organization for Economic Cooperation and Development
OECIF	Operational Energy Capability Improvement Fund
OST	Outer Space Treaty
OTA	Office of Technology Assessment (United States)
OTA	Other Transaction Authority
PDCO	Planetary Defense Coordination Office (NASA)
PLA	People's Liberation Army (China)
PLASSF	PLA Strategic Support Force (China)
PNT	Precision, Navigation, and Timing
PPP	Purchasing Power Parity
PRC	People's Republic of China
PRL	Physical Research Laboratory (India)
PSLV	Polar Satellite Launch Vehicle (India)
PTI	Press Trust of India
PwC	Price Waterhouse Cooper
QUESS	Quantum Experiments at Space Scale
RBA	Royal Bhutan Army
R&D	Research and Development
RLV	Reusable Launch Vehicle
ROI	Return on Investment
SASTIND	State Administration on Science, Technology, and Industry for National Defense (China)
SBSP	Space-based Solar Power
SCS	South China Sea
SDA	Space Development Agency (United States)
SDA	Space Domain Awareness.
SDI	Strategic Defense Initiative
SDSC	Space Development Steering Committee
SEOB	Strategic Earth Orbital Base
SEDS	Students for the Exploration and Development of Space
SES	Société Européenne des Satellites (Luxembourg)
SFF	Space Frontier Foundation
SIPRI	Stockholm International Peace Research Institute

SLOCS	Sea Lines of Communication
SLS	Space Launch System (NASA)
SMC	Space and Missile Center / Space Force Materiel Command (USSF)
SPD	Space Policy Directive (United States)
SPS	Solar Power Satellites
SRI	Space Resources Initiative
SSCG	Space Security Coordination Group
SSD	Space System Department
SSP	Space Solar Power
SSPIDR	Space Solar Power Incremental Demonstrations and Research
STM	Space Traffic Management
SWCNT	Single Wall Carbon Nano Tube
TES	Technology Experiment Satellite
TPCR	Technology Perspective and Capability Roadmap (India)
UAE	United Arab Emirates
UAVs	Unmanned Aerial Vehicles
UCS	Union of Concerned Scientists
UNOOSA	United Nations Office for Outer Space Affairs
U.S.	United States
USAF	United States Air Force
USAID	United States Agency for International Development
USCC	United States-China Economic and Security Review Commission
USCG	United States Coast Guard
USD	United States Dollar
USGS	United States Geological Survey
USSF	United States Space Force
USSR	Union of Soviet Socialist Republic
VSE	Vision for Space Exploration
VSSC	Vikram Sarabhai Space Centre (India)

Acknowledgments

It has been our privilege and honor to receive a research grant and support of several others along the path toward the writing of this book. First and foremost, we would like to thank the Minerva Research Grant for selecting our research proposal on contemporary great powers in space for a research grant. This grant enabled us to conduct field work in China and India, offering us deep insights into both countries' space policies, programs, elite discourse, and strategic culture. We thank Air Command and Staff College (ACSC), especially Dr. Paul Springer for supporting our grant proposal, his mentorship with the DoD Institutional Review Board, and his continuous support throughout the research.

Our first thank you is to those administering MINERVA, Erin Fitzgerald, David Montgomery without which this work would never have begun. A special thank you to those in finance and contracting Mark Honaker, Gladys Johnson, and Justin Holmly, and Thomas Taylor of the Alabama small business administration.

We are thankful for the early opportunities to showcase our work, including on the International Studies Association (ISA) annual conference panels organized by Dr. Dan Deudney whose framing on Space Expansionism inspired our use of the term. George Popp, and Hriar Cabayan who included us as a part of the Strategic Multilayer Assessment, Pat Cronin and Hudson for including us in their Asia-Pacific Space Strategy event, Joel Mozer for the opportunity to present at Air Force Space Command, and Alexander Bowe for including our views in testimony to the USCC. Joseph Parry, who solicited us at ISA to turn our research into a book.

Before we ever began travel and fieldwork, we are grateful for the generous insights, contacts, and suggestions, especially to Ken Allen, Steve Bachowski, Dr. Robert Baker Jr, David Brin, Steve Burgess, Michael Chase,

Dean Cheng, Elbridge Colby, Andrew Erickson, Richard Fisher, John Geis, Paul Giarra, Bonnie Glaser, Mike Green, Scott Vander Hamm, Guocheng Jiang, Ralph Jodice, Kenny Johnson, Phil Karber, John Lewis, Doug Loverro, Debra Luker, Oriana Mastro, Brandon Mila, Dawn Murphy, Paul Nelson, Robert Norton, Scott Pace, Kevin Pollpeter, Nilanthi Samaranayake, Joanie Simon, Peter Singer, Robert Spalding, Dave Stilwell, Mark Stokes, Carl Rehberg, Joe Rouge, Chris Stone, Paula Thornhill, Panos Yannakogeorgos, Larry Wortzel, and Mariana Wray.

Official travel is a difficult process, our thanks to those who helped us: Angelia Adams, Jonathan Artis, Eric Chan, Ashley Goodman, Eric Kiss, Timothy Kelly, Kevin Lackey, Nicole McCauley, Andrea E. Robles-Olson, Steven Perdue, Heidi Robinson, Julie Rottier, Lilibeth Taylor, SA Russell Taylor, Jeffrey Waldroop, and Martin Wesley, and the U.S. Embassy (China) and the U.S. Embassy (India).

Our field work engagements were generously supported by Colonel Terence J. Vance, Assistant Air Attaché, U.S. Embassy in Beijing, and for our meetings in Shanghai by Steven Angel, Defense Liaison Officer, U.S. Consulate General, Shanghai. We offer them our heartfelt thanks for their time and efforts.

We thank Dr. Tong Zhao, Carnegie Tsinghua, Dr. Han Hua, Peking University, Dr. Guo Xiaobing, China Institutes of International Contemporary Relations' Institute of Arms Control and Security Studies, Wu Chian, Renmin University, Zan Lu, Tsinghua University, Lui Ching Electronic, Industry Consortium, Robert Manning, Atlantic Council, Dr. Bharath Gopalaswami, Atlantic Council, Dr. Gaurav Kampani, Ben Polsky, Atlantic Council, Pralay Kanungo, Indian Council of Cultural Relations, Chair of Contemporary Indian Studies, Group Captain Ashish Srivastava, Indian Air Attaché, Indian Embassy in Beijing for some very stimulating discussions on the research topic. We are thankful to Dr. Qiu Yonghui, Chinese Academy of Social Science (CASS), Dr. Lin Minwang, Pangoal Institution, Fudan University, Dr. Wang Dong, Pangoal Institution, Dr. Li Linfei, Pangoal Institution, Xiao He, CASS, Professor Zhongying Pang, Renmin University School of International Studies, Dr. Liang Yabin, Associate Professor of the Institute for International Strategic Studies Party School of the Central Committee of the Communist Party of China, Dong Lu, Director of China Aerospace Science & Technology Consulting Co (CASC) for offering us insights into China's space program that has proven valuable as we continue work on this field.

We thank Ma Jiali, China Reform Forum, Executive Deputy Director Center for Strategic Studies, Sun Yuchen, Dr. Li, Assistant Program Officer, and Dr. Li Zheng CICIR. We thank Chen Zonghai, Dr. Li Shouping from the Beijing Institute of Technology for their time with us. We are grateful to Dr. Wu Chunsi, Dr. Xue Chen, and Dr. Feng Shuai from the Shanghai Institute of

International Studies (SIIS) for their time and sharing their academic insights with us. We also thank Zhao Gancheng and Dr. Zhou Shixin from SIIS. Our thanks to Professor Xia Liping, Tongji University for his insights on China's strategic culture. We thank Dr. Wu Xueming, Dr. Zhang Ming, from the Shanghai Academy of Social Sciences (SASS) and Professor HE Qisong, Shanghai University of Political Science and Law for their interactions with us and research inputs for our work. We thank Professor Xin Qiang, Fudan University, Deputy Director, Center for American Studies, CAS, Professor Wei Zongyou, Fudan University, (CAS) and Dr. Wang Xiaofeng, Fudan University, CAS for their insightful conversations on our topic.

We thank Dr. Jabin Jacob and Dr. Jagannath Panda for their kind help in guiding us on who to contact for our China field work.

For our India field trip, we would like to thank Gateway House, Mumbai for holding a discussion with us, and Dr. Nagnath R. Sonkavday, a former member of Maharashtra Electricity Regulatory Commission, Electricity Department and now working on space-based solar power with a published book titled *Space Based Solar Power*, Golden Page Publication, Pune, 2016. We thank the Tata Institute of Fundamental Research (TIFR) for hosting us for a presentation and interaction therein. We thank C-Step, Bangalore, Dr. Abhijit Singh, Assistant Professor on Space Law from the National Law School (NLS), Bangalore, the National Institute of Advanced Studies, Bangalore; SATSURE founder, Ashok, ReBeam, a new-space Indian private company, Pratik Basu, CEO, SATSURE, Rohan from Bellatrix, Astrome, a new space Indian private company, and Team Indus, the Indian private space company selected amongst six teams worldwide to compete for the GOOGLELUNARXPRIZE; SE Engineering Innovations for meeting with us and sharing their insights on India's space industry and policies, M. Matheswaran (Air Marshall-Retd). Rahul Narayan, CEO of Team Indus was very generous with his time. We thank the Indian Space Research Organisation (ISRO) headquarters for meeting with us and sharing their thoughts with us on India's space program. We thank Dr. Susmita Mohanty, CEO, Earth2Orbit (E2O), and Dr. Vasudevan Mukunth, Science Editor, *The Wire* for their conversations and insights; Dr. Gopal N. Raj, Freelance Journalist, and Space expert, Thiruvananthapuram, Kerala for his time and generous sharing of ideas. We also thank Vikram Sarabhai Space Center Engineers for meeting with us and discussing our research and ideas.

We thank Narayan Prasad, New Space entrepreneur for organizing several of our meetings, spending his precious time, and offering us an insightful analysis on India's space strategy, culture, and its new space organizations.

We thank Dr. Rajeswari Pillai Rajagopalan, Distinguished Fellow, Observer Research Foundation (ORF), for her generous help in setting up interviews and a panel discussion on our topic. We thank Dr. Medha Bisht,

Assistant Professor, South Asian University, Prakash Chandra, Science Writer, New Delhi, Dr. Jabin Jacob, Shruti Pandalai, Dr. Ranjana Kaul, leading space lawyer, Dua Associates, and Mr. Sashwant, DPS R K Puram, a school team that won the NASA Ames Research Center Prize 2017 for their project on Dead Sea underwater settlement that could be replicated in space for meeting with us during our field trip in Delhi. We thank the Defence Research and Development Organisation (DRDO) for making time to meet with us as well.

We extend our thanks to Col. P.K. Gautam (Retd.), Dr. Ajey Lele, Senior Fellow, MP-IDSA and one of India's known space experts; Vasundhara Sirnate Drennan, Co-Founder of the Polis Project, Stanly Johny, The Hindu and Dr. Roland Wittje, Associate Professor, Department of Humanites and Social Sciences, Indian Institute of Technology, for sharing ideas and thoughts with us.

In general, we would like to thank the leadership, faculty, and students of the ACSC and Air University. We thank Dr. Jim Forsyth, Brian Hastings, and Steve Kwast (Lt Gen, USAF-Ret) for their support. The support and stimulation from the faculty of Auburn University and the American Foreign Policy Council (AFPC) are also appreciated. A special thanks to the Space Horizons and Schriever Scholar students who enriched our perspective, and Schriever Scholar Faculty, Dr. Everett Dolman, Dr. Andrea Harrington, and especially Dr. Brent Ziarnick (Lt Col, USAF-Reserve), and Dr. "Coyote" Smith (Colonel, USAF-Ret) for their generous support, conversations, suggestion of valuable sources and feedback during the course of our study. The insights and mentorship of Dr. APJ Kalam, Dr. Klaus Heiss, Dr. Ed McCullough, Dr. Paul Spudis, Dr. David Criswell, Dr. Peter Glaser, now all passed, were formative and deeply appreciated. We are indebted to the insights and mentorship of Air Cmde Raghavan Gopalaswami, Buzz Aldrin, Dennis Wingo, Charles Miller, Howard Bloom, Bruce Pittman, Mark Cohen, Rick Tumlinson, Greg Autry, John Mankins, former Chief of NASA Advanced Concepts, and Dr. Simon "Pete" Worden (Brig Gen, USAF-Ret.).

Last but not the least, we thank our families for their continuous support during the course of our academic work, especially Peter Walter Garretson, Jerri Garretson, and Sanjay Goswami. Their support for our research has been constant and we would not have accomplished what we have today, without their continuous love and guidance over the years.

If we have missed any names, we apologize. Any fault in the book is our own.

Namrata Goswami and Peter A. Garretson
Montgomery, Alabama

Chapter 1

Introducing the Concept of Great Power Competition for Space Resources

In 2012, a U.S. company, Planetary Resources, went public with its ambitions to mine asteroids. Backed by billionaires from the world's richest company—Google—it made global headlines.[1]

By 2015, U.S. president Barack Obama signed into law the U.S. Commercial Space Launch Competitiveness Act (CSLCA). The law stated,

> A United States citizen engaged in commercial recovery of an asteroid resource or a space resource under this chapter shall be entitled to any asteroid resource or space resource obtained, including to possess, own, transport, use, and sell the asteroid resource or space resource obtained in accordance with applicable law, including the international obligations of the United States.[2]

After review, the International Institute of Space Law (IISL) issued an opinion paper stating,

> Therefore, in view of the absence of a clear prohibition of the taking of resources in the Outer Space Treaty one can conclude that the use of space resources is permitted. Viewed from this perspective, the new United States Act is a possible interpretation of the Outer Space Treaty.[3]

The U.S. move was soon followed by Luxembourg with its own law in 2017,[4] and an investment fund of $227 million to lure global companies.[5] These were the opening moves of a new chapter both in space and in international relations.

There is a new space race just beginning—though one of a vastly different character. It is not, as before, a race for prestige and honor among nations, or a contest between ideologies. It is rather a race to secure the determinants of economic and military power between states. It is a race not just with two Western superpowers—the United States and the Union of Soviet Socialist Republics (USSR) during the Cold War period—but now joined by the great powers of Asia: China and India. Small financial states and middle powers have also joined in such as the United Arab Emirates (UAE) and Luxembourg. It is a race to industrialize the inner solar system through access to its billion-fold greater resources. It is a race to write the rules of a new global order. The absolute gains of such a venture could uplift humanity and secure the future of our planet and our species for centuries to come. It could provide a level of riches to Earth that parallel the economic development of the first industrial revolution. Competition among states is sure to accelerate such a process. But even as the resources of the universe are vast, such a competition will bring with it pique and conflict. Our world stands, therefore, at the precipice of a new scramble for resources. This time it is not on the shores of the New World, in Asia or in Africa, but within the heavens themselves. Its outcome will be no less consequential. Despite such a far-reaching impact of space resources on Earth, the topic of space resources itself, and the emerging great power competition has yet to be treated, with little serious academic theory or policy analyses. This book is aimed to fill that gap in the literature.[6]

OUR AIM AND INTENDED AUDIENCE

This book aims to provide an analysis to inform space policy for the century ahead. We foresee a need for analysis that can anticipate foreign policy challenges. State behavior is of interest, both to students of international relations and practitioners of public policy. Our interest is to inform the debate about space policy in the context of great power competition. We provide case studies through which to understand an emerging area of competition anticipated to become significant to the accumulation of wealth and power in the international system. We provide thoughts on a theoretical framework as an "afterword" (in Appendix A) that may prove useful for further research and analysis. The book is written both for students and practitioners of international relations and U.S. public policy. It is vital, in our view, that both students and practitioners of policy making understand the contours of this scramble, this new race, so we can properly set our sails for the century ahead.

A KEY TO OUR ORGANIZATION

The book is organized to help the reader become an expert on the topic of space resources and be able to offer an informed opinion on the subject.

First, we provide a general introduction to the topic of space resources, its scale, and its importance.

Second, we provide the reader with individual chapters on each of the states who are the major contenders. These include the great powers, the United States, China, and India, and the active middle powers, Luxembourg and UAE.

In each chapter, we attempt to answer the commonsense questions a reader would want to know: Where exactly does a contender stand with regard to power and space capability? What ambitions have they expressed, and what are they actually doing in terms of law, policy, and programs? What do we know about their strategic and policy tendencies and what do we actually expect them to do?

Thus, individual chapter subheadings follow this format: "Current and Forecast Power to 2050," "Space Capability," "Elite Discourse," "Policies and Law," "Evidence of Space Programs," "Strategic Culture," "Strategic Trauma," and finally, "Expected Behavior." The theory is that before we would observe a funded space program, we would see an articulated policy, and before we would see a law or policy, we would expect to see elite discourse that attempts to shape that policy and law.[7] We would likewise expect that history has conditioned individual states to act and react in certain ways to an opportunity or a threat.[8] In our concluding chapter, we provide several scenarios to help the reader understand how the various trends and uncertainties might play out in the domain of space and space-based resources.

Because of length, we have chosen to break up the United States into two chapters. The first chapter examines the beginning of an epistemic community and its slow journey into elite discourse. The second chapter continues with the current and forecast power of the United States and the instantiation of that elite discourse in the policy.

We provide an afterword, offering our thoughts on a theoretical framework and causal mechanism suggested by our case studies. This framework emerged in the process of recording our cases and so is not reflected in the structure of the book. Nevertheless, we have provided it in the hopes it might be useful for future scholars who wish to engage in further research on this exciting subject. The afterword also attempts to make plain the interpretive stance of the researchers.

GENESIS

This book began as a research project proposed by the authors in 2015 for the MINERVA Social Science Research fund[9] titled "Contemporary Great Power Attitudes Toward Space Expansionism, Territoriality and Resource

Nationalism in Space" which resulted in fieldwork in China and India. It was our intuition at the time that we were witnessing a change in orientation of state-space programs toward exploitation and development of space resources for economic development. Since then, China and India have articulated such ambitions, followed by Luxembourg and UAE.[10] Our intuition appears vindicated as a result.

LITERATURE REVIEW

To provide a novel contribution at the nexus of space policy and international relations requires a review of a diverse set of literature including scientific facts about space resources, technical realities regarding the possibility of their recovery and development, international relations theory regarding great power competition (especially for resources), and how strategic culture influences individual state responses.

An extensive literature exists on state competition for power in the international system. Authors, including Kenneth Waltz,[11] John Mearsheimer,[12] Alexander Wendt,[13] and John Ikenberry,[14] have explored how states are generally expected to behave pursuing relative gains.[15]

A similarly extensive literature on *conflict theory* has explored the competitive drive for resources and dominance within a global political economy. Thucydides,[16] Karl Marx,[17] Rosa Luxemburg,[18] Max Weber,[19] V. I. Lenin,[20] and Immanuel Wallerstein[21] have explored how this competition leads to external acquisition, and the social and state conflict this generates within the global political-economic system. However, models that explain competition in the international system or world political-economic system or that explain scrambles for terrestrial resources have not been extended into space.[22]

States also choose to cooperate or to at least moderate their competition via institutions.[23] Various authors, including Paul Dembling & Daniel Arons,[24] Carl Everett Dolman,[25] Joan Johnson-Freese,[26] and Clay Moltz[27] have examined the role of cooperative institutions in space, but these frameworks have not been specifically extended to address space resources, and no general theory provides a causal logic of how states will behave specifically with regard to space resources.

Strong general models that explain competitive behavior have been elaborated by John Boyd[28] and Frans Osinga.[29] John Kindon[30] has explained how particular problems and opportunities become part of a state's agenda. Individual states, however, exhibit variation in their responses to various stimuli. A developed literature exists to suggest elites within nation-states perceive threats and opportunities differently,[31] including specific studies on the strategic cultures of major spacefaring states including the United States,[32] China,[33] and India.[34]

Space-based resources present a significant opportunity to advance national power and relative power. Several authors in the space science community have called attention to the scale and value of space-based resources, including Herman Kahn,[35] David Gump,[36] John Lewis,[37] Dennis Wingo,[38] and Paul Spudis.[39] Others have called attention to the increasing feasibility of recovery, including Vide Hellgren,[40] Joseph Pelton,[41] Tom James,[42] as well as the Keck Institute,[43] and International Academy of Astronautics.[44] The relevance of such claims becomes more important with the entry of the wealthiest individuals in society investing personal resources toward space exploitation and industrialization.[45]

A developed literature exists on spacepower theory and competition. Carl Everett Dolman,[46] Simon "Pete" Worden & John Shaw,[47] Joan Johnson-Freese,[48] Clay Moltz,[49] Brent Ziarnick,[50] Chris Stone,[51] and John Klein[52] were major contributors to the development of spacepower theory. Some authors concerned with the national and military space strategy and policy have observed the value of such resources to national power and positional strategies.[53] Researchers across the disciplines of international law and international relations have explored the legality and regulation of space resources, including Ricky Lee,[54] Viorel Badescu,[55] Ram Jakhu,[56] Scot Anderson,[57] and Annette Froehlich[58] and the IISL.[59] Researchers have articulated how such ambitions might lead to specific conflict scenarios, including Martin Elvis, Tony Milligan, and Alanna Krolikowski,[60] Namrata Goswami,[61] and Dan Deudney.[62] However, a discourse on space resources has yet to become a mainstream security concern within international relations literature.

No authors within the field of international relations have specifically put forth a theory that explains and predicts how nations are likely to behave with respect to the specific opportunity of space resources. Neither have they theorized on how these opportunities or threats are specifically perceived within the strategic cultures of the elites of the great powers nor how these differ or interact.

A large body of work exists that examines the current and historical space programs of the major spacefaring powers, including work by David Spires,[63] Walter McDougal,[64] Lalitendra Kaza,[65] Clay Moltz,[66] Gurbir Singh,[67] and Narayan Prasad.[68] However, this literature focuses on military and prestige competition and does not address the specific ambitions or programs motivated by the opportunity of space resources to enhance national economic power. Despite an increasing flurry of public interest stories in the media,[69] no in-depth academic case studies were found that specifically address what spacefaring states are doing to exploit space resources.

KEY CONCEPTS AND TERMS

Throughout the book, we use several terms that may be new to the reader. The two most important are *space resources* and *strategic culture.*

By *space resources*, we mean the mineral or energy resources extant (and available for exploitation) in outer space, including mineable substances from planetary bodies such as the Moon[70] and asteroids,[71] as well as energy resources present both on planetary bodies (such as Helium-3)[72] or in free space (such as space-based solar energy).[73]

By *strategic culture*,[74] we mean the shared paradigms within a state of how elites understand threats and opportunities. Often this involves "ranked grand strategic preferences derived from central paradigmatic assumptions about the nature of conflict and the enemy, and collectively shared by decision makers."[75] Throughout the book, we assert that states' strategic culture influences how they orient their policies and programs toward space resources, in particular, whether they take a passive or proactive approach and when and how they decide to do so. We introduce the concept of *strategic trauma*, which are formative experiences (usually in the form of a crisis) where states are conditioned to perceive and react to stimulus in particular and predictable ways.[76]

In our view, space resources are a subcategory of strategic resources,[77] by which we mean resources that are necessary for states to secure or improve their position relative to other states in the international system. This book assumes that states compete with each other for relative power in the international system. One of the things they compete over are strategic resources. States have options with respect to how they behave relative to strategic resources. They might seek an open trading system. Alternatively, they might choose to exploit such resources in some exclusive or patronage system of managed trade. We employ two concepts to discuss exclusivity. The first is *resource nationalism*, by which we mean anti-competitive behavior designed to restrict the international supply of a natural resource.[78] The second is *territoriality*, by which we mean the claiming of space by persons or groups[79] and the behavior of animals or people that try to keep others away from an area that they use or control.[80]

When we talk of the *global order* or *international system*, we are referring to the pecking order or leadership order that disciplines, and structures power relationships and the distribution of economic and status spoils among states.[81] Within the international system are a diversity of states, all seeking to improve their security and relative position.[82] When we speak of *great powers*,[83] we are referring to the most powerful states in the international system, capable of projecting military power overseas.

When we speak of *middle powers*,[84] we are referring to states who are not great powers, but who play important roles in the international system by legitimizing alternate proposals for institutions and norms that constitute the global order, moderating conflict and ensuring a balance between the great powers.

We use the term *space expansionism*,[85] by which we mean an ideological commitment that favors the expansion of humanity and life into the cosmos as a worthy end or means to perpetuate the survival and prospering of life. Those with an ideological commitment to space expansionism typically are interested in expanding the habitat for Earth life, including space settlements or colonies, and see industrialization and development in space as a necessary means to enable habitat expansion. They typically believe that the vast resources of space can and should be exploited to improve the quality of life on Earth.

We use the term *space development*, by which we mean the economic development of space—of a vibrant in-space economy. Space development typically imagines "people living and working in thriving communities beyond the Earth, and the use of the vast resources of space for the dramatic betterment of humanity" to promote "social, economic, technological, and political change in order to expand civilization beyond Earth, to settle space and to use the resulting resources to build a hopeful and prosperous future for humanity."[86] Included in this concept are the development of industries, infrastructure, and transportation infrastructure that grows the size and diversity of the in-space economy, improves the economic well-being and quality of life of communities and settlements in space. The aim of space development is economic development, and the establishment of a permanent, and broad-based activity. The term is used in contrast[87] to the more narrow "space exploration" whose activity is transient, and whose motivation is curiosity or status seeking.

The pursuit of space development for national power would seem to accord with Neo-Marxist and Neo-Realist insights; Dr. Carl Everett Dolman, author of *Astropolitik* (2002) notes that the path to hegemony begins by dominating the production of the most valuable commodities, using the carrying trade to become the shipper of choice, then using the profits from bulk trade to become the financial/banking leader and lender of last resort.[88] Moreover, there is an important tie between economic development and military power that is likely to hold true for space development and military spacepower as well. The level of economic development strongly predicts military effectiveness,[89] and developed nations excel in battle.[90] This has an effect even independent of military spending, and over and above the increase in wealth available to developed nations to buy weapons.[91] Economically developed states derive several advantages from their broader economies.[92] Historically, this has led to very lopsided victories.[93] In summary, according to Michael Beckly,

> The essence of economic development is efficiency of production. The higher a state's level of economic development, by definition, the more efficiently its workers produce goods and services. *There may be a natural tendency to view civilian and military realms as separate entities, but militaries are actually embedded within economic systems. Thus, countries that excel in producing civilian goods and services also tend to excel in producing military force* [emphasis added].[94]

If this is true, then the military dominance of today's Western democracies is likely just a dominance of the most economically developed states.[95] The People's Republic of China (PRC) appears to agree, valuing economic development highly in its concept of comprehensive national power. Author Michael Pillsbury notes that China's military and intelligence services measure global strength and national progress via precise quantitative assessments where military strength comprises less than ten percent of the ranking.[96] Those nations who become accomplished in space development will likely also be best equipped for space battle.

SPACE RESOURCES: AN INTRODUCTION

In the ongoing contest for standing among nations, a factor many fail to consider is space. The vast resources of space could tip the balance in favor of whichever nation proves most successful in exploiting them and ensuring the security of the celestial lines of commerce.

States pay close attention to economic opportunities that do not even amount to a single percent of their GDP. States play hardball over energy resources that are mere years of future domestic supply. But within space lies not a percent, not 10 percent, not 100 percent of a nation's GDP, but many, many times the value of the entire global GDP in resources.[97] Within space is not a small energy source, but a sufficient supply of constant solar energy to light a fully developed world six to seven times over.[98] As poetically explained in Robert Zubrin's *The Case for Space*,[99] nothing is a resource until the cleverness of humanity makes it so. Today's global order features a rapidly rising China, soon to pass the United States in real GDP. Behind it, another spacefaring power, India,[100] is expected to eclipse China circa 2050. But space resources could make all the difference to that power differential. Accessing multitrillion-dollar asteroids[101] or multitrillion-dollar power markets[102] could accelerate China's rise, prolong U.S. hegemony, or enable a late-blooming India to overtake China. At midcentury, it could permit China to hang on to its number one spot, allowing it to stay rich long after it grows old.[103] Or it could enable India with its young and vibrant population to surpass China[104] far sooner than either expects.

Today, nations consider individual percentage points of GDP growth to be strategically significant. Tiny changes in relative growth—of barely one percent at the time of the first industrial revolution—determined the relative hierarchy and fate of nations. Nations still fight over relatively small areas of land. Certainly, an opportunity that could change the balance of economic power between the first and second power, or between the second and third, would be significant. But the scale of resources in the inner solar system so far exceeds this that even the scale of colonial expansion during the Age of Imperialism appears an insufficient analogy. We are not talking about resources that would give the United States or China an extra 10 percent of productivity; we are not talking about gaining resources that would merely double a nation's land and resources—such as the capture of all of Antarctica; we are not talking about multipliers of 10-fold, or 100-fold. We are talking about resources that allow the equivalent of 3,000 Earths worth of livable surface area,[105] with a carrying capacity to support a million times our current global population, [106] with in-space habitats[107] supported by a billion times the mineral resources,[108] and a billion times more energy.[109] Surely opportunity at that scale deserves greater policy study and academic focus.

ANALOGIES

The last time technology enabled states to seek wealth and power far from their shores began about 1419 when Portugal developed sailing and navigation technologies that enabled seagoing vessels to make oceanic voyages.[110] Thereafter followed an Age of Exploration or Age of Discovery, which lasted until the early seventeenth century (200 years).[111] Then followed the Age of Old Imperialism (200 years). While journey times between continents by sail ranged between twenty and seventy-five days, the time to consolidate gains and find a power equilibrium took hundreds of years.

In this period, the theater of the European state competition expanded from the tiny European continent to the entire globe. The wealth enabled by accessing new resources, new markets, and new ideas was a significant determiner of the destinies of nations. The size and scale of their economies grew dramatically. Then a new technology, the steamship, which began development in the early 1800s, opened up the *Age of New Imperialism* in 1870 that lasted until about 1914 (forty-four years),[112] which included the *Scramble for Africa* (1881–1914).[113] This global scramble for resources, markets, and strategic position to command them created the power configuration and conflict points that set the stage for two world-spanning wars (1914–1945). It then took at least until the end of Decolonization (1945–1960) to arrive at some semblance of a stable demarcation of such territory. But now humanity has a new transportation technology, *reusable space vehicles*, which again enables states to seek wealth and power far from their shores. And the rewards are every bit as consequential as was the New World to the powers of Europe in the first Age of Exploration.

SPATIAL AND TEMPORAL SCOPE

While in principle, the topic of space resources might span the entire universe until the end of time, the scope of this book is considerably narrower. We are considering only the opportunities that appear available to the present states within our current understanding of physics. This examination of space resources is temporally limited to what could occur within the next 200 years using existing technology and spatially limited to the inner solar system.

We are most interested in the initial (and highly consequential) steps that can be taken between now (2020) and in 2050, specifically with respect to the energy and material wealth available from small bodies, such as Earth's Moon, the near-Earth asteroids (NEAs), the Martian moons, and the asteroid belt.[114]

Why confine things to the next 200 years? First, this seems to be within the timescales of modern states. The oldest surviving sovereign state in the world is San Marino, now 1,718 years old.[115] At the time of writing, the United States has persisted as a political entity for 243 years. If the PRC, established 1949 and Republic of India, established 1947, are as durable, then in 200 years, they will be barely older than the United States is now. It seems reasonable that currently powerful states such as the United States, PRC, and India might still be important actors across this timespan. Additionally, as noted above, 200 years' timescale appeared to be a politically important human epoch in earlier waves of exploration and consolidation.

Why confine things to the asteroid belt? It's a matter of travel time and exploitability. While the Moon, Mars, and Martian moons are all closer, the travel time between Earth and the largest asteroids in the belt is less than a year-and-a-half using the most efficient trajectory and current technology.[116] This is not an unreasonable time in terms of normal human commerce or travel (Christopher Columbus' journeys to the New World lasted between one and three years,[117] Marco Polo's journey lasted twenty-four years). However, it is likely with near-term nuclear propulsion (such as is being developed by the PRC and the National Aeronautics and Space Administration (NASA)), the journey to the asteroid belt might be only ten months.[118] As nuclear propulsion improves, this time could be reduced to seven to eight months using less-efficient trajectories, and perhaps to less than two months with constant acceleration profiles.[119] Given the proximity of these resources, it seems appropriate (assuming no unforeseen breakthrough in technology) to anticipate this will be the key theater of human geostrategic competition on the timescales of existing nation-states.

THE POTENTIAL TO IMPACT RELATIVE ECONOMIC POWER: A NEW WEALTH OF NATIONS?

Kenneth Waltz laid out five criteria by which to rank great powers: "size of population and territory, resource endowment, economic capability, military strength, political stability and competence."[120] The resources of the solar system offer states the opportunity to change the size of their population and territory, their resource endowment, their economic capability, and as a result their military strength. Central to the balance of power is relative economic growth.

"Economic growth can be defined as a positive change in the level of goods and services produced by a country over a certain period of time."[121] A state's production capacity is typically conceptualized as a production possibility curve, depicting the maximum possible production assuming a fixed amount of resources and a static level of technology; "an outward shift of the production possibilities frontier is only possible if the country discovers new resources or there is a change in technology."[122] Typical factors of production include land (including natural resources), labor (including population and human capital), and capital (machinery, tools, buildings, and infrastructure). An economy also needs the energy to transform it, and markets that create demand for production. A final factor of production often listed are entrepreneurs; those individuals willing to take calculated risks to bring about change or growth. Entrepreneurs cope with significant uncertainty, including uncertainty of returns.

What then is the production–possibility frontier for the inner solar system? It is hard to tell. But we can begin with a survey of the scale of resources available and consideration of what a state could do with such vast resources in space.

THE SCALE OF RESOURCES

The scale of these resources is truly vast. To understand the scale of such resources, let's consider what are considered big numbers at the time of the writing of this book. At present, the richest individual on planet Earth is Jeff Bezos, the CEO of Amazon, with a net worth of $160 billion (B).[123] Apple was the largest company in the world with a market capitalization of

$1 Trillion (T).[124] The largest state economy on planet Earth, the United States was worth $18.6T.[125] The largest state budget, the U.S. Federal Budget totaled $3.8T, out of which $686B (18 percent of the federal budget) was spent on defense,[126] $19.65B on NASA; and the U.S. national debt was $19.9T.[127] The entire annual world economy, the World GDP, was still under $100T ($84.38T),[128] and the total population of the planet Earth was still under 8B people (7.7B).[129] 7.7B people, using all forms of energy, required only 18 terawatts (TW) to power their civilization.[130]

There are estimates that by 2040 (twenty years from writing), including by the U.S. Secretary of Commerce,[131] that the space economy will grow from approximately $400B to exceed $1T without any access to space resources, just through the growth of present satellite services and associated manufacture and launch (Goldman Sachs $1.1T,[132] Morgan Stanley $1.1T,[133] U.S. Chamber of Commerce, $1.5T[134]), and to $2.7T within thirty years (Bank of America Merrill Lynch)[135] assuming only the beginnings of such capabilities. But to speak of a trillion-dollar space economy is to vastly understate its potential.

MINERAL WEALTH OF THE SOLAR SYSTEM

The valuation of asteroids below is made based on the current market exchange prices for the materials. There are two reasons to question this approach. First, we cannot know what is the actual cost of recovery of metals such as platinum or titanium from an asteroid. Second, we cannot know what future customers would be willing to pay for such metals. Third, an immediate overabundance of supply would result in significant devaluing by price. But for our purposes—which is to appreciate scale for human activity, not trade in commodities—these objections are not helpful. On the contrary, it is useful to understand what it would cost us in today's currency to acquire materials on this scale.

Individual asteroids vastly exceed $1T, eclipsing the wealth of the wealthiest individuals, companies, and many countries. In fact, Asterank, a website that maintains a catalog of known asteroids and their value, lists over 830 asteroids that are valued over $1T.[136]

Speaking plainly, Howard Bloom, Founder of the Space Development Steering Committee puts an individual asteroid in perspective:

> Just one asteroid, Amun 3554, has platinum, nickel, and cobalt worth $20 trillion, a figure bigger than the gross domestic product of Japan, Germany, the United Kingdom, France, and India combined . . . worth 200 times more than all the gold that America mined in California in the gold rush . . . and Amun 3554 is rated as "one of the smallest" asteroids in its class.[137]

Not only is Amun 3554 small for its class, $30T does not touch the upper limit of individual asteroid value. According to Asterank, at least five hundred of those are valued in excess of $100T each.[138] Among the top ten most massive asteroids in the asteroid belt is 16 Psyche, a potato-shaped metallic asteroid ("M-type"), approximately 210 kilometers across[139] reportedly valued at $10,000 quadrillion[140] (a quadrillion is 1,000 trillion). But what matters more than any single asteroid is the aggregate value of all of them. And the combined value of all the asteroids in the asteroid belt is truly staggering. According to Business Insider,

> The value of many asteroids are measured in the quintillions of dollars, which makes the market for Earth's annual production of raw metals—about $660 billion per year—look paltry in comparison . . . NASA estimates this belt to hold $700 quintillion of bounty, [a quintillion is a million trillion[141]] . . . That's about $100 billion for each person on Earth.[142]

The nearest abundance of mineral space resources is Earth's Moon, just three days by rocket, with as much surface areas as Africa and a wealth of important resources for an expanding industrial civilization. The next most accessible resources are the NEAs. According to NASA discovery statistics, about, 20,000 have been discovered, about 8,500 larger than 140m, and nearly 1000 a kilometer or larger.[143] The overall population of NEAs larger than 10–20 meters is likely tens of millions.[144] Beyond the NEAs is Mars and its two Moons, Phobos and Deimos. Just beyond Mars lies the asteroid belt, containing between 1.1 and 1.9 million asteroids larger than a kilometer, and millions of smaller ones.[145]

SPACE INDUSTRIALIZATION AND ECONOMIC DEVELOPMENT USING SPACE RESOURCES

Industry and economic growth require power. If we hope for a future of continued prosperity, even modest levels of growth in our energy requirements will drive us into space. For example, there are limits to the energy available on planet Earth. Many imagine that we could power our civilization entirely using solar power. However, this does not consider growth. As per Tom Murphy, our energy usage compounds with a growth rate of just 2.3 percent per year. At that modest growth rate, in a mere 400 years, "we hit the solar wall at the Earth's surface," our civilization would essentially require the entire land surface area of the Earth for solar panels.[146] Already, the nations of the Earth are looking to access the vast energy wealth of the solar system.

ENERGY WEALTH OF THE SOLAR SYSTEM

If the mineral wealth of the solar system is vast, the energy resource available is equally astounding. To understand the scale of the energy wealth available, it is important to put it in the context meaningful to human activity. As discussed above, the whole of human civilization and industry produces and consumes about 18TW to power an economy of $100T. The total energy required to warm our Earth and power our ecosystem is 173,000 terrawatts,[147] but this amounts to only one half of one-billionth of the Sun's total energy,[148],[149] which is 384 trillion TW (or 384.6 yottawatts).[150] Consider that per Dr. Criswell's estimates, every 1TW of electric power supports $42T of productivity.[151] Earth's geostationary belt alone could support 331TW[152] of delivered electrical power, enabling an increase in economic productivity of an additional $13,944 Trillion per annum (an economy over 166 times the size of our economy today). Clearly, there is vast energy available to transform the raw materials of space into human artifacts—mines, factories, power stations, habitats—to power the economies of states in the international system.

REAL ESTATE WEALTH AND THE LIMITS OF HABITAT EXPANSION

Population has always been an important component of the wealth of nations, and therefore national power. Population supplies labor to power industry, consumers to power markets, soldiers for the military. The most powerful states are often the most populous, with the largest labor forces and internal markets. Space development offers states possibilities to expand population.

Likewise, territory has been a determinant of national power—both its extent and location. Nations with large territories often have vast resources and can support large populations. Some nations enjoy geographic boundaries that shield them from invasion or expose them to invasion; that connect them over land to their neighbors or required them to be seafaring. Space development offers states possibilities to expand territory.

What may not be obvious to most international relations thinkers is the opportunity to create new land, new real estate by creating free-flying orbital habitats or space colonies to expand a state's population base in space. It turns out that asteroids can be turned into giant free-floating space stations housing thousands to millions.[153] As explained by Robert Walker, "asteroid resources could create space habs for trillions; land area of a thousand Earths."[154] Decades ago, Princeton Physicist Gerald K. O'Neill calculated that:

> Even if we limited ourselves to the resources of the asteroid belt (merely the most convenient source of materials, and not the only one), we could still build, in the form of orbital habitats, over 3,000 times the livable surface area of the Earth.[155]

More recently, estimates by Dr. John Lewis, a planetary geologist, indicate that the asteroid belt "could support a population of 10 million billion people[156]—a million times the ultimate carrying capacity of Earth," depending solely on the Sun for power.[157]

Such thoughts animate the richest man in the world, Jeff Bezos, and the vision of his company *Blue Origin* to enable "millions of people living and working in space"[158] who states,

> It seemed very clear that planetary surfaces were not the right place for an expanding civilization inside our solar system . . . There are a lot of other problems with planetary surfaces. But the main one is that they're not big enough. We have the resources to build room for a trillion humans in this solar system, and when we have a trillion humans, we'll have a thousand Einsteins and a thousand Mozarts. It will be a way more interesting place to live . . . we'll move all heavy, dirty industry off Earth—where, by the way, we'll be able to do it much more effectively with 24/7 solar power . . . The Earth is not a very good place to do heavy industry. It's convenient for us right now, but in the not-too-distant future, I'm talking decades, maybe 100 years, it'll start to be easier to do a lot of the things that we currently do on Earth in space, because we'll have so much energy. And then we can send the vitamins down to Earth . . . That's going to be the Great Inversion. The beginning is, we'll get bulk materials in space and we'll have to send all the vitamins up, integrated circuits and things like that. We'll have to send all of those up into space, but eventually that will invert, and we will send the vitamins down to Earth.[159]

Elon Musk, the founder of *Space X* has expressed similar ambitions. To turn human beings into space colonizers is his life's stated purpose.

> I would like to die thinking that humanity has a bright future . . . If we can solve sustainable energy and be well on our way to becoming a multiplanetary species with a self-sustaining civilization on another planet—to cope with a worst-case scenario happening and extinguishing human consciousness—then . . . I think that would be really good.[160]

Toward this end, on September 29, 2019, Musk unveiled his *Starship Mk1* project, and its heavy booster, the latest version of his Interplanetary Transport System,[161] capable of carrying 100 people to the Moon, Mars and beyond. Musk announced that "this is the fastest path to a self-sustaining city on Mars."[162]

THE ROLE OF IDEOLOGY AND INDUSTRIALIST AMBITIONS

It is in light of the above that the reader should consider the present ambitions of today's billionaire industrialists. Within fifty years, Elon Musk hopes to have a city of 1 million people on Mars.[163] Within a century, Jeff Bezos hopes to move all industries off Earth using space-based solar energy and to have Earth zoned as a park.[164] He hopes to enable millions of people living and working in space, and then billions and finally a trillion people in space.[165] The various asteroid mining firms hope to access the tremendous wealth of the millions of free-floating space rocks.

A MASSIVE RETURN ON INVESTMENT (ROI)?

But the wealth of nations is built on more than mineral wealth. Many assume that the geostrategic/geopolitical impact of space-based resources will accrue due to wealth brought from space to Earth, but this is wrong. The tremendous disparity between the wealth brought back from the New World and the wealth produced in the New World is staggering.

Consider University of Houston's Dr. David Criswell's insights regarding the economic ROI of Columbus' 1492 voyage that "discovered" the New World:

> Of the total $300 [thousand] invested (Christopher Columbus invested $100,000 (U.S.) [via a loan] and convinced Queen Isabella to invest an additional $200,000 (U.S.) to fund his expedition to "India"), by 1515 when Columbus died, there had been approximately $40,000,000 . . . in New World Gold Transported to Spain, a Return on Investment (ROI) of 130 to 1. By 1600, $30,000,000,000 . . . in Gold had been transported to Spain, an ROI of 100,000 to 1. In the time from 1500 to 1820, the sum of the Gross "New-World" product was $7,500,000,000,000 . . . which amounted to 4% of the total World GDP for that period, and by summing the contribution from 1500 to 2005 yields an amazing $1,100,000,000,000,000 ($1,000 trillion) or 30% of the cumulative World GDP.[166]

Basically, the wealth generated in the New World itself vastly exceeded the (comparatively tiny) mineral wealth brought back to Europe. Is it possible to imagine an ROI on this scale for space resources?

Much of the wealth generated in the New World accrued from what is called long-term exogenous growth; production increased due to an expansion of key inputs. Drawing attention to the strategic magnitude of the opportunity before us, Brian Wang of Next Big Future highlights the bottom line of a paper by Metzger, Muscatello, Mueller,[167] and Mantovani stating,

> The asteroid belt has everything necessary for it: water, carbon, silicates, metals, oxygen, solar energy (with much larger collecting arrays), etc. The ices in the lunar poles are a limited resource so it will be important to move the center of industry to the asteroids as quickly as possible. *There, the billion-fold greater resources could allow the industry to expand exponentially until it dwarfs that of the entire Earth within just a few decades* . . . Multiplying this by a factor of 3 per year, it would exceed the energy usage of the US within 11 more years. *After 12 more years it would exceed the US economy by a factor of a million. After another decade it would exceed the US economy by a factor of a billion.*[168]

Bringing the subject back to the question of policy and strategy and the relationship between economic might and military investments, Wang states, "exponential industrialization of space is more important than combat lasers and hypersonic fighters."[169]

DO SUCH PROJECTIONS DESERVE TO BE TAKEN SERIOUSLY?

Is this a "real" emerging security concern? Of course, projections are just that. They depend on many things. Technology may fail to progress. Various domestic or international distractions (depression, war) could result in a slower pace outside our timelines, or perhaps not at all. One cannot know where one is on the "hype curve"[170] and early space advocates vastly overestimated where we would be by 2000, 2010, and 2020. Certainly, the following chapters in the book will provide ample evidence of both the state and private ambitions and intent with regard to space-based resources, national security, and ambition.

While there is ample reason to be skeptical of the near-term possibilities of space resources to affect the global balance, one must be equally "skeptical" of the skeptics. Skeptics have had their share of spectacular failures for foresight and imagination. As evidence, consider just two instances of predictions about transportation technology and infrastructure:

Consider the skeptics of America's Transcontinental Railroad (completed in 1869):

In 1844, Asa Whitney (cousin of cotton gin inventor Eli Whitney) proposed to the U.S. Congress that America build a transcontinental railroad. U.S. Sen. Thomas Benton of Missouri responded that it was "an imposture, a humbug; it could have emanated only from a madman . . . science was unequal to overcome the Allegheny Mountains—and now Whitney proposed to scale the Rocky Mountains, four or five times as high! Why sir, it's madness!" . . . "You are one hundred years before your time."[171]

Or the stunning skepticism of the *New York Times* in October 1903, just two months before the Wright Brothers achieved powered flight. The *New York Times* carried the following editorial (October 9, 1903):

> The flying machine which will really fly might be evolved by the combined and continuous efforts of mathematicians and mechanicians in from one million to ten million years.[172]

On December 17, 1903, Wilbur and Orville Wright achieved their first powered flight at Kitty Hawk; they had invented the first successful airplane.[173]

Many were the skeptics who did not believe that a private space company (SpaceX) could even achieve orbit, let alone vertically land a reusable rocket, recapture, and dominate global market share,[174] or become one of the most valuable companies on Earth.[175]

Many will object to such space resource extraction projections on the basis that current technology does not appear to allow economic access of such materials at present. But these same objections were lodged against deep-sea oil recovery[176] and more recently to fracking.[177] None of the ideas above requires any new physics. It is just clever engineering.

Arthur C. Clarke warned us, "if we have learned one thing from the history of invention and discovery, it is that, in the long run—and often in the short one—the most daring prophecies seem laughably conservative."[178] And that "when a distinguished but elderly scientist states that something is possible, he is almost certainly right. When he states that something is impossible, he is very probably wrong."[179] Similarly, Roy Amara has observed,

> Although society generally has a dismal record in forecasting the diffusion and effect of new technology, one generalization does seem to apply: we consistently overestimate the rate of diffusion and the impacts of technology in the short run but underestimate diffusion and impact in the long run.[180]

Technology is not something apart that we guess at, and we are either right or wrong in predicting. The speed of technology and invention is strongly influenced by policy and funding, as recounted by Elon Musk:

> Then there's becoming a multiplanet species and space-faring civilization. *This is not inevitable. It's very important to appreciate this is not inevitable.* The sustainable energy future I think is largely inevitable, but being a space-faring civilization is definitely not inevitable. If you look at the progress in space, in 1969 you were able to send somebody to the moon. 1969. Then we had the Space Shuttle. The Space Shuttle could only take people to low Earth orbit. Then the Space Shuttle retired, and the United States could take no one to orbit. So that's the trend. The trend is like down to nothing. *People are mistaken when they think that technology just automatically improves. It does not automatically improve. It only improves if a lot of people work very hard to make it better*, and actually it will, I think, by itself degrade, actually. You look at great civilizations like Ancient Egypt, and they were able to make the pyramids, and they forgot how to do that. And then the Romans, they built these incredible aqueducts. They forgot how to do it.[181]

Many are projects and destinations. To mine asteroids. To establish Moon bases. To build cities on Mars. To capture the endless green energy of the Sun. To move industry off planet. To look for life on distant icy Moons. To visit another star. The counsel of the prophets of spacepower established three imperatives: "Defend the Earth, Mine the Sky, Settle the Universe."[182] Musk's stated philosophy is, "When something is important enough, you do it even if the odds are not in your favor."[183]

Ideas and beliefs about the future can strongly tilt things one way or the other. As succinctly stated in the Thomas theorem, "if men define situations as real, they are real in their consequences."[184] Just as ideas of scarcity, the viability of the size of nation-states in the time of Mackinder,[185] or ideas about global communism influenced the behavior of states, ideas such as space expansionism, development, and ideas about interstate competition influence technology today.[186] The dreams of the space expansionists echo those of the Oregon Train Pioneers: "by the 1840's many dreamed of helping the nation achieve its destiny 'to subdue the continent—to rush over this vast field to the Pacific Ocean.'"[187] The following chapters in this book will illustrate in detail much of such thinking. Ideology and ambitions do matter, especially when coupled with means. Consider these thoughts by Elon Musk:

> Conventional wisdom said that he couldn't afford to build rockets because they were too expensive and pointed to the fact that no one had ever made a rocket that cheaply before—but like the scientists who ignored those who said the Earth was 6,000 years old and those who insisted the Earth was flat, Musk started crunching numbers to do the math himself. Here's how he recounts his thoughts: "Historically, all rockets have been expensive, so therefore, in the future, all rockets will be expensive. But actually that's not true. If you say,

> what is a rocket made of? It's made of aluminum, titanium, copper, carbon fiber. And you can break it down and say, what is the raw material cost of all these components? And if you have them stacked on the floor and could wave a magic wand so that the cost of rearranging the atoms was zero, then what would the cost of the rocket be? And I was like, wow, okay, it's really small—it's like 2% of what a rocket costs. So clearly it would be in how the atoms are arranged—so you've got to figure out how can we get the atoms in the right shape much more efficiently. And so I had a series of meetings on Saturdays with people, some of whom were still working at the big aerospace companies, just to try to figure out if there's some catch here that I'm not appreciating. And I couldn't figure it out. There doesn't seem to be any catch. So I started SpaceX."[188]

Practitioners have a broad diversity of interests and policies competing for their attention. Policies usually must be decided in the absence of sufficient data, based on risk (the product of probability and severity).

This sets up a kind of Pascal's wager[189] for a game-theory risk assessment that practitioners must be aware of: "assume there is a scramble for space resources and none emerges—at worst you were prepared; assume a scramble for space resources is ongoing, and it happens—you are prepared; assume no such scramble is imminent and it doesn't emerge—you have not wasted your time; assume no such scramble is immanent and it is—you have lost significant time with potentially vast opportunity cost and consequence."

Our ethical viewpoint begins with a desire to aid policymakers and improve statecraft. Such work may take place in states, by international governmental organizations (IGOs), nongovernmental organizations (NGOs), or private actors.

For that audience, the primary questions are: what can I anticipate will happen (in systems where I have interests), what are the forces (causes) that are

Table 1.1 Policy "Pascal's Wager" Payoffs for a Space Resource Scramble

	Assume a Probability of Scramble for Space Resources	*Dismiss Probability of Scramble for Space Resources*
No Scramble for Space Resources Emerges	Overprepared/Early to need. Small opportunity cost due to lost time preparing	No loss. No opportunity cost
A Scramble for Space Resources Emerges	State is properly prepared to take full advantage of the opportunity	State is unprepared to take advantage of the opportunity, potentially experiencing major losses in relative power

Source: Compiled by authors, October 13, 2019.

driving changes (in systems I value), and, are there relationships of cause and effect that allow me to make the best choice to maximize the things I value? To this audience, specifics are often more important than theories.

But a secondary audience are academics themselves, who may have the motive to understand and develop models for more general purposes rather than any specific applied policy situation. For them, the Appendix A on a theoretical framework may constitute a foundation on which they seek to build or counter some other model or debate.

For the primary audience, we will now give the digest version starting with: what do we think is going to happen.

WHAT WE EXPECT TO HAPPEN

Is space more like cyber or nuclear? No, it is more like the South China Sea. We conceptualize space as a geography and expect behavior very similar to colonial powers seeking advantage from their neighborly rivalries in wealth and resources abroad. Nations will perceive (at different speeds) that space resources are strategic. Space resources will become "securitized" and perceived as strategic interests.[190] Those that have the ability to compete will compete. This competition will primarily be benign and there will be attempts to avoid conflict, except where they cannot because of extraordinary value. Here, we can expect concerns of loss of exceptional advantage and security dilemmas to create conflicts. Here, we can also expect significant activity from liberal institutionalists and middle powers. They would try to negotiate obstacles with an aim to keep the larger stable international order intact.

Those nations that cannot compete, or perceive themselves to be definitely lagging behind or losing, will follow a classic "weak power strategy" attempting to restrain (through lawfare or coercion in the earthly domain)[191] those ahead from getting further ahead. This push will be most acute from those who perceive their relative position declining in such a way that they fear losing the status of a great power or contender for hegemony, to the rank of middle powers. They will play spoilers with regard to new institutions. They will propose alternate norms that restrain the exploitation of space resources. They will make common cause with the masses of less empowered sovereign states, with socialists and environmentalists, and seek to undermine increasing concentration of power by the few with narratives of justice, illegitimacy, and need for redistribution.

Our expectation broadly is that the United States and China will be the actual competitors. India will be a late arrival to the endeavor. Russia, Japan, France, largely due to a lack of resources and competitive industrial base, will not be independent actors but will operate within the ecosystem of norms and institutions established by the three great spacefaring powers.

Both state and private industry will play a role. This will be an intense subject of mutual measurement. Both will use the middle powers to legitimize global norms. The middle powers will have the enviable position of receiving investment from all three great powers. The two most important middle powers appear to be Luxembourg and the UAE, which are building niches for themselves in space.

Negotiated well, it is not clear that conflict needs to come within the lifetimes of the authors or even another generation. However, it is possible that equity concerns could occur even before the intended publication date, given the interests of both the PRC and the United States in the lunar poles, the current lack of any bilateral understanding regarding claims of safety and exclusivity of use, and the reality that these are locations that convey tremendous strategic advantage within the presently understood possibilities of technology.

Over time, we see the possibility that the means of production may shift off Earth and to space. We are convinced by the facts that it is only human ingenuity that stops us from accessing wealth and expanding human habitats to many times the size of Earth's economy and population. Heavy industry and energy production can be expected increasingly to shift off Earth. Just as the center of power moved from Rome to Constantinople to Amsterdam to Paris to London to Washington, D.C., and Beijing, with the cooption of space resources, the capital of political leadership is likely to shift first to the dominating power on Earth, and at some point, eventually to space. The critical geography that controls trade and production is likely to move from oil-choke points and oil-rich regions to space at key locations of resources, propellant, power production, and choke points that can constrain space between them. The book's subsequent chapters on the United States, China, India, UAE, and Luxembourg offer compelling evidence to the nature of this shift. The next chapter indicates how the strategic myths and culture of a state informs its policy in regard to space policy and behavior. Toward the end of the book, we offer the reader alternative scenarios illuminating issues that may lead to competition, especially between the spacefaring nations. We feel compelled to include an afterword (Appendix A) on a theoretical framework given our reasoned intuition that space resources will become a topic of serious academic discourse in the next decade or so.

NOTES

1. Adam Mann, "Tech Billionaires Plan Audacious Mission to Mine Asteroids," *Wired*, April 23, 2012, accessed August 5, 2019. https://www.planetaryresources.com/2012/04/wired-tech-billionaires-plan-audacious-mission-to-mine-asteroids/

2. "H.R.2262—U.S. Commercial Space Launch Competitiveness Act 114th Congress (2015-2016)," *Congress.gov*, November 25, 2015, accessed October 16, 2019. https://www.congress.gov/bill/114th-congress/house-bill/2262/text

3. International Institute of Space Law, Position Paper on Space Resource Mining Adopted by consensus by the Board of Directors, December 20, 2015, accessed August 15, 2019. http://iislwebo.wwwnlss1.a2hosted.com/wp-content/uploads/2015/12/SpaceResourceMining.pdf

4. Mariella Moon, "Luxembourg's Asteroid Mining Law Takes Effect August 1st It's the First of its Kind in Europe," *Engaget*, July 30, 2017, accessed August 15, 2019, https://www.engadget.com/2017/07/30/luxembourg-asteroid-mining-law-august-1/

5. David Z. Morris, "Luxembourg to Invest $227 Million in Asteroid Mining," *Fortune*, June 5, 2016, accessed August 15, 2019. https://fortune.com/2016/06/05/luxembourg-asteroid-mining/

6. As part of the authors' Minerva Research Grant, field trips were conducted in China and India. As part of these field trips, authors held interactions and interviews with academics, policymakers, and journalists after duly informing them that their perspectives will be utilized as a part of future publications. It was only after participants agreed that the authors conducted all such interactions. By the authors' choice, we refrain from identifying anyone by name. This holds true for those authors interacted with, within the United States as well.

7. Randall L. Schweller offers an interesting perspective on how elite discourse and social cohesion can shape policy. Though we have not adopted his model, his article nonetheless offers an interesting perspective. See Randall L. Schweller, "Unanswered Threats: A Neoclassical Realist Theory of Understanding," *International Security* 29, no. 2 (Fall 2004): 159–201.

8. Stephen A. Kocs, "Explaining the Strategic Behaviour of States: International Law as System Structure," *International Studies Quarterly* 38, no. 4 (December 1994): 535–56.

9. "Minerva Research Initiative", accessed February 17, 2020, https://minerva.defense.gov/

10. Jeff Foust, "Lunar Exploration Providing New Impetus for Space Resources Legal Debate," *SpaceNews*, September 7, 2019, accessed September 11, 2019. https://spacenews.com/lunar-exploration-providing-new-impetus-for-space-resources-legal-debate/; Louis Brennan, "How Luxembourg is Positioning itself to be the Centre of Space Business," *The Conversation*, July 16, 2019, accessed September 11, 2019. https://theconversation.com/how-luxembourg-is-positioning-itself-to-be-the-centre-of-space-business-120436; Ajay Lele, "Space Activities Bill: India's Great Galactic Leap," *The Financial Express*, July 2, 2019, accessed September 11, 2019. https://www.financialexpress.com/lifestyle/science/space-activities-bill-indias-great-galactic-leap/1625980/; Vikram Mansharamani, "China Wants to Mine the Moon for "Space Gold"," *PBS*, March 31, 2016, accessed September 11, 2019. https://www.pbs.org/newshour/economy/china-wants-to-mine-the-moon-for-space-gold

11. Kenneth Waltz, *Theory of International Politics* (New York: Random House, 1979).

12. John Mearsheimer, "Structural Realism," in Tim Dunne et al., eds., *International Relations Theories* 3rd ed. (Oxford: Oxford University Press, 2013), 77–93.

13. Alexander Wendt, “Anarchy is What States Make of It: The Social Construction of Power Politics,” *International Organization* 46, no. 2 (Spring 1992): 391–425.

14. John G. Ikenberry, “Institutions, Strategic Restraint, and the Persistence of American Postwar Order,” *International Security* 23, no. 3 (Winter 1998/1999): 43–78

15. See also John G. Ikenberry and Charles Kupchan, “Socialization and Hegemonic Power,” *International Organization* 44, no. 3 (Summer 1990): 283–316; Martha Finnemore and Kathryn Sikkink, “International Norm Dynamics and Political Change,” *International Organization* 52, no. 4 (Autumn 1998): 887–918.

16. Thucydides, *The History of The Peloponnesian War*, 431 BC translated by Crawley, R., 1852. *History of the Peloponnesian War* (No. 455) [New York: Penguin Books, 1964]. https://www.gutenberg.org/files/7142/7142-h/7142-h.htm

17. Karl Marx, *Capital: A Critique of Political Economy, Volume 1*. Originally published 1867 (London: Arkose Press, 2015).

18. Rosa Luxemburg, *Accumulation of Capital*, originally published 1913 (London: Routledge and Kegan Paul, 1951); translated by Agnes Schwarzschild and with an introduction by Joan Robinson.

19. Max Weber, *Wirtschaft und Gesellschaft (Economy and Society)*, 2 vols. (Tuebingen: J. C. B. Mohr, 1925)

20. Lenin, Vladimir Il′ič, “Imperialism: The Highest Stage of Capitalism,” *The New Imperialism: Analysis of late Nineteenth-century Expansion* (1939), available from Resistance Books, 1999, available online at: https://www.readingfromtheleft.com/Books/Classics/LeninImperialism.pdf; Nikolaĭ Bukharin and Vladimir Il′ič Lenin. *Imperialism and world economy*. Vol. 4 (New York: International Publishers, 1929).

21. Immanuel Wallerstein, *The Modern World-System I: Capitalist Agriculture and the Origins of the European World-Economy in the Sixteenth Century* (New York: Academic Press, 1974).

22. Please refer general bibliography for full citation of the literature review.

23. Robert O. Keohane, *After Hegemony: Cooperation and Discord in the World Political Economy*. (Princeton: Princeton University Press, 1984); Jeffrey W. Legro, “Which Norms Matter? Revisiting the ‘Failure’ of Internationalism,” *International Organization* 51, no. 1 (1997): 31–63.

24. Paul G. Dembling and Daniel M. Arons, “The Evolution of the Outer Space Treaty,” *Documents on Outer Space Law. 3.*, 1967, pp. 419–456. https://digitalcommons.unl.edu/spacelawdocs/3

25. Everett C. Dolman, *Astropolitik: Classical Geopolitics in the Space Age* (Portland, OR: Frank Cass, 2002).

26. Joan Johnson-Freese, *Heavenly Ambitions: America’s Quest to Dominate Space* (Philadelphia, PA: University of Pennsylvania Press, 2009).

27. James Clay Moltz, *Asia’s Space Race: National Motivations, Regional Rivalries and International Risk* (New York: Columbia University Press, 2011).

28. John R. Boyd, *A Discourse on Winning and Losing* (Maxwell AFB, Alabama: Air University Press, 2018).

29. Frans Osinga, *Science Strategy and War: The Strategic Theory of John Boyd* (Utrecht: Eburon, 2005).

30. John W. Kingdon, *Agenda's, Alternatives, and Public Policies* (New York: Longman, 2003).

31. Alastair Iain Johnson, "Thinking about Strategic Culture," *International Security* 19, no. 4 (Spring 1995): 32–64; Stephen Peter Rosen, "Military Effectiveness: Why Society Matters," *International Security* 19, no. 4 (1995): 5–31; Jack L. Snyder, *The Soviet Strategic Culture: Implications for Limited Nuclear Operations* (Santa Monica, CA: RAND Corporation, 1977); Jeannie L. Johnson, *Strategic Culture: Refining the Theoretical Construct*. SAIC (for Defense Threat Reduction Agency), 2006; Darrl Howlett, *The Future of Strategic Culture*. SAIC (Prepared for Defense Threat Reduction Agency), 2006. http://www.socsci.uci.edu/~bskyrms/bio/papers/StagHunt.pdf; Melanie Graham, *Redefining Strategic Culture* (Prince George: The University of Northern British Columbia, June 2014).

32. Carnes Lord. "American Strategic Culture," *Comparative Strategy* 5, no. 3, (1985): 269–293. doi: 10.1080/01495938508402693; Theo Farrell, "Strategic Culture and American Empire," *SAIS Review* XXV, no. 2 (Summer-Fall 2005): 3–18; Thomas G. Mahnken, "United States Strategic Culture," SAIC (Prepared for Defense Threat Reduction Agency), 2006. http://www.au.af.mil/au/awc/awcgate/dtra/mahnken_strat_culture.pdf; Leonid Savin, *US Strategic Culture and Foreign Policy*, Katehon, 2016. http://katehon.com/article/us-strategic-culture-and-foreign-policy

33. Namrata Goswami, "China in Space: Ambitions and Possible Conflict," *Strategic Studies Quarterly* (Spring 2018): 74–97.

34. George K. Tanham, *Indian Strategic Thought: An Interpretive Essay* (Santa Monica: RAND, 1992), https://www.rand.org/pubs/reports/R4207.html

35. William M. Brown and Herman Kahn, *Long-term Prospects for Developments in Space (A Scenario Approach)* [prepared for National Aeronautics and Space Administration under contract NASW-2924] (Washington, DC: The Hudson Institute, 1977).

36. David Gump, *Space Enterprise: Beyond NASA* (New York: Praeger Publishers, 1990).

37. John S. Lewis, *Mining the Sky: Untold Riches from the Asteroids, Comets, and Planets* (Reading, MA: Addison-Wesley Pub. Co., 1996); John S. Lewis, *Asteroid Mining 101: Wealth for the New Space Economy* (Mountain View, CA: Deep Space Industries, 2014).

38. Dennis Wingo, *MoonRush: Improving Life on Earth with the Moon's Resources* (Ontario, Canada: Apogee Books, 2004).

39. Paul D. Spudis, *The Value of the Moon: How to Explore Life and Prosper in Space Using the Moon's Resources* (Washington, DC: Smithsonian Books, 2016).

40. Vide Hellgren, *Asteroid Mining: A Review of Methods and Aspects* (Lund, Sweden: Lund University, 2016).

41. Joseph N. Pelton, *The New Gold Rush: The Riches of Space Beckon* (Göttingen, Germany: Copernicus, 2017).

42. Tom James, *Deep Space Commodities: Exploration, Production and Trading* (Switzerland: Palgrave Macmillan, 2018).

43. Keck Institute for Space Studies (KISS), *Asteroid Retrieval Study* (Keck Institute, 2012). http://kiss.caltech.edu/final_reports/Asteroid_final_report.pdf

44. International Academy of Astronautics (IAA), *Space Mineral Resources: A Global Assessment of Challenges and Opportunities*. 2015.

45. Christian Davenport, *The Space Barons: Elon Musk, Jeff Bezos, and the Quest to Colonize the Cosmos* (New York US: Public Affairs, 2018).

46. Dolman, *Astropolitik: Classical Geopolitics in the Space Age*.

47. Simon P. Worden and John E. Shaw, *Whither Space Power? Forging a Strategy for the New Century* (Maxwell Air Force Base: Air University Press, 2002).

48. Joan Johnson-Freese, *Space as a Strategic Asset* (New York: Columbia University Press, 2007).

49. Moltz, *Asia's Space Race*.

50. Brent Ziarnick, *Developing National Power in Space: A Theoretical Model*. Jefferson (NC: McFarland & Company, 2015).

51. Christopher M. Stone, *Reversing the Tao: A Framework for Credible Space Deterrence* 1st ed. (Scotts Valley, California: CreateSpace Independent Publishing Platform, 2016).

52. John Klein, *Space Warfare: Strategy, Principles, and Policy* (London & New York: Routledge, 2006); John Klein, *Understanding Space Strategy: The Art of War in Space (Space Power and Politics)* [London & New York: Routledge, 2019].

53. Dolman, *Astropolitik: Classical Geopolitics in the Space Age*; Peter A. Garretson, "The Next Great White Fleet: Extending the Benefits of the International System into Space." *Astropolitics* 6, no. 1 (2008): 50–70; Peter Garretson, "Solar Power in Space?" *Strategic Studies Quarterly*, Spring 2012. http://www.au.af.mil/au/ssq/2012/spring/garretson.pdf; Ziarnick, *Developing National Power*.

54. Ricky Lee, *Law and Regulation of Commercial Mining of Minerals in Outer Space*. Vol. 7 (New York: Springer Science & Business Media, 2012).

55. Viorel Badescu, ed., *Asteroids: Prospective Energy and Material Resources* (New York: Springer Science & Business Media, 2013).

56. Ram S. Jakhu, Joseph N. Pelton, and Yaw Otu Mankata Nyampong. *Space Mining and Its Regulation* (New York: Springer International Publishing, 2017).

57. Scot W. Anderson, Korey Christensen & Julia LaManna, "The Development of Natural Resources in Outer Space," *Journal of Energy & Natural Resources Law*, pp. 1–32, 2018. https://www.hoganlovells.com/~/media/hogan-lovells/pdf/2018/the_development_of_natural_resouces_in_outer_space_august_2018.pdf

58. Annette Froehlich, ed. *Space Resource Utilization: A View from an Emerging Space Faring Nation*. Vol. 12 (New York: Springer, 2018).

59. International Institute of Space Law, *Does International Space Law Either Permit or Prohibit the Taking of Resources In Outer Space And On Celestial Bodies, And How Is This Relevant For National Actors? What Is the Context, and What Are The Contours And Limits Of This Permission or Prohibition?* (Background Paper), International Institute of Space Law, 2016. http://iislweb.org/docs/IISL_Space_Mining_Study.pdf; International Institute of Space Law. *Position Paper on Space Resource Mining*, December 20, 2018. https://iislweb.org/iisl-position-paper-on-space-resource-mining/

60. Martin Elvis, Tony Milligan, Alanna Krolikowski. "The Peaks of Eternal Light: A Near-term Property Issue on the Moon," (Postdoctoral NSF Grant#1066293), 2016. https://arxiv.org/ftp/arxiv/papers/1608/1608.01989.pdf

61. Goswami, "China in Space: Ambitions and Possible Conflict".

62. Daniel H. Deudney, *Dark Skies: Space Expansionism, Planetary Geopolitics, and the Ends of Humanity* (New York: Oxford University Press, 2020).

63. David Spires, *Beyond Horizons: A Half Century of Air Force Space Leadership*, Revised Edition (Montgomery, AL: Air University Press, 1998).

64. Walter McDougall, *The Heavens and the Earth: A Political History of the Space Age* (Washington, DC: Johns Hopkins University Press, 1997).

65. Lalitendra Kaza, *Militarisation of Space* (New Delhi: Center for Air Power Studies, 2010).

66. Moltz, *Asia's Space Race.*

67. Gurbir Singh, *India's Space Programme: India's Incredible Journey from the Third World Towards the First* (Manchester: Astrotalkuk Publications, 2017).

68. Narayan A. Prasad, *Geopolitical and Technological Survey of Potential Use of Space Weapons* (Bangalore: NIAS, 2018).

69. Leonard Davis, "The US Geological Survey is Getting Serious about Space Resources and Mining," *Space.com*, September 04, 2018, accessed September 12, 2019. https://www.space.com/41707-space-mining-usgs-resource-survey.html; Alex Gilbert and Morgan D. Bazilian, "We Need a Space Resources Institute," *Scientific American*, April 19, 2019, accessed September 12, 2019. https://blogs.scientificamerican.com/observations/we-need-a-space-resources-institute/

70. Wingo, *Moonrush: Improving Life.*

71. Lewis, *Mining the Sky.*

72. J. Sved, G. L. Kulcinski, and G. H. Miley, "A Commercial Lunar Helium 3 Fusion Power Infrastructure," *Journal of the British Interplanetary Society* 48 (1995): 55–61.

73. John Mankins and Nobuyuki Kaya, eds., *Space Solar Power: The First International Assessment of Space Solar Power: Opportunities, Issues and Potential Pathways Forward* (France: International Academy of Astronautics, 2011).

74. For an excellent discussion of how the term has been used in the literature, see the critique by Rosen, "Military Effectiveness".

75. Johnston, "Thinking about Strategic Culture."

76. Stephen M. Walt has an insightful perspective on how states respond to issues based on fear. See Stephen M. Walt, "Alliance Formation and the Balance of World Power," *International Security* 9, no.4 (Spring 1985), 3–34. Also Stephen M. Walt, *The Origin of Alliances* (Cornell Studies in Security Affairs) [Ithaca, New York: Cornell University Press, 1990].

77. David Stephen, et al., *The New Politics of Strategic Resources* (Washington DC: The Brookings Institution Press, 2015).

78. UK Government, *Resource Nationalism: A Horizon Scanning Research Paper by the Resources Demand and Supply Resource Nationalism Community of Interest*, December 2014, accessed August 14, 2019. https://assets.publishing.service

.gov.uk/government/uploads/system/uploads/attachment_data/file/389085/Horizon_Scanning_-_Resource_Nationalism_report.pdf

79. David Storey, "Territory and Territoriality," *Oxford Bibliographies*, July 26, 2017, accessed August 27, 2019, https://www.oxfordbibliographies.com/view/document/obo-9780199874002/obo-9780199874002-0076.xml

80. Merriam-Webster, "Territoriality," accessed August 27, 2019. https://www.merriam-webster.com/dictionary/territoriality

81. See Hedley Bull, *The Anarchical Society A Study of Order in World Politics* (New York: Columbia University Press, 2012). Barry Buzan and Richard Little, *International Systems in World History Remaking the Study of International Relations* (Oxford: Oxford University Press, 2000).

82. Such a view is compatible with much of conflict theory, discussed in the literature review.

83. John Mearsheimer states, "Great powers are determined largely on the basis of their relative military capability. To qualify as a great power, a state must have sufficient military assets to put up a serious fight in an all-out conventional war against the most powerful state in the world. The candidate need not have the capability to defeat the leading state, but it must have some reasonable prospect of turning the conflict into a war of attrition that leaves the dominant state seriously weakened, even if that dominant state ultimately wins the war." See John J. Mearsheimer, *The Tragedy of Great Power Politics* (New York: WW Norton & Company, 2001): 1.

84. Namrata Goswami, "The Competition for Outer Space Resources and the Role of Middle Powers," *Live Encounters Magazine*, June 2019, accessed September 11, 2019. https://liveencounters.net/2019-le-mag/06-june-2019/dr-namrata-goswami-the-competition-for-outer-space-resources/?fbclid=IwAR0DUohKoqUMf2fZZMPl-dyxXDP_4P1YVfnGpmfL7EUPRO6zIG8HdGm3h_w

85. Dan Deudney is the first scholar to coin and use this term, "Expectations of the human future in space have been dominated by a body of thought which I refer to as 'space expansionism.' The core of space expansionism is the claim that human expansion into space is desirable, perhaps even inevitable. Space expansion advocates also claim that large-scale space activities can solve a variety of important Earth problems. Perhaps most importantly, space expansionists hold that making humanity a multi-world species can help insure the survival of humanity from a variety of catastrophic and existential threats, ranging from nuclear war to asteroidal collision" as quoted in "An Interview with Dan Deudney," *World Government Research Network (WGRN)*, January 4, 2016, accessed October 11, 2019. http://wgresearch.org/an-interview-with-daniel-h-deudney/ which he elaborates further in his book, Deudney, *Dark Skies*. The authors were in a panel with Dan at the International Studies Association Annual Conference, Baltimore, 2018 in which his draft manuscript was discussed.

86. "NSS Statement of Philosopy," *NSS*, August 3, 2017, accessed October 11, 2019. https://space.nss.org/nss-statement-of-philosophy/

87. Peter A. Garretson, "US Space Command: A Vision for the Final Frontier," *The Hill*, August 28, 2019, accessed October 11, 2019. https://thehill.com/opinion/technology/459066-us-space-command-a-vision-for-the-final-frontier

88. Per Dr. Carl Everett Dolman (personal exchange on April 16, 2013),

In dominating the political economy of the inner solar system, I think the model established in global trade by the capitalists (as explained by neo-Marxist Immanel Wallerstein and neo-Realist Robert Gilpin) is valuable. Whenever a state rises to hegemonic status, it does so in a precise sequence. First, it dominates in the production of the most valuable commodities. In the space context this is the launch vehicles and satellites early on, as they are the expensive and exotic keys to further exploitation. After achieving premier status in the production of these highest value (what is generally the forefront of innovation) the dominant state will achieve economies of scale that make routine space production cost-effective. Right now, surveillance satellites in LEO, to include Earth resources satellites, and the communications satellites in GEO are the most profitable, but the US has a mighty advantage in providing a public good from space with its GPS constellation. In the near-to-mid future, the goods with the highest value for space will be from exotic resources on the Moon and solar power. When these become routine, the bulk trade in them will be fantastically profitable.

Second, the hegemonic state dominates trade by becoming the carrier or shipper of choice. You put this as your first maxim and while I agree it is vital it is in fact in the second position of importance, and is one the US is letting slip away. It can be captured, but will take an extraordinary investment of resources and political will to do so at this point. Unlike the first key of leading innovative and highest-value production, which the US managed and can still effect through its efforts to spread space commerce to its national corporations, domination of the carrying trade requires a political support structure for the equivalent of harbors, refueling stations, police or customs capacities, and the like. We're not there yet, but if we wish to dominate space—to include, and via, the space economy—we have to start pushing this.

Third, the profits made from the transfer of bulk trade in the system (through dominating shipping and movement of goods) allows the hegemonic state to become the financial or banking leader of the world. Lending capital and operating as both the lender of last resort And the international counter-cyclical lender (similar to what the Fed does domestically: loosening credit in times of economic hardship and tightening it during boom times) stabilizes the system in favor of the dominant hegemon . . . hegemon declines in the same order. First, it loses control of the most profitable production; Second, it no longer carries the bulk of trade; and Third, it is no longer the financial center of the system. But during the period that it is dominant IN ALL THREE, it is the global hegemon, and can rewrite the international governing institutions in whatever manner it chooses (so long as it pays for the bulk of them) . . . Wallerstein pointed out there have been three hegemonic powers in the last 500 years." (used with permission)

89. Michael Beckley, "Economic Development and Military Effectiveness," *The Journal of Strategic Studies* 33, no. 1 (2010): 43–79.

90. Beckley, "Economic Development and Military Effectiveness." "economically developed states significantly outperform less developed states in battle."

91. Beckley, "Economic Development and Military Effectiveness.": "the benefits of superior economic development *go beyond merely having more money to invest*

in the military. In particular, economically developed states are *more likely to have robust technological infrastructures, large-scale production capacities*, skillful personnel, and stable political environments, which *may enhance effectiveness independent of the level of defense spending*."

92. Beckley, "Economic Development and Military Effectiveness.": "economically developed states are *more capable of generating highly skilled military units and producing, maintaining, and modernizing sophisticated military equipment*. Part of this advantage stems from *a greater surplus of wealth, which allows developed states to sustain large military investments without undermining long-term economic growth. But economically developed states also derive military benefits from their technological infrastructures, efficient production capacities*, advanced data analysis networks, stocks of managerial expertise, and stable political environments. In short, military effectiveness cannot be bought; it must be developed."

93. Beckley, "Economic Development and Military Effectiveness": "Developed economies also possess the financial capital to fund technological innovation, and the production capacity to manufacture equipment in large quantities. As industries within an economy become more advanced, they increasingly benefit from economies of scale that diminish the unit costs of production and, therefore, increase the likelihood that new innovations, including military innovations, will be developed and adopted. This capacity for sustained technological progress translates into technological supremacy in battle, as evidenced most clearly by the history of colonialism. Consider that in 1850, Great Britain's GDP was only half that of India's, yet Britain's fourfold advantage in economic development (as measured by per capita income) translated into vastly superior weapons and organization, which in turn facilitated political domination. Throughout this period technological asymmetries resulted in extremely one-sided military outcomes."

94. Beckley, "Economic Development and Military Effectiveness."

95. Beckley, "Economic Development and Military Effectiveness."

96. "Chinese military and intelligence services use quantitative measurements to determine how China compares with its geopolitical competitors, and how long it will be before China can overtake them. When I read these quantitative measurements in a difficult-to-obtain book written by Chinese military analysts, I was surprised to see how precisely China measures global strength and national progress. The most startling revelation was that *military strength comprised less than 10 percent of the ranking*." As quoted in Michael Pillsbury, *The Hundred-Year Marathon: China's Secret Strategy To Replace America As The Global Superpower* (New York: Henry Holt and Company, 2015).

97. Harry Pettit, "NASA Headed Towards Giant Golden Asteroid that Could Make Everyone on Earth a Billionaire," *FoxNews*, June 27, 2019, accessed August 15, 2019. https://www.foxnews.com/science/nasa-headed-towards-giant-golden-asteroid-that-could-make-everyone-on-earth-a-billionaire

98. Garretson, "Solar Power in Space?"

99. Greg Autry, "Robert Zubrin Makes a Strong Case for Space Development," *Forbes*, May 16, 2019, accessed August 15, 2019. https://www.forbes.com/sites/gregautry/2019/05/16/robert-zubrin-makes-a-strong-case-for-space-development/

100. Namrata Goswami and Peter Garretson, "Critical Shifts in India's Outer Space Policy," *The Diplomat*, April 16, 2019, accessed August 15, 2019. https://thediplomat.com/2019/04/critical-shifts-in-indias-outer-space-policy/

101. Asterank, accessed August 15, 2019, http://www.asterank.com/

102. Kris Holland, "Space Based Solar Power: A Solution to the Carbon Crisis," *You Tube*, April 20, 2016, accessed August 15, 2019, https://www.youtube.com/watch?v=zrcoD_vHzxU

103. Karim Raslan, "Ignore Silk Road Hubris, China Will Grow Old Before It Gets Rich," *South China Morning Post*, May 23, 2017, accessed August 15, 2019. https://www.scmp.com/week-asia/opinion/article/2095326/opinion-ignore-silk-road-hubris-china-will-grow-old-it-gets-rich

104. Ganeshan Wignaraja, "India Could Overtake China—One Day," *Nikkei Asian Review*, January 19, 2018, accessed August 15, 2019. https://asia.nikkei.com/Economy/India-could-overtake-China-one-day

105. Richard D. Johnson and Charles H. Holbrow, eds., "Space Settlements: A Design Study," 413, Scientific and Technical Information Office, National Aeronautics and Space Administration, 1977, available online at https://settlement.arc.nasa.gov/75SummerStudy/Chapt7.html "Assuming 13 km of total area per person, it appears that space habitats might be constructed that would provide new lands with a total area some 3,000 times that of the Earth." (under Section 7, "Limits of Growth")

106. John S. Lewis, "'Demandite' and Resources in Space," Dr. John S. Lewis website (author of *Mining the Sky* and *Asteroid Mining 101* and Chief Technical Advisor Deep Space Industries (DSI)), January 13, 2011, accessed January, 21, 2019, http://www.johnslewis.com/2011/01/demandite-and-resources-in-space.html

107. Robert Walker, "Asteroid Resources Could Create Space Habs For Trillions; Land Area Of A Thousand Earths," *Science 2.0*, September 17, 2018, accessed October 11, 2019, https://www.science20.com/robert_walker/asteroid_resources_could_create_space_habs_for_trillions_land_area_of_a_thousand_earths-116541

108. Philip T. Metzger, Anthony Muscatello, Robert P. Mueller, and James Mantovani, "Affordable, Rapid Bootstrapping of Space Industry and Solar System Civilization," *ARVIX* (preprint), December 10, 2016, accessed February 3, 2019, https://arxiv.org/pdf/1612.03238v1.pdf (paper is also available at https://ascelibrary.org/doi/10.1061/%28ASCE%29AS.1943-5525.0000236)

109. "Looking at Earth from Space," NASA, August 1994: 62, accessed February 3, 2019, https://er.jsc.nasa.gov/seh/earth_glossary.pdf

110. A. J. R. Russel Wood, *The Portuguese Empire, 1415-1808: A World on the Move* (Washington, DC: John Hopkins University Press, 1998).

111. Kenneth Pletcher, *The Age of Exploration: From Christopher Columbus to Ferdinand Magellan* (New York: Britannica Publications, 2013).

112. "The Age of Imperialism (1870-1914)," Tamaqua K12, July 15, 2007, accessed January 21, 2019, https://www.tamaqua.k12.pa.us/cms/lib07/PA01000119/Centricity/Domain/119/TheAgeofImperialism.pdf

113. Daniel R. Headrick, "The Tools of Imperialism: Technology and the Expansion of European Colonial Empires in the Nineteenth Century," *The Journal of Modern History* 51, no. 2 (June 1979): 231–63.

114. Authors participated in a futures study workshop for U.S. Air Force Space Command, the report of which was published. Air Force Space Command, "The Future of Space 2060 and Implications for U.S. Strategy: Report on the Space Futures Workshop," September 5, 2019, accessed October 16, 2019. https://www.afspc.af.mil/Portals/3/documents/Future%20of%20Space%202060%20v2%20(5%20Sep).pdf?ver=2019-09-06-184933-230

115. Michael Turtle, "The World's Oldest Sovereign State," *Time Travel Turtle*, n.d, accessed February 3, 2019. https://www.timetravelturtle.com/visiting-san-marino-oldest-country/

116. Winchell Chung, "Mission Delta-V and Flight Times," *ProjectRho*, June 7, 2012, accessed February 3, 2019. http://www.projectrho.com/public_html/rocket/appmissiontable.php

117. Christopher Columbus, *The Four Voyages of Christopher Columbus,* translated by J. M. Cohen (New York: Penguin, 2004); also see "The Four Voyages of Christopher Columbus," Constant Contact, May 17, 2018, accessed, February 3, 2019. https://myemail.constantcontact.com/The-4-Voyages-of-Columbus.html?soid=1108762609255&aid=c0r8lhGZxC4

118. "Adjusted for a distance of 1.2 AUs, that means that a ship equipped with a NTP/NEC propulsion system could make the trip in about 293 days (about nine months and three weeks)." See Matt Williams, "How Long Does it Take to get to the Asteroid Belt?," *Universe Today*, August 10, 2016, accessed February 3, 2019. https://www.universetoday.com/130231/long-take-get-asteroid-belt/

119. Chung, "Mission Delta-V and Flight Times."

120. Kenneth N. Waltz, "The Emerging Structure of International Politics," *International security* 18, no. 2 (1993): 44–79.

121. Nitisha, "5 Factors that Affect the Economic Growth of a Country," *Economic Discussion Net*, January 10, 2015, accessed February 3, 2019. http://www.economicsdiscussion.net/economic-growth/5-factors-that-affect-the-economic-growth-of-a-country/4199

122. Prateek Agarwal, "Production Possibilities Frontier," *Intelligent Economist*, December 25, 2017, accessed February 3, 2019. https://www.intelligenteconomist.com/production-possibilities-frontier/

123. No author mentioned, "Richest People in the World," *CBS News*, n.d., accessed January 30, 2019. https://www.cbsnews.com/pictures/richest-people-in-world-forbes/21/

124. Elvis Picardo, "Eight of the World's Top Companies Are American," *Investopedia*, October 20, 2018, accessed January 30, 2019. https://www.investopedia.com/articles/active-trading/111115/why-all-worlds-top-10-companies-are-american.asp

125. No author mentioned, "List of Countries by Projected GDP," *Statistics Times*, May 6, 2018, accessed January 21, 2019. http://statisticstimes.com/economy/countries-by-projected-gdp.php

126. "DoD Releases Fiscal Year 2019 Budget Proposal," *Department of Défense*, February 12, 2018, accessed January 21, 2019. https://www.defense.gov/News/News

-Releases/News-Release-View/Article/1438798/dod-releases-fiscal-year-2019-budget-proposal/

127. Anzish Mirza, "What Are the National Debt, Debt Ceiling and Budget Deficit?," *US News and World Report*, April 13, 2017, accessed January 21, 2019. https://www.usnews.com/news/economy/articles/2017-04-13/what-are-the-national-debt-debt-ceiling-and-budget-deficit

128. No author mentioned, "List of Countries by Projected GDP."

129. Current World Population, Worldometers, accessed January 30, 2019, http://www.worldometers.info/world-population/

130. Christopher J. Rhodes, "Solar Energy: Principles and Possibilities," *Science Progress* 93, no. 1 (2010): 37–112; "These are hot topics these days . . .," The World Counts, Accessed February 20, 2020, https://www.theworldcounts.com/stories/current_world_energy_consumption

131. Thomas Barrabi, "'New Space Race' Could Yield $1 Trillion Industry: Wilbur Ross," *Fox Business News*, November 29, 2018, accessed January 30, 2019, https://www.foxbusiness.com/business-leaders/new-space-race-could-yield-1-trillion-industry-wilbur-ross; See also, Kevin O'Connell, "Remarks on the Trillion Dollar Space Economy," Office of Space Commerce, November 27, 2018, accessed January 30, 2019, https://www.space.commerce.gov/remarks-on-the-trillion-dollar-space-economy/

132. "Space: Investing in the Final Frontier," *Morgan Stanley*, November 7, 2018, accessed January 30, 2019, https://www.morganstanley.com/ideas/investing-in-space

133. Rich Smith, "The $1.1 Trillion Space Industry Prediction You Can't Afford to Miss," *The Motley Fool*, February 16, 2018, accessed March 2, 2020, https://www.fool.com/investing/2018/02/16/the-11-trillion-space-industry-prediction-you-cant.aspx

134. Brian Higgenbotham, "The Space Economy: An Industry Takes Off," *U.S. Chamber of Commerce*, November 11, 2018, accessed January 30, 2019, at https://www.uschamber.com/series/above-the-fold/the-space-economy-industry-takes

135. Michael Sheetz , "The space industry will be worth nearly $3 trillion in 30 years, Bank of America predicts," *CNBC*, October 31, 2017, accessed January 30, 2019, https://www.cnbc.com/2017/10/31/the-space-industry-will-be-worth-nearly-3-trillion-in-30-years-bank-of-america-predicts.html

136. "Query of Most Valuable Asteroids," *Asterank*, accessed January 21, 2019. http://www.asterank.com/ a database of over 600,000 asteroids, ranks them according to their composition and accessibility. Asteroid Ryugu tops the list as the most cost-effective with a value of $82.76 billion, and an estimated profit of $30.07 billion. But there are nearly 1000 asteroids in the database whose profits exceed a trillion each, and over 600 whose estimated profits would exceed $100 trillion.

Already NASA has plans to visit 16 Psyche, a rare all-metal asteroid, estimated to be worth $10,000 quadrillion all on its own. Parnell Brid-Aine, "NASA will Reach Unique Metal Asteroid Worth $10,000 Quadrillion Four Years Early," *Forbes*, May 26, 2017, accessed January 21, 2019. https://www.forbes.com/sites/bridaineparnell/2017/05/26/nasa-psyche-mission-fast-tracked/#506d52db4ae8

137. Howard Bloom, "China's New Silk Road vs. America's Platinum Highway in the Sky," *Over the Horizon Journal*, September 24, 2018, accessed January 30, 2019. https://othjournal.com/2018/09/24/chinas-new-silk-road-vs-americas-platinum-highway-in-the-sky/

138. "Query of Most Valuable Asteroids"; Brid-Aine, "NASA will Reach Unique Metal Asteroid."

139. "16 Psyche," NASA, May 16, 2019, accessed February 2, 2019, https://solarsystem.nasa.gov/asteroids-comets-and-meteors/asteroids/16-psyche/in-depth/

140. Brid-Aine, "NASA Will Reach Unique Metal Asteroid".

141. A quadrillion (1 × 10^15) is thousand trillion (1,000T), a quintillion (1 × 10^18) is a million trillion (1,000,000T). A quintillion is a 1000 quadrillion.

142. Jeff Desjardins, "There's Big Money to be Made in Asteroid Mining," *Business Insider*, November 3, 2016, accessed January 21, 2019. http://www.businessinsider.com/the-value-of-asteroid-mining-2016-11

The original NASA source cited appears to be: "NEAs as Resources," *NASA*, May 2018. https://cneos.jpl.nasa.gov/about/nea_resource.html, accessed via archive January 30, 2019 at https://web.archive.org/web/20180518005833/https://cneos.jpl.nasa.gov/about/nea_resource.html

143. "Discovery Statistics," NASA, accessed February 3, 2019. https://cneos.jpl.nasa.gov/stats/totals.html

144. Mike Wall, "About 17,000 Big Near-Earth Asteroids Remain Undetected: How NASA Could Spot Them," *Space.com*, April 10, 2018, accessed February 3, 2019. https://www.space.com/40239-near-earth-asteroid-detection-space-telescope.html

145. "Asteroids," NASA, accessed February 3, 2019. https://solarsystem.nasa.gov/asteroids-comets-and-meteors/asteroids/in-depth/

146. Tom Murphy, "Galactic-Scale Energy," *Do the Math (UCSD)*, July 7, 2012, accessed February 3, 2019. https://dothemath.ucsd.edu/2011/07/galactic-scale-energy/ Murphy goes on to mention that at such a constant compounding growth rate, we would require the entire energy of the sun in 1,400 years, and the entire galaxy in 2,500 years.

147. David Chandler, "Shining Brightly," *MIT News*, October 26, 2011, accessed February 3, 2019, http://news.mit.edu/2011/energy-scale-part3-1026

148. NASA, "Looking at Earth from Space," *NASA*, August 1994, page 62, accessed February 3, 2019, https://er.jsc.nasa.gov/seh/earth_glossary.pdf

149. For a more in-depth discussion with calculations, see Question 27 of NASA, "Amount of Energy the Earth Gets from the Sun," Ask Us: Sun, NASA GFC Cosmicopia, Accessed February 3, 2019, https://cosmicopia.gsfc.nasa.gov/qa_sun.html

150. The Sun's Energy, University of Tennessee Institute of Agriculture, accessed February 3, 2019, https://ag.tennessee.edu/solar/Pages/What%20Is%20Solar%20Energy/Sun's%20Energy.aspx

151. David R. Criswell, "Lunar Solar Power System and Lunar Exploration," *Workshop on Moon Beyond*, 2002, p. 1, accessed 15 February 2008. http://www.1pi.usro.edu/meetings/moon2002/pdf/3004.pdf

152. James M. Snead, "Energy, SSP, and Jumpstarting America's Spacefaring Future: Why America's Pro-Human Spacefaring Community Should Strongly

Advocate Starting Commercial Development of Space Solar Power," *Presentation at the 2009 International Space Development Conference (ISDC)*, June 10, 2009, accessed October 11, 2019. https://www.mikesnead.com/resources/spacefaring/presentation_energy_ssp_and_jumpstarting_america's_spacefaring_future_version_2_for_distribution.pdf. According to Snead, geosynchronous orbit could, in theory, hold as many as 66,200 5-GW SPSs at 100-percent slot use, supplying as much as 331 TW of power against a 2100 projected requirement of only 55 TW to supply all of humanity's energy needs.

153. Gerard K. O'Neill, *The High Frontier: Human Colonies in Space* (New York: William Morrow and Company, 1977).

154. Walker, "Asteroid Resources Could Create Space."

155. Mike Combs, "The Case for Space," *The Space Front*, Fall 1999, accessed January 30, 2019. https://space.nss.org/settlement/MikeCombs/case_spc.htm

156. 10 million billion could also be expressed as 10,000 Trillion, or 10 Quadrillion people. Earth's population is currently 7.7 billion according to United Nations Department of Economic and Social Affairs, https://www.un.org/development/desa/publications/world-population-prospects-2019-highlights.html

157. John S. Lewis, "'Demandite' and Resources in Space," Dr. John S. Lewis website (author of *Mining the Sky* and *Asteroid Mining 101* and Chief Technical Advisor Deep Space Industries (DSI)), January 13, 2011, accessed January, 21, 2019, http://www.johnslewis.com/2011/01/demandite-and-resources-in-space.html

158. Blue Origin, "Our Vision," accessed February 3, 2019. https://www.blueorigin.com/

159. Alan Boyle, "Jeff Bezos: 'We Will Have to Leave this Planet . . . and it's Going to Make this Planet Better,'" *GeekWire*, May 29, 2018, accessed February 3, 2019. https://www.geekwire.com/2018/jeff-bezos-isdc-space-vision/

160. Ashlee Vance, *Elon Musk: Tesla, SpaceX, and the Quest for a Fantastic Future* (New York: HarperCollins, 2017).

161. "SpaceX Interplanetary Tranport System (ITS)," You Tube, September 27, 2019, accessed October 06, 2019, https://www.youtube.com/watch?v=0qo78R_yYFA

162. Kenneth Chang, "Elon Musk Sets Out SpaceX Starship's Ambitious Launch Plan," *The New York Times*, September 28, 2019, accessed October 06, 2019. https://www.nytimes.com/2019/09/28/science/elon-musk-spacex-starship.html

163. Sarah Knapton, "Elon Musk: We'll Create a City on Mars with a Million Inhabitants," *The Telegraph UK*, June 21, 2018, accessed January 21, 2019. https://www.telegraph.co.uk/science/2017/06/21/elon-musk-create-city-mars-million-inhabitants/

164. "Jeff Bezos discusses space flight and his vision for Blue Origin" (at 12 min 01 seconds into talk), filmed October 23, 2016, at the 2016 Pathfinder Awards at Seattle's Museum of Flight," *YouTube video*, posted October 23, 2016, 13:14, accessed January 21, 2019. https://www.youtube.com/watch?v=VNwE3sRWxHw#action=share

165. Catherine Clifford, "Jeff Bezos Dreams of a World with a Trillion People Living in Space," *CNBC*, May 1, 2018, accessed January 21, 2019. https://www

.cnbc.com/2018/05/01/jeff-bezos-dreams-of-a-world-with-a-trillion-people-living-in-space.html; also see Davenport, *The Space Barons*.

166. David R. Criswell, "Return on Investment (ROI) from 'New Worlds & Lunar Solar; Jeff Bezos, Power,' (7pp.) International Conference: Moon Base—A Challenge for Humanity," Washington Workshop: Jamestown on the Moon, Washington Academy of Sciences, Washington, DC, October 12, 2005.

167. Metzger, Muscatello, Mueller, and Mantovani, "Affordable, Rapid Bootstrapping."

168. Metzger, Muscatello, Mueller, and Mantovani, "Affordable, Rapid Bootstrapping."

169. Brian Wang, "Exponential Industrialization of Space is More Important than Combat Lasers and Hypersonic Fighters," *Next Big Future*, October 26, 2017, accessed February 3, 2019. https://www.nextbigfuture.com/2017/10/exponential-industrialization-of-space-is-more-important-than-combat-lasers-and-hypersonic-fighters.html

170. The hype curve charts the inflated public expectations after the introduction of a new technology, disillusionment as it initially underperforms hyped expectations, only to ultimately pass and surpass early expectations. See: Scott Brinker, "One thing everybody forgets about Gartner's hype cycle: The gaps between hype and reality are opportunities," Think Growth, April 2, 2018, accessed August 15, 2019. https://thinkgrowth.org/one-thing-everybody-forgets-about-gartners-hype-cycle-ecfe7e9de8ff

171. Charles Miller, "Ex-NASA exec: Gingrich Moon Colony Lost in the Laughter," *CNN*, February 7, 2012, accessed August 15, 2019. https://www.cnn.com/2012/02/07/opinion/miller-gingrich-space-policy/index.html

172. Aviation Quotations, Great Aviation Quotes: Predictions, accessed August 15, 2019. http://www.aviationquotations.com/predictions.html; also see Glenn Firebaugh, *Seven Rules for Social Research* (Princeton and Oxford: Princeton University Press, 2008), 27.

173. The Wright Brothers, Inventing a Flying Machine, Smithsonian National Air and Space Museum, accessed September 11, 2019. https://airandspace.si.edu/exhibitions/wright-brothers/online/fly/1903/

174. Jay Bennett, "One Chart Shows How Much SpaceX Has Come to Dominate Rocket Launches. The Space Exploration Technologies Corporation is starting to take the lion's share of global launches," Popular Mechanics, July 13, 2017, accessed August 15, 2019, https://www.popularmechanics.com/space/rockets/a27290/one-chart-spacex-dominate-rocket-launches/

175. Jackie Wattles, "SpaceX, One of the World's Most Valuable Private Companies, Just Got More Valuable," *CNN Business*, January 3, 2019, accessed August 15, 2019. https://www.cnn.com/2019/01/03/business/spacex-valuation-500-million-fundraising-round/index.html

176. Ali Somarin, "Unconventional Oil Exploration, Part 3: Ultra-deepwater Oil," *Thermofisher*, October 21, 2014, accessed August 15, 2019, https://www.thermofisher.com/blog/mining/unconventional-oil-exploration-part-3-ultra-deepwater-oil/

177. Matt Egan, "America is Now the World's Largest Oil Producer," *CNN Money*, September 12, 2018, accessed August 15, 2019. https://money.cnn.com/2018/09/12/investing/us-oil-production-russia-saudi-arabia/index.html

178. Arthur C. Clark, *The Exploration of Space* (New York: Harper & Brothers Publishers, 1951), 111.

179. SF Encyclopedia, "Clarke's Laws," August 2, 2016, accessed August 15, 2019. http://www.sf-encyclopedia.com/entry/clarkes_laws

180. Quoted in Joseph F. Coats and Jennifer Jarratt, *What Futurists Believe* (Bethesda, MD: World Futures Society, 1989), 66, accessed July 6, 2020. http://www.laprospective.fr/dyn/francais/memoire/texte_fondamentaux/what-futurists-believe-j-f-coates.pdf

181. Elon Musk, "The Future We're Building—And Boring," *TED*, 2017, accessed February 3, 2019. https://www.ted.com/talks/elon_musk_the_future_we_re_building_and_boring/transcript?language=en

182. Valentine Lee, "A Space Roadmap: Mine the Sky, Defend the Earth, Settle the Universe," *Space Studies Institute (SSI)*, May 2002, accessed January 21, 2019. http://ssi.org/reading/papers/space-studies-institute-roadmap/

183. Anthony Galli, "Elon Musk Life Purpose Explained," *Mission.org*, October 10, 2017, accessed September 11, 2019. https://medium.com/the-mission/elon-musk-life-purpose-explained-d08e7f37ce28

184. Robert K. Merton, "The Thomas Theorem and The Matthew Effect," Social Forces, 74/2 (December 1995), 379–424, available at http://garfield.library.upenn.edu/merton/thomastheorem.pdf

185. Halford Mackinder, "The Geographical Pivot of History," *The Geographical Journal* (Royal Geographical Association), April 1904. https://rgsibg.onlinelibrary.wiley.com/doi/abs/10.1111/j.0016-7398.2004.00132.x

186. Il'ič Lenin, "Imperialism: The Highest Stage of Capitalism".

187. Authors' visit to the Oregon Trail Interpretive Center, Plaque, Baker City Oregon, photo taken July 26, 2019.

188. Tim Urban, "The Cook and the Chef: Musk's Secret Sauce," *Wait but Why*, November 6, 2015, accessed February 3, 2019. https://waitbutwhy.com/2015/11/the-cook-and-the-chef-musks-secret-sauce.html

189. Matheson Calum, "The Sublime Rhetoric of Pascal's Wager," *Argumentation and Advocacy*, 53/4, 2017: 271–86. doi: 10.1080/00028533.2017.1375737

190. Stéphane J. Baele and Catarina P. Thomson, "An Experimental Agenda for Securitization Theory," *International Studies Review* 19, no.4 (December 2017): 646–66.

191. The term lawfare was first utilized by two PLA officers in 1999 in a publication called *Unrestrictred Warfare*, that referred to "a nation's use of legalized international institutions to achieve strategic ends." See Col. Qiao Liang and Col. Wang Xiangsui, *Unrestricted Warfare: China's Master Plan to Destroy America* (Panama City, Panama: Pan American Publishing Company, 2002). The term was later popularized by Air Force Colonel (and later General) Charles Dunlap, in a 2001 paper titled, "Law and Military Interventions: Preserving Humanitarian Values in 21st Conflicts," prepared for the Humanitarian Challenges in Military Intervention Conference Carr Center for Human Rights Policy, Kennedy School of Government, Harvard University Washington, DC, November 29, 2001. https://people.duke.edu/~pfeaver/dunlap.pdf, accessed January 7, 2020.

Chapter 2

The Role of Myths, History, and Strategic Culture on Space-Based Resources

Societies have always been inspired by the myths and historical achievements of their ancestors. Especially, those myths that have played an instrumental role in offering a perspective of who we are, what should we be aspiring to, how could our policy choices be informed by those ancient wisdoms, and what can be avoided as a consequence.[1] We define myth as a traditional story explaining the worldview of a group of people. While these stories are mythical (meaning they are constructed as part of an imagination and passed down), generations enjoy these stories and accept them as part of their roots. Myths are thought not to be based on facts and are perceived as creative imaginations of the human mind. The origin of the word comes from the Greek word *mythos*, meaning story or word. Myths occupy a special position in societal memory and are a common set of stories shared by a group.

According to Mary Magoulick, "myths are symbolic tales of the distant past (often primordial times) that concern cosmogony and cosmology (the origin and nature of the universe), may be connected to belief systems or rituals, and may serve to direct social action and values."[2] Myths express certain values and are instrumental narratives that explain human nature. The setting for myths is thrown back to primordial times, connected to the origin of the universe. Most of the characters within a mythology are proto-human or deific. Myths reflect the particular foundational practices of a society, especially informing its epistemology of how to organize thoughts. Myths are such that they appear impossible in present culture but are constructed stories of a distant past when civilization just started, and the mythical protagonists exhibit extraordinary personality traits, capabilities, and feats. Significantly, countries value their ancient myths as they represent who, as a civilization, they were, how they were formed, and where they have arrived. Myths offer us a window into the soul of our ancestors and their complex intellectual engagement with the world; what they valued or feared. Consequently, understanding the context, the individual as well as the historical period within which myths first originated offers a perspective of how and why they were constructed in the first place.

Myth is the opposite of science in that they cannot be tested for their truth, and yet, it is rather interesting that societies choose to name scientific missions, especially those that leave Earth and journey into outer space based on characters from mythical stories. This can be explained as myths being represented as a metaphor, as a narrative that has symbolic appeal across a civilization, and what it implies and messages to the world. Myths often represent what intrigues and plagues human civilization: the paradoxes of life. Characters within these myths express their preferences and how they deal with those paradoxes, offering the descendants of the originator of these myths, a metaphor to the soul of their ancestors. The character of these myths are often gods and goddesses and evokes a sense of mystery and awe. They represent a reflective narrative and can work as a call for social action. Myths can be interpreted as being representative of the "context reality" in which they are conceived, the narrator's perception of that society, as well as reflective of a rational, theoretical, or analytical perspective of that time. It reflects a particular context as a metaphor that utilizes the writer's imagination or a society's shared perspective of a myth that takes hold. These are not factual narrations like a historical text, for instance, Thucydides' *Peloponnesian War*,[3] but definitely address societal conversations and offer a record of the time. The writer is thought of by society as inspired by something divine or higher; therefore, myths can be related within historical documentation of that societal context as well as religious documentation. Contrary to perception, however, poets who wrote myths or authors who offered prose to the construction of myths had to stick to a particular guideline. But that did not limit several versions of the same myth. Myths are inspirational storytelling but that cannot substitute history. Myths were utilized to explain natural occurrences in a context where science was not the norm. Myths have and are utilized to lay claim to land.

Philosophers of ancient tradition often invoked myths to give expression to their worldview and values. It was often more effective to put it in the form of a myth to their audiences than to offer rather abstract ideas, without any mythically constructed character, that would take a life of its own, embodying these values. One of the most interesting embodiments of such a technique was by the Greek philosopher Zeno and his use of Greek mythical hero Achilles to demonstrate that space and time are divisible; that no matter how fast Achilles ran, he would never catch up with the tortoise since it would have crawled further from its earlier position.[4] Zeno was perhaps the first Greek philosopher that moved away from poetry to prose and offered a dialectic method of writing; he would present an argument that he opposed, and then offer a contradictory argument or conclusion so that the first argument is negated. This, he conducted, by highlighting certain assumptions in an argument that could be abandoned.[5]

THE ROLE OF MYTHS IN OUTER SPACE

Significantly, modern society including democracies have peered deep into their past to utilize names from their civilizational ethos for their space missions. While the paradoxical nature of that *naming enterprise* is clear to all, myths and mythical characters can never be tested for their worth and represent an interpretative metaphor for their civilizations, and while space missions purely depend on science and replicability, such names represent a society, its belief system, and its value system as well.

THE U.S. MOON MISSIONS

The Greek Olympian God *Apollo* was the guardian of the city of Delphi. He was known as the God of the Light and the Sun, of oracles, archery, plague, and knowledge in Greek mythology.[6] Most remembered and represented by the statue of Apollo in the temple of Zeus in Olympia, those who revered Apollo in ancient Greece hoped to build qualities of courage, tact, moderation, wisdom, and self-reflection, and develop a unitary constitutive self. Apollo's connection to the U.S. space program is interesting. The National Aeronautics and Space Administration (NASA) had conceived of a manned circumlunar mission in 1960 that would go beyond project Mercury. Project Mercury was named after the Roman God Mercury, his Greek equivalent being *Hermes*, the Olympian God of interpreters. He was also the God of athletes, as well as swift interpretation. The Roman God Mercury was also known as the God of the Winds, due to his speed. In continuation of naming space projects after Roman and Greek Gods, Director of Space Flight Development at NASA, Abe Silverstein, proposed the name Apollo, mostly because of the Olympian God's connection as God of the Sun. In a NASA history page on Apollo, the following is highlighted in relation to the name:

> Most significantly he was the god of the sun.
> In his horse-drawn golden chariot, Apollo pulled the sun in its course across the sky each day.[7]

The name was approved by NASA and publicly announced on July 28–29, 1960, at the NASA/Industry Program Plans Conference in Washington. The mission for Project Apollo changed when President John F. Kennedy's proposal to Congress for a manned lunar landing on May 25, 1961, was approved. Similarly, when President Donald Trump's administration called for a refocus on the Moon in his Space Policy Directive 1,[8] and which was publicly reiterated by Vice President Mike Pence, during his March 2019 speech in Huntsville, Alabama during the 5th meeting of the National Space Council, when he stated that the United States will send a man and woman back to the Moon's south pole by 2024,[9] NASA went ahead and established *Project Artemis*. The strategic significance of the Moon was, of course, highlighted by Pence.

> To reach the Moon in the next five years, we must select our destinations now. NASA already knows that the lunar South Pole holds great scientific, economic, and strategic value. But now it's time to commit to go there. But in order to accomplish this, NASA must transform itself into a leaner, more accountable, and more agile organization. If NASA is not currently capable of landing American astronauts on the Moon in five years, we need to change the organization, not the mission. To continue to build a world-class workforce, NASA needs the authority to recruit, train, and motivate the world's best and brightest scientists, engineers, and managers, and to remove any barriers standing in their way. And that includes building new and renewed partnerships with America's pioneering space companies and entrepreneurs.[10]

Project *Artemis* is named after another Olympian goddess, *Artemis*, twin sister of Apollo. *Artemis* is the goddess of wild animals, chastity, vegetation. Artemis was the daughter of Zeus and Leto.[11] As per NASA, "Artemis was the twin sister of Apollo and goddess of the Moon in Greek mythology. Now, she personifies our path to the Moon as the name of NASA's program to return astronauts to the lunar surface by 2024, including the first woman and the next man. When they land, our American astronauts will step foot where no human has ever been before: the Moon's South Pole."[12] Artemis is sometimes associated as the Goddess of the Moon. Homer referred to Artemis as the mistress of wild animals.[13] She is one of the major twelve Olympian deities.

CHINA'S LUNAR EXPLORATION PROGRAM (CLEP)

CLEP is based on the philosophy and strategy of ensuring that China's space program equips and enhances its status as a great power. This implies that its space program is viewed as a critical contributor to its national development goals, based on indigenous technical capacity. As per its 2016 White Paper on space,

> China takes independent innovation as the core of the development of its space industry. It implements major space science and technology projects, strengthens scientific exploration and technological innovation, deepens institutional reforms, and stimulates innovation and creativity, working to promote rapid development of the space industry.[14]

As per the White Paper, the CLEP is one of the key projects of China's deep space exploration program. This lunar probe began with the launch of *Chang'e 1*, riding on a *Long March 3A* rocket on October 24, 2007. The *Chang'e 1* surveyed the lunar surface for resources such as *Helium-3*. At a cost of $180 million, the mission launched China's ambitious lunar mission. The *Chang'e 1* carried with it the Chinese national anthem, as well as thirty other folk tunes from China to the Lunar orbit. The objectives of that first mission were clearly stated in a paper published by scientists from the Centre for Space Science and Applied Research (CSSAR) and the Xi'an Institute of Optics and Precision Mechanics, Chinese Academy of Sciences (CAS) in the *Journal of Earth System Science*. These objectives were to "obtain a three-dimensional stereo image of the lunar surface; determine the distribution of elements and estimate their abundance on the lunar surface, survey the thickness of the lunar soil and access the existence and quantity of Helium-3."[15] The *Chang'e 2* (launched in October 2010) achieved the difficult feat of not only flying to the lunar orbit but then changing orbits and flying to the Earth-Sun Lagrange-L2 point. It then departed the L2 point and flew by the *Asteroid Toutatis*, at a distance of 2 miles (3.2 km). At that time, Wu Weiren, the chief designer of China's lunar program, stated, "The success of the extended missions also embodies that China now possesses spacecraft capable of interplanetary flight."[16] For the first time for China, it soft-landed the *Chang'e 3* on the *Mare Imbrium*, an ancient volcanic plain. This mission also carried the Yutu rover, named after the mythical jade rabbit, a friend of the Chinese Moon goddess, *Chang'e* after whom the lunar mission is named.

The legend of *Chang'e* is critical to understanding the motivations and ambitions of China's Moon mission. Legend has it that she stole the drug of immortality from her husband, *Hou Yi*, and to escape his wrath, sought refuge on the Moon. Hou Yi was impeded in his attempts to get to *Chang'e* by Yutu. On the 15th day of the eighth lunar month, *Chang'e* is celebrated across China during the autumn festival known as *Zhongqiu Jie*. This festival is also celebrated in Vietnam as a festival of abundance, harmony, and good luck. *Hou Yi* is credited with shooting down nine of the ten stars that circled the Earth. This resulted in a cooler Earth. It is expressed that *Hou Yi* was tyrannical and *Chang'e* partook of the immortal drug to save people from his tyrannical rule.[17] There is another legend that believes that *Hou Yi* was given the excelsior of life for shooting down the nine suns. In this legend, after he shot down the suns, *Hou Yi*, a common hunter and an expert archer, was made king. Once he was made king, *Hou Yi* married *Chang'e*. It was in that context that *Chang'e* is believed to have swallowed the immortality pill, that *Hou Yi* gave her for safekeeping, to share it with her later.[18] The symbolism of *Chang'e* is that the Moon is her permanent abode. Given that context, the CLEP is aimed at long-term sustainable presence on the Moon, to make it a permanent home for human settlement.

INDIA'S SPACE PROGRAM

India's space program can be seen as reflecting a similar pattern of behavior, if not from myths but its ancient past. In April 1975, when India launched its first satellite, the *Aryabhata*, it was named after ancient Indian astronomer and mathematician, Aryabhata, from the fifth century CE. Aryabhata lived in Kusumapura, close to present-day Patna in Bihar. He composed two pieces of work, *Aryabhatiya* and the *Aryabhatasiddhanta,* the latter of which has been lost. Aryabhata's work is one of the earliest works in the ancient world, where the start of the day was assigned to the stroke of midnight. His connection to astronomy is eternalized by his work on predicting eclipses, and his contribution to the world of mathematics.[19] At the time of his life, the Gupta Empire was ruling those parts of ancient India and their capital; Pataliputra was known as the center of learning. The Indian lunar mission is called *Chandrayaan*, the translation of which means, a craft to the Moon. *Gaganyaan* (sky craft) is the name of the Indian human spaceflight mission to be launched by December 2021. The Indian Mars mission is called *Mangalyaan*, or craft to Mars. India's planned Venus mission is called *Sukrayaan*, or craft to Venus.

India's mission to the Sun is called *Aditya*, or the Sun God, in Indian mythology. *Aditya* or the Sun was born to the goddess *Aditi* and the sage, *Kasyapa*. *Aditya* is viewed as the creator of everything in the universe, offering warmth to Earth. *Vishnu*, another Indian God, is also viewed as an embodiment of *Surya* (Sun) under the name Surya-Narayana. The first embodiment of *Aditya* is in the *Rig Vedas*, composed between 1500 BCE and 1000 BCE. In the Vedic world, *Surya* or *Aditya* is viewed as the chief of the nine planets; this includes the Moon, the Sun, Mars, Jupiter, Venus, and Saturn, as well as the solar and lunar shadows of *Rahu* and *Ketu*, collectively known as the *Navagraha* or the nine planets.[20] So, here too you see how societal imagination, ancient myth, and history have been utilized to name space missions of today.

In all these missions across different countries, there is something that does stand out. Each country's perception of who they are, historically and from a societal point of view, is inspired by their own societal myths as well as meanings drawn from their history. What do these imply from the point of view of their strategic culture and outer space? While chapters on individual countries will further elaborate upon each country's strategic culture, it is useful to offer a general perspective of the strategic culture and outer space within the context of societal history, perceptions, and myths.

STRATEGIC CULTURE AND OUTER SPACE

In terms of strategic culture and outer space, states' ability to perceive the meaning of space societally and politically is conditioned to a significant extent upon narrative. Among spacefaring states, and those who aspire to be space faring, the topic of space has played, and continues to play, an important role in a society's intellectual life and self-image. Inspired by ancient myths in their societies or of those who inspired these myths, achievements in space were viewed as the highest form of success. The United States and the Union of Soviet Socialist Republics (USSR) projected space power as a form of political power projection of their states' attractive political ideologies and societies. When *Sputnik* burst into the skies in 1957, followed by Yuri Gagarin, as the first man to get to space on April 12, 1961, the way humanity viewed space changed forever: that now, we can go there; it's no longer inaccessible. The success of the Soviet Union was followed by the race to the Moon, inspired and instigated by John F. Kennedy's famous May 25, 1961, speech before a Special Joint Session of Congress. In that speech, Kennedy argued that

> We choose to go to the moon in this decade and do the other things, not because they are easy, but because they are hard, because that goal will serve to organize and measure the best of our energies and skills, because that challenge is one that we are willing to accept, one we are unwilling to postpone, and one which we intend to win, and the others, too.[21]

At that time, the strategic cultures of both the United States and the USSR were informed by the logic of competition over two divergent ideologies, that of democratic capitalism or communist socialism led by a single party. Consequently, over the years, within the United States, there are several strands of strategic culture, one inspired by realism where the focus is on national security and state interest. The second strand is an inspiration-based strategic culture of liberalism where the focus is on building institutionalized norms and collective interest. The third is where space is viewed as a means to an end, as constitutive of societal identity and values, to channel something wider than just the state. Strategic culture mostly offers a conceptual understanding of the culture of the society as well as the national preferences/idiosyncrasies that go into shaping and informing national security. Strategic culture is based on a set of beliefs, history, societal narratives, perceptions, and normative preferences, that shape the contours of a state's foreign policy, the choices it makes with regard to its preferences, and its position in international relations. In this regard, the U.S. space policy definitely prioritizes the United States being the lead actor in space, as well as utilize cooperative frameworks to maintain that U.S. primacy. As such, after the turn from competition with the Soviet Union, the United States went ahead and collaborated with Russia in space, primarily to take advantage of Russian space technology especially launches of astronauts, but also to ensure that the Russian space program does not challenge U.S. primacy in space, in the future.

The U.S. strategic culture very clearly prioritizes a future space order in which the United States is in the lead, be it for space exploration or utilization of space for peaceful purposes. This is affected in the manner in which it crafts its National Space Policy, as well as articulates its ambitions on outer space. There are several key factors that could affect the way it visualizes its space ambitions, chief among which is to articulate ambitions of leadership in the final frontier. The U.S. spirit has been to always be the leader and at the forefront of something which challenges the spirit of its soul; to rise to a challenge that is beyond oneself. This can be seen in the articulation of its space policy of exploration of something unknown, Mars, and beyond, with deep space exploration missions. You see this glorification of exploring something unknown in the final frontier in popular science fiction series such as *Star Trek*. Space is deeply intertwined with the idea of American progress and innovation. Consequently, there are schools of strategy that emerges within the American conception of space. If located within the prism of realism, which focuses on the competitive nature of human beings focused on the state and its national interest (gain-lose), based on the distribution of power, the United States must maintain leadership in space for three very significant benefits: first, to ensure that space not only remains a free enterprise but also upholds American citizens' rights to utilize space for economic benefits; second, to ensure that space remains a military domain where the United States has both access and can dominate in case conflict breaks out; third, that the United States can demonstrate presence, by which is meant quick and cost-effective access to space, as states with access are the only ones that have the most influence in shaping the rules of engagement. While realists are skeptical about the holding power of norms itself, they view norms as a subset of national interest backed by power. While acknowledging the role of ethics in international relations and strategic culture, realists remain highly skeptical about the ability of ethics to influence state behavior, if it is not backed by power (economic and military) located within its national interest. Some of those Realist philosophers who have acknowledged the role of ethics and morality in state power include Thucydides,[22] Kautilya,[23] Reinhold Niebuhr,[24] and Hans Morgenthau.[25] Strategic culture from a Realist perspective is, therefore, most successful when states can accumulate power (military and economic) that can support their worldview and create an order beneficial to them. The absence of an international governing body that can enforce the rules and hold rogue states accountable only further necessitates the accumulation of power. It is only then that states can enjoy some form of security.

According to Liberal Institutionalists, power, however, has to be constrained and in moderation, imbibed by a sense of justice. Thereby, based on such a justice-oriented ethic, the United States aspires to maintain leadership in space so that it can shape that normative order. This can be witnessed by its push for cooperation in space as well, so that NASA can lead with space exploration programs, namely with projects such as getting to Mars, and the *Ultima Thule,*[26] now renamed *Arrokoth.*[27] Excelling in a space exploration project, which others have not done, ensures that the United States maintains the lead in setting the space exploration norms. This ties into the liberal project of ensuring that U.S. space leadership will build in a cooperative framework in which American values such as liberty, freedom, and free commerce continues the U.S. international project since the end of World War II.[28]

In this context, the idea of a U.S. Space Force, specifically forwarded by its most committed advocates, is based on a similar logic; that it would not and should not be a mission directed to become simply a war-fighting service,[29] but that it will be a military service specifically tasked to ensure that outer space remains free, especially when it becomes bedecked with economic activities, that will take advantage of the multi-trillion economy that awaits to be harvested. Hence, it will and should replicate the role of the U.S. Navy on the High Seas, that of protecting commerce.[30] Today, the United States, not dissimilar to British naval thinking, is the foremost economic and military power in the world, due to its ability to demonstrate continuous presence across all domains: air, sea, land, and space. Similarly, the aims for the newly established U.S. Space Force could enlarge into a service capable of demonstrating similar power projection capabilities in the domain of outer space for the purposes of ensuring free trade and space commerce. The ambition is that: equipped with such spacefaring power projection capabilities, the U.S. Space Force will be able to not only maintain leadership but also ensure that its allies/partners have free access to outer space.[31]

CHINA'S STRATEGIC CULTURE

The ideational influences on China's decision-making with regard to outer space is that it aims to become the lead player by 2045 and utilize space achievements for its centennial celebrations (2049) of the founding of the PRC. Toward achieving this goal, President Xi Jinping's directive is rather clear; invest time, talent, and resources to make that happen. In this, the role of Chinese strategic culture and its influencing power is critical in understanding where China draws its decision-making process. Within China, several streams of thought have a pull, to include communist thought, which is dominant despite being an alien idea, not indigenous to China. Chinese decision-makers are inspired by the work of Karl Marx and his insistence on how the economic forces of society matters in explaining competition over scarce resources. For Marx, who coined the term "historical materialism," human history was determined by the modes of production, and the class struggle that ensues as a result. Once a polity can overcome and remove class struggles, communism, implying equality of all classes, is established. For Marx, capitalist societies made profits from exploited and unpaid for surplus labor of the poorer classes, which made the owners (the bourgeois powerful and rich) and the proletariat (working classes) weak and poor. Such a context would lead to revolt, and establish a communist state. Marx refused to speculate as much on what communism was like or should be; instead, he believed that the historical process would determine that process. For Marx, political emancipation and human emancipation may not necessarily follow as they are distinctly separate. Marx believed that human beings cannot hope for political and economic emancipation in isolation but require a community. Such communities depend on social and political networks. Marx specified that to break through the hold of a capitalist and exploitative mode, the workers have to transform themselves and revolt guided by the founding philosopher.[32]

Mao Tse Tung and the Communist Party of China (CPC) formed and led by him, saw itself as a guardian of China as it fought against the Chinese Nationalist Party forces led by Chiang Kai-shek for two decades (1927–1949). When Mao emerged victorious, he established a communist state inspired by Marxist principles. That tradition continues despite arguments that China has accepted capitalist modes of productions. Sure, but to the extent that it creates wealth for the state and its citizens, as guided under the watchful eyes of the CPC, the upholder of ideology and Marxism. The establishment of communist China registered a move away from indigenous philosophies such as Confucianism to adopting a foreign political ideology inspired by Marx and the Soviet Union. Confucianism that had resulted in political and social hierarchy was viewed by Mao as resulting in the exploitation of the peasant classes.[33] Significantly, President Xi Jinping, while ensuring the rule of communism, is diverting from Mao, and including age-old Chinese philosophers such as Confucius who placed enormous emphasis on social stability and harmony.[34] By self-selecting himself President for life, Xi makes the point that Confucius believed in: that a single authoritarian ruler is required to maintain that balance. Xi indicates that the only way to maintain Chinese stability and prosperity is to ensure that the CPC is in the lead. Combining a heady mix of an alien political concept on which the CPC is based and organized with recollections of China's past glory, President Xi has now coined the phrase, "the great rejuvenation of the Chinese nation."[35] This is indicated by the CPC's claim as an inheritor of Chinese tradition, history, and culture. Showcasing themselves as protector of Chinese history and culture ensures the support of the Chinese population for a self-appointed group of leaders.[36]

Based on this clear need to build legitimacy for its survival, the CPC aims to utilize its economic resources to build leadership in space, which, in turn, offers that kind of legitimacy. The critical concern however is, what will China do with that kind of leadership, once it assumes it. Given the authoritarian Chinese regime, one cannot hope for an international order based on peace and freedom, when at home the CPC behaves completely otherwise offering no political freedom to its own people. Moreover, the Confucius harmony loving nature may be a shield that could hide the deeper manifestations of a propensity to violence within Chinese strategic culture. The historical assessment of Chinese behavior reveals that the country was willing to use force if and when the opportunity offered itself. This is further more entrenched when the use of force is viewed as strategically beneficial backed by sufficient resources and when the conditions are such that the adversary is weak or at a disadvantage. And we can clearly observe this behavior with regard to Chinese political culture; as China has got stronger and more resourceful, it has stopped behaving harmoniously but instead adopted a more assertive path.

> As China has become stronger, it has increasingly been abandoning non-provocative postures, and seems today ever more willing to act like exactly the self-interested hegemon that official propaganda has denied it is culturally or even "genetically" possible for China to be. There should be nothing too surprising in this, but unless we pierce the conceit of Beijing's self-Orientalizing narrative of benevolent pacifism, we will continue to be dangerously misled in dealing with China.[37]

Even more worryingly, there is a self-righteous narrative involved in justifying several of China's violent interventions; describing its occupation of Tibet as liberating Tibet, for instance. As a consequence, rejuvenating the nation and its citizens could be harked back to earlier expositions of the greatness of the Chinese nation. The use of force to ensure the national rejuvenation of the Chinese nation will have full support as a result. Since there is no basis for population consent, the CPC has to build its legitimacy around other ideas; historicity, defender of the country, ensuring the Chinese way of life, "defender of the eternal Chinese realm," the realm that has not changed eternally, and should by right include areas such as Taiwan and Tibet.

Regimes other than China are projected as lacking in civilization and higher values, and thereby a Chinese-led international order will prove superior. This Chinese narrative questions the American strategic culture based on exceptionalism, and championing values such as democracy and liberty. Instead, what China aspires to do, based on its conception of strategic culture, is to build legitimacy and support for a space order where it is in the lead. This to China undoes several levels of historical wrongs, one being the West on the lead, when China with the superior civilizational ethos, should be leading. Within Chinese strategic culture is this belief that its 5000 years of civilization has offered its citizens with superior thinking and logic, compared to upstart nations such as the United States. The Chinese official narrative has been to project China as a humiliated nation that has suffered immensely from the 100 years of Western nations imposing "unequal treaties on it."[38]

China has cleverly constructed a "Self-Orientalizing Narrative of Pacificism"[39]—that Eastern/oriental societies are peace-loving, or create mechanisms to ensure peace.[40] Much of the debates do not account for millions of deaths within, due to both interstate war and internal political oppression. For instance, China itself engaged in numerous internal wars, interstate wars with Japan (several), India (1962), Vietnam (1979), USSR (1969), seized the independent nation of Tibet (1950), and has current militarized conflicts on its borders with India, and over disputed islands with Japan, Philippines, Vietnam, Brunei, Taiwan, and Indonesia. Several millions died in Mao's Great Leap Forward and Cultural Revolutions and purges, and continue to die in Tibet and Xinjiang Province. Such conflict narratives afflict within the broader Asian framework as well, for instance, Association of South East Asian Nations (ASEAN) founding member states in 1967, to include Indonesia under the brutal Suharto regime, the violence in South Philippines, or the rise of the military dictatorship in Myanmar. The 1960s and 1970s were brutal for the populations of Laos, Cambodia, Vietnam, with millions killed during the Vietnam War. Vietnam (1995), Laos (1997), and Cambodia (1999) joined ASEAN in the 1990s. One of the most influential debates of the 1990s, early 2000s was the peaceful rise of China discourse.[41]

Anyone mindful of history will know that China utilized force whenever it felt it was to its advantage or when the conditions favored it. As Sun Tzu, one of China's most brilliant strategists, during the Warring States Period (476 BC–221 BC) had written: whenever conditions change, one must be flexible and adapt.[42] Most significantly, never ever dispel adversary perception of who they think you are. Sun Tzu, however, advised that to have a strategy for success, a Commander must be aware of five important elements that lead to a successful strategy: these included an understanding of the power of norms (moral legitimacy), heaven, earth (physical conditions), leadership, and finally, method, and discipline (assessment of military capability, context, relative power potential/difference, logistics, resources). Once all elements come together, a state can benefit from a grand strategy for success. This strategy has an element of deception in it so that the adversary is unprepared for who you actually are.[43]

However, lurking behind this pacifist pretense is a state that believes itself to be morally superior, equipped with a superior strategy; a strategy that can defeat an adversary without actually engaging in bloodshed. Chinese leaders have a sense of virtuousness and grander, elaborate displays of power, as could be seen from Mao's treatment of former U.S. National Security Advisor, Henry Kissinger, during the Nixon opening up to China in the 1970s, or their celebration of the founding day (October 1, 1949) of the PRC. The recent seventieth-anniversary celebrations (October 1, 2019) showcasing China's military and cultural power, is a case in point.[44] Such PRC efforts at impressing the other side are visible in Henry Kissinger's book *On China*.[45] In that book, Kissinger expressed awe of Mao, especially as he alluded to the aphorisms that Mao uttered, and the gratitude Kissinger felt, even to have an audience with Mao.[46] It was clear that Mao believed he was doing Kissinger a favor by even granting him an audience at the Zhongnanhai Palace in Beijing. After all, Mao was given to such theatrics where he deliberately created the image of a virtuous leader. This can be encapsulated in his famous swim in the Yangtze River in 1966 after which he returned to Beijing and started his great cultural revolution. That swim symbolized the reawakening of a nation, after decades of lull.[47]

The moral legitimacy that the CPC draws upon is to quote and draw linkages to the age-old concept of having a "mandate of heaven." This "mandate of heaven" concept somehow instills in the top-level CPC leadership virtues that every other Chinese citizen and lower-level CPC members must emulate. This itself creates a penchant for righteous violence since violence utilized in such circumstances is viewed as an eternal feature of Chinese strategic culture that has historical continuity.[48] This is the kind of strategic culture, coupled with its strategic myths that will influence Chinese behavior in space. For CPC leaders aspire to be the lead actor in space, so that other nations would then like to emulate, not only their capacity to lead, but also for their superior civilizational ethos.

INDIA'S STRATEGIC CULTURE

India's strategic culture has undergone changes since its establishment as a democratic republic in 1947. At that time, its first prime minister, Jawaharlal Nehru, forwarded the concept of nonalignment as well as envisioned a role for India that showcased its moral and civilizational values to the world.[49] Significantly, under Nehru, India chose the path of nonalignment with two of the Great Powers of the day, namely, the United States and the Soviet Union, unlike, say, China or Pakistan. China entered the Soviet bloc, whereas Pakistan joined the U.S. bloc. Nehruvian culture, as it is called, is a strategic culture that is cautious in its foreign policy and counsel restraint with regard to the use of military power. Significantly, Nehru believed in working within the United Nations to establish India as a moral force to be reckoned with. Those who fall within this school is careful to build India's reputation abroad as well as ensure that military power is squarely within civilian control. Nehruvians believe in the concept of strategic restraint, especially when it comes to acquiring military power.[50] As a consequence, the acquisition of space capability was located by Vikram Sarabhai, the founder of the Indian space program, within the context of national development and as benefitting India's economic goals. It was the close relationship between Nehru, Sarabhai, and Homi Bhabha, the founder of India's nuclear program, that led to the development of India's space program, as well as the establishment of the Atomic Energy Commission.[51] Both Sarabhai and Bhabha understood the significance of space research for a country's future. The context was also significant. It was happening at a time when the International Geophysical Year was being celebrated in 1957–1958, and the Committee for Space Research was established.[52]

Nehru, on the urging of Sarabhai, had established the Physical Research Laboratory in 1954. In 1962, the Department of Atomic Energy under Bhabha and now with Sarabhai on the board went ahead and established the Indian National Committee for Space Research. Sarabhai was made its chairman and was given the responsibility to establish India's space program. As history will tell us, India's space program continued to grow under that tutelage and within a Nehruvian strategic culture. That strategic culture ensured that India's space program remained civilian, focused on economic and social development. This was also the result of the continuance of Nehru's party, the Congress, in power for several decades, after his death, led by his daughter, Indira Gandhi, his grandsons, Sanjay and Rajiv Gandhi, his granddaughter-in-law, Sonia Gandhi, and great-grandson, Rahul Gandhi. That legacy of constraint and wariness with projecting a more militant image impacted how India's space organization utilized space. Nehru, despite being a democrat, behaved in an authoritarian manner, in the sense that he selected diplomats and internal bureaucrats who were wedded to his way of thinking and ensured that only those loyal to him got promoted. This was possible because Nehru was prime minister, external affairs minister, and defense minister, all at once, thereby conglomerating much of state power to himself. This made it possible for Nehru to shape and instill a strategic elite that understood and wedded themselves into his way of strategic thinking.[53]

Interestingly, one can perhaps see similarities between Nehru's nonalignment strategy and Deng Xiaoping's strategy of keeping a low profile. Nehru believed that alignment with a particular superpower will create obstacles for India's development, democracy, and identity at a time when it was poor and fragmented. Consequently, to remain outside of those Great Power ideological struggles gave India the opportunity to develop itself. An outlier to such a point of view is the fact that Nehru felt closer to the Soviet Union than he did to the United States; more importantly, after the 1962 China-India border war, in which China suddenly attacked India and imposed a devastating defeat on its border forces, Nehru wrote to then U.S. President John F. Kennedy for U.S. military support in case China attacks other parts of India.[54]

Recently, there have been other significant changes that have brought about India's change in posture with regard to the international reputation and how it is viewed. In March 2019, India went ahead and tested *Mission Shakti*, an Anti-Satellite Weapons Test (ASAT) that killed one of India's orbiting satellites. The Defense Research and Development Organisation utilized its Ballistic Missile Defense Interceptor and conducted the test in Low Earth Orbit (LEO). Significantly, this test demonstrated a significant change in India's strategic culture, where the present Narendra Modi government was willing to stake international reputation, to demonstrate India's national security space capacity. While stating that the test was not directed at any country, India's Ministry of External Affairs specified that "the Anti-Satellite missile test provides credible deterrence against threats to our growing space-based assets from long-range missiles, and proliferation in the types and numbers of missiles."[55] Such moves, including the willingness of India to target Pakistan with the use of air-power in response to a terror attack in February 2019, as well as unilaterally revoking the Indian controlled part of Kashmir's autonomy status, without consultation with its local population is unique and new.[56] Both actions had brought about international condemnation, but India did not seem to care much.[57] Consequently, the Nehruvian ethos of being cautious with regard to India's reputation as a civilizational and moral giant has been set aside in favor of internal and external behavior that demonstrates the power and a focus on national security. Consequently, India is now upping its space game as well, demonstrating its willingness to become a comprehensive space power. The competition to build capacity and showcase national willingness to take advantage of what space has to offer has only become more insistent with the establishment of a Defense Space Agency and a Defense Space Research Organisation.

THE STRATEGIC CULTURE OF UAE AND LUXEMBOURG

Both UAE and Luxembourg are middle powers with resources that can throw a serious challenge to some of the major spacefaring nations. Both have been utilizing their unique geographic location and economic power to shape the space discourse. The UAE is a federation that includes seven states as part of that federation, with an authoritarian political structure, where the leadership is hereditary. Oil exports are its chief source of income along with tourism. Ruled by a Supreme Council of Rulers, all seven states are represented. The kingdom does not have a free press; neither does it allow free expression of opinion.[58] Yet, in terms of broadcasting its power and influence in the Middle East, UAE hopes to play an inspirational role for not only Emirati citizens but also for the larger Middle East through its success and its strategic ambitions. One such ambition is with regard to its space program. A later chapter in the book will elaborate on the UAE space program. However, it is useful to understand some of the strategic thinking that has animated the UAE strategic space discourse. For the UAE, investments in space imply that the country can emerge as one of the key technology hubs in the Middle East. It indicates that its leaders can peer into the future and think of interplanetary missions and habitation.

In February 2017, UAE's prime minister, Sheikh Mohammed bin Rashid Al Maktoum, tweeted about Mars 2117 plan (100 years ahead), which aspires to build a city on Mars with the help of international cooperation, fitting in well with Elon Musk's plans for Mars as well. Its geographic location offers UAE a distinct advantage to establish a spaceport for future space activities and its ability to draw talent from around the world. Its vast deserts and extensive coastline offer it a distinct advantage to bring forth such space activities, and its closeness to the equator implies that it could launch into space with less fuel. A Martian environment could feel like the UAE's desert environment, which the Emirates have changed by using climate control technologies into lush water parks and skiing resorts. A similar landscape shift can be tried in space as well. We know that *Space X*'s Elon Musk aspires to nuke Mars to vaporize its ice caps, which would warm up the planet making it habitable for humans. For Musk, establishing *Space X* in 2002 was a path to his ultimate goal to colonize Mars. The idea is to establish a million-person Martian colony by 2075.[59] In this, we can see the similarity of Musk's Martian vision and UAE's vision for Mars. As mentioned earlier, the UAE aims to establish a Martian colony by 2117. The aim is to have a 100-year plan that will incrementally build the capacity required to meet such a goal. To prepare for a Martian environment, the UAE aims to build the Mars Scientific city, in which a complex set of buildings are to be constructed, replicating what it is like to be on Mars. 3D printing, heat and radiation insulation are going to be utilized to create that harsh Martian environment. Very similar to China's Moon Lab experiment where eight students from Beihang University lived in a simulated lunar lab for a year, UAE aims to have a team of scientists and astronauts live in a Martian environment for a year.[60] Within four years, this Mars Scientific city is expected to be in operation. Adnan Al Rais, Mars 2117 Program Manager at the Mohammad Bin Rashid Space Centre (MBRSC) stated that "We're developing the Mars Science City where we will have the analogue facilities and simulation facilities, where we'll simulate life on Mars either for humans or also for robotics—how robotics function on the surface of Mars."[61]

Drawing talent is the key which the UAE with its vast wealth has accomplished, a nation with 1.4 million Emiratis but nearly 8.6 million visitor workers.[62] UAE oil wealth stands at 97.8 billion barrels (2015), with Abu Dhabi holding the majority of those barrels (92.2 billion barrels), followed by Dubai with 4 billion barrels, Sharjah, 1.5 billion barrels, and Ras al Khaimah with 500 million barrels.[63] With a GDP of $460 billion and a GDP per capita of $43, 000,[64] UAE has the resources to support its strategic shift to make itself a hub for outer space in the Middle East, and the only Arab nation to have a Mars program.

Similar to UAE is Luxembourg's investments in outer space. With a GDP of USD 62.4 billion and a GDP per capita income of $115.203,[65] Luxembourg is one of the richest countries in the world.[66] In September 2018, Luxembourg established its space agency, and a Luxembourg Space Fund, aimed at developing its space industry sector.[67] Luxembourg, like UAE, aims to become a European hub for space innovation and develop future space technologies that can tackle the world of space-based resources. In a strategically clever and anticipatory move, Luxembourg drafted a law on space resources that enables those who extract a space resource the legal mechanism to own it.[68] Such moves indicate a strategic culture that is future oriented, hoping to be in the lead when such space resources become viable and benefit from the profit therein. And as Robert Zubrin states in his book, *The Case for Space*, it's not a matter of how, but when.[69]

The ability to aim for space depends on three very important factors; resources, talent, and technology. Space is becoming a networked infrastructure where we depend on it for our modern technologies, including the internet, GPS, ATM transactions, online marketing, and so much more. Those with an understanding of how a networked space enterprise will grow, to include not just critical infrastructure but also the multi-trillion-dollar space economy that will benefit future generations will profit.[70]

As we know, John Locke, the famous English philosopher wrote the *Two Treatises of Government* and published it anonymously in 1689. Locke never acknowledged that he was the author of the *Two Treatises* during his lifetime, and only his will acknowledged that he was the author. Though he did not live to see or assess the impact of his philosophical works, the *Two Treatises* is considered an important work of the Enlightenment and influenced both the French and the American revolutions. Perhaps, policymakers that are crafting ambitions for extracting space-based resources in a 100-year timeframe, are not looking to benefit from those investments in their lifetime, but with the hope that future generations will benefit. In this, Ye Peijian, the head of CLEP and an aerospace engineer aptly stated that:

> The universe is an ocean, the moon is the Diaoyu Islands, Mars is Huangyan Island. If we don't go there now even though we're capable of doing so, then we will be *blamed by our descendants* [emphasis added]. If others go there, then they will take over, and you won't be able to go even if you want to. This is reason enough.[71]

In the spirit of that quote, countries such as China, UAE, Luxembourg are focused on long-term space resource exploitation goals because it enables them to ensure that the multi-trillion-dollar economy that awaits, will benefit the generations that will follow. After all, the history of the world will inform you that those ancestors who took bold steps to take over territory and ensure that their descendants benefit from their frontier spirit are indeed better off than those who did not. Most of the Americas were taken over by Europeans, and their descendants today have indeed benefitted, with resources far higher for their civilization to enjoy from territories that once belonged to the indigenous population. Space is going to see a similar scramble for resources once methods to extract those resources become commonplace. Hence, countries that channel and craft laws to take advantage of that future to come will benefit first. Those policies and the ability to anticipate the future will depend on the strategic cultures that countries base their choices and preferences on, as well as lessons they imbibe from their own history.

NOTES

1. Andreas Gofus, eds., et al., Sage Handbook of *History, Philosophy and Sociology of International Relations* (Los Angeles: Sage, 2018).

2. Mary Magoulick, "What is a Myth?," accessed July 17, 2019, https://faculty.gcsu.edu/custom-website/mary-magoulick/defmyth.htm
3. Hunter R. Rawlings III, "Thycididian Epistemology: Between Philosophy and History," *Rheinisches Museum für Philologie* Neue Folge, 153, no. 4 (2010): 247–90.
4. Zeno's Paradoxes, Internet Encyclopedia of Philosophy, accessed July 17, 2019, *https://www.iep.utm.edu/zeno-par/#SH1b.*
5. Zeno's Paradoxes, Internet Encyclopedia of Philosophy.
6. G. S. Kirk. "Greek Mythology: Some New Perspectives," *The Journal of Hellenic Studies* 92 (1972): 74–85, accessed July 19, 2019, https://www.jstor.org/stable/629974?seq=1#page_scan_tab_contents; Cara Leigh Sailors, "The Function of Mythology and Religion in Ancient Greek Society," accessed July 19, 2019, https://dc.etsu.edu/cgi/viewcontent.cgi?article=3471&context=etd
7. Origins of NASA Names, "Manned Spaceflight Section IV," accessed July 17, 2019, https://history.nasa.gov/SP-4402/ch4.htm
8. "Presidential Memorandum on Reinvigorating America's Human Space Exploration Program," *The White House*, December 11, 2017, accessed on July 17, 2019, https://www.whitehouse.gov/presidential-actions/presidential-memorandum-reinvigorating-americas-human-space-exploration-program/.
9. "Remarks by Vice President Pence at the Fifth Meeting of the National Space Council Huntsville, Al," *The White House*, March 26, 2019, accessed July 17, 2019, https://www.whitehouse.gov/briefings-statements/remarks-vice-president-pence-fifth-meeting-national-space-council-huntsville-al/
10. "Remarks by Vice President Pence."
11. "Artemis Greek Goddess," *Encyclopedia Britannica*, accessed July 17, 2019, https://www.britannica.com/topic/Artemis-Greek-goddess
12. "What is Artemis?," *NASA*, June 28, 2019, accessed July 17, 2019, https://www.nasa.gov/feature/what-is-artemis/
13. "Artemis Greek Goddess of the Hunt, Forests and Hills, the Moon, Archery," accessed July 17, 2019, https://greekgodsandgoddesses.net/goddesses/artemis/
14. The State Council, People's Republic of China, "China's White Paper on Space Activities 2016," December 28, 2016, accessed on January 19, 2019, http://english.gov.cn/archive/white_paper/2016/12/28/content_281475527159496.htm
15. Sun Huixian, et al., "Scientific Objectives and Payloads of Chang'e-1 Lunar Satellite," *Journal of Earth System Science*, 114, no. 6 (December 2005): 789–94, accessed July 22, 2019, https://link.springer.com/article/10.1007%2FBF02715964
16. "China Space Probe Flies by Asteroid Toutatis," China Academy of Sciences, December 18, 2012, accessed July 22, 2019, http://english.cas.cn/news/ric/201212/t20121218_97190.shtml
17. J. Gordon Melton, ed., *Religious Celebrations: An Encyclopedia of Holidays, Festivals, Solemn Observances, and Spiritual Commemorations* (Santa Barbara: ABC-CLIO, 2011).
18. "The Legend of Chang'e," accessed July 22, 2019, http://www.moonfestival.org/the-legend-of-chang-e.html
19. "Aryabhata," accessed July 22, 2019, https://www.britannica.com/biography/Aryabhata-I

20. Mark Cartright, "Surya," June 23, 2016, accessed July 22, 2019, https://www.ancient.eu/Surya/

21. Mike Wall, "JFK's 'Moon Speech' Still Resonates 50 Years Later," *Space.com*, September 12, 2012, accessed August 1, 2019, https://www.space.com/17547-jfk-moon-speech-50years-anniversary.html

22. Thucydides, *The History of the Peloponnesian War*, translated by Richard Crawley, accessed October 08, 2019, http://classics.mit.edu/Thucydides/pelopwar.html

23. R. P. Kangle, *The Kautilya Arthasashtra An English Translation with Critical and Explanatory Notes* (New Delhi: Motilal Banarsidass Publ., 1986).

24. Reinhold Niebuhr, *Moral Man and Immoral Society: A Study in Ethics and Politics* (Louisville: Westminster John Knox Press, 2013).

25. Hans Morgenthau, *Politics Among Nations The Struggle for Power and Peace* (New York: Knopf, 1988).

26. Nayaf R. F. AL. Rodham, "U.S. Space Policy and Strategic Culture," *Journal of International Studies*, April 16, 2018, accessed October 8, 2019, https://jia.sipa.columbia.edu/online-articles/us-space-policy-and-strategic-culture

27. Leah Crane, "Distant Space Rock Known as Ultima Thule Renamed to Avoid Nazi Links," *New Scientist*, November 12, 2019, accessed January 8, 2020, https://www.newscientist.com/article/2223176-distant-space-rock-known-as-ultima-thule-renamed-to-avoid-nazi-links/

28. John G. Ikenberry, *Liberal Leviathan: The Origins, Crisis, and Transformation of the American World Order* (Princeton, NJ: Princeton University Press, 2012).

29. "Trump: Space is the World's Newest War-Fighting Domain," *BBC*, December 21, 2019, accessed January 8, 2020, https://www.bbc.com/news/av/world-us-canada-50875940/trump-space-is-the-world-s-newest-war-fighting-domain; Video, "Vice President Mike Pence Calls Space a War Fighting Domain," *The Washington Post*, October 23, 2018, accessed January 8, 2020, https://www.washingtonpost.com/video/politics/vice-president-mike-pence-calls-space-a-war-fighting-domain/2018/10/23/2f7e5f06-d6d1-11e8-8384-bcc5492fef49_video.html

30. Peter Garretson, "The Purpose of a Space Force is a Space Faring Economy," *The Hill*, June 26, 2019, accessed August 5, 2019, https://thehill.com/opinion/technology/450519-the-purpose-of-a-space-force-is-a-spacefaring-economy. Also see Brent Ziarnick, "The Space Corp Needs Naval Rank," *Politico*, July 26, 2019, accessed August 5, 2019, https://www.politico.com/story/2019/07/26/space-corps-naval-rank-1433541

31. Steve Kwast, "There Won't Be Many Prizes for Second Place," *Politico*, August 10, 2018, accessed August 5, 2019, https://www.politico.com/story/2018/08/10/space-race-kwast-768751

32. David McLellan ed., *Karl Marx: Selected Writings* (Oxford: Oxford University Press, 2000).

33. A. James Gregor and Maria Hsia Chang, "Anti-Confucianism: Mao's Last Campaign," *Asian Survey* 19, no. 11 (November 1979): 1073–92.

34. Wang Xiangwei, "Xi Jinping Endorses the Promotion of Confucius," *South China Morning Post*, September 29, 2014, accessed January 9, 2020, https://www.scmp.com/news/china/article/1603487/xi-jinping-endorses-promotion-confucius

35. "Xi Focus: Xi Jinping and China's New Era," *Xinhua*, September 30, 2019, accessed January 9, 2020, http://www.xinhuanet.com/english/2019-09/30/c_1384 37362.htm

36. Mitchell Chan, "Communism with Chinese Characteristics," Confucius and China's Future," *Penn Political Review*, February 16, 2018, accessed August 5, 2019, https://pennpoliticalreview.org/2018/02/communism-chinese-characteristics-confuc ius-and-chinas-future/

37. The National Bureau of Asian Research, "Behind the Official Narrative China's Strategic Culture in Perspective," National Bureau of Asian Research, November 1, 2016, accessed August 7, 2019, https://www.nbr.org/publication/behind -the-official-narrative-chinas-strategic-culture-in-perspective/

38. Zheng Wang, *Never Forget National Humiliation: Historical Memory in Chinese Politics and Foreign Relations* (New York: Columbia University Press, 2014).

39. The National Bureau of Asian Research, "Behind the Official Narrative China's"

40. Timo Kivimaki, "The Long Peace of ASEAN," *Journal of Peace Research* 38, no. 1 (January 2001): 5–25, accessed October 4, 2019, https://www.jstor.org/stable/4 25780?seq=1#page_scan_tab_contents

41. "China's Peaceful Rise: Speeches of Zheng Bijian 1997–2004," *The Brookings Institution*, accessed October 4, 2019, https://www.brookings.edu/wp-content/upl oads/2012/04/20050616bijianlunch.pdf

42. Sun Tzu, *The Art of War*, Accessed August 6, 2019, http://classics.mit.edu/Tzu /artwar.html

43. Namrata Goswami, "China's Grand Strategy in Outer Space: To Establish Compelling Standards of Behavior," *The Space Review*, August 5, 2019, accessed January 9, 2020, https://www.thespacereview.com/article/3773/1

44. "The 70th Anniversary of the Founding of the People's Republic of China," *Xinhua*, October 1, 2019, accessed October 4, 2019, http://www.xinhuanet.com/engli sh/special/2019-10/01/c_138438666_181.htm

45. Henry Kissinger, *On China* (New York/London: Penguin/Random House, 2012).

46. Kissinger, *On China*, 31.

47. Stuart Heaver, "Chairman Mao's Historic Swim-Glorified in China but Ridiculed by the Rest of the World," *South China Morning Post*, August 4, 2016, accessed October 4, 2019, https://www.scmp.com/magazines/post-magazine/long-reads/article/1999098/chairman-maos-historic-swim-glorified-china; Shuk Waah Poon, "Embodying Maoism: The Swimming Craze, the Mao Cult, and Body Politics in Communist China, 1950s-1970s," *Modern Asian Studies* 53, no. 5 (September 2019): 1450–85, accessed October 4, 2019, https://www.cambridge.org/core/jour nals/modern-asian-studies/article/embodying-maoism-the-swimming-craze-the-mao -cult-and-body-politics-in-communist-china-1950s1970s/955A833A02B8053585F 2B1D8631698B0

48. Donovan C. Chau and Thomas M. Kane, eds., *China and International Security History, Strategy and 21st Century Policy* (Santa Barbara: Praeger, 2014), 160.

49. Ministry of External Affairs, Government of India, "History and Evolution of Non-Aligned Movement," August 22, 2012, accessed October 4, 2019, https://mea.gov.in/in-focus-article.htm?20349/History+and+Evolution+of+NonAligned+Movement; M.S. Rajan, *Nonalignment and Non Aligned Movement: Retrospect and Prospect* (New Delhi: Vikas Publications, 1990).

50. Kanti Bajpai, "Indian Conceptions of Order and Justice: Nehruvian, Gandhian, Hindutva, and Neo-Liberal," in *Order and Justice in International Relations* Rosemary Foot, John Gaddis, and Andrew Hurrell, ed. (Oxford: Oxford University Press, 2003), 236–61.

51. Amrita Shah, "Nehru's Belief in Science Fuelled India's Audacious Space Program," *The Print*, November 14, 2017, accessed August 8, 2019, https://theprint.in/opinion/nehrus-belief-fuelled-indias-space-atomic-programmes/16224/

52. "Shaping the Space Age: The First International Geophysical Year," NASA, Appel News Digest, 1/12, December 2008, accessed October 4, 2019, https://appel.nasa.gov/2010/02/23/aa_1-12_f_shaping-html/

53. The National Bureau of Asian Research, "The Legacy of Nehruvianism and the Implications for India's Strategic Culture," National Bureau of Asian Research, November 21, 2016, accessed August 8, 2019, https://www.nbr.org/publication/the-legacy-of-nehruvianism-and-the-implications-for-indias-strategic-culture/

54. "India: Subjects: Nehru Correspondence, November 1962: Nov 1–19," John F. Kennedy Presidential Library and Museum, accessed October 4, 2019, https://www.jfklibrary.org/asset-viewer/archives/JFKNSF/111/JFKNSF-111-016

55. "All You Need to Know about Mission Shakti," *Hindu Business Line*, March 27, 2019, accessed October 4, 2019, https://www.thehindubusinessline.com/news/all-you-need-to-know-about-mission-shakti/article26652887.ece

56. Kelsey Davenport, "India's ASAT Test Raises Space Risks," *Arms Control Today*, May 2019, accessed October 4, 2019, https://www.armscontrol.org/act/2019-05/news/indian-asat-test-raises-space-risks; Sean Gallahar, "India ASAT Test Debris Poses Danger to International Space Station, NASA Says," *Ars Technica*, April 2, 2019, accessed October 4, 2019, https://arstechnica.com/tech-policy/2019/04/india-asat-test-debris-poses-danger-to-international-space-station-nasa-says/

57. "The UN Can't Ignore Kashmir Anymore," *The New York Times* Editorial Board, October 2, 2019, accessed October 4, 2019, https://www.nytimes.com/2019/10/02/opinion/editorials/kashmir-india-pakistan-un.html; "Indian Judges are Ignoring the Government's Abuses in Kashmir," *The Economist*, October 5, 2019, accessed October 4, 2019, https://www.economist.com/asia/2019/10/05/indias-judges-are-ignoring-the-governments-abuses-in-kashmir

58. Andrea B. Rugh, *The Political Culture of Leadership in the United Arab Emirates* (New York: Palgrave, MacMillan, 2007).

59. Hanneke Weitering, "How SpaceX's Starship will Help Establish a Mars base," *Space.com*, April 15, 2019, accessed October 04, 2019, https://www.space.com/spacex-starship-mars-transportation-plans.html

60. UAE, "2030-2117," accessed August 22, 2019, https://government.ae/en/more/uae-future/2030-2117

61. Janice Ponce de Leon, " Mars Science City to be Fully Operational in 4 Years," *Gulf News*, January 22, 2019, accessed August 22, 2019, https://gulfnews.com/uae/science/mars-science-city-to-be-fully-operational-in-4-years-1.61596317

62. Kate Greene, "Why the United Arab Emirates is Building a Space Program from Scratch," *Slate*, March 30, 2017. accessed August 9, 2019, https://slate.com/technology/2017/03/why-the-united-arab-emirates-is-building-a-space-program.html

63. "The UAE and the Global Oil Supply," Embassy of the United Arab Emirates, Washington, DC, accessed August 22, 2019, https://www.uae-embassy.org/about-uae/energy/uae-and-global-oil-supply

64. The World Bank, "GDP Per Capita UAE," accessed August 22, 2019, https://data.worldbank.org/indicator/NY.GDP.PCAP.CD?locations=AE

65. The World Bank, "GDP Per Capita (Current US$)," accessed September 14, 2018, https://data.worldbank.org/indicator/ny.gdp.pcap.cd

66. The World Bank, "GDP per capita, PPP (current international $) – Luxembourg," 2018, accessed January 9, 2020, https://data.worldbank.org/indicator/NY.GDP.PCAP.PP.CD?locations=LU

67. Jeff Froust, "Luxembourg Establishes Space Agency and New Fund," Spacenews, September 13, 2018, accessed January 9, 2020, https://spacenews.com/luxembourg-establishes-space-agency-and-new-fund/

68. "Space Resources," Luxembourg Space Agency, n.d, accessed January 9, 2020, https://space-agency.public.lu/en/space-resources.html

69. Robert Zubrin, *The Case for Space How the Revolution in Spaceflight Opens up a Future of Limitless Possibilities* (New York: Prometheus Books, 2019).

70. Lt. Gen. Steve Kwast, "The Real Stakes in the New Space Race," *War on the Rocks*, August 19, 2019, accessed August 22, 2019, https://warontherocks.com/2019/08/the-real-stakes-in-the-new-space-race/

71. Brendon Hong, "China's Looming Land Grab in Outer Space," *Newsweek*, June 22, 2016, accessed October 4, 2019, https://www.thedailybeast.com/chinas-looming-land-grab-in-outer-space

Chapter 3

The Epistemic Community and the Foundations of Discourse in the United States

The centrality of the United States to the global space dialogue is so deep and rich that we have decided to break it into two chapters. This chapter provides an overview of the U.S. epistemic community, important works of thought leadership, and history of elite discourse. To enable the reader to understand the elite debate, we have chosen to include significant segments of key public documents and speeches. Unless specifically noted otherwise, all italics are ours for emphasis, and not in the original.

Since its entrance into Apollo, the United States has been the central node in a broader transnational epistemic community interested in all aspects of space expansionism, including space settlement (previously space colonization) and space resources. Importantly, most of these discussions have remained within the epistemic communities, consisting of advocacy groups, professional associations, science fiction communities, and scientists.[1] Serious consideration of space resources and settlement has only recently entered into the broader public imagination policymaking discourse among elites. This chapter examines the discourse in the broader community, while the next chapter will examine its instantiation in the policy.

THE OUTER SPACE EPISTEMIC COMMUNITY

The broad U.S. outer space epistemic community that includes scientists, engineers, astronauts, political advocates, and enthusiasts of all walks of life cohered around certain societies and the conferences and information systems they established. These organizations helped the community cohere and focus its influence. It is important to note that very different visions animate the broader space community. The space development community—those components that share a definite development, commerce, industrial, or settlement vision—form a distinct pole from other visions, such as those that animate the 1980 Planetary Society (formed in 1980 by Carl Sagan), the latter being focused on robotic science and exploration.[2] In the taxonomy put forward by Greg Klerkx, there are three distinct visions or paradigms that inform individual interest in space: these are the von Braunian (after Werner von Braun), the Saganite (after Carl Sagan), and the O'Neillian (after Gerald K. O'Neill).[3]

"Von Braunian's are deeply committed to human space travel" and believe that "space demand military imperatives of national scope," and "find their most comfortable home within NASA and Big Aerospace partners."[4] This is a statist, nationalist, and socialist approach to space. The state is the main actor. Outer space is an arena for the expression of national vibrancy and leadership. The means (rockets, budgets, space stations) are collectively owned by the government. The fundamental motivation is glory. The focus is on inspirational "firsts"—destinations such as Mars.

The Saganite is a "look but don't touch" astronomer's vision, where robotic probes economically survey a pristine universe. "Space is something akin to a virtual theme park; you can't actually go there yourself but you can look at it, admire it."

In contrast, the O'Neillian vision foresaw settlements in space, significant off-world industrialization, the use of space resources to transcend the limits of growth, and the use of solar power satellites. O'Neill's vision would inspire an entire movement as well as multiple organizations, including the L5 Society, The National Space Society (NSS), Space Frontier Foundation (SFF), Students for the Exploration and Development of Space (SEDS), the Space Development Steering Committee (SDSC), the Earthlight Foundation, and the Alliance for Space Development (ASD). It would likewise inspire the founders of the X-Prize (and the Ansari and Google Lunar X-prizes), the International Space University (ISU), Singularity University, and Jeff Bezos (of Blue Origin). Many individuals participate in more than one organization and attend each other's conferences. As with most epistemic communities, there are a few individuals—usually policy entrepreneurs—who are more central, and have founded or participated in multiple organizations, as well as companies.

First among the important organizations pushing a vision for space settlement and industrialization is the NSS, which was formed in 1987 from the merger of the National Space Institute founded by Werner von Braun in 1974 and the L5 Society, begun in 1975 to promote the ideas of Dr. Gerald K. O'Neill.[5] This is the largest space advocacy organization with the broadest mandate. Their "Statement of Philosophy" lays out their commitments[6]:

> The Vision of NSS is people living and working in thriving communities beyond the Earth, and the use of the vast resources of space for the dramatic

> betterment of humanity. . . . The Mission of NSS is to promote social, economic, technological, and political change in order to expand civilization beyond Earth, to settle space and to use the resulting resources to build a hopeful and prosperous future for humanity. Accordingly, we support steps toward this goal, including human spaceflight, commercial space development, space exploration, space applications, space resource utilization, robotic precursors, defense against asteroids, relevant science, and space settlement-oriented education.[7]

The NSS lists several rationales for its mission:

(a) Survival of the Human Species and Earth's Biosphere.
(b) Growth—Unlimited Room for expansion (New Habitats for Life; New Frontier for Humanity).
(c) Prosperity—Unlimited Resources (Improved Standards of Living; Economic Opportunity; Technological Development).
(d) Curiosity—The Quest for Knowledge.[8]

It also puts forward several guiding principles of human rights, ethics, and pragmatism, as well as lists beliefs in the following: (1) individual rights, (2) unrestricted access to space, (3) personal property rights, (4) free market economics, (5) government funding of high-risk research & development (R&D), (6) international cooperation, (7) democratic values, (8) enhancement of earth's ecology, and (9) protection of new environments.[9] The NSS holds an annual International Space Development Conference and has various print and electronic publications. It also organizes an annual Legislative Blitz in concert with the Space Frontier Foundation's March Storm, and an August Home District Blitz.[10] They maintain important online libraries[11] for space settlements[12] and space solar power.[13] Through their members, they sponsor two associated space settlement competitions, Al Globus's Space Settlement Contest (previously sponsored by NASA Ames)[14] and Anita Gale's International Space Settlement Design Competition,[15] which both annually engage hundreds of students in designing cities in space. They also provide a roadmap to space settlement specifying 31 specific milestones.[16] The NSS also sponsors an annual Space Settlement Summit, which "brings together the leading people, companies and organizations that are making space settlement a reality."[17] The NSS is among the most politically active institutions and is deeply committed to personal property rights, reflecting the influence of its predecessor, the L5 Society that prevented U.S. ratification of the Moon Treaty[18] due to its opposition to the treaty's ban on private property and its redistributive system.[19]

The second most important carrier of the space settlement and space resources vision is SFF, founded in 1988, whose credo identifies the Space

Frontier Foundation as "an organization of people dedicated to opening the Space Frontier to human settlement as rapidly as possible."[20] Their goals include "protecting the Earth's fragile biosphere and creating a freer and more prosperous life for each generation by using the unlimited energy and material resources of space."[21] Among their listed "purpose is to unleash the power of free enterprise and lead a united humanity permanently into the Solar System."[22] The Space Frontier Foundation hosts the annual NewSpace conference and helps organize an annual "March Storm" grassroots engagement of the U.S. Congress to advocate for commercial space.[23]

Other important organizations that share a space development agenda include the Space Studies Institute (1977), SSEDS (1980), of which Jeff Bezos was a member, ISU (1987), the Mars Society[24] (1998), The Moon Society (2000),[25] the American Institute of Aeronautics and Astronautics (AIAA) Space Colonization Technical Committee (2002),[26] the SDSC (2009),[27] the Earthlight Foundation (2012),[28] and the ASD (2015).[29] The X-Prize foundation also played a key role in extending the community of doers and public interest through the Ansari X-Prize that ran from 1996 to 2004[30] and achieved the first private suborbital flight, and the Google Lunar X-Prize[31] that ran from 2007 to 2018 and catalyzed work on lunar lander capability.[32]

DIALOGUE WITHIN THE EPISTEMIC COMMUNITY

John Kingdon has provided a model to understand agenda setting and policymaking in which three separate streams—policy proposals, problems, and politics—each follow their own logic and interact largely by chance and opportunity.[33] The space movement's epistemic community (including its policy entrepreneurs) constitutes the first distinct stream, providing policy proposals, which at various times rose to salience in the larger political elites, depending to what extent they were perceived as relevant to problems or served an extant political mandate.

A review of the breadth of the dialogue among the U.S. epistemic community would involve hundreds to thousands of authors and is beyond the scope of this work.[34] A summary pointing out some of the more important works will be offered before providing a history of elite discourse.

MAJOR EPISTEMIC WORKS

By far, the most important work was Gerald K. O'Neill's 1977 book *The High Frontier: Human Colonies in Space,*[35] which provided a full industrial vision of colonies in space, how to build them, and an economy based upon solar power satellites. It inspired an entire movement, multiple organizations (including the L5 Society, NSS, Space Frontier Foundation, SEDS, ISU), as well as Jeff Bezos and Blue Origin.

To be noted, no less a luminary than Herman Kahn considered in-depth scenarios for space development in a 1977 work "Long-Term Prospects For Developments In Space (A Scenario Approach)" for NASA that was originally restricted by the Department of Defense.[36] In their "moderate scenario" to 2176:

> Space development: Large human colonies will be able to travel anywhere in the solar system and will be self-sufficient for at least 20 years, perhaps indefinitely. *Cost of space transportation will fall by a factor of 10 or more. At least 100 million people will have their homes in space colonies.* One of the socio-political consequences of the above developments is that human beings in space colonies, whether in orbit or on a planetary surface such as the moon or Mars, probably will become independent of earth—that is, they will become sovereign in the sense of nations on earth. Most sociologists today believe that this is an inevitable result in space—just as it has been on earth. I can only assume that these natives will be friendly—if we are.[37]

By 1978, NASA had published its first study on *Extraterrestrial Materials Processing and Construction*. In 1979, author Harry G Stine published *The Third Industrial Revolution* articulating the vast potential of the future space industrial economy.[38]

While the focuses of most of O'Neill's followers were broadly societal, libertarian, and humanitarian, the potential to affect Earthly balance of power did not go unnoticed. A second work with a similar title, *High Frontier: A New National Strategy* by General Daniel O. Graham in 1982 articulated a comprehensive national strategy that included solar power satellites, space industrialization, as well as the rationale for Reagan's Strategic Defense Initiative (SDI).[39]

While the concept of harvesting energy in space has been attributed to both Konstantin Tsiolkovsky (1925)[40] and Isaac Asimov (1941)[41], it was Peter Glaser who invented (1968) and patented (1973)[42] a method and succeeded in galvanizing study of the concept. Between 1977 and 1981, an extensive collection of reports was undertaken by NASA and the Department of Energy, including the use of lunar resources,[43] as well as published evaluations by the National Research Council[44] the Office of Technology Assessment in 1981,[45] and a book by G. Harry Stine, *Space Power*[46] that same year.

In 1995, Ralph Nansen's *Sun Power: The Global Solution for the Coming Energy Crisis*[47] reintroduced to the public the concept of Space Solar Power in a popular nonfiction book. The following year, 1996, Dr. John Lewis published *Mining the Sky: Untold Riches from Asteroids, Comets, and Planets*, which provided the public with the first popular and public survey of the potential wealth of the solar system, and especially asteroids.[48]

In 2002, Brigadier General Simon "Pete" Worden & and then Major John E. Shaw (now Major General) provided an exceptionally broad, *Whither Space Power?* which included consideration of the resources of the Moon and asteroids.[49] That same year, Dr. Carl Everett Dolman published *Astropolitik: Classical Geopolitics in the Space Age*, which considered the interaction of an in-space economy, space geography, and military power from the perspective of realism and geopolitical conceptions.[50] Dolman's was the first to create a rich theoretical conception of commerce, position, and astrostrategy, which would be later elaborated by others.

Within the space settlement and space development community, distinct camps had emerged between those who favored a Moon-first approach, those who favored a Mars-first approach, and those who favored asteroids. In 2004, Dennis Wingo's *MoonRush: Improving Life on Earth with the Moon's Resources* provided an environmental vision for how lunar resources could be applied to problems of scarcity on Earth.[51] In 2011, the founder of the Mars Society, Robert Zubrin authored (originally published 1996) and updated *The Case for Mars: The Plan to Settle the Red Planet and Why We Must*,[52] which ignited significant public interest and is likely to have had a catalytic effect on the thinking of Elon Musk.

The Pentagon Study Group's *Space Solar Power: An Opportunity for Strategic Security* in 2007 received significant attention.[53] It was followed by Ralph Nansen's *Energy Crisis: Solution from Space*[54] in 2009, though these failed to gain significant traction.

The formation of Google billionaire-backed Asteroid companies Planetary Resources (2012) and its competitor Deep Space Industries (2013), as well as an important 2012 study on asteroid mining by the Keck Institute[55] put asteroid mining in the public's mind. Zubrin's call for a focus on Mars would be followed by a new book by Dr. John Lewis in 2014, *Asteroid Mining 101: Wealth for the New Space Economy*[56] and by Paul Spudis in 2016, *The Value of the Moon: How to Explore, Live, and Prosper in Space Using the Moon's Resource*.[57] John Mankins, the former head of NASA Exploration Systems and Advanced Concepts in 2014, published *The Case for Space Solar Power*.[58] In 2017, the National Space Society also created an online library featuring the most important technical works on the topics of space settlement, space manufacturing, space solar power, lunar bases, and space resources.[59] It was the first time that such an archive was made available to the public.

Spacefaring advocates within military circles elaborated on the development paradigm. In 2015, Dr. Brent Ziarnick published *Developing National*

Power in Space: A Theoretical Approach,[60] which built on the prior naval theory of Alfred Thayer Mahan and economist Joseph Schumpeter.

More recently, the highly public and visually stunning attempts and successes of Virgin Galactic, SpaceX, and Blue Origin to fly commercial reusable rockets reignited public interest in spacefaring, and significant popular works followed, including Christian Davenport's *Space Barons*,[61] Robert Zubrin's *The Case for Space: How the Revolution in Spaceflight Opens Up a Future of Limitless Possibility*,[62] and Rod Pyle's *Space 2.0: How Private Spaceflight, a Resurgent NASA, and International Partners are Creating a New Space Age*.[63]

ELITE DISCOURSE

Below we offer an overview of the history of elite discourse in the United States on the subject of space resources. It is divided into two sections: an early period (lasting from 1957 until 2011) that developed the concepts and ideas, but generally failed both to grow roots in policy and to bear fruit programmatically; and the modern era, beginning with the formation of Planetary Resources in 2012[64] and lasting till present.

The section is meant to provide the first-ever record of primary source data of how the U.S. policy conversation has evolved chronologically, and to provide a strong case of the deep roots of U.S. space resource ambitions, including from Congressional bills, policy think-tank reports, and influential thought leaders. It is intended to be a resource for those doing future policy and legislative work. Those readers less interested in policy history can skip to the end of the chapter to read our conclusions.

What the reader should look for is a progressive elaboration of space expansionist and space development themes, which appear to become more specific and more urgent as both private actors and foreign actors show interest. Over time, the reader will see a consensus building on a policy image that seeks to elevate the importance of space commerce and legal protections for space resources, and linkage to concerns of predation by foreign powers and the need for a U.S. military power to be postured to protect these future interests.

ELITE DISCOURSE IN THE EARLY PERIOD (1957–2011)

While consideration of space expansionism in America has even earlier roots in speculative fiction and rocketry hobbyist clubs,[65] it became a subject of serious policy interest with the launch of Sputnik in 1957.[66] An early example was Cox and Stoiko's 1958 *Spacepower: What it means to you*, which gave early consideration to lunar bases, resources, and property.[67] Reacting to Sputnik, President Eisenhower passed the National Aeronautics and Space Act of 1958 establishing NASA; among its founding objectives is "the establishment of long-range studies of the *potential benefits to be gained from, the opportunities for, and the problems involved in the utilization of* aeronautical and *space activities for peaceful and scientific purposes*."[68]

The breadth of the early policy debate was captured in the 1961 Simon Ramo series, *The Peaceful Uses of Outers Space* where Ralph Cordiner first laid out a compelling vision for space development.[69] Dennis Wingo recounts the industrial vision of space he proposed:

> In 1960 the Chairman of the Board of the General Electric Corporation, Ralph Cordiner, asked three questions regarding space policy and these questions are still relevant to the development of a new 21st century space policy.
>
> - How can we utilize our dynamic system of competitive private enterprise in space, as on earth, to make *newly discovered resources useful to mankind*?
> - How can private enterprise and private capital make their maximum contribution? What projects will necessarily require government chairmanship and support for their execution?
> - What must be done to preserve a free society while competing in the international race for space? How can we assure that when the space frontier is developed, it will be an area of freedom rather than regimentation?[70]
>
> Cordiner did not see space policy as separate, or as an appendage to national policy, but a natural growth of the aspirations of a free people in a free market economy where space affords boundless opportunity. . . . Those who might dismiss Cordiner's vision, must consider his impactful role in American history. Beyond being the CEO of General Electric, one of the largest industrial corporations of mid-20th century America, he was also the number two man during World War II on the war production board. It was the war production board that transformed America's industrial might from civilian products to producing fleets of thousands of navy ships, tens of thousands of airplanes, tanks, and millions of military vehicles and victory in a global war. He, as few others, understood the power of government directed industry, yet he rejected that path for the American move into space and had the vision to tie that to a time where the productive abilities of our competitive free enterprise system could take us across the solar system.[71]

But the vision of Ralph Cordiner[72] would not hold sway. Instead, in its civil program, America would pursue a state-centric socialist approach to space aimed not at economic development but international prestige, to the detriment of America's would-have-been commercial space entrepreneurs. This approach would last through the end of the Apollo era.

For a brief period, the U.S. military held the banner for thought leadership on space development. Prior to President Eisenhower's decision to de-emphasize military space competition in favor of a civilian space agency[73] (NASA[74]), the U.S. military of the late 1950s and early 1960s had proposed ambitious plans for space development[75] in line with traditional military peacetime missions of frontier exploration and development.[76] These plans included a manned orbital station intended to house twenty to twenty-four persons, requiring the development of a fully reusable rocket system able to ferry 7 metric tons of people and supplies (replacement crews of nine to twelve persons) per flight, and a fleet of these vehicles able to provide 74 servicing flights per year.[77] The military was codeveloping a nuclear thermal rocket "NERVA" intended to provide reach throughout the entire solar system,[78] and an incredibly ambitious Strategic Earth Orbital Base predicated on the Advanced Research Project Agency's (ARPA) Project Orion nuclear pulse detonation propulsion system[79] that would have been capable of going to Mars and back in 150 days with 5,000 tons.[80]

Consideration for off-world bases and their commercial potential occurred very early. In 1959, just two years after Sputnik, Maj Gen Bernard A. Schriever, Commander of the Air Force Ballistic Missile Division, testifying before the Senate Committee on Aeronautical and Space Sciences stated that all three military services should be studying the possibility of a base on the Moon.[81] That same year, the Army completed its study of a Moon base, "Project Horizon" with Lt Gen Arthur G. Trudeau, Chief of Research and Development stating "I envision expeditious development of the proposal to *establish a lunar outpost to be of critical importance to the Army of the future*. This evaluation is apparently shared by the Chief of Staff." Notably, the report echoes a Lewis & Clark[82] dual-use purpose: "The primary objective is *to establish the first permanent manned installation on the moon*. Incidental to this mission will be the investigation of the scientific, *commercial*, and military potential of the Moon."[83] This extremely early plan considered space resources: "*materials on the moon itself may prove to be valuable and commercially exploitable*"[84] and asserted:

> Assuming we arrive on the moon, before or after the Russians, and *find a concentration (the only known concentration) of very valuable resources . . . if American forces are in possession of an area having resources of great and significant value, it is reasonable to anticipate that her peaceful occupancy or holding would be respected.* So, too, with respect to a like holding by the USSR.[85]

The report also noted the commercial-military nexus, "military space capabilities are technically inseparable from peaceful capabilities which are well worth pursuing in their own right. Reconnaissance for merchant-ship lane patrol and/or peaceful mapping of resources can also be used to locate military targets."[86]

Although the military was prepared to allocate a large percentage of its budget to fund its ambitious manned space program,[87] by 1963 a combination of factors had precluded any role for the U.S. military in leading manned spaceflight and space development.[88] First, concerns over atmospheric and high-altitude nuclear testing[89] led to the Limited Nuclear Test Ban Treaty, signed by the United States, Soviet Union, and Great Britain on August 5, 1963, which prohibited the testing of nuclear weapons in outer space, underwater, or atmosphere,[90] thus putting an end to the nuclear ambitions of the U.S. military in space.[91] Second, although the Secretary of the Air Force Eugene Zuckert supported the ambitious space program, he refused to request funding for it from DOD, knowing that Dr. Harold Brown, Director of Research and Engineering, and Secretary of Defense Robert McNamara would not support it. Third, in May 1961, framed in the context of "if we are to win the battle that is now going on around the world between freedom and tyranny," President Kennedy proposed the Moonshot to Congress, stating "I believe that this nation should commit itself to achieving the goal, before this decade is out, of landing a man on the moon and returning him safely to the Earth." By September 1962, Kennedy had publicly committed NASA to the endeavor,[92] thus excluding the military from its traditional peacetime exploration role. Space development was denied a powerful champion in the U.S. military. Thought leadership would now shift to other venues.

On January 27, 1967, during the height of the moon race, the United States and the Soviet Union signed the Outer Space Treaty (OST).[93] It contained the following provisions relevant to space resources.

Article II

> Outer space, including the *moon and other celestial bodies, is not subject to national appropriation by claim of sovereignty*, by means of use or occupation, or by any other means.

Article IX

> In the exploration *and use* of outer space, *including the Moon* and other celestial bodies, States Parties to the Treaty shall be guided by the principle of co-operation and mutual assistance and *shall conduct all their activities* in outer space, including the Moon and other celestial bodies, *with due regard to the corresponding interests of all other States Parties to the Treaty.* [94]

A history of the OST suggests that the overriding concern was prohibiting nuclear weapons in orbit[95] and not space resources.Apparently "Article II, which prohibits national appropriation of outer space and celestial bodies, provoked only a few minutes of debate";[96] "there was *no disagreement that the scope* of the Treaty *should include 'use' of outer space and celestial bodies*, even though potential uses of outer space and celestial bodies can be foreseen only to a limited extent at present"[97] and the United States took the position before the General assembly that "Article II, by banning national appropriation of outer space and celestial bodies, *reinforces the free access language in Article I.* If an individual nation *cannot claim sovereignty* to any particular area of outer space or of a celestial body, *it cannot deny access to that area.*"[98]

In fact, "the United States has long taken the position that Article I of the Treaty . . . *recognizes the right of exploitation*" (1980)[99] and that "this 'non-appropriation' principle applies to the natural resources of celestial bodies only when such resources are 'in place'"[100] and does not limit "ownership to be exercised by States or private entities over those natural resources which have been removed from their 'place' on or below the surface of the moon or other celestial bodies" (1979).[101] Nevertheless, Article II would have a chilling effect on humanity's journey to the stars.

As the first Space Race was won, America began to debate its follow-on space vision. The 1969 Space Task Group, chaired by Vice President Spiro Agnew expressed support for the settlement-related concepts and proposed a series of programs and options including a fifty-person space station.[102] However, contrary to the expectations of some, landing on the Moon did not result in a significant expansion of activity. An insightful view given by John Cramer is relevant to the topic of space resources and territoriality:

> Some of you may recall that *after the Apollo 11 landing on the Moon in 1969, there was an amazingly abrupt decrease in public enthusiasm for manned space flight.* I have a theory about what happened. I believe that *a principal sociological force driving the manned space program of the 1960s was territoriality, the basic human urge to explore and occupy new territory.* I think the man-in-the-street expected the Apollo astronauts to go to the Moon and claim it as our territory. *Neil Armstrong was supposed to step out of the Lunar Lander and say "I claim this territory in the name of the United States of America.*" When instead, all he said was the scripted lines about a "giant step for mankind," the aftermath was like air escaping from a balloon. The *entire space program deflated very rapidly when the taxpayers discovered that the Moon Race was not about claiming the Moon as a territory of the United States of America. The pattern of exploration and territorial expansion we had learned in history class had not occurred.* Moreover, in 1969 the Vietnam War was in full swing, diverting money and attention away from the space program.[103]

Undeterred, the epistemic community had become an activist community. In this period, beginning in 1972, a resource-centric industrial vision would again make its way into discussions among the policy elites. An excellent history chronicles that period in Michael Michaud's *Reaching for the High Frontier: The American Pro-Space Movement, 1972-84.*[104] Michaud's fourth chapter vividly recalls the engagement of the pro-space movement with Congress, in particular how Gerald O'Neill's vision for industrial development and space colonies and Peter Glaser's ideas for Space Solar Power Satellites would receive national attention. In late 1977, Representative Olin Teague even introduced this pro-High Frontier resolution (H.R. 451) which included the following text:

> As longer-range, high priority national goals, it is anticipated that by the year 2000 these explorations will have opened *the resources* and environment of extraterrestrial space to an as yet incalculable range of other positive uses, including, but not limited to, international cooperation for the maintenance of peace, the *discovery and development of new sources of energy and materials, industrial processing and manufacturing*, food and chemical production, health benefits, recreation, and conceivably, the *establishment of self-sustaining communities in space.*[105]

But America would not adopt the grand vision of the pro-space movement at that time. It did not resonate with NASA that saw itself as the star player in a destination-focused vision. It did not appeal to a president who was focused on Vietnam and a domestic agenda.

The criticism of Senator William Proxmire in 1977 on "60 Minutes" regarding O'Neill's vision was telling of the time.

> L-5 by 1995 calls for "only" a few billion dollars of the taxpayers' money to establish an idyllic colony out in space with all the amenities of life for a few thousand adventurous earthlings. This proposal gives the best argument yet for chopping NASA's funding to the bone. The cost would have to be millions of dollars annually for each earthling maintained in space. As Chairman of the Senate Subcommittee responsible for NASA's appropriations, I say not a penny for this nutty fantasy. Sen. William Proxmire.[106]

This hostile reaction ensured NASA distanced itself from the grander vision animating the pro-space and pro-industrial development community, forcing the United States to cede its multi-decade lead in the race to become the first true industrial spacefaring civilization.

A second attempt was made in 1981 by Representative Newt Gingrich with thirteen cosponsors, who introduced H.R. 4286 "The National Space and Aeronautics Policy Act of 1981," the purpose of which was to "establish a comprehensive civil space and aeronautics policy that will provide a framework for a world information system, Earth orbital facilities, exploration of the solar system, and the development of other space and aeronautical activities to preserve and expand the leadership of the United States in space and aeronautics." It forwarded an exceptionally strong industrial vision, stating,

> The future potential of permanent manned and automated *facilities in orbit* around the Earth include those aspects of our civilization that provide for research, education, *power production, manufacturing* and health care . . . the continued exploration and *utilization* of the solar system, including the Moon and Mars, is important to present and future generations of Americans and should be a long-term goal of the United States Space Program.[107]

The bill sought to establish policies toward the creation of

> *Orbital facilities during the 1990's that could consist of permanent, generally self-financing facilities for research, education, power production, manufacturing and health care* . . . a solar system exploration capability during the first decade of the twenty-first century which provides the option for *Moon bases*, manned missions to Mars, *a Moon settlement*, manned missions to Venus, and a Mars Settlement . . . an environmentally acceptable *space to Earth power capability that is economically competitive with power generation on Earth.*[108]

Moreover, it contained a "Northwest Ordinance for Space," which "offered a framework for future space settlements to join the Union, first as territories, later as states—just as Ohio and other western settlements joined the Union in the 19th Century."[109] The bill stated, "The Congress declares that the *United States is committed to the expansion of free people and free institutions into space*,"[110] and set forth the following:

Title IV—Government of Space Territories

Constitutional protection

All persons residing in any community in space organized under the authority and flag of the United States shall be entitled to the protection of the Constitution of the United States.

Self-government

Whenever any such community shall have acquired twenty thousand inhabitants, on giving due proof thereof to Congress, they shall receive from Congress authority with the appointment of time a place to call a convention of representatives to establish a permanent constitution and government for themselves.

Admission to statehood

> Whenever any such community shall have as many inhabitants as shall then be in any one of the least numerous of the United States such community shall be admitted as a State into the Congress of the United States on equal footing with the original states.[111]

H.R. 4286 showcases just how much foresight some U.S. congressmen displayed with regard to space resources and settlement during the Cold War period. But this farsighted bill would not become law.

The United States also entertained more internationalist approaches to space resources. During this same period, from 1980 to 1982, the United States would debate the Moon Treaty (formally the *Agreement Governing The Activities of States on the Moon and Other Celestial Bodies* of 1979), which specifically sought to allow utilization, though executed exclusively through an international authority. It stated the following provisions:

3. Neither the surface nor the subsurface of the moon, *nor any part thereof or natural resources in place, shall become property of any State*, international intergovernmental or nongovernmental organization, national organization or *non-governmental entity or of any natural person*.
5. States Parties to this Agreement hereby undertake to establish *an international regime, including appropriate procedures, to govern the exploitation of the natural resources of the moon as such exploitation is about to become feasible*.
7. The main purposes of the international regime to be established shall include: (a) *The orderly and safe development of the natural resources of the moon;* (b) The *rational management of those resources*; (c) The expansion of opportunities in the use of those resources; (d) *An equitable sharing by all States Parties in the benefits derived from those resources, whereby the interests and needs of the developing countries*, as well as the efforts of those countries which have contributed either directly or indirectly to the exploration of the moon, shall be given special consideration.[112]

Strong opposition arose from the space activist community, especially the L5 Society, which opposed the treaty because of its anti-private ownership, collectivist approach. Said their 1982 L5 newsletter:

> The Moon Treaty, for example, *designates the Moon and its resources as the "common heritage of mankind" and calls for the establishment of an "international regulatory regime"* to create and oversee an equitable system to share resources once the means of *lunar exploitation* becomes feasible. That, critics in the R&D industries and in the Senate charged, *essentially means private industry would be prohibited from developing outer space*.[113]

In 1986, the U.S. National Commission on Space final report reiterated the animating vision of the space advocacy community: "To lead the exploration *and development of the space frontier*, advancing science, technology, and enterprise, and building institutions and systems that *make accessible vast new resources and support human settlement beyond Earth orbit, from the highlands of the Moon to the plains of Mars*."[114] It provided justification for legislative attempts to enshrine this as a policy.

After over a decade of attempts, the space development community achieved significant legislative success. In 1988, Rep George Brown (D-Calif.) introduced the "Space Settlement Act of 1988" (H.R. 4218[115]). It stated: "the term '*space settlement*' *means any community of humans living beyond Earth's atmosphere which exists with a substantial degree of independence of resupply from Earth*," and required NASA to submit a report every two years to congress analyzing how current science and technology could be applied to the establishment of space settlements, identify scientific and technical steps that must be taken toward their establishment, alternative locations and architectures, status of technologies for resource development and the use of energy production, and economics of financing such settlements, as well as sociological factors such as psychology, political science, legal issues involved with space settlements. It sought to authorize $3M for this purpose. The bill was passed and became Public Law 100-685, amending the NASA charter to read[116],[117]:

> SEC. 217. (a) The Congress declares that the extension of human life beyond Earth's atmosphere, leading *ultimately to the establishment of space settlements*, will fulfill the purposes of advancing science, exploration, and development and will enhance the general welfare.[118]

This act was U.S. law from 1988 to 1995. However, NASA never produced such reports, and NASA (under Administrator Goldin) used[119] the Paperwork Reduction Act of 1995 to have it dropped from legislation in future amendments to the NASA charter.[120]

Coinciding roughly with the beginning of the construction of the International Space Station (ISS) in 1998, elite discourse on broader space industrial vision appears to have gone into a period of dormancy as the U.S. space program got "stuck in low Earth orbit," until the Vision for Space Exploration (VSE) was announced by President George W. Bush, January 14, 2004, which specifically mentioned resources twice:

- Use lunar exploration activities to further science, and to develop and test new approaches, technologies, and systems, *including use of lunar and*

other space resources, to support sustained human space exploration to Mars and other destinations.
- Conduct robotic exploration across the solar system for scientific purposes and to support human exploration. In particular, explore Jupiter's moons, *asteroids, and other bodies* to search for evidence of life, to understand the history of the solar system, *and to search for resources.*[121]

The clear economic motivations of the Bush policy were highlighted in a speech by the president's science advisor, Dr. John Marburger in 2004, who for the first time articulated the vision of incorporating the solar system into our economic sphere:

> As I see it, questions about the vision boil down to *whether we want to incorporate the Solar System in our economic sphere, or not.* Our national policy, declared by President Bush and endorsed by Congress last December in the NASA authorization act, affirms that, "The fundamental goal of this vision is to advance U.S. scientific, security, *and economic interests* through a robust space exploration program." So at least for now the question has been decided in the affirmative.[122]

The wording of this policy phrase is significant. It subordinates space exploration to the primary goals of scientific, security, and economic interests. This was the first time an administration had subordinated exploration to economic and security interests and may be marked a subtle but important shift in the U.S. policy stance.

An equally ambitious and long-term vision for human space settlement and its civilizational importance was articulated by President Bush's NASA Administrator Dr. Mike Griffin in a 2005 Washington Post interview;[123] there he provided an unabashedly pro-settlement vision, and linking this vision to U.S. values:

> the goal isn't just scientific exploration . . . *it's also about extending the range of human habitat out from Earth into the solar system* as we go forward in time . . . In the long run a single-planet species will not survive. We have ample evidence of that . . . [Species have] been wiped out in mass extinctions on an average of every 30 million years . . . I cannot say that multiple-planet species will survive, I think I can prove to you from our own geologic record that single-planet species don't . . . There will be another mass-extinction event. *If we humans want to survive for hundreds of thousands or millions of years, we must ultimately populate other planets* . . . I'm talking about that one day, I don't know when that day is, but *there will be more human beings who live off the Earth than on*

> *it. We may well have people living on the moon. We may have people living on the moons of Jupiter and other planets. We may have people making habitats on asteroids.* We've got places that humans will go, not in our lifetime, but they will go there . . . I don't know the date—but I know that *humans will colonize the solar system and one day go beyond.* And *it is important for me that humans who carry—I'll characterize it as Western values—are there with them.* You know, I think we know the kind of society we would get if you, for example, carry *Soviet values. That means you want a gulag on Mars. Is that what you're looking for?*[124]

Dr. John Marburger spoke again in 2006, this time clearly subordinating space exploration to grander achievements of occupation, infrastructure, and routine access to resources for economic incorporation of the solar system into the Earth's economic sphere:

> The Moon is the most massive near-Earth object—massive enough to have a useful surface gravity, but substantially out of Earth's gravity well and therefore of great interest to deep space operations. *I think it is inevitable that the Moon will eventually become a space station and a source of mass for space applications.* Between Earth and Moon there are interesting places: low Earth orbits, geostationary orbits, Lagrange points. *Whatever operations we perform in these places have to be conducted from platforms we construct and launch from Earth or Moon* . . . As I put it in my speech two years ago, "*Questions about the vision boil down to whether we want to incorporate the Solar System in our economic sphere, or not.*" If we are serious about this, then our objective must be more than a disconnected series of missions, each conducted at huge expense and risk, and none *building a lasting infrastructure* to reduce the expense and risk of future operations. If we are serious, we will build capability, not just on the ground but in space. And our objective must be to make the use of space for human purposes a routine function . . . *They ignore the very likely possibility that operations on the Moon "and other intermediate destinations" will "serve national and international interests" other than science*, but including science as an important objective. Our current experience with space, dramatically portrayed by the existence today of a commercial space industry, is that it is useful in ways not imagined even by the early visionaries. *To me this is an important point. Exploration by a few is not the grandest achievement. Occupation by many is grander. Not necessarily in the sense of permanent human occupation, but in the sense of routine access to resources.* The future I look for in the human space enterprise is one in which exploration has long since ceased *and our successors reap the benefits of the new territories.*[125]

Space solar power briefly received national-level attention following a 2007 Pentagon Study *Space-Based Solar Power: an Opportunity for Strategic Security*, which had been jointly undertaken between the Pentagon's National Security Space Office (NSSO), the National Space Society and the Space Frontier Foundation. It concluded that "space-based solar power does present a strategic opportunity that could significantly advance U.S. and partner security, capability, and freedom of action and merits significant further attention on the part of both the U.S. Government and the private sector" and that "SBSP requires a coordinated national program with high-level leadership and resourcing commensurate with its promise, but at least on the level of fusion energy research or International Space Station construction and operations."[126]

The NSSO report came at the end of the second Bush administration and was not acted upon. However, following the election of Barack Obama in 2008, a white paper appeared on the Obama-Biden Transition Project website Change.gov titled, "Space Solar Power (SSP)—A Solution for Energy Independence & Climate Change." It stated, "we urge the next President of the United States to include SSP as a new start in a balanced federal strategy for energy independence and environmental stewardship, and to assign lead responsibility to a U.S. federal agency."[127] The transition team website invited public comment and it became among the top five most-discussed papers.[128] SBSP appeared to have reached a new level in the policy discourse. SBSP was strongly supported by the advocacy community, and the mood of the space advocacy community was decidedly pro-commercial in its approach to the government, as evidenced by the statement from the Space Frontier Foundation spokesman.[129]

The concept continued to gain support via the Obama Open Government Plan, where the concept received significant attention from the public. As reported in an open letter from citizens for space-based solar power to Obama's Office of Science and Technology Policy in 2010, "The idea for convening a Space Based Solar Power Conference was the number one idea supported on OpenOSTP, OpenEnergy and OpenNASA" and provided the supporting website participation statistics.[130] Reportedly, both George Whitesides[131] and Lori Garver, the NASA deputy were space solar power advocates. Space solar power was also nominated as a top-line "deliverable" for President Obama's first visit to India by the Department of State but failed to gain sufficient interagency consensus. Why the Obama administration did not act on space-based solar power given its consistency with its energy and environmental goals remains both a mystery and a lost opportunity for the United States.

Consistent with President Kennedy's framing of space as a forum to showcase civilizational vibrancy through its symbolic and inspirational power, President Obama sought to refocus the U.S. space program away from what it saw as a redo of Apollo and toward new destinations such as Mars.

MODERN ERA (2012–PRESENT)

In 2012, the public was excited to hear of a company, called Planetary Resources, which intended to mine asteroids. It was soon followed by the formation of Deep Space Industries in 2013, and TransAstra in 2015.

In 2014, Tom Kalil in the Obama Office of Science and Technology Policy published a conversation on the White House official blog with Dr. Phillip Metzger, formerly NASA KSC, on the subject of "rapidly bootstrapping a solar system civilization."[132] It appeared to be a clear administration endorsement of a grand space industrial vision to instantiate a complete supply chain in space for the benefit of future generations, and signaled administration openness to commercial space exploitation:

> If we want to create a robust civilization in our solar system, more of the energy, raw materials, and equipment that we use in space has to come from space. Launching everything we need from Earth is too expensive. It would also be too expensive to send all of the factories required to manufacture everything necessary to support a solar system civilization.
>
> Ultimately what we need to do is to evolve a complete supply chain in space, utilizing the energy and resources of space along the way. We are calling this approach "bootstrapping" because of the old saying that you have to pull yourself up by your own bootstraps. Industry in space can start small then pull itself up to more advanced levels through its own productivity, minimizing the cost of launching things from Earth in the meantime. Obviously, this isn't going to happen overnight, but I think that it is the right long-term goal.
>
> *Why do you think that developing the capability for a self-sustaining space industry would be a desirable goal?*
>
> We need to realize we live in a solar system with literally billions of times the resources we have here on Earth and if we can get beyond the barrier of Earth's deep gravity well then the civilization our children and grandchildren will build shall be as unimaginable to us as modern civilization once was to our ancestors.[133]

That same year, a bill was introduced, H.R.5063 Asteroids Act 2014, which was a first attempt to assert Americans could own space resources.[134] While it would not pass, it laid the groundwork for future efforts that would, ultimately, become law.

On January 10, 2015, in his annual State of the Union Address, President Obama would include in his remarks his wish for America to win the race for new discoveries of all kinds, including "pushing out into the solar system, not just to visit but to stay. Last month, we launched a new spacecraft as part of a re-energized space program that will send American astronauts to Mars."[135] Despite the focus on new "flags and footprints" initiatives, the remarks hinted at a muted, if pro-settlement disposition.

A renaissance was also underway to instantiate space settlement as the overarching goal of the U.S. space program. Led by entrepreneur and space advocate pioneer Rick Tumlinson, in 2015, Pioneering Space National Summit brought together "over 100 leaders in human spaceflight from academia, government and industry" in an effort to forge a consensus to guide the future of America's human spaceflight initiatives, and agreed on the following:

> The long term goal of the human spaceflight and exploration program of the United States is to expand permanent human presence beyond low-Earth orbit and to do so in a way that will *enable human settlement and a thriving space economy*. This will be best achieved through public-private partnerships and international collaboration.[136]

The community efforts at consensus building were followed up with attempted legislation the same year (2015). Representative Dana Rohrabacher (R-Calif.) introduced[137] "The Space Exploration, Development and Settlement Act of 2016" (H.R.4752). The bill contained several provisions that sought to establish development and settlement as U.S. policy, including:

> "To require the National Aeronautics and Space Administration to investigate and promote the exploration and *development of space leading to human settlements* beyond Earth, and for other purposes." It would have amended the National Aeronautics and Space Act to read, "(d) *Exploration, Development, And Settlement Of Space.—The Congress declares that expanding permanent human presence beyond low-Earth orbit in a way that enables human settlement and a thriving space economy will enhance the general welfare of the United States* and requires the Administration to encourage and support the development of permanent space settlements" and defines "the term '*space settlement*' *means any community of humans living beyond Earth's atmosphere that is able to economically sustain its population through a neutral or positive balance of trade of goods and services*, and is able to expand its habitable real estate as need and desire of the community may warrant and international law permits." It also stated, "*It is the sense of Congress that the President should conduct a review of national space policy to incorporate as a long-term goal of the human*

> *spaceflight and exploration program to expand human presence beyond low-Earth orbit in a way that will enable human settlement and a thriving space economy.*[138]

The bill was not to become law in 2016, but it set the stage for continuing efforts.

Space solar power would receive a boost in legitimacy within the U.S. executive branch in August of 2016 when a multi-agency-industry team, won the Secretary of State, Secretary of Defense, and United States Agency for International Development (USAID), Director's Development Diplomacy and Defense "D3" Innovation Challenge.[139]

In April of 2016,[140] Rep Jim Bridenstine (R-OK) also introduced "H.R.4945—American Space Renaissance Act" in the 114th Congress (2015–2016) aimed to create a comprehensive (military, civil, commercial) space policy to "permanently secure the United States as the preeminent spacefaring nation, and for other purposes."[141] While silent on the specific topics of space resources or settlements, it featured a strong development agenda. The bill sought to amend the NASA charter to make clear their role in facilitating non-NASA activity, "(1) The expansion of the human sphere of influence throughout the Solar System. (2) To be among those who first arrive at a destination in space and to open it for subsequent use and development by others. (3) To create and prepare infrastructure precursors in support of the future use and development of space by others."[142]

It sought to define a new role for NASA: "the Administration also be directed toward the pioneering of space. The objectives of such pioneering shall be to *increase access to destinations* in space, *explore the possible options for development at these destinations, demonstrate the engineering feasibility of such development, and transition* those activities to Federal agencies outside of the Administration or persons or entities outside of the Federal Government."

The bill also attempted to frame how NASA was to view the Moon and cislunar space: "NASA should *utilize the Moon and cislunar space* in order to accomplish the goal of sending American astronauts to Mars."[143] It also sought to push NASA to embrace commercial development: "NASA should utilize commercial assets, when practicable and available, to support exploration beyond Earth orbit, including to Mars,"[144] that NASA "establish a Commercial Habitat Pilot Program to demonstrate the viability of using commercially built on-orbit habitats" and that "NASA shall enter into not less than 1 competitively bid agreement with a private sector entity to demonstrate the viability and capabilities of crewed commercial low Earth orbit platforms."[145] The bill sought to establish certain prizes for space-related activities including operation of space stations, lunar missions, asteroid missions,

Mars missions, debris clean up and salvage, and point-to-point missions on Earth.[146]

Its author, Rep Jim Bridenstine (who subsequently became the NASA Administrator in 2018[147]) gave a speech in November of 2016 (two days before the Presidential Election), directly addressing the subject of this study. Showing clear continuity with and influence by the epistemic community, in his opening, he referenced Dr. Paul Spudis and his book, *The Value of the Moon*, Dennis Wingo and his book *MoonRush*, and Lunar entrepreneur Bob Richards of Moon Express. The bill drew a clear link between the presence of water resources, the interest of the PRC, and the need for both commercial exploitation and military protection, going so far as to call for a cislunar power projection capability to comply with a constitutional duty to protect commerce. This publicly available speech is so rich in strategic thought that major portions are provided below[148]:

> Later experiments by NASA and other space agencies provided more evidence suggesting 10 billion tons of water ice at each lunar pole. Many in this room have been involved in these experiments.
>
> This single discovery should have immediately transformed America's space program. Water ice not only represents a critical in situ resource for life support (air and water); it can be cracked into its components, hydrogen and oxygen, to create the same chemical propellant that powered the Space Shuttle. Even better, this chemical propellant sits at the poles of the moon, which receive almost constant sunlight for photovoltaic power, which is necessary crack the water into hydrogen and oxygen.
>
> All of this is available on a world that has no atmosphere and a gravity well that is 1/6 that of earth. In other words, standard aerodynamic limitations do not apply.
>
> From the discovery of water ice on the moon until this day, the American objective should have been a permanent outpost of rovers and machines at the poles with occasional manned missions for science and maintenance. The purpose of such an outpost should have been to utilize the materials and energy of the moon to drive down the costs and increase the capabilities of cis-lunar space.
>
> To be clear, satellite servicing and assembly requires a lunar program that is permanent to include long term human habitation, machines, rovers, and resource production.
>
> Let's talk about China. In 2007 China used a direct ascent anti-satellite missile to shoot down one of its own weather satellites creating thousands of pieces of

orbital debris. Since then, they have been testing such missiles all the way out to geostationary orbit. Recently, the *Chang'E 2* spacecraft orbited and mapped the moon. It then travelled to the Sun-Earth L-2 point before travelling to a near-earth asteroid. Such devices could be used for a kinetic intercept path from above GEO or other spoofing, dazzling, or jamming activities. Attacks from above GEO would be very difficult if not impossible to detect. In 2014, the Chinese hacked into the U.S. National Weather Service and compelled us to stop collecting space-based weather data for a period of three days. That significant if you're from Oklahoma. They currently have two astronauts (taikonauts) on their domestically produced space station: the Tiangong-2. The Taikonauts flew there on domestic Chinese rockets and capsules: the Shenzhou Program. They currently have rovers and machines on the moon, and they have developed unhackable, unjammable, unspoofable quantum communications. It is clear the Chinese understand the geopolitical value of space operations.

The U.S Government understands that in the future, and even today, it will be a customer of routine space services, not a provider of routine space services. One of those services could also someday be in situ resource utilization from the Moon to fuel and power missions to locations deeper into our solar system, such as Mars.

This is only possible because of all the risk that the government has already retired for these capabilities. Now, the U.S. government should play a part in developing the tools for lunar energy resource development, cis-lunar satellite servicing, and maintenance. The U.S. government must work to retire risk, make the operations routine, and once again empower commercial companies. This has already worked to an extent in low Earth orbit, and now we should apply this model to cis-lunar space. This is not only appropriate for economic development and to improve the human condition on Earth, but to provide for national security, which is now entirely dependent on space-based capabilities. Every domain of warfare today depends on space.

Once the cis-lunar market develops to service and maintain our traditional space-based military and commercial capabilities, other opportunities will naturally follow. The surface of the moon is composed mainly of oxides of metals: iron, magnesium, aluminum, silicon, titanium and others. While these oxides can be used to produce oxygen for life support and metals for additive manufacturing in situ, they will not likely be exported to earth. However, it is possible, if not likely, that highly valuable platinum group metals are much more available on the moon from astroblemes than they are on earth. *Such a discovery with cis-lunar transportation capabilities would fundamentally transform American commercial lunar development and could profoundly alter the economic and*

> *geopolitical balance of power on Earth.* This could explain the Chinese interest in the moon. The question is: What are WE, the United States, doing to make sure the free world participates economically in such a discovery? The U.S. government has a role to play here.
>
> *Competition for locations on the moon (the poles) and resources is inevitable. It must be stated that constitutionally, the U.S. government is required to provide for the common defense. This includes defending American military assets in space AND commercial assets in space*, many of which have and will have a dual role of providing commercial and military capabilities. President Kennedy said, "Whatever men shall undertake, free men must fully share." The U.S. government must establish a legal framework and be prepared to defend private and corporate rights and obligations all within keeping the Outer Space Treaty. *And to enable freedom of action, the United States must have cis-lunar situational awareness, a cis-lunar presence, and eventually must be able to enforce the law through cis-lunar power projection. Cis-lunar development will either take the form of American values with the rule of law, or it will take the form of totalitarian state control.* The United States can decide who leads.
>
> The United States of America is the only nation that can protect space for the free world and responsible entities, and preserve space for generations to come.
>
> This is our Sputnik moment. America must forever be the preeminent spacefaring nation and the moon is a path to being so.[149]

As the second term of the Obama administration came to a close, the preceding eight years showcased a mixed record for space development. At its close, President Obama held the White House Frontiers Conference in October of 2016[150] that featured an interplanetary frontiers track for "the next stage of space exploration, including the journey to Mars." In a speech and later in an OpEd, President Obama asserted "we have set a clear goal vital to the next chapter of America's story in space: sending humans to Mars by the 2030s and returning them safely to Earth, with the ultimate ambition to one day remain there for an extended time,"[151] though the proposed NASA budget for 2017 featured a 20 percent cut from deep space exploration programs.[152]

On the one hand, the Obama administration failed in its larger ambitions to move human exploration beyond low earth orbit, and failed to act on space solar power. Early on, the Obama White House abandoned the Bush Vision for Space Exploration, rejecting a focus on developing the Moon and cislunar space as "been there"[153] [done that] and attempted to de-fund President Bush's Constellation program.[154] It hoped to instead fund enabling technologies in favor of achieving new "firsts" such as human visits to an asteroid (the Asteroid Redirect Mission (ARM)[155]) and then Mars. Congress, however, had other ideas, forcing the continuation of major portions of Constellation to remain funded, and failed to show strong interest in the ARM,[156] and ARM was ultimately canceled.[157]

On the other hand, two very important developments did occur under Obama's leadership. First, the Obama administration championed commercial space access, bringing to fruition the NASA Commercial Orbital Transportation System ("COTS program")[158] begun under George W. Bush and NASA Administrator Griffin. COTS provided an example of a successful and cost-effective public-private partnership and enabled the interplanetary vision of Elon Musk and SpaceX.[159] Second, President Obama signed the Commercial Space Launch Competitiveness Act of 2015, perhaps the most consequential change in U.S. space policy related to space resources, which will be discussed in the subsequent chapter.[160]

Prior to the election, military spacepower theorists concerned with the strategic potential of space resources and favorable to a Space Force (called the "Blue Water school of spacepower"[161]) had hoped to influence an incoming administration with a series of publications in popular media to highlight the new strategic environment,[162] and suggest proactive policies both in long[163] and short[164] formats that emphasized the importance of taking leadership in the exploitation of space resources.

The Presidential Campaign would put in office President Donald J. Trump. His advisors would articulate a Trump space agenda that included the revival of the National Space Council to improve interagency collaboration, attention to the emerging military space threats from China and Russia, human exploration of the entire solar system, and open embrace of commercial and public-private partnerships.[165]

Just after the election (November 2016), as the new administration took office, one think tank, the Center for New American Security (CNAS), published a monograph titled *From Blue to Black*, where it anticipated the possibility of rivalries to control key strategic positions within cislunar space. This was the first time a major think tank had paid serious attention to space resources since the publication by Heritage/High Frontier in 1982.[166]

> A historical parallel could be considered is the Dutch East India Company, chartered in 1602 by the Dutch government and granted an exclusive monopoly over the spice trade from Southeast Asia. The East India Company was first to issue stock as a means of raising capital to cover its expenses, a move that revolutionized the world economy. SpaceX's innovative, low cost, competitive approach to spaceflight, including the improvement of landing and reusing first stage boosters, promises to change mankind's approach to space exploration. Today the challenge posed by the mining of minerals from the moon and the asteroid belt requires economic innovations of the current generation. Certainly such actions are coming into the realm of the possible from both technological

> and economic perspectives, but whither goes trade, so goes national competition. Low Earth orbit, geosynchronous orbit, Lagrange points, and the Moon each represent key strategic positions that might trigger rivalries to control them. Adapting the writings of Mahan and Corbett as a starting point for future conversations regarding a strategic approach offers firm ground upon which to build. Although the future remains uncertain and the vastness of space offers ambiguity, the lessons of the past can chart a good course forward.[167]

In early 2017, the Air University published *FAST SPACE: Leveraging Ultra low-cost Space Access for 21st Century Challenges.*[168] The report was important because it was the first articulation of a broader U.S. military strategy linked to private space development interests, and contextualized the role of spacepower within a framework of space resources and the rule of law. For example, it asserted, "it is in the national security interests of the United States to establish the Western principles of free trade and commerce, including free enterprise development and use of space resources, in international common law,"[169] and laid an agenda of significant scope:

> The United States needs to lead in utilizing space and space resources, in order to establish common international operating principles based on Western values and law. Ceding the "high ground" to other nations to develop this nascent area of the law would enable precedents that directly conflict with Western beliefs and ethics, which could have devastating long-term repercussions to national security, economic policy, and fundamental American freedoms and liberties.

> Much larger markets for clean electricity exist for baseload power, which would justify development of larger SSP systems in the Gigawatt class. However, those systems are unlikely to be enabled by an initial 3X reduction in launch costs from first generation RLVs, so it was beyond the scope of our study to assess them. In addition to first generation RLVs, these larger systems are likely to require in-situ resource development from the Moon and asteroids, advances in automated manufacturing, on-orbit construction, and satellite servicing. The required capabilities are likely to be within reach if smaller SSP prototypes are developed and a virtuous economic cycle in space transportation is jump-started. After US industry develops and demonstrates the first generation of fully-reusable LVs, the USG would be reasonably well-positioned to consider development of an SSP pilot demonstration system.[170]

The Trump administration more fully embraced a space development policy than administrations before it. Prior to the election, it had pledged to reestablish the National Space Council headed by the vice president, and make public-private partnerships the foundation of its space efforts, and cited the military-focused initiatives of China and Russia,[171] further stating, "increased reliance on the private sector will be a cornerstone of Trump space policy," and "no space goals will be more important to Donald Trump than defense of our nation and that a freedom-loving people will lead the way to the heavens above."[172] As the administration took office with its transition team, the Wall Street Journal reported "Trump in Space: Transition Focuses on Private-Public Initiatives."[173]

Just after the U.S. presidential election in 2016, the CNAS authored *A Space Policy for the Trump Administration,* which gave significant prominence to the topic of space resources. It urged the Trump administration to "consider the express authorization and incentivization of the exploitation of resources from space"[174] including tax-incentives, time-limited permissions and provided a historical analogy to the U.S. westward expansion.

> Just as the Lincoln administration *encouraged the building of railroads across the U.S. West, the Trump administration should develop strategies that incentivize commercial expansion into space*. Where Lincoln used land grants, the Trump administration could *leverage tax incentives or time-limited permissions, such as those granted to the British and Dutch East Indies Companies, limited to celestial bodies such as asteroids, or to coordinates such as LaGrange points*, to encourage investment and innovation in space. A core component of the commercial space sector is the capacity to take advantage of the many economic opportunities outside Earth's atmosphere, including mining and tourism. These endeavors are currently untapped markets but have significant investor interest.[175]

Moreover, the report took a skeptical position with regard to the OST. It likened the OST to "Pope Alexander VI's 1496 Papal Bull dividing the outer, unexplored world (from a European viewpoint) between Portugal and Spain" and nineteenth-century treaties between the United States and European powers regarding the interior of the North American continent—"aspirational and practical tools to avert or delay conflicts"—which "carried full legal authority" until some power "decided that it did not, and began acting outside its guidance."[176]

The document was bullish on the potential to access space resources[177] but expressed concern regarding the OST language on non-appropriation of space resources, stating,

> *"Without the capacity to lay claim to space-based resources, the commercial sector will be limited in its economic potential" and "no economic entity will fully invest in an enterprise involving the moon (which has many rare mineral resources) or asteroids (some of which are extraordinarily valuable) without having some claim of mineral rights or ownership—which the Outer Space Treaty makes difficult at best."*[178]

The report suggested the OST might need to be reviewed for modification[179] and the United States should clearly establish an enabling interpretation, reasoning that "promoting commercial development is only the first reason that the 1967 *treaty should be reviewed and profit incentivizing judicial interpretations established. If it is not, the United States will likely fall behind near-peer competitors who are eager to reap the domain's many economic rewards.*"[180]

Echoing such sentiments, Senator Ted Cruz (R-TX) held hearings in 2017,[181] "Reopening the American Frontier: Exploring How the Outer Space Treaty Will Impact American Commerce and Settlement in Space," where the importance of property rights and concerns over the potential limitations of the OST—including withdrawal to enable exploitation—were discussed. The commercial panelists, however, felt that the OST was sufficiently liberal so as not to limit their ambitions.[182]

In June 2017, H.R.2809—American Space Commerce Free Enterprise Act, was introduced in an attempt to elevate the place of commercial space in U.S. policy, stating,

> "It is the policy of the United States that—(1) United States *citizens* and entities *are free to explore and use space, including the utilization of outer space and resources contained therein, without conditions or limitation*" asserting that "the *private* exploration and *use of outer space by nongovernmental entities* will *further the national security, foreign policy, and economic interests* of the United States.[183]

The bill sought to the elevate Office of Space Commerce within the Department of Commerce, stating, "the Director shall be the Assistant Secretary of Commerce for Space Commerce and shall report directly to the Secretary of Commerce." It hinted at a similar elevation for the Office of Commercial Space transportation, requiring the "Comptroller General of the United States shall submit to Congress a report on removing the Office of Commercial Space Transportation from under the jurisdiction of the Federal Aviation Administration and reestablishing the office under the jurisdiction of the Secretary of Transportation."[184]

Moreover, the act contained the following provision:

The President shall—

(1) *protect the interests of United States entity exploration and use of outer space, including commercial activity and the exploitation of space resources, from acts of foreign aggression and foreign harmful interference*;

(2) protect ownership rights of United States entity space objects and obtained space resources; and
(3) ensure that United States entities operating in outer space are given due regard.[185]

Dr. Mineiro, Former Staff Director and Senior Counsel of U.S. House of Representatives Science, Space, and Technology Space Subcommittee asserted its import,

> If you look at the *2015 Commercial Space Launch Competitiveness Act, it established a right under U.S. courts for people to maintain rights over resources they got from outer space*. And it allowed them to avail themselves to the court system. If you look at the HR 2809, the Free Enterprise Act, there's a provision in there *which directs the president to use the ambit of national power*—it's a soft directive—but the ambit of national power, essentially, to protect and promote U.S.-flagged equity interests in there.[186]

H.R.2809 passed the U.S. House but failed that year to convince the Senate (the bill would be reintroduced in 2019[187]). Nevertheless, the attempt signals an intent by a significant number of U.S. policymakers to elevate the importance of space commerce in U.S. policy and raise the standing and the power of those offices whose primary concern is commercial space development.

In April 2018, the Aerospace Corporation issued a report *Cislunar Development: What to Build—and Why*. The report advocated for cislunar infrastructure development, stating "the Trump administration and subsequent U.S. leadership may determine that the *best way to achieve human expansion into space while building a space economy is by focusing on cislunar development* through a *combination of government programs and industry partnerships*."[188] The report argued:

> It is reasonable to project that in the next 20 to 30 years, global efforts in cislunar space will . . . use the unique characteristics of space—such as microgravity, vacuum, high-intensity solar exposure, and isolation from Earth—to produce useful knowledge and products; *harvest and process extraterrestrial materials and energy resources; build sophisticated structures in Earth orbit and in the vicinity of the moon*; build installations on the moon, constructed to the greatest extent possible *with local materials*.[189]

It envisioned "activities such as *extraction, processing, and use of extraterrestrial resources, primarily from the moon at first but eventually from asteroids as well; demand for propellant storage depots in various cislunar locations* (e.g., to fuel on-orbit servicing vehicles or deep-space missions) *using propellants derived from lunar ice deposits*"[190] would make the Moon the natural hub of activities.[191] It asserted that "employment of large multipurpose orbiting platforms that would benefit from the use of lunar materials in their construction or resupply (e.g., *solar power satellites*, lab/manufacturing/habitat '*industrial parks*')."[192]

The 2018 Aerospace report is an important milestone because Aerospace, as a Federally Funded Research and Development Center, is a trusted source of advice for conservative federal agencies. While organizations such as the U.S. Space Force, U.S. Space Command, National Reconnaissance Office, and NASA might dismiss the space advocacy organizations, it is more likely to give credence to Aerospace Corp.

Former Speaker of the House Newt Gingrich who played a prominent role in efforts to give space development a central place in the 1980s reemerged as an influential advisor to the Trump administration. In a memo titled "Space Force Planning" dated July 17, 2018, circulated by former speaker of the House Newt Gingrich, he provided a clear vision of the future of space development, colonization, and the associated need for a space force, stating, "A space-faring nation will inevitably require a Space Force"[193]:

> *The Trump-Pence plans for space call for a dramatic shift from a very small number of highly trained astronauts exploring in very expensive small vehicles to a large number of Americans pioneering and colonizing.* By 2040 there should be *a permanent American presence on the Moon and Mars, asteroid mining will be a viable business*, and space tourism will be available for a surprising number of Americans. The cost crashes from second and third generation reusable rockets, 3D printing and breakthroughs in material and propulsion technology will make space dramatically more accessible and more profitable. The government will play a major role in the early investments just as it did with aviation, computing and the internet. As commercial activities grow and costs decline the government's role will decline. Aggressively developing public private partnerships will increase both the resources available and the rate of innovation as it did with military aviation. *The economic engine will emerge as the United States government makes the initial non-recurring investments (ideally with private partners). This will inevitably increase both the value of space assets and the need to protect them* from other players.
>
> By 2040 there will be autonomous activities and assets around and on the Moon, around and on Mars, and on some asteroids and on the path between those asteroids and Earth. The dramatic expansion of time and space as an arena for space warfare will be larger than the shift from defending the English Channel

to operating in the Pacific (note the tremendous differences in British and American thinking in World War 2 about the scale of logistics and the nature of ships needed between closed naval environments and blue water campaigns). A critical requirement for a genuine Space Force will be the unique characteristics of war beyond LEO. *This dramatic change in the nature of the challenge also explains why the Earth-based services will inherently be under-prepared for deep space warfare. This new type of warfare will require new technologies, new doctrines, and new realistic training. The traditional services will always underfund this effort in favor of their more traditional and culturally more comfortable roles.* Exploring the potential for space warfare beyond LEO while maintaining our dominance in near space is the third step toward planning a Space Force.

The Trump-Pence commitment to dramatically more Americans in space *creates a parallel to 19th century American history. In that parallel NASA encourages, facilitates and sometimes organizes the space equivalent of the wagon trains. It is NASA's job to maximize the development of pioneers and colonizers.* It is NASA's job to help develop the new technologies and to encourage the commercial activities (including space tourism) which will make civilian space sustainable and expandable with minimum taxpayer support. *The Space Force in a sense is parallel to the role of the United States cavalry in opening the West. Properly designed, the Space Force itself will be supporting US government actions that are mandated for an accepted societal purpose.* It was a societal decision to build a Panama Canal. The US government ran the job but all the major work was done by contractors. But it was not done as a commercial venture; they would have gone broke. *As a provider of security and a rescue system (in this aspect the Space Force has a little of the Coast Guard's functionality) the Space Force will accelerate the American evolution as a spacefaring nation. Part of the design of the Space Force has to be synergistic with and supportive of civilian activities,* just as it is the U.S. Navy that, for most of the world, guarantees the freedom of the high seas. Similarly, as we better understand the requirements for dominance in space, some supporting elements may be built into the commercial and civilian activities and structures. Thinking through the requirements for this synergistic approach is the fourth step in designing a Space Force.

Because the Space Force *must have continuous presence to establish security and must interact with and help grow the non-military American activities in space, its real forerunner is the Navy* rather than the Air Force, the Air Force currently has the bulk of space oriented technologies but the culture of the Navy is almost certainly a better fit for the civilian supporting and synergistic requirements of civilian space. Furthermore, the autonomous role of the ship's captain

> is much closer to the requirements of space systems that will require on the spot, instantaneous command in response to unforeseen challenges. There is a sound psychological reason the original 1966 Star Trek was built around a ship (the USS Enterprise) commanded by a Captain.[194]

Like the 2016 Bridenstine speech, the 2018 Gingrich Space Force memo is remarkable in showcasing how thought leaders in the American political elite linked to commerce, geopolitical competition and the need for military protection, and connected them to earlier American experiences with its westward and naval expansion.

Such thinking had also permeated military academia. Students in the Space Horizons Task Force, begun under Lieutenant General Steven Kwast, published a series of monographs in the summer of 2018, which examined these strategic aspects of spacepower.

In *Movement and Maneuver in Deep Space: A Framework to Leverage Advanced Propulsion*[195] (2018), authors Brian Hans, Christopher Jefferson, and Joshua Wehrle picked up where the Fast Space Study of 2017 left off. The report was important because it further elaborated on the need to have deep space access commensurate with commerce, acknowledged the logistical potential of in-space resources for propellant, and proposed future military requirements.

Movement and Maneuver in Deep Space also contained the following (2016) quote from Dr. Simon "Pete" Worden, former NASA Ames Center Director, and Brigadier General, USAF (Retired), which first used the term "Blue Water":

> *True space operations will spread across the solar system in the decades ahead* and *the nation that controls them will dominate the planet. Focusing on low Earth orbit (LEO) is akin to having a Navy that never leaves sight of the shore. The US Military needs to focus on "blue-water" space operations*—GEO and above. US military space operations need to be in deep space, *initially all of cislunar space, with an eye upon the entire inner solar system.* To operate in deep space one needs to use the resources there, starting with *fuel from asteroids. Once this is recognized, the military-economic imperative of identifying and protecting these assets becomes clear.*[196]

In 2018, U.S. Air Force Lawyer Maj Paul Gesl argued "as the United States and humanity expand away from the surface of the Earth, *we must create a legal framework, or we will be left with frontier justice*"[197] and that "*it is in the best interest of the United States to lead the effort to provide laws that govern space colonies*"[198] and "instead of passive supervision, the United States should pursue laws and policies that both encourage and direct these activities."[199] He recounts that "Mr. Musk envisions the full colonization to take 40–100 years. *Even if this timeline misses its ambitious deadline by a decade, humanity will be a multi-planetary species in many readers' lifetimes*" and that "*space colonization is more important to our species than the economic benefits of a space economy and the conquests of exploration.*"[200] He argued:

> *If colonization is going to happen, then it is in the United States' best interest to develop a legal framework that supports the efforts and protects our citizens who will travel to and live in these habitats.* This is important for several reasons. First, private corporations appear to have an interest in colonizing space, so it is in humanity's future whether the government is involved nor not. However, governments can take actions that will accelerate things. Second, it is in the best interest of the United States' economy to support commercial companies that are expanding into space. Third, if the United States does not create a favorable legal framework for space colonization, someone else will. Finally, as humanity expands away from the surface of the Earth, it is important to create a free society based on the principles of the Rule of Law rather than some other form of government, or an anarchistic company town.[201]

Other U.S. Air Force lawyers, Major Dustin Grant and Maj Matt Neil offered a strong argument for an independent U.S. Space Force (USSF) based in part on future space commerce requirements. The report was important because it offered a fully developed draft legislation and justification for a U.S. Space Force shortly after the Trump Administration began to use the term, and the paper was discussed by Congress, White House Office of Management and Budget (OMB), and Air Force Headquarters.

Grant and Neil argued, "Rather than the supporting role space served up to this point, *warfare in the future is likely to extend or even begin within the space domain.* Additionally, *as the burgeoning commercial space economy continues to grow, so too will the need for security and protection of lines of commerce grow.*"[202] The paper advocated for an independent Department of the Space Force as well as offered fully developed legislation before any such legislation had been initiated by the Administration.[203] The authors argued,

> *In the space domain, perhaps more than any other domain, the line between civil and military is blurred, and there is no civil law enforcement infrastructure in space. As a result, the USSF will have significant responsibility to enforce, and assist in the enforcement of the laws of the United States.* For this reason, perhaps more than any other, the *USSF needs to be an independent force as opposed to a force organized under the Department of the Air Force, to avoid potential conflicts with the Posse Comitatus Act* . . . of the key functions, roles,

> and responsibilities a USSF must assume, some of the *most critical will be the roles and responsibilities the USCG [U.S. Coast Guard] currently assumes in the maritime domain* . . . the space domain needs a *DOD agency more akin to the United States Navy than a hybrid agency similar to the USCG* . . . "; that "much as the *United States*" *Navy assumed responsibility for protecting lines of commerce on the high seas, only a military force will be equipped to protect lines of commerce in space.* With commercial space activities growing exponentially, and the expressed NSS of the United States, such a force is needed, and the USSF should be that force.[204]

While the draft legislation itself would not be the actual Space Force legislation (which is unfortunate, because it contained many innovative authorities), it made the prospect of Space Force legislation appear more tractable.

This vision, linking a great power race, commerce, space resources, and dedicated military protection would find its first expression by a serving senior military officer with Lieutenant General Steve Kwast, Commander of Air Education and Training Command. In an August 2018 opinion piece in *Politico* (authored just prior to Vice President Pence's Space Force guidance at the Pentagon), he articulated for the American people the complete vision, including asteroid mining, in-space manufacturing, space solar power, and the need for a military, which played both a frontier development and guardian role:

> Rapid technological progress makes space more of an opportunity for exploration and discovery than ever before. The *commercial opportunities are simply staggering*. Visionary investors see new markets across a variety of industries, to include space tourism, space-based wireless internet across the globe, *space-based solar power, asteroid mining, and space-based manufacturing*. Imagine a world where materials harvested from asteroids are stored in space, used to 3D-print new satellites or space vehicles, without any need for a gravity-defeating launch vehicle to put them into space. *Imagine the impact on humanity if space-based solar power and internet access could fuel and connect the whole world* for free. This vision is not a 100-year pipe dream. The components and vision are already here. We only require the national vision, will, and commitment. *As human development of space increases rapidly in the coming years, the U.S. military certainly understands and appreciates its expanded role*. As a first step, the military recognizes that our reliance on space creates vulnerabilities that must be actively defended . . . *But as humanity moves to space in more committed and comprehensive ways, the next phase of military activity in space will likely be as the government's frontier development arm*. Just as the Navy and Army have been involved in frontier development since our

> colonial days-opening new markets and infrastructure across the seas or across the American west-so too today's U.S. military needs to be prepared for similar activity in space. *The U.S. military can expect to be a frontier development service once again, establishing the cosmic equivalent of navigational aids, charts, lighthouses, dams, search and rescue services, forts, supply depots, and coaling stations.*[205]

Clearly, as of 2018, there was a new policy image taking shape that prioritizes space commerce, takes seriously the development of space resources, and anticipated the need to protect those interests with dedicated military presence. The advocates of this linked space-development/security paradigm are far from achieving a policymaking monopoly though.[206] Many thinkers were skeptical of the promise of space resources, fearing its distraction from urgent problems. Nor was such advocacy without risk and professional casualties. Among those most opposed to the new policy image that included an independent U.S. Space Force with a mission to protect commerce was U.S. Air Force senior leadership. General Kwast, rather than being rewarded for his insight commissioning the Fast Space Report, Space Horizons, and his own public advocacy, would be sidelined by the Air Force.[207] After opposing President Trump regarding the U.S. Space Force, the administration would find leadership for the Air Force more amenable to the new policy image.[208]

U.S. policymakers seemed to exhibit a level of shock when Chang'e 4 landed in January 2019 on the far side of the Moon. This was the first feat by the Chinese space exploration program that was a distinct "first" and not a repeat of something the United States had previously accomplished. It was a potential signal that the United States was losing its position of leadership, and introduced a much broader audience to consideration of the importance of cislunar space.

U.S. intelligence officials expressed a certain degree of concern about the PRC lunar program "building steam" and noting that "*putting a satellite at L2 could also enable Chinese attack spacecraft to zoom past the moon—about a quarter-million miles away—and then sneak up on critical U.S. intelligence and communications satellites in geosynchronous orbit, just 28,300 miles up.*"[209] But the larger Department of Defense remained skeptical. The prevailing attitude might be summarized as "let China waste their money" and "let the billionaires have their mid-life crises." It denies the probability there is any wealth to be had and seeks to avoid any risk or embarrassment by adopting the role of watcher and follower. This conservative bias has resulted in a more passive response by the Department of Defense and appears increasingly at odds with the new policy image.

In April of 2019, the U.S.-China Economic and Security Review Commission held a hearing on "China in Space: A Strategic Competition?" where three of the individuals giving testimony—Admiral Cartright, Dr. Roper, and Dr. Goswami[210]—expressed cislunar concerns.[211]

In June of 2019, a joint report by Air Force Research Lab and the Defense Innovation Unit concluded that:

> The U.S. space industry is vulnerable and the government must protect it from "manipulation, distortion, penetration and domination by our adversaries, allies and neutral countries . . . The breadth and depth of Chinese malfeasance with regard not only to our technology, but also to our larger economy and our nation is significant and intentional." The *accelerating space race carries high economic stakes*, the report says, including *dominance of the cislunar domain and its economic exploitation through space manufacturing, space power and resource extraction*. "The foundation for a sustainable space economy, such as *cislunar infrastructure*, strategically depends on close collaboration with national commercial capabilities and the maintenance of a strong space industrial base."[212]

Throughout 2019, a series of articles associated with the "Blue Water School" reflected the new policy image, arguing for "A Space Service in support of American grand strategy,"[213] how "The Space Force has gone from joke to reality,"[214] that "The Purpose of a Space Force is a Spacefaring Economy,"[215] that "The First Duty of a Space Force is to Protect Space Commerce,"[216] that "The Future of the Asian Order Will be Decided in Space,"[217] that this necessitates a distinctive space culture able to mount a peacetime strategic initiative to aid cislunar development,[218] that this requires Naval Rank for the Space Force,[219] that the Department of Energy must take leadership on space resources, that "In Space Force debate, the military's space experts are missing in action,"[220] and that the "Space Force defense must stretch to the Moon."[221] General Kwast, now retired, would make additional speeches at Hillsborough College, and write on "The Urgent Need for a United States Space Force."[222]

The space advocacy community continues its efforts to enshrine space settlement within the U.S. legislation. The community has updated "The Space Exploration, Development and Settlement Act of 2016," (H.R.4752), and it is now "The Space Development Act of 2019."[223] It was listed as the number-one priority for the ASD.[224] The updated bill is more directive to the President, stating, "The Administration shall encourage, facilitate, promote and support the *development of permanent space settlements*." It also would redesignate the Office of Space Commerce as the Office of Space Commerce and Settlement, and require its director to "*encourage, facilitate, promote and support the development of space settlement.*" A parallel effort has taken place in the Senate. From 2015 to 2018, an active campaign by the ASD engaged over twenty-five members of Congress, ten aerospace companies, and twenty nonprofit organizations, ultimately resulting in a bill [S.584] introduced by Senators John Cornyn and Gary Peters, which would add enabling space settlement to NASA's mission.[225] The effort features prominently in the top three priorities of the ASD's 2020 Congressional Blitz:

(1) Support planetary defense by funding NASA's Near-Earth Object Surveillance Mission.
(2) Start developing and demonstrating SSP.
(3) Make space development and settlement part of NASA's, the Department of Transportation's, and the Office of Space Commerce's official mission statement.[226]

Outside the circle of government elites, military academia and Washington policy circles, among the most important developments in elite dialogue were interviews and long-form presentations made by Elon Musk and Jeff Bezos. The most important of these was Elon Musk's 2016 presentation "Making Humans a Multiplanetary Species"[227] and 2017 "Making Life Multiplanetary"[228] at the International Astronautical Congress (IAC), and Jeff Bezos' "Going to Space to Benefit Earth," which became "viral" hits on YouTube.[229] These established ideologically committed agency outside of government who were "putting their money where their mouth is" and championing the concepts of space settlement and space development. The cultural importance of the star quality of very successful individuals (billionaires) articulating multigenerational societal ambitions and plans has generated significant public interest, reduced skepticism, and attracted government interest.

Inspired by the prominent successes and positive press of Musk and Bezos, policy entrepreneurs within the space advocacy community continue to submit proposals they hope will catalyze space development and settlement. The NSS has offered several recent position papers[230] calling for a Space Guard,[231] a Space Salvage Entity,[232] an Outer Space Private Investment Corporation,[233] and a Public/Private COTS-Type Program to Develop Space Solar Power.[234] Independent voices have explored a Space Commodities Exchange,[235] a Space Commodities Futures Exchange,[236] a multinational joint-stock company,[237] and a U.S. Government Space Strategic Minerals Reserve at L1 (to purchase 100,000–250,000 tons of asteroid material over a thirty-year period),[238] and a concept for an international Lunar Development Cooperative, designed to incentivize cooperative economic development of the Moon.[239]

Perhaps, the most significant recent proposal is an extremely comprehensive piece of draft legislation, the *Space Pioneer Act* (written to augment international law and further consolidate U.S. protections for private property) by space lawyer and space Advocate Wayne White.[240] The act would augment the OST, by executing the Article VI "the activities of nongovernmental entities in outer space, including the Moon and other celestial bodies, shall require authorization and continuing supervision by the appropriate State Party to the Treaty."[241] The act would accomplish this by establishing that the

> United States—(a) *exercises its jurisdiction over United States entities and Space Objects on its Registry, and foreign persons and Space Objects subject to its jurisdiction due to their presence in a safety zone around a U.S. Entity's Preliminary or Perfected Claim*, consistent with the freedom of exploration and use enumerated in Article I of the Outer Space Treaty, and the international law principle of non-interference; but (b) *does not thereby assert territorial sovereignty, or exclusive rights, or permanent jurisdiction over, or the permanent ownership of, any areas of Outer Space or Celestial Bodies.*[242]

The draft legislation covers nearly every conceivable aspect of regulating and facilitating pioneer activities, creating an Office of Outer Space Claims under the Secretary of the Interior, establishing a claims registry, and establishing procedures for certification of space objects, for registration of claims of homesteads, outposts, safety zones, as well as surveys of real property on the Moon and Mars by the United States Geological Survey. It establishes the principle of first come, first served limited by noninterference, defines "space object," "celestial body," "claimant," "owner," "areas of ongoing activity," "beneficial use," "harmful contamination," "preliminary claims," "perfected claims," and jurisdiction. It establishes a process for approval of safety zones by the Office of Outer Space Claims, limitations on size, and sets forth principles of operational control by the claimant or owner, covers emergency transit and rescuer liability. It also covers titles, transfer of titles, easements, condominium associations, multiple retail locations, spaceports, and historical sites, as well as a procedure by the U.S. State Department to designate or revoke a foreign nation as a reciprocating state. If passed, this would be a major step forward, reducing significant uncertainty and regulatory risk.

This chapter has explored important discourses in the United States from their beginnings in the epistemic community and the dawn of the space age (1957) until the present time. Because the United States is an open system and because the United States is both the leading space power and global leader, its discourse has set the stage both domestically and internationally. The reader should get a sense of how interest in space resources has waxed and waned but over time has been elaborated and gained support. In the next chapter, we will examine how these ideas and themes have come to be reflected in policy, law, and programs.

NOTES

1. Andrew Chaikin, *A Man on the Moon: The Voyages of the Apollo Astronauts* (New York: Penguin, 2007).
2. The Planetary Society, "About Us," n.d., accessed July 27, 2019, http://www.planetary.org/about/
3. Greg Klerkx, *The Fall of NASA and the Dream of a New Space Age* (New York: Vintage Books, 2004, 2005): 63–90.
4. Klerkx, *Lost in Space,* 63–90.
5. National Space Society, "About the National Space Society," n.d., accessed July 29, 2019, https://space.nss.org/about-national-space-society/
6. National Space Society, "NSS Statement of Philosophy," n.d., accessed July 27, 2019, https://space.nss.org/nss-statement-of-philosophy/
7. National Space Society, "NSS Statement of Philosophy".
8. National Space Society, "NSS Statement of Philosophy".
9. National Space Society, "NSS Statement of Philosophy".
10. NSS, "NSS Legislative Blitzes," accessed August 8, 2019, https://space.nss.org/nss-legislative-blitzes/
11. National Space Society, "NSS Library," n.d., accessed July 27, 2019, https://space.nss.org/national-space-society-library/
12. National Space Society, "NSS Library".
13. National Space Society, "NSS Library," accessed July 27, 2019, https://space.nss.org/sun-power-the-global-solution-for-the-coming-energy-crisis/
14. "Space Settlement Contest," n.d., accessed July 27, 2019, https://space.nss.org/settlement/nasa/Contest/
15. "International Space Settlement Design Competition," n.d., accessed July 27, 2019, https://spaceset.org/
16. National Space Society, "NSS Roadmap to Space Settlement (3rd Edition 2018–2019)," accessed July 29, 2019, https://space.nss.org/about-national-space-society/
17. National Space Society, "Space Settlement Summit," n.d., accessed August 6, 2019, https://space.nss.org/space-settlement-summit/; see also https://spacesettlementsummit2019.nss.org/
18. L5 News, "UN Moon Treaty Falling to US Opposition Groups," March 1982, accessed August 7, 2019, https://space.nss.org/l5-news-un-moon-treaty-falling-to-us-opposition-groups/
19. H. Henson and Arel Lucas, "Star Laws," Reason, August 1982, accessed August 7, 2019, https://reason.com/1982/08/01/star-laws
20. Space Frontier Foundation, "About the Space Frontier Foundation: Who We Are," n.d., accessed July 27, 2019, http://newspace.spacefrontier.org/about/;

Facebook, March Storm 2019, accessed August 8, 2019, https://www.facebook.com/events/351030189012751/

21. Space Frontier Foundation, "About the Space Frontier Foundation".

22. Space Frontier Foundation, "About the Space Frontier Foundation".

23. Space Frontier Foundation, "About the Space Frontier Foundation".

24. The Mars Society, "Founding Declaration," August 1998, accessed July 27, 2019, https://www.marssociety.org/founding-declaration/

25. "The Moon Society, About Us," n.d., accessed August 7, 2019, https://www.moonsociety.org/about

26. Lunarpedia, "Space Colonization Technical Committee," December 7, 2019, accessed February 10, 2020, http://carbondate.cs.odu.edu/#http://lunarpedia.org/w/American_Institute_of_Aeronautics_and_Astronautics

27. Space Development Steering Committee, "About the Space Development Steering Committee," accessed August 7, 2019, http://spacedevelopmentsteeringcommittee.org/about-the-space-development-steering-committee/

28. The Earthlight Foundation, "About," n.d., accessed July 29, 2019, http://earthlightfoundation.org/about/

29. Alliance Objectives for 2019, "Alliance for Space Development," n.d., accessed August 1, 2019, http://allianceforspacedevelopment.org/about-us/

30. NASA, "Ansari X-Prize: A Brief History and Background," n.d., accessed August 7, 2019, https://history.nasa.gov/x-prize.htm

31. X-Prize Org, "Google Lunar X-Prize: The New Space Race," n.d., accessed August 7, 2019, https://www.xprize.org/prizes/google-lunar

32. Mike Wall, "Ex-Prize: Google's $30 Million Moon Race Ends with No Winner," Space.com, January 23, 2018, accessed August 7, 2019, https://www.space.com/39467-google-lunar-xprize-moon-race-ends.html

33. John W. Kingdon. *Agenda's, Alternatives, and Public Policies* (New York: Longman, 2003).

34. There is also a significant, and largely uncatalogued, counter-narrative. Much of the space movement thinking was motivated by the "limits of growth" ideology that saw the earth as a closed system with a limited environmental carrying capacity where the only route to survival was control of human liberty, reproduction, economic growth, and rationing through a global and collectivist approach. That narrative remains active and largely independent. New attacks are also emerging from the political left as well, including from those opposed to space expansionism since it would split humanity and undoes the "closure" toward a one-world government. Others oppose it based on an ideology that humanity can and must first be perfected. Still others see it as an extension of privilege, whether that privileged identity might be industrialized/powerful nations (colonialism), male (feminist misanthropy), ethnic (critical theorists). Donella H. Meadows, et al., Limits of Growth, 1972, accessed September 13, 2019, https://www.clubofrome.org/report/the-limits-to-growth/

35. Gerald K O'Neill, *The High Frontier: Human Colonies in Space*, 3rd edition (Ontario: Apogee books in conjunction with the Space Studies Institute, 1976) [3rd edition republished in 2000].

36. William M. Brown and Herman Kahn, *Long-term Prospects for Developments in Space (A Scenario Approach)* [prepared for National Aeronautics and Space Administration under contract NASW-2924] (Washington, DC: The Hudson Institute, 1977), accessed July 27, 2019, https://apps.dtic.mil/dtic/tr/fulltext/u2/b310563.pdf

37. Brown and Kahn, *Long-term Prospects for Developments in Space.*

38. Harry. G. Stine, *The Third Industrial Revolution* (New York: Ace Books, 1979).

39. Daniel O. Graham, *High Frontier: A New National Strategy* (Washington DC: High Frontier, 1982). Also see: Daniel O. Graham, *High Frontier: A Strategy for National Survival* (New York: Tom Doherty Associates, 1984).

40. Solar Power Satellites – Selected References, accessed July 29, 2019, https://www.spaceacademy.net.au/library/refs/spsrefs.htm

41. Kirk Nankievell, "Why the Future of Solar Power Is from Space," Singularity Hub, December 31, 2018, accessed July 29, 2019, https://singularityhub.com/2018/12/31/why-the-future-of-solar-power-is-from-space/

42. Leonard David, "Peter Glaser, Father of Solar-Power Satellite Idea, Dies at 90," Space.com, June 9, 2014, accessed July 29, 2019, https://www.space.com/26175-peter-glaser-solar-power-satellite-obituary.html

43. These studies are collected and available at the NSS Solar Power Library, https://space.nss.org/space-solar-power-library/

44. National Research Council, Electric Power From Orbit: A Critique of a Satellite System, National Academy Press, 1981, accessed July 29, 2019, https://space.nss.org/media/1981-NRC-Electric-Power-From-Orbit-1-Report.pdf

45. Office of Technology Assessment (OTA), *Solar Power Satellites*, August 1981, accessed July 29, 2019, https://space.nss.org/media/1981-OTA-Solar-Power-Satellites.pdf

46. G. Harry Stine, "Space Power: The News the Department of Energy Tried to Bury!," 1981, accessed July 29, 2019, https://space.nss.org/space-power-by-harry-stine/

47. Ralph Nansen, *Sun Power: The Global Solution for the Coming Energy Crisis* (London: Ocean Press, 1995).

48. John S. Lewis, *Mining the Sky: Untold Riches from the Asteroids, Comets, and Planets* (Reading, MA: Helix Books Addison-Wesley Publishing, Co, 1996).

49. Simon P. Worden and John E. Shaw, *Whither Space Power? Forging a Strategy for the New Century* (Maxwell Air Force Base: Air University Press, 2002).

50. Everett C. Dolman, *Astropolitik: Classical Geopolitics in the Space Age* (Portland, OR: Frank Cass, 2002).

51. Dennis Wingo, *MoonRush: Improving Life on Earth with the Moon's Resources* (Ontario, Canada: Apogee Books, 2004).

52. Robert Zubrin and Richard Wagner, *The Case for Mars: The Plan to Settle the Red Planet and Why We Must.* Rev. and updated., 1st Free Press trade pbk. Ed (New York, NY: Free Press, 2011).

53. National Security Space Office (NSSO), *Space-Based Solar Power As an Opportunity for Strategic Security*, October 10, 2007, accessed September 13, 2019, https://space.nss.org/space-based-solar-power-as-an-opportunity-for-strategic-security/

54. Ralph Nansen, *Energy Crisis: Solution from Space* (Ontario, Canada: Apogee Books, 2009).

55. Keck Institute for Space Studies (KISS), *Asteroid Retrieval Study* (Keck Institute, 2012), accessed September 13, 2019, http://kiss.caltech.edu/final_reports/Asteroid_final_report.pdf

56. John S. Lewis, *Asteroid Mining 101: Wealth for the New Space Economy* (Mountain View, CA: Deep Space Industries, 2014).

57. Paul D. Spudis, *The Value of the Moon: How to Explore, Life, and Prosper in Space Using the Moon's Resources* (Washington, DC: Smithsonian Books, 2016).

58. John Mankins, *The Case for Space Solar Power* (Houston, TX: Virginia Edition Publishing, 2014).

59. National Space Society's "Space Settlement Library," *NSS*, August 3, 2017, accessed February 11, 2020, https://space.nss.org/space-settlement-library/

60. Brent Ziarnick, *Developing National Power in Space: A Theoretical Mode*l (Jefferson, NC: McFarland & Company, 2015).

61. Christian Davenport, *The Space Barons: Elon Musk, Jeff Bezos, and the Quest to Colonize the Cosmos* (New York: Public Affairs, 2018).

62. Robert Zubrin, *The Case for Space: How the Revolution in Spaceflight Opens Up a Future of Limitless Possibility.* (New York: Prometheus, 2019).

63. Rod Pyle, *Space 2.0: How Private Spaceflight, a Resurgent NASA, and International Partners are Creating a New Space Age* (Dallas: BenBella Books, 2019).

64. We have chosen the founding of Planetary Resources, Incorporated (PRI) as the beginning of this era, as it was the first time that there was a broad public and international interest given to Asteroid resources. It was followed shortly thereafter by the founding and public announcement by Deep Space Industries in 2013. Alternate (earlier) dates might include the founding of *Blue Origin* by Jeff Bezos in 2000, or of *SpaceX* in 2002, or Shackleton Energy in 2007.

65. For example, the American Interplanetary Society (AMS) was founded in 2030, a contemporary of Verein für Raumschiffahrt (VfR), the GIRD and even predated the British Interplanetary Society (BIS). It became the ARS and ultimately merged with another organization to become the AIAA. See Barron Hilton, "Rocket Societies,". Pioneers of Flight, July 2, 2014, accessed February 10, 2020, https://pioneersofflight.si.edu/culture/rocket_societies

66. NASA, "Sputnik and The Dawn of the Space Age," April 24, 2001, accessed February 10, 2020, https://history.nasa.gov/sputnik/

67. Donald William Cox and Michael Stoiko, *Spacepower: What it Means to You* (Philadelphia and Toronto: Winston, 1958).

68. "National Aeronautics and Space Act of 1958 (Unamended)," *NASA*, April 24, 2001, accessed February 10, 2020, https://history.nasa.gov/spaceact.html

69. Ralph J. Cordiner, "Competitive Private Enterprise in Space," in Simon Ramo and Lloyd V Berkner, eds., *Peacetime Uses of Outer Space* (New York: McGraw-Hill, 1961).

70. Dennis Wingo, "The Early Space Age, The Path Not Taken Then, But Now? (Part 1)," Dennis Wingo Blog, February 16, 2015, accessed October 12, 2019, https

://denniswingo.wordpress.com/2015/02/16/the-early-space-age-the-path-not-taken-then-but-now/

71. Dennis Wingo, "Creating a Bold and Courageous 21st Century American Space Policy," International Institute of Space Commerce, August 2, 2017, accessed July 27, 2019, https://iisc.im/wp-content/uploads/2017/08/Creating-a-Courageous-21st-Century-American-Space-Policy-_v4c.pdf

72. Ralph J. Cordiner, Chairman of the Board General Electric Company, "Competitive Private Enterprise in Space," accessed March 9, 2019, https://denniswingo.files.wordpress.com/2014/04/cordiner-article-1961.pdf

73. For an excellent critique of this decision, "Currently, NASA is the agent of US space exploration. This is counter to the traditional American approaches to exploring and exploiting new territory. It is also counter to common sense. NASA is a research and technology organization. It has little incentive to develop, open or protect new areas for commerce" in: Simon P. Worden and John E. Shaw, *Whither Space Power? Forging a Strategy for the New Century* (Maxwell Air Force Base: Air University Press, 2002), 110–116.

74. For a history of NASA legislation, see John M. Logsdon, "Legislative Origins Of The National Aeronautics And Space Act Of 1958," Proceedings of an Oral History Workshop, April 3, 1992, accessed February 20, 2020, https://history.nasa.gov/40thann/legislat.pdf

75. This thinking incorporated developed ideas of spaceflight that came with the former NAZI "Operation Paperclip" scientist which had its antecedents in the VfR. For more on the connection between Operation Paperclip and the US Space Program see, "Vanguard-A History: Background Of Space Exploration," NASA, April 24, 2001, accessed February 10, 2020, https://history.nasa.gov/SP-4202/chapter1.html

76. For example, the Lewis and Clark expedition, the use of the Cavalry to protect infrastructure and settlement on the frontier, the use of Navies and Air Forces to map far reaches.

77. Paul Czysz Berndt Chudoba, and Christopher Rahaim, "Low Earth Orbit Infrastructure Configuration and Requirements Translated into Support Vehicles Requirements," in 15th AIAA International Space Planes and Hypersonic Systems and Technologies Conference, 2008, p. 2543.

78. James A. Dewar, *To the End of the Solar System: The Story of the Nuclear Rocket* (Ontario, Canada: Apogee Books, 1971 (second edition in 2002)).

79. George Dyson, *Project Orion: The True Story of the Atomic Spaceship* (New York: Henry Holt and Co., 2002).

80. Brent Ziarnick and Peter Garretson, "Starfleet was closer than you think," The Space Review, March 16, 2015, accessed February 10, 2020, https://www.thespacereview.com/article/2714/1

81. U.S. Congress Senate Committee on Aeronautical and Space Sciences Investigation of Governmental Organization for Space Activities Hearings 86th Congress 1st Session 1959: 483

82. The Lewis and Clark expedition was a military expedition to survey the newly acquired Louisiana Territory, the primary objective of which was "to explore . . . the most direct and practicable water communication across the continent, for the

purpose of commerce." From: "Lewis And Clark Expedition," *History.net*, May 6, 2013, accessed February 10, 2020. https://www.historynet.com/lewis-and-clark-expedition

83. US Army, *Project Horizon: Volume I: Summary and Supporting Considerations*, 1959, accessed February 14, 2020. https://nsarchive2.gwu.edu/NSAEBB/NSAEBB479/docs/EBB-Moon01_sm.pdf

84. U.S. Army, *Project Horizon: Volume I*

85. U.S. Army, *Project Horizon: Volume I*

86. U.S. Army, *Project Horizon: Volume I.*

87. Brent Ziarnick, *Tommy's Space Force: General Thomas S. Power and the Air Force Space Program*, Drew Paper No. 34 (Maxwell AFB, AL: Air University Press, April 2019), 80–83.

88. These factors included 1) choice of both Presidents Eisenhower and Kennedy to emphasize civilian competition through NASA (NASA was created in 1958 and Kennedy gave his Moon speech at Rice in September of 1962); 2) The limited vision of Secretary of Defense Robert McNamara and Director of Research and Engineering, Dr. Harold Brown who refused to fund the Air Force Proposal; 3) The 1963 Limited Test Ban Treaty which had the effect of killing space nuclear propulsion research; and 4) The Outer Space Treaty of 1968 which prohibited the possibility of a space-based nuclear deterrent.

89. Louis de Gouyon Matignon, "Starfish Prime Or The Legality Of High-Altitude Nuclear Explosions," Space Legal Issues, February 14, 2019 , accessed February 20, 2020, https://www.spacelegalissues.com/space-law-starfish-prime-or-the-legality-of-high-altitude-nuclear-explosions/

90. "Nuclear Test-Ban Treaty," *History.com*, August 21, 2018, accessed February 20, 2020, https://www.history.com/topics/cold-war/nuclear-test-ban-treaty

91. Both Dyson's *Project Orion* and Dewar's *To the End of the Solar System* (cited above) chronicle the direct detrimental effects of this treaty on nuclear propulsion programs.

92. NASA, "John F. Kennedy Moon Speech—Rice Stadium," (September 12, 1962), NASA, January 7, 2004, accessed February 20, 2020, https://er.jsc.nasa.gov/seh/ricetalk.htm

93. "2222 (XXI). Treaty on Principles Governing the Activities of States in the Exploration and Use of Outer Space, including the Moon and Other Celestial Bodies," United Nations Office of Outer Space Affairs (UNOOSA), May 26, 2015, accessed February 19, 2020, https://www.unoosa.org/oosa/en/ourwork/spacelaw/treaties/outerspacetreaty.html & https://www.unoosa.org/pdf/publications/STSPACE11E.pdf

94. Bureau Of Arms Control, Verification, and Compliance, "Treaty on Principles Governing the Activities of States in the Exploration and Use of Outer Space, Including the Moon and Other Celestial Bodies," U.S. State Department, January 21, 2017, accessed February 19, 2020, https://2009-2017.state.gov/t/isn/5181.htm

95. "The hearings of the Committee on Foreign Relations in the United States

Senate also provide accurate assessments of the United States intent in signing the Treaty. The statements of the persons who spoke before this Committee indicate concern about three basic problems: an arms race in space, the placement of nuclear devices in orbit as offensive weapons and the impact of the Outer Space Treaty on ABM and other technology." From: George Bernhardt, Sandra M. Gresko, and Thomas R. Merry, "Star Wars versus Star Laws: Does SDI Conform to Outer Space Law; The Reagan Legacy and the Strategic Defense Initiative, " *Journal of Legislation* Volume 15 | Issue 2 Article 14, May 1, 1989, accessed February 19, 2020, https://scholarship.law.nd.edu/cgi/viewcontent.cgi?article=1319&context=jleg

96. Paul G. Dembling and Daniel M. Arons, "The Evolution of the Outer Space Treaty," *Documents on Outer Space Law 3*, 1967, https://digitalcommons.unl.edu/spacelawdocs/3

97. Dembling and Arons, "The Evolution of the Outer Space Treaty".

98. Dembling and Arons, "The Evolution of the Outer Space Treaty".

99. Brian J. Egan, "The Next Fifty Years of the Outer Space Treaty" [to Galloway Symposium on Critical Issues in Space Law in Washington, DC on December 7, 2016], State Department, December 7, 2016, accessed February 19, 2020, http://carbondate.cs.odu.edu/#https://2009-2017.state.gov/s/l/releases/remarks/264963.htm

100. Egan, "The Next Fifty Years of the Outer Space Treaty".

101. Egan, "The Next Fifty Years of the Outer Space Treaty".

102. NASA, "Report of the Task Force on Space," *The Evolution of US Space Policy and Plans* 1969, accessed February 21, 2020, https://history.nasa.gov/SP-4407/vol1/chapter3-3.pdf

103. John G. Cramer, "2001—Then and Now," *Analog Science Fiction & Fact Magazine*, July, 2001, available online at: https://www.npl.washington.edu/AV/altvw107.html

104. Michael A. G. Michaud, *Reaching for the High Frontier: The American Pro-Space Movement*, 1972–84 (Westport, CT: Praeger, 1986), accessed July 27, 2019, https://space.nss.org/reaching-for-the-high-frontier-the-american-pro-space-movement-1972-84/

105. Michaud, *Reaching for the High Frontier.*

106. Full Text of Proxmire's Letter to "60 Minutes," *L-5 News*, Vol 2 No. 12, December 1977, accessed March 9, 2019, https://space.nss.org/media/L5-News-1977-12.pdf

107. "H.R. 4286 97th Congress," SCRIBD, July 28, 1981,accessed August 1, 2019, https://www.scribd.com/doc/79483990/H-R-4286-97th-Congress

108. "H.R. 4286 97th Congress," SCRIBD.

109. "The Northwest Ordinance For Space," Citizens in Space, n.d., accessed August 1, 2019, http://www.citizensinspace.org/2014/04/the-northwest-ordinance-for-space/

110. "The Northwest Ordinance For Space," *Citizens in Space*

111. "The Northwest Ordinance For Space," *Citizens in Space*

112. *Agreement Governing The Activities of States on the Moon and Other Celestial Bodies,* UNOOSA, December 5, 1979, accessed February 19, 2020, https://

www.unoosa.org/pdf/gares/ARES_34_68E.pdf; *Agreement Governing The Activities of States on the Moon and Other Celestial Bodies,* NTI, October 26, 2011 accessed February 19, 2020, *https://media.nti.org/documents/moon_agreement.pdf*

113. L5 Society, "UN Moon Treaty Falling to US Opposition Groups," L5 News, March 1982, accessed February 21, 2020,: https://space.nss.org/l5-news-un-moon-treaty-falling-to-us-opposition-groups/

114. US National Commission on Space. *Pioneering the Space Frontier; The Report of the National Commission on Space*. New York: Bantam, 1986.

115. H.R. 4218, NSS, accessed August 1, 2019, https://space.nss.org/media/Space-Settlement-Act-Of-1988.pdf

116. PUBLIC LAW 100-685, November 17, 1988, GovInfo, accessed August 1, 2019, https://www.govinfo.gov/content/pkg/STATUTE-102/pdf/STATUTE-102-Pg4083.pdf

117. National Aeronautics and Space Act of 1958, Pub. L. No. 85-568, Stat. 426 (1958).

118. PUBLIC LAW 100-685

119. Personal Text Message from a Space Frontier Foundation advocate (anonymous), August 31, 2019; Jessa Gamble, "How do you build a city in space?," *The Guardian*, May 16, 2014, accessed August 7, 2019, *https://www.theguardian.com/cities/2014/may/16/how-build-city-in-space-nasa-elon-musk-spacex*

120. Jessa Gamble, "How Do you Build a City in Space?," *The Guardian*, May 16, 2014, accessed August 7, 2019, *https://www.theguardian.com/cities/2014/may/16/how-build-city-in-space-nasa-elon-musk-spacex*

121. NASA, "The Vision for Space Exploration," February 2004, accessed July 27, 2019, https://www.nasa.gov/pdf/55583main_vision_space_exploration2.pdf

122. John Marburger, "Keynote Address," 44th Robert H. Goddard Memorial Symposium Greenbelt, Maryland March 15, 2006, accessed July 27, 2019, https://www.lpi.usra.edu/lunar/strategies/WhiteHouseDocuments/marburger_goddard06.pdf

123. Editors, "NASA's Griffin: 'Humans Will Colonize the Solar System'," *The Washington Post*, September 25, 2005, accessed July 30, 2019, http://www.washingtonpost.com/wp-dyn/content/article/2005/09/23/AR2005092301691.html?noredirect=on

124. Editors, "NASA's Griffin: 'Humans Will Colonize the Solar System.'"

125. John Marburger, Keynote Address, 46th Robert H. Goddard Memorial Symposium Greenbelt, Maryland March 6, 2008, accessed July 27, 2019, https://www.nasa.gov/pdf/376588main_03%20-%2003-06-08%20jhm%20Goddard%20Symposium.pdf

126. National Security Space Office (NSSO), *Space-Based Solar Power*, accessed February 21, 2020, https://space.nss.org/space-based-solar-power-as-an-opportunity-for-strategic-security/

127. "Obama-Biden Transition Project: Space Solar Power (SSP)—A Solution for Energy Independence & Climate Change," *SpaceRef*, December 6, 2008, accessed July 29, 2019, http://www.spaceref.com/news/viewsr.html?pid=30044 original available via archive at: https://web.archive.org/web/20160712094225/http://change.gov

/open_government/entry/space_solar_power_ssp_a_solution_for_energy_independence_climate_change/

128. Space Frontier Foundation, "Obama Space Team Seeks Public Comment On Space Solar Power White Paper," accessed July 29, 2019 via archive, https://web.archive.org/web/20150929070923/https://spacefrontier.org/2008/12/obama-space-team-seeks-public-comment-on-space-solar-power-white-paper-submitted-by-sff/

129. SFF: "While we support a national initiative for Space Solar Power, we do not support, nor can the taxpayers afford, another massively expensive 'White Elephant' government space program. Only real 'Change' in how we pursue national space objectives can make SSP competitive with other energy sources. We believe the private sector will eventually develop SSP-the only questions are how long it takes and which country will lead. The government cannot economically develop SSP on its own, but it can assist the U.S. private sector by funding basic R&D, creating the right investment incentives, and buying SSP for its own needs. Such an unprecedented collaboration between the private and public sectors could build not just another program, but a new, green industry that would create large numbers of high-paying jobs for American citizens. Someday, well into this century, the SSP industry could even turn America into a net energy exporter."

From: Space Frontier Foundation, "Obama Space Team Seeks Public Comment."

130. BillionYearPlan Blog, "Citizens for Space Based Solar Power Open Letter to OSTP," April 10, 2010, accessed July 30, 2019, https://billionyearplan.blogspot.com/search?q=openenergy

131. David Shiga, "Will Obama Pursue Space-based Solar Power?," *New Scientist*, December 22, 2008, accessed July 30, 2019, via archive, https://web.archive.org/web/20160731181015/https://www.newscientist.com/blogs/shortsharpscience/2008/12/will-obama-pursue-space-based.html

132. Tom Kalil, "Bootstrapping a Solar System Civilization," *The White House Blog*, October 14, 2014, accessed July 27, 2019, https://obamawhitehouse.archives.gov/blog/2014/10/14/bootstrapping-solar-system-civilization

133. Kalil, "Bootstrapping a Solar System Civilization."

134. "H.R.5063—ASTEROIDS Act," 113th Congress, *Congress.gov*, introduced July 10, 2014, accessed March 11, 2019, https://www.congress.gov/bill/113th-congress/house-bill/5063

135. CNN, "State of the Union 2015: Full transcript," *CNN*, January 20, 2015, accessed February 18, 2020, https://www.cnn.com/2015/01/20/politics/state-of-the-union-2015-transcript-full-text/index.html; NASA, "State of the Union and NASA," *Youtube*, January 20, 2015, accessed February 18, 2020, https://www.youtube.com/watch?v=xhyVEbZls54

136. Rick Tumlinson, "The Summit, the Solution, the Possibility," *SpaceNews*, May 27, 2015, accessed July 29, 2019, https://spacenews.com/op-ed-the-summit-the-solution-the-possibility/

137. Jeff Foust, "Proposed Legislation Would Make Space Settlement a National Goal," *SpaceNews*, May 1, 2015, accessed August 1, 2019, https://spacenews.com/proposed-legislation-would-make-space-settlement-a-national-goal/

138. Foust, "Proposed Legislation Would Make Space."

139. D3SSP, D3 Space Solar Power Proposal, accessed August 7, 2019, http://d3ssp.org/ see also YouTube link: https://www.youtube.com/watch?v=M9dQsRv1XDg

140. Colin Clark, "'This Is Our Sputnik Moment:' Rep. Bridenstine Offers Sweeping Space Bill," *Breaking Defense*, April 12, 2016, accessed August 1, 2019, https://breakingdefense.com/2016/04/this-is-our-sputnik-moment-rep-bridenstine-offers-sweeping-space-bill/

141. "H.R.4945—American Space Renaissance Act" in the 114th Congress (2015-2016), Congress.gov, January 6, 2017, accessed August 1, 2019, https://www.congress.gov/bill/114th-congress/house-bill/4945/text

142. "H.R.4945—American Space Renaissance Act".

143. "H.R.4945—American Space Renaissance Act".

144. "H.R.4945—American Space Renaissance Act".

145. "H.R.4945—American Space Renaissance Act".

146. "H.R.4945—American Space Renaissance Act".

147. Kenneth Chang, "Trump's NASA Nominee, Jim Bridenstine, Confirmed by Senate on Party-Line Vote," *The New York Times*, April 19, 2018, accessed August 1, 2019, https://www.nytimes.com/2018/04/19/science/jim-bridenstine-nasa.html

148. Note: this extensive quote is part of the public record by a U.S. official.

149. Jim Bridenstine, "This is Our Sputnik Moment," *Bridenstine.house.gov*, November 6, 2016, accessed November 7, 2016, https://bridenstine.house.gov/blog/?postid=758 archived (as of August 1, 2019) at: https://web.archive.org/web/20171126063553/https://bridenstine.house.gov/blog/?postid=758; also available here Jim Bridenstine, "This is Our Sputnik Moment," speech to the Lunar Exploration Group, OKG News, November 6, 2016, accessed August 1, 2019, https://okgrassroots.com/?p=642815

150. Hanneke Weitering, "Barack Obama Declares U.S. on Track to Mars," *Scientific American*, October 11, 2016, accessed February 18, 2020, https://www.scientificamerican.com/article/barack-obama-declares-u-s-on-track-to-mars/

151. Barack Obama, "America will take the giant leap to Mars," *CNN*, October 11, 2016, accessed February 18, 2020, https://www.cnn.com/2016/10/11/opinions/america-will-take-giant-leap-to-mars-barack-obama/index.html

152. Ledyard King, "President Obama proposes $19 billion for NASA in fiscal 2017 budget," *USA TODAY*, February 9, 2016, accessed February 18, 2020, https://www.usatoday.com/story/news/politics/2016/02/09/president-obama-proposes-19-billion-nasa-fiscal-2017/80053964/

153. NASA, "President Obama's Speech at Kennedy," April 15, 2010, accessed February 16, 2020, https://www.nasa.gov/about/obama_ksc_pod.html

154. Tariq Malik "Obama Budget Scraps NASA Moon Plan for '21st Century Space Program'," Space.com, February 01, 2010, accessed February 16, 2020, https://www.space.com/7849-obama-budget-scraps-nasa-moon-plan-21st-century-space-program.html

155. NASA, "What Is NASA's Asteroid Redirect Mission?," December 11, 2017, accessed February 16, 2020, https://www.nasa.gov/content/what-is-nasa-s-asteroid-redirect-mission

156. Sarah Fecht, "U.S. Lawmakers Want To Kill NASA's Mission To Send Astronauts To An Asteroid: Instead they favor going back to the moon," Popular Science, May 24, 2016, accessed February 16, 2020, https://www.popsci.com/house-bill-would-provide-zero-funding-for-nasas-asteroid-redirect-mission/

157. Jeff Foust, "NASA closing out Asteroid Redirect Mission," *Space News*, June 14, 2017, accessed February 16, 2020, https://spacenews.com/nasa-closing-out-asteroid-redirect-mission/

158. NASA, "COTS: Final Report," May 2014, accessed August 7, 2019, https://www.nasa.gov/content/cots-final-report

159. Greg Autry, "Commercial Orbital Transportation Services: A Case Study for National Industrial Policy," NewSpace, September 2018, accessed August 7, 2019, https://www.liebertpub.com/doi/10.1089/space.2017.0043

160. "H.R.2262—U.S. Commercial Space Launch Competitiveness Act 114th Congress (2015-2016)," Congress.gov, November 25, 2015, https://www.congress.gov/bill/114th-congress/house-bill/2262/text

161. Phrase coined by Dr. Brent Ziarnick in 2019 to describe a group of thinkers who saw military spacepower primarily in the context of economic power, and the imperative to seek and protect space resources in deep space Brent Ziarnick, "The Space Force Strategic Vision Emerges," *The Hill*, September 09, 2019, accessed October 14, 2019, https://thehill.com/opinion/national-security/460469-the-space-force-strategic-vision-emerges

162. Peter Garretson, "Guess What Could Be Totally Missing From the New U.S. President's Intel Briefing: Oh, just the entire outer-space thing," War is Boring, September 29, 2016, August 1, 2019, https://warisboring.com/guess-what-could-be-totally-missing-from-the-new-u-s-presidents-intel-briefing/

163. Peter Garretson, Brent Ziarnick, M. V. "Coyote" Smith, and Everett Dolman, "A National Space Policy for this Century," *Wired*, November 7, 2016, accessed August 1, 2019, http://www.thespacereview.com/article/3099/1

164. Brent Ziarnick, Peter Garretson, Everett Dolman and Coyote Smith, "Dear President Trump: Here's How to make Space Great Again," *The Space Review*, December 14, 2016, accessed August 1, 2019, https://www.wired.com/2016/12/dear-president-trump-heres-make-space-great/

165. Robert S. Walker and Peter Navarro,"Op-ed | Donald Trump's 'peace through strength' space vision," Space News, October 24, 2016, accessed February 18, 2020, https://spacenews.com/op-ed-donald-trumps-peace-through-strength-space-doctrine/; Bruce Dorminey, "Trump Space Policy To Aim For Mars And Beyond," Forbes, Nov 14, 2016, accessed February 18, 2020, https://www.forbes.com/sites/brucedorminey/2016/11/14/trump-space-policy-to-aim-for-mars-and-beyond/#2756651a250d

166. Graham, *High Frontier*. Also see Graham, *High Frontier: A strategy for National Survival*.

167. Jerry Hendrix and Michelle Shevin-Coetzee, "From Blue To Black: Applying the Concepts of Sea Power to the Ocean of Space, " Center for New American Security (CNAS), November 2016, accessed March 11, 2019, https://s3.amazonaws.com/files.cnas.org/documents/CNAS-Report-FromBluetoBlack-2016.pdf?mtime=20161118095339

168. *FAST SPACE: Leveraging Ultra Low-Cost Space Access for 21st Century Challenges*, Air University, January 13, 2017, accessed August 1, 2019, https://www.airuniversity.af.edu/Portals/10/Research/Space-Horizons/documents/Fast%20Space_Public_2017.pdf

169. *FAST SPACE: Leveraging Ultra low-cost Space Access.*

170. *FAST SPACE: Leveraging Ultra low-cost Space Access.*

171. Robert S. Walker and Peter Navarro, "Trump's space policy reaches for Mars and the stars," *Space News*, October 19, 2016, accessed August 1, 2019, https://spacenews.com/trumps-space-policy-reaches-for-mars-and-the-stars/

172. Robert S. Walker and Peter Navarro, "Trump's space policy reaches for Mars and the stars".

173. Andy Pasztor, "Trump in Space: Transition Focuses on Private-Public Initiatives: Regarding NASA's future, Steve Bannon meets with SpaceX's Elon Musk and others," Wall Street Journal, January 18, 2017, accessed February 18, 2020, https://www.wsj.com/articles/trump-in-space-transition-focuses-on-private-public-initiatives-1484781119

174. Hendrix and Routh: "One initiative for the U.S. government to consider is the express authorization and incentivization of the exploitation of resources from space. As long as a commercial entity does not claim the entirety of a celestial body (e.g., the moon or an asteroid), but rather confines itself to simply gathering resources contained within or on that body, then the entity and the government itself are within the confines of the Outer Space Treaty. The key is to pass the minimum number of laws governing authorization and supervision of space operations, in order to encourage the minimally fettered development of space by the commercial sector."

Jerry Hendrix and Adam Routh, "A Space Policy for the Trump Administration," *Center for New American Security (CNAS)*, November 2016, accessed March 11, 2019, https://s3.amazonaws.com/files.cnas.org/documents/CNASReport-SpacePolicy-Final5.pdf?mtime=20171023151801

175. Hendrix and Routh, "A Space Policy for the Trump Administration."

176. Hendrix and Routh, "A Space Policy for the Trump Administration.": "Unfortunately, as one analyst has observed, this prohibition has essentially the same level of wisdom behind it as Pope Alexander VI's 1496 Papal Bull dividing the outer, unexplored world (from a European viewpoint) between Portugal and Spain. The bull carried full legal authority until England and the Netherlands decided that it did not, and began acting outside its guidance" and likened the situation to "During the 19th century, the United States entered into a series of treaties with European powers regarding the interior of the North American continent. These were both aspirational and practical tools to avert or delay conflicts with larger countries until the United States was in a stronger position to protect and promote its interests. In its time, the 1967 Outer Space Treaty served a similar aspirational purpose, providing the basis for the entire current body of international law as it applies to the peaceful use of the space environment."

177. Hendrix and Routh, "A Space Policy for the Trump Administration.": "As more affordable access to LEO, medium earth orbits, and geo-stationary orbits becomes a reality and activities related to science, technology, and, most important,

the economic incentive outside Earth's atmosphere, humanity can begin to make use of the resources of space" justifying their prescription that, They note, "Goldman Sachs recently produced a 98-page document for investors detailing the feasibility and lucrativeness of asteroid mining."

178. Hendrix and Routh, "A Space Policy for the Trump Administration": "The 1967 Outer Space Treaty may still offer some benefit today, however, technology and economic opportunities are proving challenging for the treaty. Problematically, Article II states: "Outer space, including the Moon and other celestial bodies, is not subject to national appropriation by claim of sovereignty, by means of use or occupation, or by any other means." . . . In business and in government, the bottom-line matters, and in the case of space, the bottom line is that the Outer Space Treaty has dis-incentivized nations and companies from commercially developing the domain, a troubling consequence. "We Came in Peace for All Mankind" may have been a beautiful sentiment for the Apollo 11 crew as the first humans to step foot on the moon, *but no economic entity will fully invest in an enterprise involving the moon (which has many rare mineral resources) or asteroids (some of which are extraordinarily valuable) without having some claim of mineral rights or ownership—which the Outer Space Treaty makes difficult at best.*"

179. Hendrix and Routh, "A Space Policy for the Trump Administration".: "The high seas are currently part of the global commons where actors are free to harvest resources but prohibited from laying sovereign claim. The same perspective should be adopted for space and the resources within, thereby incentivizing the exploitation of space without encroaching on the idea of a celestial commons. Understandably, the economic exploitation of space may require additional modifications to the 1967 treaty" and "Without the capacity to lay claim to space-based resources, the commercial sector will be limited in its economic potential. The 2015 Commercial Space Launch Competitiveness Act provided legal authority for the United States to claim space-based resources, but U.S.-based commercial companies will need a broader international consensus regarding ownership or controlling interests"

180. Hendrix and Routh, "A Space Policy for the Trump Administration."

181. Peter Garretson, "Re-opening the American Frontier: Recent Congressional Hearings on space," *The Space Review*, July 3, 2017, accessed August 7, 2019, http://www.thespacereview.com/article/3275/1

182. NPS Homeland Security Digital Library, "Reopening the American Frontier: Exploring How the Outer Space Treaty Will Impact American Commerce and Settlement in Space," May 23, 2017, accessed August 7, 2019, https://www.hsdl.org/?view&did=807259

183. "H.R.2809—American Space Commerce Free Enterprise Act," 115th Congress, introduced June 7, 2017, accessed March 11, 2019, https://www.congress.gov/bill/115th-congress/house-bill/2809

184. "H.R.2809—American Space Commerce Free Enterprise Act."

185. "H.R.2809—American Space Commerce Free Enterprise Act."

186. Scott Pace, Executive Secretary, National Space Council, "Space 2.0: U.S. Competitiveness and Policy in the New Space Era," Hudson Institute, Keynote Address, May 17, 2018, accessed March 9, 2019, https://www.hudson.org/re

search/14341-full-transcript-space-2-0-u-s-competitiveness-and-policy-in-the-new-space-era

187. Cal Biesecker, "Commercial Space Bill Reintroduced In House," *Defense Daily*, July 3,2019, accessed February 18, 2020, https://www.defensedaily.com/commercial-space-bill-reintroduced-house/congress/

188. James A. Vedda, Cislunar Development: What To Build—And Why, The Aerospace Corporation, April 2018, accessed March 11, 2019, https://aerospace.org/sites/default/files/2018-05/CislunarDevelopment.pdf

189. Vedda, Cislunar Development: What To Build—And Why.

190. Vedda, Cislunar Development: What To Build—And Why.

191. Vedda, Cislunar Development: What To Build—And Why.

192. Vedda, Cislunar Development: What To Build—And Why.

193. Newt Gingrich, "Space Force Planning," July 17, 2018. Used with the permission of the author.

194. Gingrich, "Space Force Planning."

195. Brian Hans, Christopher Jefferson, and Joshua Wehrle, *Movement and Maneuver in Deep Space: A Framework to Leverage Advanced Propulsion* (Maxwell Air Force Base: Air University Press, 2018), accessed August 1, 2019, https://apps.dtic.mil/dtic/tr/fulltext/u2/1042207.pdf updated here: https://media.defense.gov/2019/May/23/2002135778/-1/-1/0/WF_0067_HANS_JEFFERSON_WEHRLE_MOVEMENT_AND_%20MANEUVER_IN_DEEP%20SPACE_A_FRAMEWORK_TO_LEVERAGE_ADVANCED_PROPULSION.PDF

196. Hans, Jefferson, and Wehrle, *Movement and Maneuver in Deep Space.*

197. Paul M. Gesl, "Preparing For The Next Space Race: Legislation and Policy Recommendations for Space Colonies," Air Command and Staff College Maxwell AFB United State, April 1, 2018, accessed August 13, 2019, https://apps.dtic.mil/dtic/tr/fulltext/u2/1053024.pdf

198. Gesl, "Preparing For The Next Space Race".

199. Gesl, "Preparing For The Next Space Race".

200. Gesl, "Preparing For The Next Space Race."

201. Gesl, "Preparing For The Next Space Race."

202. Dustin L. Grant and Matthew J. Neil, "The Case For Space: A Legislative Framework For An Independent United States Space Force," Air Command and Staff College Maxwell AFB United States, April 1, 2018. https://apps.dtic.mil/dtic/tr/fulltext/u2/1053020.pdf

203. Grant and Neil, "The Case For Space."

204. Grant and Neil, "The Case For Space."

205. Steve Kwast, "There won't be Many Prizes for Second Place," *Politico*, August 10, 2018, accessed July 29, 2019, https://www.politico.com/story/2018/08/10/space-race-kwast-768751

206. Policy monopolies were first discussed by Baumgartner and Jones' punctuated equilibrium theory. James L. True, Bryan D. Jones, and Frank R. Baumgartner, "Punctuated-Equilibrium Theory: Explaining Stability and Change in Public Policymaking," in Paul Sabatier, ed., *Theories of the Policy Process*, 2nd Edition

(Boulder, Colorado: Westview Press, 2006), accessed August 13, 2019, https://fbaum.unc.edu/teaching/articles/True_Jones_Baumgartner_2006_chapter.pdf

207. Peter Garretson, "Air Force Suppressed Space Force Debate; Lt. Gen. Kwast Spoke Truth To Power," Breaking Defense, August 08, 2019, accessed February 18, 2020, https://breakingdefense.com/2019/08/air-force-suppressed-space-force-debate-lt-gen-kwast-spoke-truth-to-power/

208. Peter Garretson, "A Speech For The Next Secaf To Launch A New Era Of Spacepower," *War on the Rocks*, September 20, 2019, accessed February 18, 2020, https://warontherocks.com/2019/09/a-speech-for-the-next-secaf-to-launch-a-new-era-of-spacepower/

209. Patrick Tucker, "China's Moon Missions Could Threaten US Satellites: Pentagon," *Defense One*, October 16, 2018, accessed March 9, 2019, https://www.defenseone.com/technology/2018/10/chinas-moon-missions-could-threaten-us-satellites-pentagon/152084/

210. U.S.-China Economic and Security Review Commission (USCC), "Statement of Dr. Namrata Goswami Independent Senior Analyst and Author 2016–2017 Minerva Grantee Before the U.S.-China Economic and Security Review Commission Hearing on 'China in Space: A Strategic Competition?'" April 25, 2019, accessed February 18, 2020, https://www.uscc.gov/sites/default/files/Namrata%20Goswami%20USCC%2025%20April.pdf; U.S.-China Economic and Security Review Commission, Hearing on "China in Space: A Strategic Competition?," April 25, 2019, accessed July 27, 2019.

211. U.S.-China Economic and Security Review Commission, Hearing on "China in Space: A Strategic Competition?," April 25, 2019, accessed July 27, 2019, https://www.uscc.gov/sites/default/files/Namrata%20Goswami%20USCC%2025%20April.pdf

212. Sandra Erwin, "Air Force, DoD Research Organizations Call for Space Industrial Policies to Counter China," June 17, 2019, accessed August 8, 2019, https://spacenews.com/air-force-dod-research-organizations-call-for-space-industrial-policies-to-counter-china/

213. Lamont Colucci, "A Space Service in support of American grand strategy," The Space Review, February 25, 2019, accessed August 13, 2019, http://www.thespacereview.com/article/3664/1

214. Mark R. Whittington, "The Space Force has gone from joke to reality," The Hill, December 16, 2018, accessed February 21, 2020, https://thehill.com/opinion/technology/474713-the-space-force-has-gone-from-joke-to-reality

215. Peter Garretson, "The purpose of a Space Force is a spacefaring economy," The Hill, June 26, 2019, accessed August 5, 2019, https://thehill.com/opinion/technology/450519-the-purpose-of-a-space-force-is-a-spacefaring-economy

216. Peter Garretson, "The First Duty of a Space Force is to Protect Space Commerce," POLITICO, June 21, 2019, accessed August 5, 2019, https://www.politico.com/story/2019/06/21/opinion-space-force-commerce-1374229

217. Peter Garretson, "The Future of the Asian Order Will Be Decided in Space," The Diplomat, July 3, 2019, accessed August 5, 2019, https://thediplomat.com/2019/07/the-future-of-the-asian-order-will-be-decided-in-space/

218. Peter Garretson, "Space Force's Jupiter-Sized Culture Problem," War on the Rocks, July 11, 2019, accessed August 5, 2019, https://warontherocks.com/2019/07/space-forces-jupiter-sized-culture-problem/

219. Brent Ziarnick, "The Space Corps needs naval rank," POLITICO, July 26, 2019, accessed August 5, 2019, https://www.politico.com/story/2019/07/26/space-corps-naval-rank-1433541

220. Tim Cox, "In Space Force debate, the military's space experts are missing in action," The Hill, February 12, 2019, accessed February 21, 2020, https://thehill.com/opinion/national-security/429665-in-the-space-force-debate-the-militarys-space-experts-are-missing

221. Peter Garretson, "Space Force defenses must stretch to the moon," The Hill, August 5, 2019, accessed August 5, 2019, https://thehill.com/opinion/national-security/456214-space-force-defenses-must-stretch-to-the-moon

222. Steven L. Kwast, "The Urgent Need for a United States Space Force," *Imprimis* 49, No. 1 (January 2020), accessed February 18, 2020, https://imprimis.hillsdale.edu/urgent-need-united-states-space-force/

223. "The Space Development and Settlement Act of 2019," Alliance for Space Development, accessed August 1, 2019, http://allianceforspacedevelopment.org/wp-content/uploads/2019/01/Space-Development-and-Settlement-Act-of-2019.pdf

224. "Alliance Objectives for 2019," Alliance for Space Development

225. For an excellent history of this campaign, see Aaron Oesterle, "The Viability of the Space Settlement Narrative in Politics," (IAC-19.E3.2.9x52875), International Astronautical Congress, November 1, 2018.

226. Alliance for Space Development, "March Storm 2020 Congressional Visits," accessed February 11, 2020, http://allianceforspacedevelopment.org/blitzes/

227. SpaceX, "Making Humans a Multiplanetary Species" presentation to IAC 2016, *YouTube*, https://www.youtube.com/watch?v=H7Uyfqi_TE8&t=10s

228. SpaceX, "Making Life Multiplanetary" presentation to IAC 2017, *YouTube*, September 29, 2017, accessed August 7, 2019, https://www.youtube.com/watch?v=tdUX3ypDVwI&t=8s

229. Jeff Bezos, "Going to Space to Benefit Earth," *YouTube*, May 9, 2019, accessed July 29, 2019, https://www.youtube.com/watch?v=GQ98hGUe6FM

230. All NSS position papers can be found here: National Space Society, "National Space Society Position Papers," n.d., accessed February 11, 2020, https://space.nss.org/national-space-society-position-papers/

231. NSS, "NSS Position Paper: A Space Guard to Enable, Regulate, and Protect National Civil and Commercial Space Activities," January, 2019, accessed February 14, 2020, https://space.nss.org/media/NSS-Position-Paper-Space-Guard-2019.pdf

232. NSS, "NSS Position Paper: Space Debris Removal, Salvage, and Use: Maritime Lessons," October, 2019, accessed February 14, 2020, https://space.nss.org/media/NSS-Position-Paper-Space-Debris-Removal-2019.pdf

233. NSS, "Position Paper: Outer Space Private Investment Corporation (OSPIC)," December 2018, accessed February 14, 2020, https://space.nss.org/national-space-society-position-papers/

234. NSS, "Position Paper: A Public/Private COTS-Type Program to Develop Space Solar Power," February 2020, accessed February 14, 2020, https://space.nss.org/media/NSS-Position-Paper-COTS-Type-Space-Solar-Power-2020.pdf

235. Bruce Cahan, "Space Commodities Exchange – a new frontier," Urban Logic, October 8, 2016, accessed August 5, 2019, https://urbanlogic.org/space-commodities-exchange/

236. Bruce Cahan, "A Space Commodities Futures Trading Exchange to Grow the Lunar Economy," 2017 Annual Meeting of the Lunar Exploration Analysis Group, held October 10–12, 2017 in Columbia, Maryland. LPI Contribution No. 2041, id.5008

237. Trevor Brown, "A Space Joint Stock Company," *The Space Review*, April 30, 2012, accessed March 9, 2020, https://www.thespacereview.com/article/2073/1: Brown's proposal includes space debris removal, space solar power, colonization, finance and even private self-defense. Brown asserts that annual predicted returns between $3.5 and $5 trillion would create a company valuation of $50—$100T, and that a mere $10T in company stock would create 10,000 billionaires. He notes "if enterprising members of the military and NASA took the initiative to form the first ever Space Joint Stock Company, a wealth of resources could be obtained with which to create a flourishing civilization in space."

238. Scott Phillips, "The New Frontier Playbook: The Political Economic Plan to Win the Frontier of Space, 2017, accessed August 13, 2019, https://newfrontier-playbook.com/ Phillips states: "a vision and strategy based on exploration is deeply flawed. At best, it gets us back to a future before we cancelled Apollo and scuttled the massive Saturn V rockets that once lofted our astronauts into the heavens, but does not resolve the reasons why we abandoned them in the first place. A more fundamental test of strategy and vision is whether a proposed effort can gain the financial support required to make it sustainable. . . Space exploration, as a vision and strategy, are wrong for our time. We need a vision that corresponds to what the public wants and is willing to fund . . . My alternative point of view on the future of America in space can be articulated in three simple words: Scale is possible. . . I will argue the paradox that it may be easier to increase NASA's budget by 10x than 2x. . . The Proposal: An L1 Strategic Materials Reserve: I am proposing that the US Congress authorize and the US Government move forthwith to implement a program of open market acquisition of asteroid material to be purchased upon delivery at Lagrange 1 (L1) or other suitable delivery point in near Earth space. The delivery of said material should be contracted with the private sector, primarily, but not necessarily exclusively American entities, which will be responsible for delivering in unprocessed form at a fixed price per unit of measure a specific quantity of materials at a specific location. . . Upon delivery and payment, said material will become the sole property of the US Government and the American people. The goal for this strategic material fund would be to acquire in the range of 100,000–250,000 tons of asteroid material over a 30 year period beginning as soon as can be operationally initiated. I will demonstrate with this proposal that the net economic benefit of the material acquired is greatly in excess of the cost of its acquisition and that moving to purchase at scale will create both an industry of high paying jobs as well as an asset for the

American people. In blunt terms, this program will create jobs at home on Earth for Americans, demonstrate American technological leadership, and open up the frontier of near-earth space for both development and exploration." (used with permission of the author)

239. Michael Castle-Miller, "The Lunar Development Cooperative (LDC)," February 2020, forthcoming.

240. Wayne White, "The Space Pioneer Act," October 28, 2019, draft legislation distributed among members of Congress. White sent a copy to the author, Peter Garretson.

241. Bureau Of Arms Control, Verification, And Compliance, "Treaty on Principles Governing the Activities of States," U.S. State Department, January 27, 1967, accessed February 18, 2020, https://2009-2017.state.gov/t/isn/5181.htm

242. White, "The Space Pioneer Act," October 28, 2019.

Chapter 4

U.S. Strategy and Space Resource Ambitions

Since the landing of Apollo 11 in 1969 until the present day, the United States maintained its position as the dominant space power. On nearly every metric of national power, the United States has reigned supreme. The United States pioneered the ideational use of space to portray civilizational vibrance, and established the standard that any state which would contest global leadership must meet or exceed, in the theater of space. It has inspired the world with unmatched feats of landing humans on the Moon, sent probes to every planet, the farthest reaches of our solar system,[1] and provided the wonders of the Hubble telescope to the world. Technologically, it remains the envy of the world, with the most vibrant education system, the most extensive research infrastructure, and the most advanced satellites and launch vehicles. Economically, it maintains the largest share of revenues of the global space sector, the global launch market, the most vibrant commercial space sector, and the largest fleet of satellites. Militarily, it maintains the largest space forces, the largest fleet of military satellites, and has pioneered the use of space to support global military operations. The United States currently plans to land the first woman and next man on the South Pole of the Moon by 2024 and to establish a base for resource utilization there. In the contest for hegemony in space, the United States is the power to beat. This chapter offers an in-depth perspective on U.S. spacepower and expressed ambitions on space resources in policy, law, programs, and the expected behavior of the United States in the century ahead.

UNITED STATES: CURRENT AND FORECAST POWER

The United States finds itself at the top of the world order. In 2019, it was ranked #1 of the most powerful countries by U.S. News and World Report.[2] Global Firepower ranks it as #1 in military power.[3] The Stockholm International Peace Research Institute (SIPRI) estimates its military spending at $609,758 Billion (Bn).[4] The International Institute of Strategic Studies (IISS) estimates it at $602.8Bn and places it as #1 in the military balance.[5]

Today, the United States has a population of 329,093,110, growing at 0.71 percent,[6] with a Human Development Index (HDI) score of 0.924 (ranked #13/189 countries).[7] Its total energy consumption was 2,201 million tons of oil equivalent (MTOE)[8] (47.7 barrels of oil (BOE) per capita), growing at 0.27 percent, with a carbon output of 5,073 MtCO2.[9] Its electrical consumption is 3,808 TWh growing at –2.1 percent,[10] with a per capita electrical usage of 12,984 Kw-hr/annum.[11]

According to International Monetary Fund (IMF) figures, its GDP (real) was $19,390.6 Bn, and $19,390.6 Bn at purchasing power parity (PPP), with a per capita income of $59,501. The United States spent 35.6 percent of its GDP on government expenditures, its estimated gross national savings was 17.4 percent of GDP, and the United States had a current account balance of $ –466.246 Bn (–2.4 percent of GDP).

However, despite continued growth, the United States is expected to be outgrown by China and India, and to move from a position of economic dominance to a position of second or third place depending upon metric. According to PricewaterhouseCooper, its economy is expected to grow from $20.1 trillion (T) in 2020 to $34.1T in 2050 at market exchange rates (MER; real) and from $20.1T to 34.1T in PPP, suggesting it will be the #2 economy in 2050 at MER and the #3 economy (after China and India) measured by PPP, with 12 percent share of the total global economy (down from 16 percent in 2016).[12] The Carnegie Endowment for International Peace (CEIP) similarly estimated it to be the #2 economy in 2050 at $38.646 T.[13] The Economist Intelligence Unit (EIU) placed the United States #2 at MER, with a projected GDP rising from $17.4T in 2014 to $70.913T in 2050.[14]

As a consequence, if the United States does nothing to significantly grow its economy (such as to access the vast resources of space), it will decline in relative economic power, and therefore total power.

UNITED STATES: SPACE CAPACITY

At the time of writing, the United States is the acknowledged leading space power. It leads in almost all areas: civilian space spending, military space spending, the share of the global space sector, the share of the global launch market, and the number of satellites.

According to the World Economic Forum, it spends $39.332Bn on its space program (0.23 percent of GDP)[15], whereas Euroconsult estimated its space spending at $35.947Bn.[16] The most recent (and somewhat dated) Futron Space competitiveness report gave it an overall space competitive score of 90.32 (ranked #1),[17] with an overall military space capability score of 99.5 (ranked #1).[18] According to the Space Foundation's report released in 2018, the U.S. share of global government spending was 57 percent, its share of global launch activities was 33 percent, and its share of global spacecraft was 65 percent.[19]

As of November 30, 2018, of the 1,957 active satellites cataloged in the Union of Concerned Scientists (UCS) Satellite Database, 849 belonged to the United States (significantly more than the next three space powers combined, China with 284, Russia with 152, India 57).[20] The United States articulated the most ambitious plans for the expansion of satellite constellations. In March 2019, four satellite companies (SpaceX: 12,000; OneWeb: 4,560; Boeing 2,956; and Spire Global: 972) are seeking to place in orbit over 20,000 satellites (20,488),[21] over 10-fold as many as are on orbit today, with the U.S. Federal Communication Commission already having approved SpaceX's plans to launch nearly 12,000 internet satellites.[22] SpaceX would submit paperwork for 30,000 more satellites (total of 42,000) in October 2019.[23] An indication of the speed and vibrancy of the commercial U.S. space sector, within a single year (Mar 2019–Feb 2020), SpaceX would successfully deploy 300 of its Starlink satellites, deploying 60 satellites each time on five missions.[24]

The United States leads the world in space exploration. It is the first, and to date, the only nation to have visited every planet in the Solar System,[25] having sent probes to Mercury (1), Venus (2), Mars (10), Jupiter (5), Saturn (3), Uranus (1), Neptune (1), Pluto (1), as well to Ceres (1), other asteroids (3), Comets (4), and Kuiper belt objects (1)—no other nation has yet independently reached Jupiter, Saturn, Uranus, Neptune, or Pluto.[26]

The U.S. spacecraft has achieved the fastest speeds of any human-made objects, holding the top five speed records: #1 the Parker Solar Probe on October 29, 2018,[27] estimated to achieve 430,000mph on its closest orbit to the Sun,[28] #2 Juno @165,000[29], #3 Helios 2 @157,078mph, #4 Helios 1 @142,000mph, #5 Voyager 1 @38,610.[30] Voyager 1 holds the record for being the first human spacecraft to have reached the edge of our solar system and is the farthest any human artifact has ventured from Earth, over 21.2 billion kilometers.[31] The United States holds the record for having visited the most distant object, Ultima Thule, now renamed Arrokoth, as mentioned earlier.[32]

The United States remains the first and only nation to have been able to land a human being on the Moon and return them safely to Earth.[33] It is at present, the first and only nation that has demonstrated operational reusable launch technology.[34] The United States has the largest, best capitalized, and most vibrant private space industry. Culturally initiated by an epistemic and activist community, which gave rise to the cultural phenomenon of the Ansari X-Prize and later Google Lunar X-Prize, the United States is enjoying great success though the high-visibility efforts of ideologically driven billionaire philanthro-capitalists such as Jeff Bezos and Elon Musk with their broad visions for spacefaring. A broad commercial space ecosystem has now emerged with significant and growing private capital. As of 2019, Space Angels reported that sixty-seven entrepreneurial space companies had received a total of $7.2Bn in investments from the government between 2000 and 2018. This was dwarfed by the broader 374 companies within the private space industry that combined received $19Bn in private funding since 2008. In 2018 alone, over 120 venture capital firms invested in the space industry, resulting in space companies receiving $3.9Bn in private investment in a single year.

U.S. ELITE DISCOURSE

The previous chapter covered U.S. elite discourse from the first Space Age beginning 1957 to the modern era beginning circa 2012.

U.S. POLICIES AND LAW

This section catalogs the instantiation of space expansionist and space development ideas into official policy statements and law.[35] What the reader should look for is a progressive elaboration of a policy image that links the promise of future space commerce and resources with concerns of constriction and predation by foreign powers and the need for a dedicated force to protect these future interests.

The most globally significant event happened in 2015. The U.S. Congress passed[36] and former president Barack Obama signed into law[37] the U.S. Commercial Space Launch Competitiveness Act (CSLCA) of 2015 (formerly H.R. 2262 now Public Law No.:114-90), in which Title IV, section 51303 states:

> A United States citizen engaged in commercial recovery of an asteroid resource or a space resource under this chapter shall be entitled to any asteroid resource or space resource obtained, including to possess, own, transport, use, and sell

> the asteroid resource or space resource obtained in accordance with applicable law, including the international obligations of the United States.[38]

While this initially drew significant global controversy, as discussed in chapter 1, it is now the consensus opinion that this is a valid interpretation of the Outer Space Treaty, and has been followed by similar legislation in Luxembourg and draft legislation in UAE. As remarked by Rep Bridenstine in the last chapter, this means that the U.S. Congress has put the full force of American law behind the claims of private industry to mine asteroids. The law directed the president to

> (1) *facilitate commercial exploration for and commercial recovery of space resources by United States citizens*; (2) discourage government barriers to the development in the United States of economically viable, safe, and stable industries for commercial exploration for and commercial recovery of space resources . . . ; and (3) *promote the right of United States citizens to engage in commercial exploration for and commercial recovery of space resources free from harmful interference*, in accordance with the international obligations of the United States and subject to authorization and continuing supervision by the Federal Government.[39]

The following year (2016), President Donald J. Trump won the election and took office in 2017. The Trump administration more fully embraced a space development policy than administrations before it. Below are excerpts of policies and policy statements of the Trump administration that offers the readers an understanding of the key policy shifts. What is remarkably consistent across this administration is the repeated linking of economy, partnership with the commercial sector, and military protection of the economy. The language strongly echoes observations of American strategic culture where identity and motivation are sought in foundational myths referencing a pioneer or a pioneering spirit,[40] and power must be couched within a crusade of values and freedom. As Bernard L. Brock, et al. stated:

> Democracy requires political philosophy a rhetorical framework for contextualizing action. A political actor can announce an action or a policy, but democracy requires that the actor contextualize the policy, justify it, and orient it to the character of the political culture. Political actors must find the language, images, and interpretations, to connect significant decisions with their context in political philosophy.[41]

Shortly after taking office, in June 2017, President Trump revived the National Space Council via executive order to "advise and assist the President regarding national space policy and strategy" including the direction to "review . . . long-range goals, and develop a strategy for national space activities."[42] In October 2017, Vice President Pence stated, "The President has directed us to relaunch the National Space Council's advisory group to foster close coordination, cooperation, technology information."[43] The Trump administration was to set a significant pace of space-related announcements, ceremonies, and high-profile events by the reconstituted National Space Council,[44] and National Space Council Users' Advisory Group.

Before the end of its first year, on December 11, 2017, the Trump administration would release Space Policy Directive 1 (SPD-1) tasking NASA to:

> Lead an innovative and *sustainable program* of exploration *with commercial* and international partners to *enable human expansion across the solar system* and to bring back to Earth new knowledge *and opportunities*. Beginning with missions beyond low-Earth orbit, the United States will lead the *return of humans to the Moon* for long-term exploration and *utilization*, followed by human missions to Mars and other destinations.[45]

The prominence of commercial partners and utilization is clear evidence of the new policy image discussed in the last chapter.

On December 13, 2017, the secretary of the National Space Council, Dr. Scott Pace, delivered remarks to the International Institute of Space Law where he provided a clear policy statement showcasing that the Trump administration had given significant thought to space resources, and their legal and military protection. He stated:

> U.S. private sector must have confidence that it will be able to profit from capital investments made to *develop and utilize in-situ resources, commercial infrastructure*, and facilities in outer space. Furthermore, certain types of rights and obligations typically associated with exclusive use and private property are needed. In 2015, the United States took an important step with the enactment of the Commercial Space Launch Competitiveness Act. This Act provides that U.S. citizens are *entitled to own, as private property, asteroid and space resources they have obtained* in accordance with applicable law, including our international obligations. . . . The U.S. Government, working with its space partners and the private sector, should use legal and diplomatic means to create a stable, peaceful environment not only for governmental activities, but also for commercial ones. These legal and diplomatic means include *efforts to minimize and mitigate harmful interference to our space systems*, whether from terrestrial actors or from space actors. In addition to the UN Charter and other applicable law, *such as the right of self-defense*, several provisions of the Outer Space Treaty provide legal principles that would be applied toward these ends.[46]

The subsequently released National Security Strategy (December 18, 2017) laid out three priority actions: *Advance Space as a Priority Domain, Promote Space Commerce, Maintain Lead in Exploration*. Notice that within the National *Security* Strategy, commerce precedes exploration. The document took a clear stand that space was of vital interest and interference would generate a military response. It noted that:

> Others believe that the ability *to attack space assets offers an asymmetric advantage* and as a result, *are pursuing a range of anti-satellite (ASAT) weapons* . . . The United States considers *unfettered access to and freedom to operate in space to be a vital interest. Any harmful interference with or an attack upon critical components of our space architecture that directly affects this vital U.S. interest will be met with a deliberate response* at a time, place, manner, and domain of our choosing.[47]

While muscular in tone, this was not a departure from the previous policy. But the following line suggested a shift to conform to the emerging policy image. It stated, "we will also *consider extending national security protections to our private sector partners* as needed."[48] The inclusion of protecting private sector partners is significant and provides evidence that the new policy image discussed in the previous chapter was beginning to take hold.

In February 2018, Secretary of Commerce Wilbur Ross would draw headlines, and for the first time, place a U.S. secretary of commerce in a prominent role in the shaping of space policy. Ross told CNBC, "we want to turn the Moon into a gas station." CNBC reported that "Ross, a former private-equity investor with more than 55 years experience, has *emerged as the point man for promoting commercial space projects*," and reported him stating:

> I think a *lot depends upon how successful we are in turning the moon into a kind of gas station for outer space* . . . the plan is to break down the ice [there] *into hydrogen and oxygen, use those as the fuel propellant* . . . at the moon, you have very low gravity so you don't need so much thrust to go from the moon to Mars, for example, or another asteroid.[49]

This was the first time that a cabinet member of a Presidential administration had articulated an unambiguous space resource ambition.

In March 2018, the administration announced a new National Space Strategy. The new strategy continued the clear linking of economic and security interests. It emphasized the importance of utilization, is evidence of space expansionist thinking, and clearly showcases America's strategic desire for leadership and primacy. The public White House fact sheet[50] stated the strategy sought to place "America first among the stars," emphasizing the "dynamic and cooperative *interplay between the national security, commercial, and civil space sectors*" and stating "the United States will partner with the commercial sector to ensure that American companies remain world leaders in space technology." Calling to mind America's past and space expansionist thinking, the strategy "builds on America's *pioneering* spacefaring tradition."

The mostly defense-focused document called attention to its civil and commercial ambitions and SPD-1, restating,

> By signing [SPD-1] . . . which instructed [NASA] to return American astronauts to the moon for long-term exploration *and utilization*, followed by human missions to Mars and other destinations. In signing the directive, the *President ordered action to work with commercial* and international partners *to enable human expansion across the solar system.*[51]

The document asserted that "our competitors and *adversaries have turned space into a warfighting domain*," and "while the United States would *prefer that the space domain remain free of conflict*, we will prepare to meet and overcome any challenges that arise." As a result, the United States would pursue "peace through strength" and would "*strengthen deterrence and warfighting options*." President Trump would use the term "Space Force" for the first time in a speech on March 13, 2018.[52]

On May 1, 2018, President Trump would again mention Space Force, stating,

> And *we're actually thinking of a sixth, and that would be the Space Force*. Does that make sense? The Space Force, General. You probably haven't even heard that. I'm just telling you now. This is perhaps—because *we're getting very big in space, both militarily and for other reasons*, and we are seriously thinking of the Space Force.[53]

Those other reasons would soon be made clear.

The following day, on May 2, 2018, Commerce Secretary Wilbur Ross would directly link space resources, colonization, and Space Force. The *Washington Post* reported that Secretary of Commerce Wilbur Ross asserted that "space is the new frontier for the country and for the world," and "called for the United States to aim for a dominant role in space when it comes to *security, commerce and, yes, even colonizing Mars*."[54] The Post reported:

> He began by saying travel to Mars is coming, but in order to get there with any large payload, "for *colonization or mining or anything else*," the dark side of the moon would be the staging ground. The moon has low gravity for easy launch

> and *the dark side has ice, which could be broken down into hydrogen and oxygen for rocket fuel,* he explained. "I'm intrigued with the notion that the man in the moon is pretty soon going to be a gas station attendant." He reportedly "predicted *a huge increase in asteroid mining*, which Ross explained is quite different than the terrestrial mining he had experience with in the past. All the very valuable minerals—*gold, silver, platinum, rare minerals—are right on the surface*, said Ross. 'Somebody just has to scoop them off. You just have to get there. That's going to be a big activity.'"[55]

That such a vision was coming not from a NASA administrator but a secretary of commerce indicates a level of interest and confidence within the commercial sector. Importantly, in the same talk, Ross echoed President Trump's first suggestion the day prior on adding Space Force as a branch of the military, stating "operating in space will be enormously costly and that *our interests in space are enormously vulnerable. We are going to see amazing things in space much sooner than we thought possible*, he said. 'It's going to be the most determinative and the most exciting new development . . . and *space is going to be a very big battleground.*'"[56] Here you can see the clear linkage between resources and military concerns.

Dr. Scott Pace, Secretary of the National Space Council, spoke at the Hudson Institute on May 17, 2018.[57] His remarks evidenced a very strong space development emphasis, including anticipated changes due to entrepreneurial firms, and specific mention of prospecting, in-space assembly and manufacture, and new commercial and military competition.

> The United States needs strong and innovative space industries. And if we limit, however, space activities—the activities of governments, we would handicap ourselves and *cede a key American strength, which is the private sector. If we unleash private enterprise in partnership with government, there are really no limits* to what the U.S. can accomplish. And this is why the Space Council has prioritized the development of policies at the interface of public and private sector interests. These include policies on space traffic management, new space technologies, regulatory reform, ensuring there are opportunities for private enterprise and space exploration, and improving space resilience. So, again, as you might have heard recently, discussions about *lunar exploration*—one of the parts of that story, which is still not fully appreciated, is the range of commercial opportunities that are being provided for sending instruments and so forth *to the lunar surface using private partners*. Today, space sector revenues continue to be dominated by large aerospace and telecom firms. However, an increasing number of entrepreneurial firms are seeking nontraditional market sectors. And these sectors are experiencing revolutionary change: space launch, remote sensing, global broadband via large constellations, new services, such as

> suborbital and orbital human flight, *prospecting,* satellite service and *assembly and manufacturing* . . . the challenge of space leadership today is really to manage in the face of rapid change. And while we're *facing new commercial and military competition*, the most important factor affecting the U.S. in space is not really what other countries do but what the United States chooses to do or fails to do. The National Space Council, with Vice President Pence's leadership, is committed to taking the initiatives necessary to sustain and expand a U.S. position of leadership in space.[58]

Dr. Pace's Hudson remarks presaged SPD-2, released thereafter, on May 24, 2018. This policy directive was intended to streamline commercial regulations to put the United States in a more competitive position for the NewSpace economy. SPD-2 sought to elevate the importance of the Department of Commerce, and provided a tight timeline to develop legislation: "within 30 days of the date of this memorandum, the Secretary of Commerce shall transmit to the Director of the Office of Management and *Budget a legislative proposal to create within the Department of Commerce an entity with primary responsibility for administering the Department's regulation of commercial space flight activities*."[59]

Barely a month later, on June 18, 2018, the Trump administration released SPD-3, which transferred Space Traffic Management (STM) responsibility to the Department of Commerce (till that point it had been conducted by the Air Force Space Command). Importantly, its justification anticipates space development activities including space manufacturing. SPD-3 stated:

> Emerging commercial ventures such as *satellite servicing, debris removal, in-space manufacturing, and tourism*, as well as new technologies enabling small satellites and very large constellations of satellites, are increasingly outpacing efforts to develop and implement government policies and processes to address these new activities. *To maintain U.S. leadership in space, we must develop a new approach to space traffic management* (STM) that addresses current and future operational risks . . . Through this national policy for STM and other national space strategies and policies, the United States will enhance safety and ensure continued leadership, preeminence, and freedom of action in space.[60]

The president gave an accompanying speech to the release of SPD-3, which is a strong evidence of space expansionist and space development thinking, giving priority to commerce, industry, and economy.

> In December, I signed a historic directive that will *return Americans to the moon for the first time since 1972*, if you can believe that. (Applause.) Always remembering it's about that, but *it's also about jobs and the economy*. This is a

> great thing we're doing. *This time, we will do more than plant our flag and leave our footprints. We will establish a long-term presence, expand our economy,* and build the foundation for the eventual mission to Mars—which is actually going to happen very quickly . . . I am instructing my administration to *embrace the budding commercial space industry*. We are modernizing out-of-date space regulations. They're way out of date. They haven't been changed in many, many years. And today we're taking one more step to unleash the power of American ingenuity.[61]

The assembled audience assumed the occasion was about SPD-3 and primarily commercial. President Trump then surprised almost everyone[62] by delivering an order to the Pentagon to begin the process to establish the Space Force. His justification weaves together destiny, identity, American Presence, and taming frontiers all in the broader context of economic interests suggested by SPD-3.

> My administration is reclaiming America's *heritage as the world's greatest space-faring nation*. The essence of the American character is to explore new horizons and to tame new frontiers. But *our destiny, beyond the Earth, is not only a matter of national identity, but a matter of national security*. So important for our military. So important. And people don't talk about it. When it comes to *defending America, it is not enough to merely have an American presence in space. We must have American dominance in space*. So important. Very importantly, I'm here by *directing the Department of Defense and Pentagon to immediately begin the process necessary to establish a space force as the sixth branch of the armed forces*. That's a big statement. We are going to have the Air Force and *we are going to have the Space Force*—separate but equal.[63]

In early August 2018, NASA Administrator Bridenstine articulated a view that reflected a space development policy image; one that prized a reusable architecture and the success of commercial rockets. He stated,

> We *want the entire architecture between here and the Moon to all be reusable* . . . if we can *take advantage of commercial industry that can develop a reusable rocket, we want them to be successful* . . . if 10 years from now, 20 years from now, there's a commercial capability that's successful, we're going to use it. And we want them to be successful. In fact, we're partnering with those companies today on commercial crew and other things.[64]

Because previous NASA exploration vehicles had been government owned and expendable, this signaled a change in the policy image to reflect a space development paradigm.

On August 13, 2018, Vice President Pence visited the Pentagon to hear their implementation plan for Space Force and provide them with more "hands-on" guidance. Pence's speech is notable for its cultural references to the frontier, liberty, its connection between the Space Force and prosperity, and the Space Force as a vehicle to carry American commitment to freedom, private property, and the rule of law into a new frontier.

> As the President will discuss in further detail in the days ahead, the United States Space Force will strengthen our security, *it will ensure our prosperity*, and it will also *carry American ideals into the boundless expanse of space*. While other nations increasingly possess the capability to operate in space, *not all of them share our commitment to freedom, to private property, and the rule of law*. So as we continue to carry American leadership in space, so also will we *carry America's commitment to freedom into this new frontier*.[65]

Pence asserted space had become a warfighting domain, and "the time has come to establish the United States Space Force,"[66] but he framed it in a commercial context. He stated, "President Trump stated clearly and forcefully that space is, in his words, 'a warfighting domain, just like . . . land, [and] air, and sea'" and provided firm guidance "the President made it clear that our ultimate objective is *to create a new branch of our military that is separate from, and equal to, five other branches*."[67] But the context was not just military threats, it was prosperity, expanding commercial interests, way of life, and liberty on the new frontier. Pence reiterated the administration had revived the National Space Council, signed three new space policy directives to "reorient our space program toward human exploration, *unleash America's burgeoning commercial space companies, and safeguard our vital space assets* with a new space traffic management policy" and "since day one of our administration, this President has kept his promise to restore America's proud legacy of leadership in space, believing that space is essential to our nation's security, *prosperity, and our very way of life*."[68] Hinting even at a future need to protect human life and freedom (perhaps in the context of space settlement) in the space domain, he stated, the United States "will meet the emerging threats on this new battlefield with American ingenuity and strength to defend our nation, protect our people, and *carry the cause of liberty and peace* into the *next great American frontier*."

That same day, public support for a Space Force would come not from the Department of Defense—which had yet to embrace a space development or space expansionist paradigm—but instead from NASA Administrator Bridenstine.[69]

On February 19, 2019, the Trump administration released SPD-4 that provided direction for establishing the Space Force. Important to note was its linking of way of life, economy, lawful use, and in particular the priority of, "ensuring unfettered *use of space* for United States national security purposes, *the United States economy, and United States persons,* partners, and allies."[70] The inclusion of persons is interesting, as it anticipates presence by U.S. persons in the space domain. The broader context is repeated below.

> *Space is integral to our way of life, our national security*, and modern warfare. Although United States space systems have historically maintained a technological advantage over those of our potential adversaries, those potential *adversaries are now advancing their space capabilities and actively developing ways to deny our use of space in a crisis or conflict.* It is imperative that the United States adapt its national security organizations, policies, doctrine, and capabilities to deter aggression and protect our interests . . . Under this proposal, the United States Space Force would be authorized to organize, train, and equip military space forces of the United States to *ensure unfettered access to, and freedom to operate in, space,* and to provide vital capabilities to joint and coalition forces in peacetime and across the spectrum of conflict . . . The United States Space Force shall be organized, trained, and equipped to meet the following priorities: (a) *Protecting the Nation's interests in space and the peaceful use of space for all responsible actors, consistent with applicable law, including international law*; (b) *Ensuring unfettered use of space* for United States national security purposes, *the United States economy, and United States persons, partners, and allies*;(c) Deterring aggression and defending the Nation, United States allies, and United States interests from hostile acts in and from space; (d) Ensuring that needed space capabilities are integrated and available to all United States Combatant Commands; (e) *Projecting military power in, from, and to space in support of our Nation's interests*; and (f) Developing, maintaining, and improving a community of professionals focused on the national security demands of the space domain.[71]

Those economic interests would take center stage again when on February 28, 2019, the U.S. Commerce Secretary Wilbur Ross wrote an op-ed in the *Orlando Sentinel* titled "Launching toward a $1 trillion space economy."[72] Ross highlighted the ongoing innovation,

> Thanks to the combined innovation and ingenuity of private companies and NASA, the United States' space industry is transitioning from public sector dependence to private sector dominance. The global space economy is nearing $400 billion. And commercial enterprise now accounts for the majority of space activity.[73]

He went on to give voice and legitimacy to industry forecasts stating,

The *Trump Administration is continuing to advance policies that will eliminate regulatory barriers to space commerce and ignite steady economic growth in the industry. Morgan Stanley projects the global space economy will be $1.1 trillion by 2040. Bank of America's forecast is even more bullish, projecting the industry could reach $3 trillion.*[74]

This clearly shows that the administration was anticipating significant new interests. Ross also called upon financial interests to participate in U.S. ambitions "if America is going to return to the moon and get to Mars—and fully realize the unprecedented business ideas in between—we will need strong participation from all levels of the capital structure."[75]

On March 26, 2019, the National Space Council approved several recommendations for President Trump, including a consequential recommendation that NASA should focus its efforts on the lunar South Pole and include resource utilization.

Recommendations Approved by the National Space Council to President Trump.

Recommendations on Human Space Exploration

1. Consistent with the overall goals of SPD-1, the *United States will seek to land Americans on the Moon's South Pole by 2024*, establish a sustainable human presence on the Moon by 2028, and chart a future path for human Mars exploration. NASA's lunar presence will focus on science, *resource utilization*, and risk reduction for future missions to Mars.[76]

Vice President Pence would address the fifth National Space Council at NASA Marshall and establish as U.S. policy the lunar South Pole as the destination and lunar mining as a focus.

> *To reach the Moon in the next five years, we must select our destinations now. NASA already knows that the lunar South Pole holds great scientific, economic, and strategic value.* But now it's time to commit to go there. And today, the National Space Council will recommend that when the first American astronauts return to the lunar surface, that they will take their first steps on the Moon's South Pole . . . And in this century, we're going back to the Moon with new ambitions, *not just to travel there*, not just to develop technologies there, *but also to mine oxygen from lunar rocks that will refuel our ships; to use nuclear power to extract water from the permanently shadowed craters of the South Pole*; and to fly on a new generation of spacecraft that will enable us to *reach Mars not in years but in months.*[77]

The next month, Secretary Ross would again iterate the administration's anticipated future economic interests in space. In April 2019, Gazzette reported that Secretary of Commerce Wilbur Ross,

> expects the *commercial space industry* to grow in value from $24 billion today to $3 trillion during the next 20 years as U.S. companies continue to develop innovative products and service, including *space tourism*. He said department leaders plan to meet soon with financial industry executives to find ways to finance innovation while reducing the risk of funding such projects.[78]

These remarks suggest the salience of the topic among industry executives. He would also express concerns about international competition as well as international regulatory uncertainty. He stated he was "committed to making sure the United States remains the flag of choice for innovative space companies" but that he feared moon landings by multiple countries could trigger a 'Wild West' scenario with few rules to keep key discoveries from being stolen."[79] The Daily Beast reported the same event as "Wilbur Ross: We've Gotta *Get Back to the Moon—to Exploit the Hell Out of It*" quoting him as saying "As more countries land on the moon, we risk *a Wild West situation without clarification of ownership rights*," and "We must *make sure the United States captures the lion's share of burgeoning space markets*."[80] Ross was clearly contextualizing the importance of the U.S. lunar program not in the context of inspiration or exploration, but in the context of resource exploitation for economic advantage.

In June of 2019, NASA Administrator Bridenstine provided similar framing, justifying the lunar program on both resource grounds and as a pathfinder for future commercialization. Bridenstine stated, "We learned in 2009 that there's *hundreds of millions of tons of water ice at the south pole of the moon*. Water ice represents water to drink. It also is hydrogen and oxygen, which is rocket fuel. It's the same rocket fuel that powered the space shuttles."[81] He then discussed NASA's efforts to commercialize Low Earth Orbit (LEO), build markets, capital, and commercial architectures with the Moon:

> then we can take all of these *resources* and we can *take them to the moon for the development of a sustainable lunar architecture, and in fact, a lunar outpost with the intent, again, to drive down costs, to commercialize, to prove the market, and then ultimately let commercial industry take over there*. Then we can take the resources, again, to Mars . . . What NASA does is, we go first, we prove technology, we *retire risk, and then we allow commercial industry to follow*.[82]

In mid-July of 2019, Breaking Defense outlined new concerns about the nexus of commercial satellite vulnerability and roles and missions for the military. Defense Secretary nominee Mark Esper told the Senate during his confirmation hearing, "we anticipate that *adversary nations are unlikely to discriminate between U.S. military satellites and commercial satellites* providing services to the U.S. Government, in the event of a conflict."[83] The same week, Shawn Barnes, Air Force deputy principal assistant for space reflected the policy concern, "does the Defense Department have the responsibility to protect commercial space, if that commercial space becomes part of our national [security] infrastructure?"[84] and provided his own thinking:

> Initially I think we would want to focus on 'where are those commercial capabilities providing a clear, distinct national security activity'? . . . Then I think if you look larger, you start to think about what does the nation and our economic vitality depend on? And if there is a dependence on space capabilities, is there then a responsibility for the Department of Defense or Homeland Security or others to provide some element of security and safety?[85]

The article highlighted the disconnect between the development paradigm of the administration and their clear policy statements (expressed above), and the thinking of the Department of Defense, which had yet to embrace the new policy image. The article also showcased the distance between the thinking of the academic military intelligentsia discussed in the previous chapter (the "blue water school of spacepower") and the Department of Defense leadership. The author of the article, Theresa Hitchens noted that:

> *There is a cadre of future-minded space proponents within the Air Force and national security space circles that long have championed the concept that 'flag follows trade' and the concomitant idea of a Coast Guard for space. Strong advocates of a separate Space Force*, these officers and former government *officials are particularly interested in the role of the military in cislunar space as commercial firms seek to exploit on-orbit resources such as water ice and minerals on asteroids and the Moon.*[86]

But gave her judgment that "Air Force and DoD leadership are nowhere near that advanced in their planning."[87]

In late July of 2019, Bridenstine further expanded on the economic rationale for space development, employing a new term, "industrialization." Bridenstine stated "we want to see commercialization. I love the word '*industrialization*,'" and that "industrialization builds nations, and *if we want to build space, we need to industrialize*."[88] He provided this ultimate justification,

> We want to prove capabilities and prove technologies, prove markets, that ultimately enable us to do more in space and have *more people in space, more*

> *projects in space, commercialize space, industrialize space and, of course, have this be a major driver for the United States of America economically* when it comes to the *balance of payments* and exports. *That's the ultimate objective* of the United States of America leading in space.[89]

He stated that NASA is *building upon its commercial cargo and crew programs* and that "the next step after that, of course, is to make sure we are *commercializing habitation* in low Earth orbit. In order to get there, we have to *have the industrialization take place*."[90] Bridenstine asserted a mandate, "we have a mandate to commercialize low Earth orbit, to industrialize low Earth orbit, to the point where NASA becomes a customer rather than the owner and operator."[91]

In August 2019, the United States reestablished the U.S. Space Command,[92] with the mission to "The USSPACECOM mission is to deter aggression and conflict, *defend U.S. and allied freedom of action*, deliver space combat power for the Joint/Combined force, and develop joint warfighters to *advance U.S. and allied interests in, from, and through the space domain*."[93]

In September 2019, Air Force Space Command released its report of its Space Futures Workshop titled, *The Future of Space 2060 and Implications for U.S. Strategy*, which clearly showed that space expansionist and space development thinking was beginning to penetrate the conservative Department of Defense. The document itself reflected the new policy image linking economic development with protective military presence, and an unabashed space expansionist agenda, proposing an ambition for the United States based on space resources and space settlement. It concluded:

> The *U.S. should establish space settlement and human presence as a primary driver of the nation's civil space program to determine the path for large-scale human space settlement* and ensure America is the foremost power in achieving that end. Accordingly, civil space programs must be assessed as to their utility to further space settlement goals. *The Departments of Commerce, Energy, and Transportation, along with NASA, must execute a coordinated strategy to develop U.S space commerce and to incorporate space commodities and resources into the larger U.S and global economy.* If entrepreneurs can achieve large returns from space commercial enterprises, the U.S. government must enact policies that ensure the U.S. captures the dominant position in that market. The U.S. must continue to lead in developing a rules-based, democratic international order for space. The U.S. *must commit to having a military force structure that can defend this international space order and defend American space interests, to include American space settlements and commerce.*[94]

And:

> The U.S. must recognize that *in the world of 2060, space will be a significant engine of national political, economic, and military power* for whichever nations or nation best recognize(s) the potential of space and organizes and operates to exploit and maximize that potential. The U.S. faces growing competition from allies, rivals, and adversaries to remaining the leading nation in the exploration and exploitation of space as an expanded domain for human endeavor. *China is executing a long-term civil, commercial, and military strategy for exploration and economic development of the cislunar domain, to include the settlement of the Moon*, with the explicit aim of displacing the U.S. as the leading space power. Other nations are developing similar national strategies. *A failure to remain the leading space power will place U.S. national power at risk*. The U.S. and its allies must promote and optimize the combined civil, military, and commercial exploitation of space that best serves the nation's interests . . . *The U.S. military must define and execute its role in promoting, exploiting, and defending the expanded commercial, civil, and military activities and human presence in space* driven by industry, NASA, and other nation-states.[95]

Air Force Space Command's *The Future of Space 2060 and Implications for U.S. Strategy* will likely be seen as a watershed document, where the strategic agenda of the new Space Force was first expressed in an official military document.[96]

On November 14, 2019, the U.S.-China Economic and Security Commission (USCC) released its annual report. The report was important because it was the first time that China's lunar industrial ambitions became a central part of public discourse outside of military academia. The USCC expressed concern that:

> Beijing has specific plans not merely to explore space, but to *industrially dominate the space within the moon's orbit of Earth*. China has invested significant resources in exploring the national security and economic value of this area, including *its potential for space-based manufacturing, resource extraction, and power generation*.[97]

Moreover, the USCC recommended Congress direct the National Space Council to develop a strategy to ensure the United States remains the pre-eminent space power in the face of growing competition from China and Russia including by addressing U.S. space resource and cislunar interests. The USCC recommended the National Space Council develop:

> *A long-term economic space resource policy strategy*, including an assessment of the viability of extraction of space-based precious minerals, onsite

> exploitation of space-based natural resources, and space-based solar power. It would also include a comparative assessment of China's programs related to these issues. *An assessment of U.S. strategic interests in or relating to cislunar space . . . Ensure U.S. Space Command and any future space-oriented service are responsible for protecting freedom of navigation* and keeping lines of communication open, safe, and secure in the space domain, as the U.S. Navy does for U.S. interests in the maritime commons.[98]

These recommendations showcase an ever-wider consensus of the perceived importance of space resource and cislunar space, and the need for the future Space Force to be responsible for protecting the freedom of navigation and lines of communication akin to the Navy.

On November 15, 2019, Dr. Mir Sadat of the National Security Council staff gave remarks at the Hudson Institute, again asserting the new paradigm and policy image, tying together economic projections, national interests, strategic threats, and the need of a space force to stand watch to ensure freedom of navigation.

> economic impact estimates project that the *U.S. space economy will grow between three- to eightfold over the next few decades in a approximately $2.7 trillion space economy*. American efforts in the commercialization of space are vital. *Let me explain to you what vital means—to our national interest because it reflects U.S. security investments in space*. The future space operating environment is expected to be exponentially more voluminous and diverse. Emerging commercial ventures such as satellite servicing, debris removal and *space manufacturing, tourism* and cutting-edge technologies enabling smallsat and expansive satellite constellations are outpacing efforts to develop and implement government policies and processes to address these new activities . . . *On the other hand, the Chinese are actively working to undermine any such efforts and have laid out a 30-year cislunar economic plan that will supposedly generate $10 trillion by 2050—three times the size of our estimated economy*. If true, we must create a collaborative, conscientious setting in which our nations—which other nations choose to align themselves and their space activities with us, the United States, as opposed to those without any regard to space—the space operating environment. We need to think about space before we lose space. Right now, we have the competitive advantage, but that advantage is rapidly eroding as peer and near-peer competitors ascend to space without any regard for space, as they are aided by advanced technologies which they might have stolen from others. *A recent Air Force futures report claimed that by 2060 space will be a significant engine of national, political, economic and military power and that the United States must commit to having a military force structure that can defend this international space order and defend American space interests, to include American space settlements and commerce. It will be the U.S. Space*

> *Force which will provide the necessary expertise for the U.S. Space Command to ensure unfettered access to and the freedom to operate within space.* U.S. Space Command will provide vital effects and capabilities to joint and coalition partners during peacetime and across the spectrum of conflict, *just as . . . the U.S. Navy stands to watch to ensure that we freely navigate the world's seas*, such as those in the Indo-Pacific. America must now ensure the freedom to navigate through space. If we continue to delay the creation of a U.S. Space Force until it is too late, we jeopardize the comforts of our American life, the lifeline of our economy, the enabling of our national security assets and the execution of priorities, especially in the Indo-Pacific and elsewhere around the world . . . let me conclude by drawing your attention to the start of my talk. While the Cold War is over, *great power competition is back, and the great game ensues for cislunar leadership . . . To win the great game in space*, we must ensure our economy, innovative technology, and national security are acquiring.[99]

On December 20, 2019, the Congress presented, and the president signed into law the creation of the U.S. Space Force,[100] with the functions to organize, train, and equip to provide "(1) freedom of operation for the United States in, from, and to space; and (2) prompt and sustained space operations" with specified duties to: "(1) *protect the interests of the United States in space*; (2) deter aggression in, from, and to space; and (3) conduct space operations."[101] This act was important because it elevated the status of military space, giving it equal voice to the Air Force, Navy, Army, and Marines on the Joint Chiefs of Staff, before the president, and before the Congress. It would create a constituency of interests, and make the Space Force a focal point for the protection of U.S. commercial ambitions in space.

In January 2020, Secretary of Commerce Wilbur Ross gave a speech, "A New Space Race: Getting to the Trillion-Dollar Space Economy World Economic Forum, Davos, Switzerland," where he outlined,

> Government activity in space is growing, but over 80 percent of the $415 billion space economy is commercial. *We believe the future of space is overwhelmingly commercial in nature, and will no longer be dominated by government agencies and their priorities. Trump Administration initiatives will encourage economic growth from space activities, and encourage like-minded nations to do the same.* To this end, President Trump reestablished the National Space Council in 2017 to coordinate national space policy and prioritize economic development and technological advancement . . . One assessment done by Bryce Space and Technology found that *the economic impact of space in 2018 had grown to $5 trillion annually . . . Current industry projections place the 2040 global space economy at between $1 and $3 trillion.* And I think we will certainly get to a

> trillion before 2030 . . . This June, the Departments of Commerce and State will again co-host the Space Enterprise Summit in Washington to discuss the nature of space partnerships in pursuing our mutual goals.[102]

At his February 2020 State of the Union Address, President Trump invoked a clear expansionist metaphor. The term used, manifest destiny, "a phrase coined in 1845, is the idea that the United States is destined—by God, its advocates believed—to expand its dominion and spread democracy and capitalism across the entire North American continent."[103]

> In reaffirming our heritage as a free nation, we must remember that America has always been a frontier nation. Now we must embrace the next frontier, America's manifest destiny in the stars. I am asking Congress to fully fund the Artemis program to ensure that the next man and the first woman on the Moon will be American astronauts—(applause)—using this as a launching pad to ensure that America is the first nation to plant its flag on Mars.[104]

In February 2020, President Trump amended the membership of the National Space Council to add the Assistant to the President for Economic Policy, Assistant to the President for Domestic Policy, and the Secretary of Energy.[105] This change reflects the increased importance of the economic component of space, and the desire to mobilize the national labs within the Department of Energy (DOE) to speed space industrialization. Statements by the Secretary of Energy, which mention development, industry, commerce clearly indicate a space development focus:

> Under President Trump's leadership, we will unlock the mysteries of space and *empower our commercial partners to see the untapped potential in space*. The Department of Energy, through our *nuclear R&D*, has long provided the capabilities to provide *fuel for spacecraft for deep space missions* . . . joining the National Space Council only amplifies our nation's commitment towards learning more about our universe, and *creating innovative technologies to assist in advancing development in space* . . . We must *recognize space as an industry of the future* that will lead to worldly discoveries in our lifetime.[106]

The reader will note the steady elevation of the importance of commerce and the proliferation of the space development paradigm since 2015, as well as its linkage in policy and law to a Space Force intended to protect emerging U.S. interests.

On April 6, 2020, President Trump issued an "Executive Order on Encouraging International Support for the Recovery and Use of Space Resources," which stated,

> Americans should have the right to engage in commercial exploration, recovery, and use of resources in outer space, consistent with applicable law. Outer space is a legally and physically unique domain of human activity, and the *United States does not view it as a global commons*. Accordingly, it *shall be the policy of the United States to encourage international support for the public and private recovery and use of resources in outer space, consistent with applicable law* [emphasis added][107]

On May 15, 2020, NASA made clear how it would be encouraged in its "Principles for a Safe, Peaceful, and Prosperous Future," where it laid out a requirement for international cooperation:

> With *numerous countries and private sector players* conducting missions and operations in *cislunar space*, it's critical to establish a common set of principles to govern the civil exploration and *use of outer space*.
>
> *International space agencies that join NASA in the Artemis program will do so by executing bilateral Artemis Accords agreements*, which will describe a shared vision for principles, grounded in the Outer Space Treaty of 1967, to create a safe and transparent environment that facilitates exploration, science, and commercial activities for all of humanity to enjoy [emphasis added].[108]

Cooperation would be to support for recovery and use of space resources:

> *Space Resources: The ability to extract and utilize resources on the Moon, Mars, and asteroids will be critical to support safe and sustainable space exploration and development*. The Artemis Accords reinforce that *space resource extraction and utilization can and will be conducted under the auspices of the Outer Space Treaty*, with specific emphasis on Articles II, VI, and XI [emphasis added].[109]

This long process of slow policy accretion seemed to culminate in the summer of 2020. Following new PRC-related threat assessments[110] and an open embrace of a competitive relationship with the PRC,[111] a new *Defense Space Strategy* was released which recognized the value of commercial space for economic opportunity but claimed "China and Russia each have weaponized space as a means to reduce U.S. and allied military effectiveness and challenge our freedom of operation in space."[112] Finally, on July 23, 2020, the White House released a groundbreaking space vision and strategy, *A New Era in Deep Space Exploration and Development*[113] which formally and for the first time broadened U.S. space strategy to include utilization of space resources and space settlement as part of an overall policy of space economic development. The White House document mentions space mining three times, and space resources more than thirty times. Stating that "this vision requires a secure international environment conducive to U.S. commercial growth," it maintained that the U.S. Space Force's responsibilities included ensuring unfettered access and that its "*activities such as space transportation and logistics, power, communication, navigation, and space domain awareness, are of dual-use value to all space sectors—civil, national security, and commercial.*"[114] The following week, the U.S. Space Force, Defense Innovation Unit, and Air Force Research Laboratory released *The State of the Space Industrial Base 2020*, which identified in-space power and space resources to be one of "the six areas most vital to over US national power in space, and the areas most likely to be at the center of gravity in great power competition" devoting an entire section to space resources.[115] It likewise called for an expanded role for the USSF, "protecting and enabling US commerce across cislunar space." The pursuit of space resources was now U.S. policy at the highest level.

U.S. EVIDENCE OF PROGRAMS

Commercial Interest

Clear strategic intent to exploit space resources in the United States began in the private sector with the formation of Planetary Resources in 2012, which "aims to be the leading provider of resources for people and products in space through its goal of identifying, extracting, and refining resources from near-Earth asteroids."[116] Its formation created a flurry of press—some seventy-eight stories in 2012 alone[117]—including "Tech Billionaires Plan Audacious Mission to Mine Asteroids,"[118] and "Google Chief Back Startup Mining Asteroids for Metals,"[119] and "Tech billionaires bankroll gold rush to mine asteroids," listing both Larry Page, Erick Schmidt of Google, and filmmaker James Cameron as backers as well as Space Adventures founder Eric Anderson and X-Prize founder Peter Diamandis.[120] It provided a powerful vision, "everything we hold of value on Earth—metals, minerals, energy, water, real-estate—are literally of near infinite quantities in space. Planetary Resources mission is to gain access to the natural resources of space by mining near earth approaching asteroids"[121] in its initial public videos and subsequent public engagement.[122] Planetary Resources was followed by the formation of Deep Space Industries (DSI) in January of 2013 with the vision, "Deep Space Industries believes the human race is ready to begin harvesting the resources of space both for their use in space and to increase the wealth and prosperity of the people of planet Earth," and a stated mission,

> DSI will build on the incredible heritage of the first age of space exploration and harness the power of a new age of information to locate, explore, harvest and utilize the vast numbers of asteroids in Earth's community. We will do so by being creative and practical—taking small steps to begin with, and giant leaps when we can—to supply our customers and provide a new and hopeful future for humanity.[123]

DSI stated, "our tiny planet sits in a vast sea of resources including millions of asteroids bathed in the Sun's free energy 24 hours a day. The same rocks that fall from our skies also contain everything we could need, both out there and in here. Its time someone seized the opportunity."[124] That, in turn, was followed by *TransAstra* in 2015,

> dedicated to building the 'transcontinental railroad of space' to open the solar system to humanity. Our vision is a future where some humans are homesteading the solar system living off the resources of the asteroids. Our mission is to turn thousands of asteroids into resources for refueling stations for NASA and commercial spacecraft. TransAstra will lead the way in supplying the 21st century industries of asteroid mining, space solar power, space tourism, space data processing industries, and manufacturing in space.[125]

A major victory for the private space community was the passage[126] of the U.S. CSLCA of 2015 that enabled them under U.S. law to "possess, own, transport, use, and sell the asteroid resource or space resource."[127] The community further succeeded in gaining international legitimacy by triggering similar interests and policies from Luxembourg and the UAE (which will be covered in subsequent chapters).

Space resources became a part of mainstream aerospace discourse when United Launch Alliance (ULA) authored its cislunar 1000 vision in 2016,[128] with a visionary video where CEO Tory Bruno begins by stating, "We now know that water is everywhere, that's huge! Not just because we can sustain life in space, but make propellant with it."[129] ULA envisioned the space economy would grow over the next thirty years (~2046) to $2.7T in annual revenues and over 1,000 humans living and working in space, and provided a shaping architecture of LEO space tourism, on-orbit manufacture, lunar and asteroid mining, and solar power satellites, all based upon in-space refueling using lunar propellant. Space.com reported that,

> As a customer, ULA is willing to pay about $1,360 per lb. ($3,000 per kilogram) for propellant in low Earth orbit. The going rate for fuel on the surface of the moon is $225 per lb. ($500 per kg), Sowers said. In talking with asteroid-mining experts, ULA would take delivery of propellant at L1 for $450 per lb. ($1,000 per kg), he said.[130]

Such statements of interest by industry were followed by action in the education establishment to create a pipeline of expertise. In 2017, the Colorado School of Mines began its "Mines Space Resources Graduate Program." This is significant because The Colorado School of Mines "has a world-renowned presence, including remote sensing, geomechanics, mining, materials / metallurgy, robotics / automation, advanced manufacturing, electrochemistry, resource economics and solar and nuclear energy," and "Colorado has the second-largest aerospace economy in the nation. This program is a great way for Mines to have a presence in aerospace."[131]

Various Lunar companies were also active in this period. The Google Lunar X-Prize (2007–2018) offered $20 million for the first privately funded team to land a robotic spacecraft on the Moon, travel 500 meters and transmit high definition video and images back to Earth. The effort galvanized significant global interest, with twenty-nine registered teams, of which sixteen were from the United States (Advaeros, Astrobotic, JURBAN, LunaTrex, Micro-Space, Moon Express, Mystical Moon, Next Giant Leap, Odyssey Moon, Omega Envoy, Penn State Lunar Lion Team, Quantum 3, Rocket City Space Pioneers, SCSG, STELLAR, Team Phoenicia).[132] Most did not survive.

Some companies had lunar ambitions beyond just winning the prize, or were not chasing the prize at all. Shackleton Energy Company (2007–present) hoped to prospect the Moon and establish a network of refueling stations in LEO to process and provide fuel and consumables for commercial and government customers, boldly stating,

> We Are Going Back to the Moon to Get Water. There are billions of tons of water ice on the poles of the Moon. We are going to extract it, turn it into rocket fuel and create fuel stations in Earth's orbit. Just like on Earth you won't get far on a single tank of gas, what we can do in space today is straight-jacketed by how much fuel we can bring along from the Earth's surface. Our fuel stations will change how we do business in space and jump-start a multi-trillion-dollar industry. Much like gold opened the West, lunar water will open space like never before.[133]

Another company, The Golden Spike Company (2010–2013) hoped to create a reliable "Cislunar Superhighway" providing private commercial transportation to the surface of the Moon and planned to offer commercial passenger flights to the Moon by 2020 for $1.5Bn each.[134]

Others that survive include Masten Space Systems (2004–Present),[135] Astrobotic (2008–Present),[136] and Moon Express (2010–Present),[137] which have developed small landers. Moon Express' ambitions clearly echo space development "Our mission is to redefine possible by returning to the Moon and unlocking its mysteries and resources for the benefit of humanity" and explain:

> Why the Moon? The *Moon is Earth's 8th continent*, a new frontier for humanity with precious resources that can bring enormous benefits to life on Earth and our future in space. Expanding Earth's economic and social sphere to the Moon is our first step in securing our future. Not long from now a new generation will look up and see lights on the Moon, and know that they are *part of a multi-world species* . . . The Moon is a new world with a total landmass approximating North and South America combined.
>
> Like the Earth, the *Moon has been enriched with vast resources through billions of years of asteroid bombardment.* Unlike the Earth, these resources are largely on or near the lunar surface, relatively accessible. We are blazing a trail to the Moon to seek and harvest these resources to support a new space renaissance, where economic trade between countries will eventually become trade between worlds . . . One of the greatest practical space discoveries of our generation is the presence *of vast quantities of water on the Moon, verified by NASA in 2009. Water (H2O) not only supports life but its constituents hydrogen and oxygen can be used as rocket fue*l. The *discovery of water on the Moon is a game changer, not just for the economic viability of lunar resources, but for the economics of reaching Mars and other deep space* destinations. *Water is the oil of the solar system, and the Moon will become a gas station in the sky.*[138]

While the initial space mining industry entrants saw an initial boom and then bust—PRI and DSI both being acquired and having to scale back ambitions,[139] and other prominent entrants falling silent—DSI continues to build water-based thrusters and PRI continues to build survey capable spacecraft.

However, the broader ecosystem—in particular, the transportation ecosystem—through cislunar space is now expanding, in part due to a belated recognition by the U.S. government of the importance of the Moon due to interest by the PRC, and in part due to significant private capital.

New entrants are backed by significant capital; *SpaceX* with its plans for starship that could service a Moon Base (with a potential landing between 2021 and 2023),[140] and *Blue Origin*'s Blue Moon lander.[141] Momentus Space hopes to provide transport services throughout the cislunar system with water-based propellant, stating, "We envision a future where humanity is expanding, thriving, and moving freely throughout the solar system. To succeed, we are building transportation technology with sustainable, safe, and scalable in situ resource utilization in mind."[142] OrbitFab "are working to establish the first gas station in space for satellite refueling" to enable "a thriving in-space market for products and services that support both existing space businesses (communications and Earth observation) and new industries like space tourism, manufacturing, and mining."[143] The Cislunar Space Development Company hopes to "provide commercial transportation from Low Earth Orbit to the Moon's surface and all points in between."[144] OffWorld, founded in 2016, is building a robotic platform that would enable off-world mining operations; the company "envision millions of smart robots working under human supervision on and offworld, turning the inner solar system into a better, gentler, greener place for life and civilization," stating, "if our offworld energy infrastructure could be built using offworld resources, we can dramatically lower the growing industrial burden on our home planet."[145]

A significant dialogue now exists on how to create a Space Commodities Exchange,[146] and Space Commodities Futures Trading Exchange.[147],[148] One Company, SCX, already exists attempting to build it, stating,

> Space is filled with the resources that we will need to sustain a growing, spacefaring civilization. Water, minerals, and other valuable commodities found on earth, are also found and in space; far exceeding those available on Earth. The developing economy of space resources will be driven by the extraction, processing, transportation and utilization of these commodities. A robust space economy will require a regulated public market where buyers and sellers can meet and transact business for space commodities and services using standardized commercial contracts. Lunar water miners and asteroid prospectors will need to sell their raw materials to space manufacturers and other consumers. The Space Commodities Exchange will provide a regulated exchange for transactions and orbiting facilities for deliveries to take place.[149]

Although none have succeeded in sufficient capitalization, there have also been no shortage of U.S., or partial-U.S. startups who expressed intent to exploit space solar power. These included, Space Island Group,[150] Heliosat,[151] PowerSat,[152] SolarHigh,[153] PlanetaryPower,[154] SpaceEnergy,[155] Mankins Space Technology, NGC[156]-Caltech,[157] and Solaren, of which only Mankins,[158] Caltech, and Solaren appear to be currently active. However, at the unveiling of the Blue Moon lander, Jeff Bezos hinted at his own interest when he emphasized the need for energy, and showcased an article where he was interviewed in his youth where he stated, "as a 'Space entrepreneur,' he would 'construct solar power satellites that would make the world peaceful and affluent through abundant, cheap energy.'"[159]

U.S. Civilian Space Program

The U.S. civilian space program primarily continues in a split between the Saganite paradigm pushing for more unmanned probes and the Von Braunian paradigm, championing prestige-seeking firsts for human spaceflight.[160] Both primarily use exploration, science, and curiosity as their major driver. It is certainly the case that since its inception, NASA has funded numerous internal studies, minor technology development projects, and minor centennial challenges. For example, in 2005, NASA announced several challenges or innovation prizes of interest to the space development community,[161] including for Power Beaming, Strong Tethers for a space elevator, and Lunar Oxygen Production or MoonROx Challenge (which expired on June of 2009).[162] NASA under President Obama developed peripheral capabilities that supported asteroid-mining, including the planning of an Asteroid Redirect Mission[163], stand-up[164] and funding increases[165] for the Planetary Defense Coordination Office, and the first-ever "National Near-Earth Object Preparedness Strategy"[166] (since updated by the Trump administration as the "National Near-Earth Object Preparedness Strategy and Action Plan").[167] The U.S. civilian space program began to evidence programmatic intent regarding space resources only in March 2019 when, in response to the Chinese lunar efforts, the National Space Council recommended,

> Consistent with the overall goals of SPD-1, the United States will *seek to land Americans on the Moon's South Pole by 2024*, establish a *sustainable human presence on the Moon by 2028*, and chart a future path for human Mars exploration. NASA's lunar presence will focus on science, *resource utilization*, and risk reduction for future missions to Mars.
>
> NASA will *unleash American industry, including public-private partnerships* and other mechanisms, to enhance innovation and sustainability of activities from low Earth orbit *to the lunar surface and beyond*."[168]

Vice President Pence then provided strong guidance to NASA. His speech emphasized "use," mentioned water resources, referencing China, stressing urgency, and displayed a candid willingness to use commercial rockets.

> With Space Policy Directive-1, President Trump finally gave NASA the clear direction and clear mission that it needs. And as President Trump said, we will return "American astronauts to the Moon for the first time since 1972 for long-term exploration *and use*."
>
> Now, make no mistake about it: *We're in a space race today*, just as we were in the 1960s, and the *stakes are even higher*. Last December, *China became the first nation to land on the far side of the Moon and revealed their ambition to seize the lunar strategic high ground and become the world's preeminent spacefaring nation* . . . We refused to accept a future in space written by the enemies of freedom.
>
> And I'm here, on the President's behalf, to tell the men and women of the Marshall Space Flight Center and the American people that, at the direction of

> the President of the United States, it is the stated policy of this administration and the United States of America to *return American astronauts to the Moon within the next five years.*
>
> In order to succeed, as the Administrator will discuss today, we must focus on the mission over the means. *You must consider every available option and platform* to meet our goals, including industry, government, and the entire American space enterprise . . . But to be clear, *we're not committed to any one contractor. If our current contractors can't meet this objective, then we'll find ones that will.* If American industry can provide critical commercial services without government development, then we'll buy them. And if commercial rockets are the only way to get American astronauts to the Moon in the next five years, *then commercial rockets it will be.*
>
> As we continue to push farther into our solar system, we'll need innovative new propulsion systems to get us there, including nuclear power.
>
> The United States must remain first in space, in this century as in the last, not just to propel our economy and secure our nation, but above all because the rules and values of space, like every great frontier, will be written by those who have the courage to get there first and the commitment to stay. And as Americans, and as heirs of this great nation dedicated to life, to liberty, and the pursuit of happiness, it's nothing less than our duty to *ensure that our most cherished values are the foundation of mankind's future in space.*
>
> To reach the Moon in the next five years, we must select our destinations now. NASA already knows that the lunar South Pole holds great scientific, economic, and strategic value. *But now it's time to commit to go there. And today, the National Space Council will recommend that when the first American astronauts return to the lunar surface, that they will take their first steps on the Moon's South Pole.*[169]
>
> And in this century, we're going back to the Moon with new ambitions, not just to travel there, not just to develop technologies there, *but also to mine oxygen from lunar rocks that will refuel our ships; to use nuclear power to extract water from the permanently shadowed craters of the South Pole*; and to fly on a new generation of spacecraft that will enable us to reach Mars not in years but in months.[170]

NASA has named the mission and its associated architecture after the Greek god and sister of *Apollo*, *Artemis*.[171] *Artemis* consists of the Space Launch System (SLS)[172], the Orion Space Capsule,[173] the Lunar Gateway,[174] and a yet-to-be named Lunar Lander. The SLS is a super-heavy launcher, capable of launching 26 metric tons to the Moon in its initial increment (SLS continues the questionable practice[175] of pursuing space launch through a government-owned expendable rocket at an estimated cost of $2Bn per launch[176]). The Lunar Gateway is a mini-space station to be in the vicinity of Earth's Moon enabling access to a greater range of the surface. The Orion capsule provides life support for four astronauts for several weeks and safe reentry.

At this point, in terms of major investments, NASA has largely re-spun its previous exploration program to attempt to accommodate the administration's direction, but it is still significant that it has been partially recontextualized for the first time in terms of resources to "*find and use water and other critical resources* needed for long-term exploration," and "establish American leadership and a *strategic presence on the Moon* while expanding our *U.S. global economic impact*."[177]

Public-Private Partnerships

The most significant aspect of *Artemis* is the associated public-private partnerships, which "Ahead of the human return, we will send a suite of science instruments and technology demonstrations to the lunar surface through commercial Moon deliveries."

The most significant programmatic public-private partnership is the NASA Commercial Lunar Payload Service (CLPS) program.[178] CLPS program is "Indefinite Delivery, Indefinite Quantity (IDIQ)" contract funded at $2.3Bn. In November 2018, NASA selected nine companies for the program, these included Astrobotics, Deep Space Systems, Draper, Firefly Aerospace, Intuitive Machines, Lockheed Martin, Masten Space Systems, Moon Express, and Orbit Beyond.[179] In May 2019, NASA awarded over $250 million in contracts, selecting the first three companies, "Astrobotic, Intuitive Machines and OrbitBeyond to carry up to 23 payloads to the moon on three commercial lunar lander missions scheduled for launch between September 2020 and July 2021."[180] NASA awarded OrbitBeyond $97 million for carrying four payloads on a lander that intends to touch down on Mare Imbrium (a mare is a plain formed by volcanic eruptions), while Astrobotic got $79.5 million for fourteen payloads to the crater Lacus Mortis, and Intuitive Machines $77 million for four payloads to Oceanus Procellarum or Mare Serenitatis.[181] *Orbit Beyond* later dropped out due to internal corporate challenges.[182] NASA stated, "The CLPS program continues to formulate additional requests for task order proposals to expand the scope of NASA payloads requiring transportation services to the lunar surface in advance of human return. CLPS remains strong and the project includes diverse partners helping NASA to get to the lunar surface quickly and efficiently."[183]

More recently, NASA announced nineteen new public-private partnerships (termed Space Act Agreements) with thirteen companies to advance Moon, Mars technology, including helping mature Blue Origin's Blue Moon lander to survive the lunar night with a fuel cell,[184] and mature SpaceX starship[185] Moon landing and orbital refueling tech.[186] Those are significant steps.

Why Public-Private Partnerships?

This is an approach that has long been championed by the epistemic community and its policy entrepreneurs, which militated for NASA COTS and follow-on Commercial Crew Development.[187] The NASA COTS program (because it delivered a capability at nearly 1/10th the cost of a cost-plus contract[188]) would be held up as a model[189] in multiple policy proposals relating to developing space resources, including a NASA Ames proposal for Lunar COTS (2015),[190] the Evolvable Lunar Architecture (2015),[191] the Air University FAST SPACE study (2017),[192] an Economic Analysis of Space Transportation Supplied from NEO Resources (2017),[193] and a Collaborative Lunar Propellant Architecture Study (2019).[194] These made strong claims about lower costs through public-private partnership. For instance, the NASA COTS study suggested that "Analysis of one part of the cargo & crew system, the Falcon 9, indicates the development cost was 4 to 11X times less than a traditional cost-plus acquisition."[195] The FAST SPACE study found, "The development of multiple competing fully-reusable launch vehicles could lower prices by a factor of 3X in the near-term,"[196] and that "in the Long-Term, a successful virtuous cycle can enable a launch price reduction of 10X"[197]:

> New markets and applications develop based on the availability of 3X lower launch costs, this will increase flight rates, and the availability and reliability of launch services. This will increase investor confidence, driving investments in the next generation of RLVs, with increased reliability, robustness, and operability, lowering costs and increasing flight rates even further. This cycle could achieve a 10X reduction in launch costs.[198]

These in turn would enable "new supporting space services such as propellant resupply, extraterrestrial resource extraction, on-orbit construction and assembly, and satellite servicing are now attracting significant private investment."[199] The Evolvable Lunar Architecture study stated,

> America could lead a return of humans to the surface of the Moon within a period of 5–7 years from authority to proceed at an estimated total cost of about $10 Billion (+/– 30%) for two independent and competing commercial service providers, or about $5 Billion for each provider, using partnership methods . . . America could lead the development of a permanent industrial base on the Moon of 4 private-sector astronauts in about 10–12 years after setting foot on the Moon that could provide 200 MT of propellant per year in lunar orbit for NASA for a total cost of about $40 Billion (+/– 30%).[200]

The Lunar Propellant Architecture Study "identified a near term annual demand of 450 metric tons of lunar derived propellant equating to 2,450 metric tons of processed lunar water generating \$2.4Bn of revenue annually," and that "To achieve production demand with this method, 2.8 megawatts of power is required (2 megawatts electrical and 0.8 megawatts thermal)." The Stepping Stones study stated that "NASA Business As Usual, Supply From Earth" to enable Moon-Mars mission goals over 20 years would cost \$400B and was "Unaffordable: requires doubling the NASA Budget"; however, using both "Commercial Business Practices, Asteroid Resources," the cost would be \$95B (a savings of \$305B) and would be "Affordable, Average NASA Human Exploration Budget ≈\$6B Annually, Fits Within Congressional Guidelines for HEOMD" and that, "analysis further suggests that with relatively modest initial government investment, a business case can be developed for a profitable industry in space resources."[201]

Perhaps the most ambitious proposal for a public-private partnership drew headlines: "A general, Gingrich and Michael Jackson's publicist are proposing a \$2Bn contest to return Americans to the moon"[202] and "Gingrich, Kwast, Walker and Autry: A \$2Bn prize will bring US permanent moon and Mars settlements"; the authors proposed reducing the risk of a NASA-only return to the Moon by setting up a Moon-Mars development prize commission with a \$2Bn purse.[203] The suggestion drew praise from Elon Musk of SpaceX.[204]

These statements show how increased capability in the private sector to provide cost-effective space launch makes the idea of major public-private partnerships appear increasingly viable.

Military

At the time of writing, the U.S. military had barely begun to show evidence of serious programmatic interest in space resources, either energy or materiel. There are no major system purchases or articulated formal requirements. What exists today are conceptual discussions, low-level development efforts, and a potential to use similar public-private partnerships [termed Other Transaction Authorities ("OTAs") in the Department of Defense vs SpaceAct agreements for NASA]. For example, Air Force Space Command's Space and Missile Center (SMC) [now the U.S. Space Force Space Materiel Command] was exploring an Advanced Space Testbed that might mature technologies for in-space assembly and manufacture, which considered both space logistics and space power.[205] A budget justification for the former Operationally Responsive Space Office had requested funds to "develop space-based solar power collection and transmission capability using light weight, high efficiency solar cells coupled with individual radio frequency transmitters to collect solar energy and provide uninterrupted, assured, and logistically agile power to expeditionary forces operating in unimproved areas such as forward operating bases,"[206] but was declined by the U.S. Congress as "early to need, but without prejudice"[207] in favor of a scaled-back effort at Air Force Research lab,[208] and no funds were allocated toward it in FY2020. The OSD Operational Energy Capability Improvement Fund (OECIF) which had funded a recent DoD study *Opportunities and Challenges for Space Solar for Remote Installations*[209] assessed that "*near-peer competitors are overmatching DoD in power beaming and space solar technologies* for battlefield advantage, necessitating a critical OECIF [Operational Energy Capability Improvement Fund] effort to address the shortfalls,"[210] and recommended measured, comprehensive investments in Advanced Technology Development that will be made in six focus areas: "(1) Space Solar Collection; (2) Power Beaming Transmission; (3) Power Beaming Reception; (4) Receiver Power Distribution; (5) Architecture Analytics; and (6) Integrating Technologies"[211] and requested $40M for fiscal year 2020 to "advance Space Solar Collection, Power Beaming Transmission, Power Beaming Reception, Receiver Power Distribution, Architecture Analytics, and Supporting Technologies."[212] However, as of June 2019, funding in the FY 20 federal budget for technologies for space solar was at risk, leading proponents to urge action lest the U.S. further cede leadership.

The newer acquisition arms appear more aggressive. The Defense Innovation Unit (DIU), "a fast-moving government entity that provides recurring revenue to companies to solve national security problems" appears interested in cislunar development,[213] and recently drew headlines "Pentagon Eyes Military Space Station,"[214] when it solicited solution briefs for an orbital outpost to support logistics and storage, manufacturing, in-space assembly.[215] DIU is reportedly also considering partnerships to advance cislunar access. Similarly, *Breaking Defense* reported that Fred Kennedy, the first director of the newly created Space Development Agency (SDA)

> is boldly advocating to take routine U.S. military space operations where they have never gone before: to cislunar space. "Defense follows where commercial goes," he told me in a phone interview today, echoing the old aphorism of '*flag follows trade.' As commercial activities (such as resource extraction) expand outward from Earth to the Moon, he explained, there will be "a need for the equivalent to a Navy or Air Force" to protect that region of space.* "We haven't specifically been told to go worry from GEO to the Moon" he said, "but we know emerging threats will drive us there . . . and sought "optical sensors 'looking out' toward the Moon to 'provide real-time custody of objects' in cislunar space" as well as looking to develop a "minimum of three so-called 'Advanced Maneuvering Vehicles (AMVs)' the notional architecture envisions for providing the U.S. military with *access to and rapid transit between the Earth and the Moon* in order to 'deliver effects.'"[216]

Given the discourse and plans of the PRC, which will be discussed in a subsequent chapter, U.S. Department of Defense interest is only likely to increase, as evidenced by recent Air Force Space Command "Space Futures" industry event hosted by its Chief Scientist, Dr. Joel Mozer. While the event was part of the fact-finding for the command's future visioning,[217] *The Future of Space 2060 and Implications for US Strategy*[218] (discussed above), the fact that it took place with industry signals programmatic intent. The list of topics explored suggests the first articulation of a "shopping" list for the new U.S. Space Force in the decades ahead.

> Space Logistics and Infrastructure (*base stations, propellant depots; satellite servicing, refueling*/"Maneuver without Regret," repairs, upgrades, *manufacturing*); *Lunar resources & logistics* (HE3, titanium, H2O, silicates, solar power, regolith-based construction, mass transport, etc.); Domain Sustainment (Space Traffic Management, debris removal, and mitigation); Exploitation of novel orbits & Lagrange points (e.g., space habitats at L4, L5, COMM, PNT, SDA, depot storage, etc.); New or enhanced markets/services from space (e.g., *space-based solar power, space tourism, space production*, commercial intelligence); *Asteroid-based resources* (iron, Rare Earth metals, H2O, silicates).[219]

And it appears we are already seeing its beginnings. Military interest in cislunar space development illustrates the process of elite discourse to policy to programs. Over time, military interest in cislunar has progressively become more programmatic. From early articulations by intelligence officials[220] and academics,[221] it became formalized in Air Force Space Command's 2060[222] vision and open talking points[223] and has now moved to programmatic efforts among various organizations: the SDA[224], AFWERX[225], and DIU[226] for space domain awareness and physical logistics, and AFRL for power beaming logistics.[227]

While the U.S. military programmatic investment related to space resources and space development is modest, most of it has occurred within the time it has taken to write this book. Therefore, we have every expectation that as military-led competitor states advance their industrial space programs, so too will the U.S. military to keep pace. If not driven internally by ambition, the new U.S. Space Force will be forced to meet the PRC's peacetime strategic initiative by the external U.S. political electorate who will demand primacy and protection of its celestial lines of commerce.

U.S. STRATEGIC CULTURE

All nations have a strategic culture, and the United States is no exception. It exhibits learned biases by which its elites perceive and interpret opportunities and threats.

Several aspects of U.S. strategic culture are relevant to the question of space resources. First, the United States exhibits a bias to perceive frontiers and "overseas" markets as opportunities. Second, a tendency to wish to transform the international order to reflect its own internal system of limited, representative government. Third, its strong preference for a stable, rules-based international order which limits territorial aggression, aggression against individual liberty, and closed trading system. Fourth, a tendency to perceive illiberal regimes as threats and an exceptionally strong bias to perceive constriction of trade as a vital threat. Fifth, a disinclination toward power politics or limited war necessary to maintain the system against illiberal powers in peacetime, which leads to cycles of disengagement or restraint followed by a discontinuity between normal politics and war.

America's romance for the frontier is strongly reflected in the myths they tell themselves in westerns and more recently space-westerns.[228] Seeing boundless opportunities of natural wealth[229] that can be transformed by individual effort, American strategic culture with respect to frontiers is future focused and optimistic. This mythology of the frontier biases America's strategic culture to interpret open-spaces as opportunities for wealth, along with both an individual ("Go west young man"[230]), and collective moral imperative (a "manifest destiny"[231]) to fill them.

America's preference to transform the international order[232] to more closely reflect its internal system of limited government has become a fundamental component of its grand strategy. Often seen as liberal idealism, this reflects "canny assumptions" about the kind of environment where America is likely to flourish.[233] America sees the normative order of world society as, and seeks to transform the world into a system of sovereign nations whose sovereignty which is held together by common (contractual) values and which is limited by higher values which respect human liberty.[234] Because liberals tend to see a threat to liberty anywhere as a threat to liberty everywhere,[235] America has an impulse to build a global system that limits the ability to constrain human liberty promotes representative and procedural justice. These values are not a luxury, but ensure the broader security of the American republic.[236]

The United States strongly prefers an international regime that is liberal in trade, and based on a social contract of specified duties rather than on hierarchy and power politics. With the growth of U.S. power after World War II, this preference has led to a grand strategy called "primacy" or "leadership," which has endured for over a century.[237] The strategy requires military preponderance, and seeks to contain and reassure U.S. allies using United States designed and led institutions and markets.[238] Broadly, a plethora of institutions[239] maintain a policy of economic liberalism[240] to preclude a return to closed economic empires, and provide representation to assure voice on a diversity of governance and security issues (United Nations). The strategy requires monitoring and enforcement of this order, and necessitates a global web of alliances and a global ability to project military force globally via airpower and seapower against those who would violate the order, and especially against territorial aggression as well as to prevent a roll back to a world of closed economic empires.[241]

The United States tends to perceive illiberal regimes as threats[242] and has an exceptionally strong bias to perceive constriction of trade as a vital threat. This is strongly conditioned by America's experience as a seapower state. Seapower states, with the "sea at the heart of their identity and economy . . . *are disproportionately engaged in international oceanic trade, maritime identities occupy prominent places in their culture and they will be first to respond when the peaceful use of the sea is threatened.*"[243] America's early experience with the Barbary states (detailed below)—when it was a weak and newly independent state—was a formative influence.

The United States has a disinclination toward power politics and limited war to maintain the system against illiberal powers in peacetime, which leads to cycles of disengagement and restraint followed by a discontinuity between normal politics and war. Because of the influences of insularity, "free security" and a sense of exceptionalism, the United States explicitly rejected the European tradition of power politics.[244] U.S. elites exhibit an aversion to limited war and a disinclination to use force to maintain parity or balance in power politics.[245] This leads to a view that "war is a deviation from the norm of peace."[246] The United States, perceiving itself as a principled nonaggressor nation, views war as a complete collapse of policy, making all prewar policy invalid, and to see politics and strategy as things fundamentally apart.[247]

America's tendencies to act "astrategically" can set up significant discontinuities in the international system due to its prolonged under-reaction. The U.S. discomfort with the requirements of a hegemon to engage in power politics, and its economic and reputational costs occasionally fatigues and nauseates the U.S. electorate, creating cycles of withdrawal where the United States seeks to "tend its own garden" in isolation and let the world be. While some express concern that current forces of nationalism, nativism,[248] protectionism, and anti-globalism[249] seem to be pushing the United States into a period of withdrawal from its traditionally aggressive support of the international system,[250] it is equally important to keep in mind Collin Gray's astute observation that "a United States *true to its indigenous political culture could swing from global guardian of order to self-defined virtuous (though well-armed) recluse* and still *remain firmly within the American tradition.*"[251] Other powers can misinterpret U.S. retrenchment, disengagement and restraint as license, only to find the United States will express military power should things decay sufficiently. An aggression that an autocracy might see as justifiable limited war to maintain a balance in power politics, U.S. elites are likely to interpret an armed rebellion against "universal and eternal principles of the world society."[252] Therefore, the use of force is likely to be understood or portrayed as a "crusade for values."[253] A war fought based on high moral principle demands not a limited response but total domination,[254] which can end only "*unconditional surrender* of the aggressor nation in the overthrow and transformation of its political regime."[255]

America's strategic culture is eerily reminiscent of the fictional western characters[256] with whom it identifies: the character of the rugged,[257] self-sufficient, honest-dealing cowboy making their own way, who reluctantly adopts an unwanted leadership position of defender or sheriff when cheating predators prey on honest folk. After attempts at reason or restraint, force is employed until there is submission and surrender. Having done the needful, the rugged individual wishes to tend to their own business, interacting with society only to uphold those necessary social duties for which they have given their word.

How, then, can we expect U.S. strategic culture will apply to space[258] and space resources? How will a strategic culture that values, primacy, frontiers, a liberal trading system, and a rules-based order perceive the opportunities of the space domain? How will a nation that is wary of illiberal regimes, perceives closed trade arrangements as threats, yet fails to use force early[259] to maintain balance of power, likely to behave, should it encounter such challenges in the space domain?

The elite debate cataloged in these chapters suggests an answer to this question. American elites have, and will perceive the space as a frontier and opportunity, and frame it in those terms. America will attempt to extend a rules-based international order to space, and to attempt to create an open trading regime free of closed economic empires and resource nationalism. Its desire for primacy, and decades of experience with alliances and global power projection will cause it to extend military primacy and shared security to space. Space is likely to, and there is ample evidence that space (and the Space Force) is already part of the American grand strategy[260] of primacy.[261] As detailed in the sections above, there is a strong current among the spacepower thinkers concerned with space resources who see spacepower analogous to seapower. Seapower states have an important disposition to protect their lines of commerce. By analogy, then, we can expect U.S. strategic elites to display a similar concern for the emerging celestial lines of communication (CLOCS) as they do for sea lines of communication (SLOCS).[262] Based upon what we observe about U.S. strategic culture and the ongoing discourse among elites, we can assert that U.S. elites are likely to perceive the resources of space as a strategic opportunity, and would look to extend the liberal order with allies and partners, as well as to protect that trade with military spacepower.[263] Similarly, these themes suggest that U.S. elites will perceive rival efforts by an illiberal power (such as China) to develop cislunar space, and certainly any attempt to set up an illiberal trading regime or pursue resource nationalism as a threat to its primacy. The United States is likely to show restraint and "play by the rules" rather than use force to maintain a balance of power or astrostrategic position, even if things are not unfolding in its favor, unless and until there is territorial aggression or a threat to United States or allied freedom trade or citizens, or perhaps an unconscionable human rights violation in some future space settlement. In such instance, the United States is not likely to interpret the situation as a limited war for limited aims, but more likely to interpret this as an attack of world society necessitating a significant mobilization and muscular response sufficient to curb the influence of the illiberal power.

U.S. STRATEGIC TRAUMA

Strategic culture is often conditioned by significant crisis events where a country learns formative lessons, conceptualized here as strategic traumas that condition a state's policymaking elites to be sensitive and reactive to particular stimulus. American history has conditioned it not to be dependent on the largess of foreign powers. A formative experience for the early United States was the demands of the Barbary states for tribute and safe passage.

As recounted by Ian Toll, a leading objective of America's postrevolutionary foreign policy was to secure access to new export markets.[264] Max Boot notes,

> Until 1776, American ships were protected by English tribute and the Royal Navy. As many as 100 American merchantmen made annual voyages to the Mediterranean, carrying salted fish, flour, lumber, sugar and other goods which they traded for lemons, oranges, figs, olive oil, and opium, among other valuable items. After the revolution, the enterprising merchants of New England tried to reestablish this lucrative trade but found it dangerous going.[265]

But unprotected U.S. merchant ships drew attention of predators. Toll recounts:

> Peace with Britain had removed the threat posed by the Royal Navy to American ships but it had left them without the umbrella of protection the Royal Navy had provided before 1776. *For the first time, the Stars and Stripes was seen on the high seas in foreign seaports—but the flag was seen flying only on richly laden and defenseless merchant vessels, never on ships of war.* Greedy eyes studied the ships of this new nation the way wolves study sheep. *The British let it be known that the Americans no longer enjoyed their protection. The wolves were*

> *hungry; the sheep were fat, numerous, and slow; and there was not a shepherd in sight.*[266]

The Barbary states, to finance their governments, would routinely declare war on a European state, and then send their naval or privateer vessels to seize the "enemy" shipping.[267] It was apparently a lucrative business, with captured being auctioned off to the highest bidders, captured slaves sent to slave markets, and wealthy captives ransomed for release.[268]

> The first attacks took place in the Mediterranean, where piracy had been practiced since the beginning of recorded history. The pirates in this case were from the four Barbary States of Tripoli, Tunis, Algiers, and Morocco, whose corsairs operated out of ancient seaports along the North African coast . . . Captured crew members were transported back into port in chains, where they were imprisoned, put to hard labor, or sold at slave markets. Women faced the prospect of being raped or sold into private harems. Prisoners who disobeyed or attempted to escape might be burned alive or impaled.[269]

The news of these attacks created a sensation in the United States. While Britain alone had sufficient strength in its naval squadrons to put all four Barbary states out of business, Britain chose to tolerate Barbary piracy as it served to check against economic competition from maritime rivals such as the growing United States.[270] American leaders came under increasing pressure to "strike back at the pirates, to rescue their enslaved countrymen, and to prevent further attacks."[271] The seizure of eleven American merchantment by Algiers led to George Washington's push toward a more martial approach on the seas.[272] In 1793 Algiers captured ten U.S. naval vessels, and took 100 men prisoner. All men were then forced to become slaves of the Algiers regency. As a result, American vessel that had "once been ubiquitous in the Mediterranean . . . in a manner of weeks, they vanished."[273] This led to the resolution of 1794, that the United States would require and set up "a naval force adequate to the protection of commerce of the United States, against the Algerine corsairs, out to be provided."[274] While there were some grumbling with regard to the cost of establishing a navy, it far outweighed the consequences of not establishing one. While republicans objected to the costs of building a navy, the Federalists championed the costs of not building one.[275] By the spring of 1794, the costs of insurance for transatlantic ships had risen to 25 percent of the value of ships and their cargo, a burden that would be passed on to the merchants as well as the farmers exporting the produce and the U.S. import consumers[276]—"in *the first year alone the costs would equal three to six times the total cost of the proposed squadron.*"[277]

But before the U.S. Navy ships completed construction, the U.S and Algiers agreed to peace, and the United States agreed to pay for the 100 American hostages (imprisoned merchantmen) a sum of $1 million, *one sixth of the Federal budget at that time.*[278] The U.S. policy of appeasement had not sated the demands of the Barbary rulers, merely "*whetted their appetite for more.*"[279] "the impossibly large sums demanded by one ruler would be followed by similarly large demands from others."[280] In 1800, an incident occurred that would forever alter the American attitude toward appeasement and constriction to trade, "the humiliation of the *George Washington* was the final straw; according to Secretary of State James Madison, it 'deeply affected the sensibility, not only of the President, but of the people of the United States.'"[281]

In 1801, the United States would take its first expeditionary naval action, with the *United States Ship Enterprise* (after which the *Enterprise* in the fictional Star Trek spaceship is named), taking the first successful action against the Barbary states. This early formative experience, this strategic trauma, taught the United States to beware of illiberal foreign powers, and be ready to use military power to defend its commerce.

U.S. EXPECTED BEHAVIOR

It is expected that it will be the private sector that leads the American society toward space resource exploitation, and that the U.S. government will play a supportive role by attempting to extend the international system to enable a liberal trading regime in space. The United States is unlikely to take a resource nationalist approach with respect to space resources.

However, there is significant uncertainty about the prominence of space in the U.S. grand strategy, and whether the United States will meet the PRC's peacetime strategic offensive with its own equivalent bid to seek a position of industrial and logistical advantage. Depending on whether the broader policy elites and the American electorate are convinced of the opportunity and threat, space could assume either a central or marginal role in U.S. grand strategy.

As a result, the United States faces a fork in the road. Accessing space resources could prolong its primacy. Ceding control of those resources could speed the economic power of its competitor, the PRC.[282]

One scenario, "U.S. Intent to Compete," would result in early, strong action to match the PRC offensive, and result in a coordinated national policy to help private actors achieve a broader societal purpose, extending the liberal international order and trading system into cislunar space and backing it up with military power through a U.S. Space Force.

A second scenario, "U.S. Complacency" would fail to appreciate the full scale of the strategic threat and be slow in response. The combination of a disbelief in either the opportunity or the need to take seriously the strategic threat results in the United States resting on its laurels and taking a "wait and see" approach. The effect of such inaction and lack of investment mean the United States would arrive late to the game and become serious only when there is a "Sputnik Moment" (an event of strategic surprise) or a clear threat to commercial interests or access. In this case, we would predict the United States would understand this as a significant breach requiring a significant United States overreaction. If interests were perceived to be limited, this might take a form similar to the punitive response of the Barbary Wars. However, depending on the level of interests, offense, and level of societal mobilization, this is likely to be couched as a "crusade" for values to the American people, admitting of no easy negotiated victory short of "unconditional surrender of the aggressor nation and by the overthrow and transformation of its political regime."[283] However, it is equally likely that if the United States fails to prepare a position of industrial and logistical advantage as a precursor to such a crisis, it could very well lose such a conflict.

As remarked by Bear Bryant, "*It's not the will to win that matters—everyone has that. It's the will to prepare to win that matters.*"[284] Although the United States has a proud tradition of peacetime strategic offensives in its westward expansion, its naval/maritime expansion, as well as in aviation,[285] it is yet to be demonstrated in space. As detailed above, the U.S. Department of Defense has been slow to appreciate and adopt a space expansionist paradigm with its own peacetime strategic offensive.[286] If the new Space Force continues to remain on the strategic defensive to focus on tactical excellence, it might repeat the mistake of Germany's Navy in World War I, where a myopic focus on "risk theory,"[287] and the "tactical will to battle"[288] resulted in a lack of "strategic will"[289] and a "war of coastal defense,"[290] which kept it on the strategic defensive (failing to contest control of the strategic lines of communication), and turned its tactical victories into strategic defeat.[291]

Although the dominant construction of the "value of space" among United States policymakers has focused on non-resource factors (prestige, exploration, science, espionage), the clarity of the PRC's programs is currently under examination by policy elites (e.g., see Hearing on "China in Space: A Strategic Competition?").[292] It is typical of U.S. administrations to perceive the importance of space and space access as a critical national interest worth defending. The prominence of influential high-net-worth individuals such as Jeff Bezos and Elon Musk, and the growing economic importance of the NewSpace economy are forces likely to raise the awareness of the strategic and societal importance of space. However, no U.S. administration, to include the Trump administration has yet to create a comprehensive national space strategy specifically to secure a position of industrial and logistical advantage to enable and protect space resources. The differing priorities and domestic focus of the Democratic party, as well as the partisan politicization surrounding the Space Force and PRC-related tariffs could mean that a Democratic party electoral victory might alter priorities, potentially de-emphasize U.S.-PRC competition and result in an anemic space program.

Therefore, the expected trajectory is that China will continue to secure significant positional and societal mobilization advantages before the broader American electorate "wake up" to mobilize an equivalent response. Such a societal mobilization and accompanying policy will only occur after China secures additional accomplishments and commercial and military interests grow increasingly wary of a potential restriction of space commerce.

NOTES

1. "New Horizons Spacecraft Return its Sharpest Views of Ultima Thule," *NASA*, February 22, 2019, accessed September 13, 2019, https://www.nasa.gov/feature/new-horizons-spacecraft-returns-its-sharpest-views-of-ultima-thule
2. "The Power sub-ranking is based on an equally weighted average of scores from five country attributes that related to a country's power: a leader, economically influential, politically influential, strong international alliances and strong military alliances." "Power Rankings," *US News & World Report*, January 23, 2019, accessed September 13, 2019, https://www.usnews.com/news/best-countries/power-full-list
3. "2018 Military Strength Ranking," *Global Firepower*, November 22, 2018, accessed January 19, 2019, https://www.globalfirepower.com/countries-listing.asp
4. Nan Tian, Aude Fleurant, Alexandra Kuimova, Pieter D. Wezeman, and Siemon T. Wezeman, *Trends in World Military Expenditure, 2017*. Stockholm International Peace Research Institute, 2018; "Trends in World Military Expenditure 2017," *SIPRI*, accessed January 19, 2019, https://www.sipri.org/publications/2018/sipri-fact-sheets/trends-world-military-expenditure-2017
5. "The Military Balance 2018," *IISS*, February 1, 2018, accessed January 19, 2019, https://www.iiss.org/publications/the-military-balance/the-military-balance-2018
6. "Total Population by Country 2018," *World Population Review*, 2018, accessed January 19, 2019, http://worldpopulationreview.com/countries/
7. "Human Development Index (HDI)," *United Nations Development Program (UNDP)*, October 12, 2019, accessed January 19, 2019, http://hdr.undp.org/en/data
8. Energy is expressed a number of different ways. 2,201 MTOE = 15.72 Billion Barrel of Oils Equivalent (BBOE) = 92.15 Exajoules (EJ) = 87.34 Quadrillion BTU (QUAD or PBtu) = 22,024 Megatons TNT equivalent = 25,597.63 Terawatt Hours (TWh) = 2.92 Terawatts (continuous). (Author's math)

9. "Global Energy Statistical Yearbook 2018," *Enerdata*, July 9, 2019, accessed January 19, 2019, https://yearbook.enerdata.net/total-energy/world-consumption-statistics.html

10. "Global Energy Statistical Yearbook 2018," *Enerdata*.

11. "Electric Power Consumption (kWh per capita)," *World Bank*, March 22, 2010, accessed January 20, 2019, https://data.worldbank.org/indicator/EG.USE.ELEC.KH.PC

12. "The World in 2050," *Price Waterhouse Cooper (PwC)*, December 3, 2015, accessed January 19, 2019, https://www.pwc.com/gx/en/issues/economy/the-world-in-2050.html

13. Uri Dadush and Bennett Stancil, "The World Order in 2050," *Carnegie Endowment for International Peace*, April 2010, accessed January 19, 2019, https://carnegieendowment.org/files/World_Order_in_2050.pdf

14. Brian Wang, "EIU GDP Forecasts to 2050," *Next Big Future*, June 24, 2015, accessed October 11, 2019, https://www.nextbigfuture.com/2015/06/eiu-gdp-forecasts-to-2050.html

15. "Which countries spend the most on space exploration?," *World Economic Forum*, January 11, 2016, accessed January 19, 2019, https://www.weforum.org/agenda/2016/01/which-countries-spend-the-most-on-space-exploration/

16. "Government Spending in Space Programs Reaches $62 Billion in 2016," *Euroconsult*, May 30, 2017, accessed January 19, 2019, http://www.euroconsult-ec.com/30_May_2017

17. "InSight: Futron The Space Competitiveness Index—Update, by Futron," *SatMagazine*, December 9, 2009, accessed January 19, 2019, http://www.satmagazine.com/story.php?number=1856883389

18. "INTEL: Global Military Space by Futron," *MilSat Magazine* (September 2009 Edition), September 9, 2009, accessed January 19, 2019, http://www.milsatmagazine.com/story.php?number=875113067

19. "Space Foundation Report Reveals Global Space Economy at $383.5 Billion in 2017," *The Space Foundation*, July 19, 2018, accessed March 9, 2019, https://www.spacefoundation.org/news/space-foundation-report-reveals-global-space-economy-3835-billion-2017

20. "UCS Satellite Database," Union of Concerned Scientists, November 30, 2018, accessed March 9, 2019, https://www.ucsusa.org/nuclear-weapons/space-weapons/satellite-database

21. John P. Thomas, "20,000 Satellites for 5G to be Launched Sending Focused Beams of Intense Microwave Radiation Over Entire Earth," *Health Impact News*, March 9, 2019, accessed March 9, 2019, http://healthimpactnews.com/2019/20000-satellites-for-5g-to-be-launched-sending-focused-beams-of-intense-microwave-radiation-over-entire-earth/

22. "FCC Approves SpaceX's Plan to Launch more than 7,000 Internet-Beaming Satellites," *The Verge*, November 15, 2018, accessed March 9, 2019, https://www.theverge.com/2018/11/15/18096943/spacex-fcc-starlink-satellites-approval-constellation-internet-from-space

23. Caleb Henry, "SpaceX Submits Paperwork for 30,000 more Starlink Satellites," *Space News*, October 15, 2019, accessed February 18, 2020, https://spacenews.com/spacex-submits-paperwork-for-30000-more-starlink-satellites/

24. Amy Thompson, "SpaceX launches 60 Starlink Satellites for New Megaconstellation, Misses Rocket Landing," *Space.com*, February 16, 2020, accessed February 18, 2020, https://www.space.com/spacex-starlink-4-launch-success-misses-rocket-landing.html; for an excellent overview of the significance of Starlink, see Casey Hadmer, "Starlink is a very big deal," *Casey Handmer's Blog*, November 2, 2019, accessed February 21, 2020, https://caseyhandmer.wordpress.com/2019/11/02/starlink-is-a-very-big-deal/

25. Sarah Fecht, "Now That We've Explored All The Planets, What's The Next Frontier?," *Popular Science*, July 17, 2015, accessed March 9, 2019, https://www.popsci.com/now-weve-explored-all-planets-whats-next-frontier

26. Author counts. From NASA website list excluding Hubble Telescope, NASA, Solar System Missions, No single consolidated source seems to exist other than:"List of Solar System Probes," July 3, 2013, accessed March 14, 2020, https://www.nasa.gov/content/solar-missions-list ; see also: NASA Jet Propulsion Laboratory, "JPL Mission Paths," *JPL*, August 21, 2013, accessed March 14, 2020. https://www.jpl.nasa.gov/infographics/infographic.view.php?id=11049

27. Eleanor Imster, "Parker Solar Probe Becomes Closest-Yet Spacecraft to Sun" *EarthSky*, October 20, 2018, accessed March 9, 2019, https://earthsky.org/space/parker-solar-record-closest-yet-to-sun-oct2018

28. Tim Ferholz, "The Fastest Object Made by Humans is Just Trying to Slow Down," *QZ*, August 10, 2018, accessed March 9, 2019, https://qz.com/1353429/the-parker-solar-probe-is-the-fastest-man-made-object-ever/

29. Sasha Savitsky, "NASA's Juno Spacecraft to Become Fastest Man-Made Object as it Slingshots around Earth towards Jupiter," *Fox News*, October 9, 02013, accessed March 9, 2019, https://www.foxnews.com/science/nasas-juno-spacecraft-to-become-fastest-man-made-object-as-it-slingshots-around-earth-towards-jupiter

30. "Top 10 Fastest Manmade Objects Ever," *The Mysterious World*, March 2, 2014, accessed March 9, 2019, https://themysteriousworld.com/top-10-fastest-man-made-objects-ever/

31. Elizabeth Howell, "Voyager 1: Earth's Farthest Spacecraft," *Space.com*, March 1, 2018, accessed March 9, 2019, https://www.space.com/17688-voyager-1.html

32. Sarah Kaplan, "NASA's New Horizons Spacecraft Just Visited the Farthest Object Ever Explored," *The Washington Post*, January 1, 2019, accessed March 9, 2019, https://www.washingtonpost.com/science/2018/12/31/most-distant-space-encounter-history-is-happening-now/?noredirect=on&utm_term=.54345095fc60

33. Nancy Atkinson, "How Many People Have Walked on the Moon?," *Universe Today*, August 1, 2013, accessed March 9, 2019, https://www.universetoday.com/55512/how-many-people-have-walked-on-the-moon/; Charles Murray and Catherine Bly Cox, *Apollo: The Race to the Moon* (New York: Simon & Schuster, 1989).

34. Michael Sheetz, "Elon Musk's SpaceX Becomes the First to Launch Reused Rocket," *CNBC*, December 15, 2018, accessed March 19, 2019, https://www.cnbc.com/2017/12/15/elon-musks-spacex-becomes-first-to-launch-reused-rocket-on-a-nasa-mission.html

35. For a complimentary overview of space settlement's policy history, see Aaron Oesterle, "The Viability of the Space Settlement Narrative in Politics," (IAC-19.E3.2.9x52875), International Astronautical Congress, November 1, 2018.

36. Mike Wall, "New Space Mining Legislation Is 'History in the Making'," *Space.com*, November 20, 2015, accessed February 16, 2020, https://www.space.com/31177-space-mining-commercial-spaceflight-congress.html

37. Planetary Resources, "President Obama Signs Bill Recognizing Asteroid Resource Property Rights Into Law," November 25, 2015, accessed August 1, 2019, https://www.planetaryresources.com/2015/11/president-obama-signs-bill-recognizing-asteroid-resource-property-rights-into-law/

38. "H.R.2262—U.S. Commercial Space Launch Competitiveness Act 114th Congress (2015–2016)," *Congress.gov*, November 25, 2015, accessed September 16, 2019, https://www.congress.gov/bill/114th-congress/house-bill/2262/text

39. "H.R.2262—U.S. Commercial Space Launch."

40. Bernard L. Brock, et al., *Making Sense of Political Ideology The Power of Language in Democracy* (Lanham: Rowman & Littlefield, 2005).

41. Brock, *Making Sense of Political Ideology*, 59.

42. Donald J. Trump, "Presidential Executive Order on Reviving the National Space Council," *The White House*, June 30, 2017, accessed February 19, 2020, https://www.whitehouse.gov/presidential-actions/presidential-executive-order-reviving-national-space-council/

43. "National Space Council Users' Advisory Group," *NASA*, October 5, 2017, accessed February 19, 2020, https://www.nasa.gov/content/national-space-council-users-advisory-group

44. For example, the National Space Council held highly theatric, televised events of cabinet-level officials in front of backdrops such as the Udvar-Hazy Air and Space Museum (October 5, 2017), Cape Canaveral Space Station Processing Facility (February 21, 2018), The White House (June 18, 2018), the National War College (October 23, 2018, the U.S. Space and Rocket Center in Huntsville, Alabama (March 27, 2019), and the Udvar-Hazy center (August 20, 2019).

45. Presidential Memorandum on Reinvigorating America's Human Space Exploration Program, The White House, December 11, 2017, accessed August 1, 2019, https://www.whitehouse.gov/presidential-actions/presidential-memorandum-reinvigorating-americas-human-space-exploration-program/

46. Scott Pace, "Space Development, Law, and Values" [IISL Galloway Space Law Symposium Cosmos Club, Washington, DC], December 13, 2017, August 1, 2019, https://spacepolicyonline.com/wp-content/uploads/2017/12/Scott-Pace-to-Galloway-FINAL.pdf

47. The National Security Strategy of the United States of America, *The White House*, December 18, 2017, accessed August 1, 2019, https://www.whitehouse.gov/wp-content/uploads/2017/12/NSS-Final-12-18-2017-0905.pdf

48. "The National Security Strategy."

49. Berkeley Lovelace, "Wilbur Ross on Space Race: We Want to Turn the Moon into a Gas Station," *CNBC*, February 22, 2018, accessed August 8, 2019, https://www.cnbc.com/2018/02/22/wilbur-ross-on-space-race-we-want-to-turn-the-moon-into-a-gas-station.html

50. "President Donald J. Trump is Unveiling an America First National Space Strategy," The White House, March 23, 2018, accessed August 1, 2019, https://www.whitehouse.gov/briefings-statements/president-donald-j-trump-unveiling-america-first-national-space-strategy/

51. "President Donald J. Trump is Unveiling an America."

52. Marina Koren, "What Does Trump Mean By 'Space Force'?" *The Atlantic*, March 13, 2018, accessed February 18, 2020, https://www.theatlantic.com/science/archive/2018/03/trump-space-force-nasa/555560/

53. Mike Wall, "Trump Teases 'Space Force' Idea Again for US Military," *Space.com*, May 02, 2018, accessed February 21, 2020, https://www.space.com/40471-trump-teases-military-space-force-again.html

54. Josh Rogin, "Wilbur Ross Wants the United States to Dominate Space," *The Washington Post*, May 2, 2018, accessed October 14, 2019, https://www.washingtonpost.com/news/josh-rogin/wp/2018/05/02/wilbur-ross-wants-the-united-states-to-dominate-space/

55. Rogin, "Wilbur Ross Wants the United States."

56. Rogin, "Wilbur Ross Wants the United States."

57. Scott Pace, Full Transcript: Space 2.0: U.S. Competitiveness and Policy in the New Space Era (Hudson Institute), May 17, 2018, accessed August 1, 2019, https://www.hudson.org/research/14341-full-transcript-space-2-0-u-s-competitiveness-and-policy-in-the-new-space-era

58. Scott Pace, Executive Secretary, National Space Council, "Space 2.0: U.S. Competitiveness and Policy in the New Space Era."

59. Space Policy Directive-2: Streamlining Regulations on Commercial Use of Space, *The White House*, May 24, 2018, accessed August 1, 2019, https://www.whitehouse.gov/presidential-actions/space-policy-directive-2-streamlining-regulations-commercial-use-space/

60. Space Policy Directive-3: National Space Traffic Management Policy, *The White House*, June 18, 2018, August 1, 2019, https://www.whitehouse.gov/presidential-actions/space-policy-directive-3-national-space-traffic-management-policy/

61. The White House, "Remarks by President Trump at a Meeting with the National Space Council and Signing of Space Policy Directive-3," June 18, 2018, accessed August 8, 2019, https://www.whitehouse.gov/briefings-statements/remarks-president-trump-meeting-national-space-council-signing-space-policy-directive-3/

62. Christopher Buckley, "Is Trump's 'Space Force' Really Such an Insane Idea?," *Spectator USA*, June 27, 2018, accessed June 3, 2019, https://spectator.us/is-trumps-space-force-really-such-an-insane-idea/

63. The White House, "Remarks by President Trump."

64. Eric Berger, "How Can NASA Return to the Moon? By Making Everything Reusable, Chief Says," *ArsTechnica*, August 2, 2018, accessed August 8, 2019, https://arstechnica.com/science/2018/08/nasa-chief-seeks-a-sustainable-reusable-pathway-for-lunar-return/

65. The White House, "Remarks by Vice President Pence on the Future of the U.S. Military in Space," August 9, 2018, accessed August 8, 2019, https://www.whitehouse.gov/briefings-statements/remarks-vice-president-pence-future-u-s-military-space/

66. The White House, "Remarks by Vice President Pence": "The space environment has fundamentally changed in the last generation. What was once peaceful and uncontested is now crowded and adversarial. Today, other nations are seeking to disrupt our space-based systems and challenge American supremacy in space as never before . . . As their actions make clear, our adversaries have transformed space into a warfighting domain already. And the United States will not shrink from this challenge. (Applause.) Under President Trump's leadership, we will meet it head on to defend our nation and build a peaceful future here on Earth and in space. America will always seek peace in space as on the Earth. But history proves that peace only comes through strength. And in the realm of outer space, the United States Space Force will be that strength in the years ahead . . . As the President will discuss in further detail in the days ahead, the United States Space Force will strengthen our security, it will ensure our prosperity, and it will also carry American ideals into the boundless expanse of space. While other nations increasingly possess the capability to operate in space, not all of them share our commitment to freedom, to private property, and the rule of law. So as we continue to carry American leadership in space, so also will we carry America's commitment to freedom into this new frontier."

67. See note 66.

68. See note 66.

69. Kevin McGill, "NASA Administrator Supports Trump's 'Space Force' Proposal," *Bloomberg*, August 13, 2018, accessed August 8, 2019, https://www.bloomberg.com/news/articles/2018-08-13/nasa-administrator-supports-trump-space-force-proposal

70. *Space Policy Directive-4: Establishment of the United States Space Force*, The White House, February 19, 2019, accessed August 1, 2019, https://media.defense.gov/2019/Mar/01/2002095015/-1/-1/1/SPACE-POLICY-DIRECTIVE-4-FINAL.PDF also available at https://spacepolicyonline.com/news/text-of-space-policy-directive-4-spd-4-establishing-a-u-s-space-force/

71. *Space Policy Directive-4: Establishment.*

72. Wilbur Ross, "Launching Toward a $1 Trillion Space Economy," *Orlando Sentinel*, February 28, 2019, accessed August 8, 2019, https://www.orlandosentinel.com/opinion/os-op-wilbur-ross-space-1-trillion-20190228-story.html

Also available at: https://www.commerce.gov/news/op-eds/2019/02/op-ed-commerce-secretary-wilbur-l-ross-launching-toward-1-trillion-space

73. Ross, "Launching Toward a $1 Trillion Space Economy."

74. Ross, "Launching Toward a $1 Trillion Space Economy."

75. Ross, "Launching Toward a $1 Trillion Space Economy." Also available at: https://www.commerce.gov/news/op-eds/2019/02/op-ed-commerce-secretary-wilbur-l-ross-launching-toward-1-trillion-space

76. "Recommendations Approved by the National Space Council to President Trump," *SpaceRef*, March 26, 2019, accessed February 20, 2020, http://spaceref.com/news/viewsr.html?pid=52358

77. "Remarks by Vice President Pence at the Fifth Meeting of the National Space Council | Huntsville, AL," *The White House*, March 26, 2019, accessed February 19, 2020, https://www.whitehouse.gov/briefings-statements/remarks-vice-president-pence-fifth-meeting-national-space-council-huntsville-al/

78. Wayne Heilman, "Space Symposium 2019: Commerce Secretary Warns of Moon Becoming 'Wild West'," *Gazette*, April 9, 2019, accessed August 8, 2019, https://gazette.com/business/space-symposium-commerce-secretary-warns-of-moon-becoming-wild-west/article_f24f7b62-5ad6-11e9-abd9-97c2e5f7da0f.html

79. Heilman, "Space Symposium 2019: Commerce Secretary Warns."

80. David Axe, "Wilbur Ross: We've Gotta Get Back to the Moon—to Exploit the Hell Out of It," *Daily Beast*, April 9, 2019, accessed August 8, 2019, https://www.thedailybeast.com/wilbur-ross-weve-gotta-get-back-to-the-moonto-exploit-the-hell-out-of-it

81. M. Scott Mahaskey, "We Want Access to Every Part of the Moon, At Any Time," *Politico*, June 13, 2019, accessed August 8, 2019, https://www.politico.com/agenda/story/2019/06/13/jim-bridenstine-nasa-interview-000900

82. Mahaskey, "We Want Access to Every Part of the Moon."

83. Theresa Hitchens, "Commercial Satellites: Will They Be Military Targets?" Breaking Defense, July 16, 2019, accessed February 3, 2020, https://breakingdefense.com/2019/07/commercial-satellites-will-they-be-military-targets/

84. Hitchens, "Commercial Satellites: Will They Be Military Targets?"

85. Hitchens, "Commercial Satellites: Will They Be Military Targets?"

86. Hitchens, "Commercial Satellites: Will They Be Military Targets?"

87. Hitchens, "Commercial Satellites: Will They Be Military Targets?"

88. Jeff Foust, "NASA Seeks "Industrialization" of Low Earth Orbit with ISS Commercialization Strategy," *SpaceNews*, August 1, 2019, accessed March 3, 2020, https://spacenews.com/nasa-seeks-industrialization-of-low-earth-orbit-with-iss-commercialization-strategy/

89. Foust, "NASA Seeks "Industrialization" of Low Earth Orbit."

90. Foust, "NASA Seeks "Industrialization" of Low Earth Orbit."

91. Foust, "NASA Seeks "Industrialization" of Low Earth Orbit."

92. US Department of Defense, "Department of Defense Establishes U.S. Space Command," August 2019, accessed January 31, 2020, https://www.defense.gov/Newsroom/Releases/Release/Article/1948288/department-of-defense-establishes-us-space-command/

93. USSPACECOM, "United States Space Command Fact Sheet," August 29, 2019, accessed February 3, 2020, https://www.spacecom.mil/About/Fact-Sheets-Editor/Article/1948216/united-states-space-command-fact-sheet/

94. Air Force Space Command, *The Future of Space 2060 and Implications for U.S. Strategy: Report on the Space Futures Workshop*, September 5, 2019, accessed February 10, 2020, https://www.afspc.af.mil/Portals/3/documents/Future%20of%20Space%202060%20v2%20(5%20Sep).pdf?ver=2019-09-06-184933-230 with an graphically updated version released on 3 October 2019 at: https://www.afspc.af.mil/Portals/3/The%20Future%20of%20Space%202060%20-%203Oct19.pdf

95. Air Force Space Command, *The Future of Space 2060 and Implications for U.S. Strategy.*

96. Brent D. Ziarnick, "The Space Force Strategic Vision Emerges," *The Hill*, September 9, 2019, accessed February 19, 2020, https://thehill.com/opinion/national-security/460469-the-space-force-strategic-vision-emerges; Peter Garretson, "A Historic National Vision For Spacepower," *War on the Rocks*, September 9, 2019, accessed February 19, 2020, https://warontherocks.com/2019/09/a-historic-national-vision-for-spacepower/

97. U.S.-China Economic and Security Review Commission, *2019 Report to Congress*, November 2019, accessed February 10, 2020, https://www.uscc.gov/sites/default/files/2019-11/Chapter%204%20Section%203%20-%20China%E2%80%99s%20Ambitions%20in%20Space%20-%20Contesting%20the%20Final%20Frontier.pdf

98. U.S.-China Economic and Security Review Commission, *2019 Report to Congress*, November 2019.

99. Patrick M. Cronin, Masashi Murano and H. R. McMaster, "Transcript: U.S. Space Strategy and Indo-Pacific Cooperation," *The Hudson Institute*, November 13, 2019, accessed March 1, 2020, https://www.hudson.org/research/15481-transcript-u-s-space-strategy-and-indo-pacific-cooperation (used with permission)

100. Oriana Paylyk, "It's Official: President Trump Has Signed Space Force Into Existence," *Military.com*, December 20, 2019, accessed February 3, 2020, https://www.military.com/daily-news/2019/12/20/its-official-president-trump-has-signed-space-force-existence.html

101. Congress.gov, "S.1790 - National Defense Authorization Act for Fiscal Year 2020" (Subtitle D—United States Space Force Chapter 908, section 9083 Became Public Law No: 116-92) accessed February 3, 2020, https://www.congress.gov/bill/116th-congress/senate-bill/1790/text?format=txt

102. Wilbur Ross, "A New Space Race: Getting to the Trillion-Dollar Space Economy World Economic Forum, Davos, Switzerland," *Commerce Department*, January 24, 2020, accessed February 21, 2020, https://www.commerce.gov/news/speeches/2020/01/remarks-secretary-commerce-wilbur-ross-new-space-race-getting-trillion-dollar

103. "Manifest Destiny," November 15, 2019, *History.com*, accessed February 19, 2020, https://www.history.com/topics/westward-expansion/manifest-destiny

104. Donald J. Trump, "Remarks by President Trump in State of the Union Address," *The White House*, February 4, 2020, accessed February 20, 2020, https://www.whitehouse.gov/briefings-statements/remarks-president-trump-state-union-address-3/

105. Donald J. Trump, "Amending Executive Order 13803—Reviving the National Space Council," *The White House*, February 13, 2020, accessed February

19, 2020, https://www.whitehouse.gov/presidential-actions/amending-executive-order-13803-reviving-national-space-council/

106. Full statement: "The Department of Energy is proud to become an official member of the National Space Council to further efforts toward American leadership in scientific advancement and space exploration," said U.S. Secretary of Energy Dan Brouillette. "Under President Trump's leadership, we will unlock the mysteries of space and empower our commercial partners to see the untapped potential in space. The Department of Energy, through our nuclear R&D, has long provided the capabilities to provide fuel for spacecraft for deep space missions. We look forward to working across the federal government and with industry to renew our commitment to space."; "The Department of Energy joining the National Space Council only amplifies our nation's commitment towards learning more about our universe, and creating innovative technologies to assist in advancing development in space," said Under Secretary for Science Paul Dabbar. "We must recognize space as an industry of the future that will lead to worldly discoveries in our lifetime." "President Trump Designates Secretary Brouillette a Member of the National Space Council," Department of Energy, February 14, 2020, accessed February 19, 2020, https://www.energy.gov/articles/president-trump-designates-secretary-brouillette-member-national-space-council

107. Donald J. Trump, Executive Order on Encouraging International Support for the Recovery and Use of Space Resources, *The White House*, April 6, 2020, accessed July 6, 2020, https://www.whitehouse.gov/presidential-actions/executive-order-encouraging-international-support-recovery-use-space-resources/

108. NASA, "Principles for a Safe, Peaceful, and Prosperous Future," NASA.gov, May 15, 2020, accessed July 6, 2020, https://www.nasa.gov/specials/artemis-accords/index.html

109. NASA, "Principles for a Safe, Peaceful, and Prosperous Future."

110. Mark Stokes, Gabriel Alvarado, Emily Weinstein, and Ian Easton, *China's Space and Counterspace Capabilities and Activities*, U.S.-China Economic and Security Review Commission, May 11, 2020, accessed July 31, 2020, https://www.uscc.gov/research/chinas-space-and-counterspace-activities; Peter Vincent Pry, *China: EMP Threat The People's Republic of China Military Doctrine, Plans, and Capabilities for Electromagnetic Pulse (EMP) Attack,* EMP Task Force on National and Homeland Security, June 10, 2020, accessed July 28, 2020, https://www.centerforsecuritypolicy.org/wp-content/uploads/2020/06/CHINAempTHREAT2020logo.pdf

111. See for example, the White House's *United States Strategic Approach to the People's Republic of China, and amplification given by the Secretary of State, National Security Advisor, FBI Director, and Acting Undersecretary for Arms Control and International security:* The White House, *United States Strategic Approach to the People's Republic of China*, WhiteHouse.gov, May 26, 2020, accessed July 28, 2020, https://www.whitehouse.gov/wp-content/uploads/2020/05/U.S.-Strategic-Approach-to-The-Peoples-Republic-of-China-Report-5.24v1.pdf; Michael R. Pompeo, "Communist China and the Free World's Future," State.gov, July 23, 2020, accessed July 28, 2020, https://www.state.gov/communist-china-and-the-free-worlds-future/; Robert C. O'Brien, "The Chinese Communist Party's Ideology and Global

Ambitions," The White House, June 26, 2020, accessed July 28, 2020, https://www.whitehouse.gov/briefings-statements/chinese-communist-partys-ideology-global-ambitions/; Christopher Wray, "The Threat Posed by the Chinese Government and the Chinese Communist Party to the Economic and National Security of the United States," FBI.gov, July 7, 2020, accessed July 28, 2020, https://www.fbi.gov/news/speeches/the-threat-posed-by-the-chinese-government-and-the-chinese-communist-party-to-the-economic-and-national-security-of-the-united-states; Christopher A. Ford, *Competitive Strategy vis-a-vis China and Russia: A View from the "T Suite."* Arms Control and International Security Papers Volume 11 Number 6, May 11, 2020, accessed July 28, 2020, https://www.state.gov/wp-content/uploads/2020/06/T-paper-series-6-Strategic-competition-508.pdf

112. Department of Defense, *Defense Space Strategy Summary*, Defense.gov, JUNE 2020, accessed July 28, 2020, https://media.defense.gov/2020/Jun/17/2002317391/-1/-1/1/2020_DEFENSE_SPACE_STRATEGY_SUMMARY.PDF

113. The White House National Space Council, A New Era for Deep Space Exploration and Development, WhiteHouse.gov, JULY 23, 2020, accessed https://www.whitehouse.gov/wp-content/uploads/2020/07/A-New-Era-for-Space-Exploration-and-Development-07-23-2020.pdf

114. The White House National Space Council, A New Era for Deep Space Exploration and Development, WhiteHouse.gov, JULY 23, 2020, accessed https://www.whitehouse.gov/wp-content/uploads/2020/07/A-New-Era-for-Space-Exploration-and-Development-07-23-2020.pdf

115. Steven J. Butow, Thomas Cooley, Eric Felt, and Joel B. Mozer, *State of The Space Industrial Base 2020: A Time For Action To Sustain US Economic & Military Leadership in Space Summary Report,* Department of Defense, July 2020, accessed August 3, 2020, https://cdn.afresearchlab.com/wp-content/uploads/2020/07/27223753/State-of-the-Space-Industrial-Base-2020-Report_July-2020_FINAL.pdf

116. Planetary Resources, "Timeline," n.d., accessed August 5, 2019, https://www.planetaryresources.com/company/timeline/

117. Planetary Resources, "Media Coverage," n.d., accessed August 5, 2019, https://www.planetaryresources.com/category/media/page/25/

118. Adam Mann, "Tech Billionaires Plan Audacious Mission to Mine Asteroids," *Wired*, April 23, 2012, accessed August 5, 2019, https://www.wired.com/2012/04/planetary-resources-asteroid-mining/

119. Beth Jinks, "Google Chiefs Back Startup Mining Asteroids for Metals," *Bloomberg*, April 6, 2012, accessed August 5, 2019, https://www.bloomberg.com/news/articles/2012-04-24/google-chiefs-back-startup-mining-asteroids-for-metals

120. Irene Klotz, "Tech billionaires bankroll gold rush to mine asteroids," *Reuters*, April 24, 2012, accessed August 5, 2019, https://www.reuters.com/article/space-asteroid-mining/tech-billionaires-bankroll-gold-rush-to-mine-asteroids-idUSB27208920120424

121. Planetary Resources, "Asteroid Mining Mission Revealed by Planetary Resources, Inc.," *Youtube*, April 24, 2012, accessed August 5, 2019, https://www.youtube.com/watch?v=zXXJtSZffVg

122. Planetary Resources, "Planetary Resources: Our Vision for Humanity's Future in Space," *Youtube*, December 3, 2017, accessed August 5, 2019, https://www.youtube.com/watch?v=hqW9HtPCnR0&t=9s

123. Deep Space Industries, Mission, January 24, 2013, accessed February 10, 2020, https://web.archive.org/web/20130124211904/http://deepspaceindustries.com/mission

124. Deep Space Industries, Deep Space Industries, *Youtube*, January 26, 2013, accessed August 5, 2019, https://www.youtube.com/watch?v=zXXJtSZffVg

125. Trans Astranautica Corporation, accessed August 5, 2019, https://www.trans-astracorp.com/

126. Planetary Resources, "President Obama Signs Bill Recognizing Asteroid Resource Property Rights Into Law," November 25, 2015, accessed August 1, 2019, https://www.planetaryresources.com/2015/11/president-obama-signs-bill-recognizing-asteroid-resource-property-rights-into-law/

127. "H.R.2262 - U.S. Commercial Space Launch Competitiveness Act 114th Congress (2015-2016)," *Congress.gov*, November 25, 2015, accessed February 16, 2020, https://www.congress.gov/bill/114th-congress/house-bill/2262/text

128. Barnard F. Kutter and George F. Sowers, "Cislunar-1000: Transportation supporting a self-sustaining Space Economy," ULA, AIAA Space 2016, available at AIAA, accessed March 11, 2019, https://arc.aiaa.org/doi/10.2514/6.2016-5491

129. United Launch Alliance, "Cis-Lunar 1000," *Youtube*, January 11, 2016, accessed August 5, 2019, https://www.youtube.com/watch?v=uxftPmpt7aA

130. Leonard David, "Inside ULA's Plan to Have 1,000 People Working in Space by 2045," *Space.com*, June 29, 2016, accessed August 5, 2019, https://www.space.com/33297-satellite-refueling-business-proposal-ula.html

131. "Space Resources Program Approved for Fall 2018 Launch," *Colorado School of Mines*, accessed August 11, 2019, http://www.minesnewsroom.com/news/space-resources-program-approved-fall-2018-launch; The program stated, "Post-baccalaureate certificates, master's degrees and doctoral degrees will be offered through the unique interdisciplinary program, which aims to prepare scientists, engineers, economists, policymakers and entrepreneurs to responsibly explore, extract and use resources on the Moon, Mars, asteroids and beyond to fuel future space exploration, cislunar development, as well as needs back on Earth," with "more than two dozen students living in five countries across three continents are currently enrolled in the program's pilot class, Space Resources Fundamentals, a distance learning-enabled course that was offered for the first time last fall. A second pilot class, Space Systems Engineering, debuted this semester and is being taught by recent Mines hire George Sowers, professor of practice in mechanical engineering and former chief scientist at United Launch Alliance. Other core classes will include space resources-focused design project courses and graduate seminars, with technical electives spanning prospecting, mining, extraction, power and energy, engineering systems such as robotics, and economics and policy."

132. XPrize, Competing Teams, accessed August 8, 2019, https://www.xprize.org/prizes/google-lunar/competing-teams

133. Shackleton Energy Website, accessed March 11, 2019, http://www.shackletonenergy.com/overview

134. Karl Tate, "How Golden Spike's Moon Landing Plan Works (Infographic)," *Space.com*, December 6, 2012, accessed August 8, 2019, https://www.space.com/18805-golden-spike-private-moon-landing-graphic.html; Mann, Adam, "Golden Spike Company Unveils Plans to Fly Commercial Crews to the Moon," *Wired*, December 6, 2012, accessed August 8, 2019, https://www.wired.com/2012/12/golden-spike-announcement/; Bruce Dorminey, "Golden Spike Still Aims For Human Lunar Surface Expeditions By Decade's End," *Forbes*, January 30, 2014, accessed February 16, 2020, https://www.forbes.com/sites/brucedorminey/2014/01/30/golden-spike-still-aims-for-human-lunar-surface-expeditions-by-decades-end/#54695f4d7afe .

135. Masten Space Systems, Masten Space Systems, accessed August 8, 2019, https://www.masten.aero/

136. Astrobotic, Astrobotic, accessed August 8, 2019, https://www.astrobotic.com/

137. Moon Express, "Moon Express," accessed August 8, 2019, http://www.moonexpress.com/

138. Moon Express, "Moon Express."

139. Jeff Foust, "Deep Space Industries Acquired by Bradford Space," *Spacenews*, January 2, 2019, accessed September 16, 2019, https://spacenews.com/deep-space-industries-acquired-by-bradford-space/; Brian Heater, "A Blockchain Firm Bought Asteroid Mining Company Planetary Resources," *Techcrunch*, November 3, 2018, accessed February 16, 2020, https://techcrunch.com/2018/11/03/a-blockchain-firm-bought-asteroid-mining-company-planetary-resources/

140. Jeffrey Kluger, "Elon Musk Told Us Why He Thinks We Can Land on the Moon in 'Less Than 2 Years'," *Time*, July 18, 2019, accessed August 5, 2019, https://time.com/5628572/elon-musk-moon-landing/

141. Blue Origin, Blue Moon, accessed August 5, 2019, https://www.blueorigin.com/blue-moon

142. Momentus Space, Powered by Water Plasma Propulsion, accessed August 5, 2019, https://momentus.space/

143. OrbitFab, Gas Stations in Space, accessed August 5, 2019, https://www.orbit-fab.space/

144. CSDC, "Welcome to Cislunar Space Development Company," accessed August 5, 2019, https://csdc.space/about-csdc/

145. Offworld Website vision, September 9, 2016, accessed October 11, 2019, https://www.offworld.ai/

146. Bruce Cahan, "Space Commodities Exchange – a new frontier," *Urban Logic*, October 8, 2016, accessed August 5, 2019, https://urbanlogic.org/space-commodities-exchange/

147. Bruce Cahan, "A Space Commodities Futures Trading Exchange to Grow the Lunar Economy," 2017 Annual Meeting of the Lunar Exploration Analysis Group, held October 10–12, 2017 in Columbia, Maryland. LPI Contribution No. 2041, id.5008

148. Bruce B. Cahan, Bruce Pittman, Sarah Cooper and John Cumbers, "Space Commodities Futures Trading Exchange: Adapting Terrestrial Market Mechanisms

to Grow a Sustainable Space Economy," *New Space*, September 1, 2018, accessed August 5, 2019, https://www.liebertpub.com/doi/10.1089/space.2017.0047

149. Space Commodities Exchange (SCX), "Space Commodities Exchange," n.d., accessed August 5, 2019, http://www.spacecommoditiesexchange.com/

150. Space Island Group, Clean Energy, Cheap Hydrogen and Weather Control From Space, accessed August 7, 2019, https://www.spaceislandgroup.com/solarsat.html

151. The Space Show, Heliosat, April 27, 2009, accessed August 7, 2019, https://thespaceshow.com/heliosat

152. PowerSat, Energy market drivers behind Space Solar Power (SSP), October 24, 2009, accessed August 7, 2019, http://www.powersat.com

153. The Space Show, SolarHigh.org, June 14, 2013, accessed August 7, 2019, https://www.thespaceshow.com/solarhigh.org

154. Taylor Soper, "Planetary Power raises $3.4M to provide efficient energy for off-grid telecom towers," *GeekWire*, https://www.geekwire.com/2016/planetary-power-raises-3-4m/

155. Peter Sage, Space Energy: An Open Letter From Space Energy Co-Founder Peter Sage Covering," accessed August 7, 2019, https://petersage.com/spaceenergy/ see also Peter Sage's 2009 TED talk "Space Based Solar Power," Mar 17, 2016, accessed February 16, 2020, https://www.youtube.com/watch?v=cT_xNON1_bA

156. Caltech, "Space-Based Solar Power Project Funded," April 28, 2015, accessed August 7, 2019, https://www.caltech.edu/about/news/space-based-solar-power-project-funded-46644

157. Caltech, "Space Solar Power Project," accessed August 7, 2019, https://www.spacesolar.caltech.edu/

158. Solaren, "The Solaren Mission: Cost Competitive Space Electricity," accessed August 7, 2019, https://www.solarenspace.com/

159. Jeff Bezos, "Going to Space to Benefit Earth," *YouTube*, May 9, 2019, accessed July 29, 2019, https://www.youtube.com/watch?v=GQ98hGUe6FM

160. Greg Klerkx, *Lost in Space: The Fall of NASA and the Dream of a New Space Age*, 65–80.

161. NASA STMD, STMD: Centennial Challenges, October 3, 2018, accessed August 7, 2019, https://www.nasa.gov/directorates/spacetech/centennial_challenges/history/index.html

162. NASA, Lunar Oxygen Production or MoonROx Challenge, August 3, 2017, accessed August 7, 2019, https://www.nasa.gov/directorates/spacetech/centennial_challenges/moonrox/index.html

163. NASA, "What Is NASA's Asteroid Redirect Mission?," December 11, 2017, accessed February 16, 2020, https://www.nasa.gov/content/what-is-nasa-s-asteroid-redirect-mission

164. NASA, "NASA Office to Coordinate Asteroid Detection, Hazard Mitigation," January 7, 2016, accessed February 21, 2020, https://www.nasa.gov/feature/nasa-office-to-coordinate-asteroid-detection-hazard-mitigation

165. Casey Dreier, "How NASA's Planetary Defense Budget Grew By More Than 4000% in 10 years," The Planetary Society, September 26, 2019, accessed February

21, 2020, https://www.planetary.org/blogs/casey-dreier/2019/nasas-planetary-defense-budget-growth.html

166. Executive Office of the President, "National Near-Earth Object Preparedness Strategy," December 2016, accessed February 21, 2020, available from NASA, https://www.nasa.gov/sites/default/files/atoms/files/national_near-earth_object_preparedness_strategy_tagged.pdf

167. Jeff Foust "White House releases near Earth object action plan," *Space News*, June 21, 2018, accessed February 21, 2020, https://spacenews.com/white-house-releases-near-earth-object-action-plan/ ; Tariq Malik, "How Trump's Space Force Would Help Protect Earth from Future Asteroid Threats," *Space.com*, June 20, 2018, accessed February 21, 2020, https://www.space.com/40949-trump-space-force-asteroid-defense.html (Plan available here: https://www.whitehouse.gov/wp-content/uploads/2018/06/National-Near-Earth-Object-Preparedness-Strategy-and-Action-Plan-23-pages-1MB.pdf)

168. White House, "Recommendations Approved by the National Space Council to President Trump," March 26, 2019, accessed August 8, 2019, http://www.spaceref.com/news/viewsr.html?pid=52358

169. White House, "Recommendations Approved by the National Space."

170. The White House, "Remarks by Vice President Pence at the Fifth Meeting of the National Space Council, Huntsville, AL," March 26, 2019, August 7, 2019, accessed February 16, 2020, https://www.whitehouse.gov/briefings-statements/remarks-vice-president-pence-fifth-meeting-national-space-council-huntsville-al/

171. NASA, What is Artemis?, July 25, 2019, accessed August 5, 2019, https://www.nasa.gov/what-is-artemis

172. NASA, Space Launch System, accessed August 8, 2019, https://www.nasa.gov/exploration/systems/sls/index.html and https://www.nasa.gov/sites/default/files/atoms/files/0080_sls_fact_sheet_10092018.pdf

173. NASA, Orion Overview, accessed August 8, 2019, https://www.nasa.gov/exploration/systems/orion/about/index.html

174. NASA, Q&A NASA's New Spaceship, November 13, 2018, accessed August 8, 2019, https://www.nasa.gov/feature/questions-nasas-new-spaceship

175. NASA does not deny the "over $2 billion" cost of a single SLS launch

Eric Berger, "NASA is working to bring down the cost of a single SLS launch," ArsTechnica, November 8, 2019, accessed February 23, 2020, https://arstechnica.com/science/2019/11/nasa-does-not-deny-the-over-2-billion-cost-of-a-single-sls-launch/; Hanneke Weitering, "NASA's New Human Spaceflight Chief Optimistic for Moon 2024 Goals," *Space.com*, December 04, 2019, accessed February 23, 2020, https://www.space.com/nasa-human-spaceflight-chief-douglas-loverro-moon-2024.html : "And while the development of SLS has cost more than the agency originally planned, the cost of launching an SLS rocket is also far greater than the cost of launching on a commercial heavy-lift rocket. At around $2 billion per launch, an SLS mission would cost roughly $1.5 billion more than a launch using SpaceX's Starship or Blue Origin's New Glenn rocket — both of which are reusable, unlike SLS."

176. Consider the opportunity cost to America's space development ambitions. If each SLS costs $2B+ per launch (not counting the tens of billions of development costs) "success" of each SLS launch (70–140 MT to LEO) means an opportunity cost to the U.S. taxpayers of:

=22 Falcon Heavy launches (a total of 1,452MT to LEO) @$90M & 63.8MT to LEO

=74 Falcon 9 Reusable launches (a total of 703MT to LEO) @$50M &13MT to LEO

=1,000 Starship launches (an estimated total of 100,000–150,000MT to LEO depending on actual payload) @$2M

Moreover, the U.S. sacrifices several advantages of high launch tempo: industrial learning, and all the public excitement such a launch tempo could generate. Moreover, for a development estimate of $10B through 2017 NASA through PPP's might have accelerated New Glenn, Starship, ULA's Vulcan-ACES, as well as a development program including Space Solar Power (Solaren and SPS-Alpha), as well as a NASA/DARPA's Nuclear thermal rocket. Even if SLS is successful, can US Spacepower afford such a Pyhric "success"? [Calculated by author Peter Garretson]

177. NASA, What is Artemis?

178. NASA, About CLPS, accessed August 5, 2019, https://www.nasa.gov/content/commercial-lunar-payload-services ; NASA, Commercial Lunar Payload Service CLPS, accessed August 5, 2019, https://www.nasa.gov/subject/16440/commercial-lunar-payload-services-clps/

179. Jeff Foust, "NASA Selects Nine Companies for Commercial Lunar Lander Program," *SpaceNews*, November 29, 2018, accessed August 7, 2019, https://spacenews.com/nasa-selects-nine-companies-for-commercial-lunar-lander-program/

180. Jeff Foust, "NASA Awards Contracts to Three Companies to Land Payloads on the Moon," *SpaceNews*, May 31, 2019, accessed August 7, 2019, https://spacenews.com/nasa-awards-contracts-to-three-companies-to-land-payloads-on-the-moon/

181. Narayan Prasad, "As the Race for Moon Heats Up, Chandrayaan-2 Could Give ISRO an Edge, and Spoils," *ThePrint*, June 28, 2019, accessed August 7, 2019, https://theprint.in/science/as-the-race-for-moon-heats-up-chandrayaan-2-could-give-isro-an-edge-and-spoils/254145/

182. NASA, "Commercial Lunar Payload Services Update," July 29, 2019, accessed August 5, 2019, https://www.nasa.gov/feature/commercial-lunar-payload-services-update

183. NASA, "Commercial Lunar Payload Services Update."

184. NASA, "NASA Announces US Industry Partnerships to Advance Moon, Mars Technology," July 30, 2019, accessed August 7, 2019, https://www.nasa.gov/press-release/nasa-announces-us-industry-partnerships-to-advance-moon-mars-technology

185. For an excellent overview of the strategic importance of Starship, see Casey Handmer, "The SpaceX Starship is a very big deal," *Casey Handmer's Blog*, October 29, 2019, accessed February 21, 2020, https://caseyhandmer.wordpress.com/2019/

10/29/the-spacex-starship-is-a-very-big-deal/; also see Jeffrey Becker, "Militarizing Musk," *War on the Rocks*, May 17, 2016, accessed February 21, 2020, https://warontherocks.com/2016/05/militarizing-musk/

186. Eric Ralf, "SpaceX to mature Starship Moon landing and orbital refueling tech with NASA's help," *Teslarati*, July 31, 2019, August 5, 2019, https://www.teslarati.com/spacex-starship-moon-landing-orbital-refueling-nasa/

187. National Space Society, "Position Paper: The NASA Commercial Crew Program," March 2014, accessed August 8, 2019.

188. Edgar Zapata, "An Assessment of Cost Improvements in the NASA COTS/CRS Program and Implications for Future NASA Missions," *NASA*, 2017, accessed August 7, 2019, https://ntrs.nasa.gov/archive/nasa/casi.ntrs.nasa.gov/20170008895.pdf

189. Greg Autry, "Commercial Orbital Transportation Services: A Case Study for National Industrial Policy," *NewSpace*, September 2018, accessed August 7, 2019, https://www.liebertpub.com/doi/10.1089/space.2017.0043

190. Allison F. Zuniga, Daniel Rasky, Bruce Pittman, Edgar Zapata, and Robert Lepsch, "Lunar COTS: An Economical and Sustainable Approach to Reaching Mars," AIAA 2015, *NASA*, accessed August 7, 2019, https://www.nasa.gov/sites/default/files/atoms/files/aiaa2015-4408zunigalunarcotspaper_0.pdf

191. NextGen LLC, *Economic Assessment and Systems Analysis of an Evolvable Lunar Architecture that Leverages Commercial Space Capabilities and Public-Private-Partnerships*, July 13, 2015, accessed August 7, 2019, https://space.nss.org/media/Evolvable-Lunar-Architecture.pdf

192. *FAST SPACE: Leveraging Ultra low-cost Space Access for 21st Century Challenges*, Air University, January 13, 2017, accessed August 1, 2019, https://www.airuniversity.af.edu/Portals/10/Research/Space-Horizons/documents/Fast%20Space_Public_2017.pdf

193. Joel Sercel, "Stepping Stones: Economic Analysis of Space Transportation Supplied From NEO Resources," *NASA*, August 16, 2017, accessed August 7, 2019, fiso.spiritastro.net/telecon/Sercel_8-16-17/Sercel_8-16-17.pdf Final report here: https://www.nasa.gov/sites/default/files/atoms/files/eso_final_report.pdf

194. David Kornuta, Angel Abbud-Madrid, Jared Atkinson, Jonathan Barr, Gary Barnhard, Dallas Bienhoff, Brad Blair et al. "Commercial Lunar Propellant Architecture: A Collaborative Study of Lunar Propellant Production." *Reach* (March 14, 2019): 100026.

195. Zapata, "An Assessment of Cost Improvements."

196. *FAST SPACE: Leveraging Ultra low-cost Space Access for 21st Century Challenges*, Air University, January 13, 2017, accessed August 1, 2019, https://www.airuniversity.af.edu/Portals/10/Research/Space-Horizons/documents/Fast%20Space_Public_2017.pdf

197. *FAST SPACE: Leveraging Ultra low-cost Space Access for 21st Century Challenges*, Air University, January 13, 2017, accessed August 1, 2019, https://www

.airuniversity.af.edu/Portals/10/Research/Space-Horizons/documents/Fast%20Space_Public_2017.pdf

198. *FAST SPACE: Leveraging Ultra low-cost*

199. *FAST SPACE: Leveraging Ultra low-cost*

200. NextGen LLC, *Economic Assessment and Systems.*

201. Sercel, "Stepping Stones: Economic Analysis."

202. Bryan Bender, "Newt Gingrich trying to sell Trump on a cheap moon plan: A general, Gingrich and Michael Jackson's publicist are proposing a $2 billion contest to return Americans to the moon," *Politico*, August 19, 2019, accessed October 13, 2019, https://www.politico.com/story/2019/08/19/newt-gingrich-michael-jackson-moon-1466853

203. Bryan Bender, "Newt Gingrich trying to sell Trump on a cheap moon plan: A general, Gingrich and Michael Jackson's publicist are proposing a $2 billion contest to return Americans to the moon," *Politico*, August 19, 2019, accessed October 13, 2019, https://www.politico.com/story/2019/08/19/newt-gingrich-michael-jackson-moon-1466853; Newt Gingrich, "Gingrich, Kwast, Walker and Autry: A $2 Billion Prize Will Bring US Permanent Moon and Mars Settlements," *Fox News*, August 22, 2019, accessed October 13, 2019, https://www.foxnews.com/opinion/newt-gingrich-steve-kwast-bob-walker-greg-autry-we-have-an-innovative-plan-for-getting-back-to-space

204. Chris Ciaccia, "Elon Musk Lauds Newt Gingrich's $2B Moon Competition Prize," *Fox News*, August 20, 2019, accessed October 13, 2019, https://www.foxnews.com/science/elon-musk-newt-gingrich-moon-prize

205. Roberta Ewart, "XST: On the Shoulders of Giants," November 13, 2018, accessed August 8, 2019, http://afcea-la.org/afceala-wp/wp-content/uploads/XST-Dr-Ewart-AFCEA-15-Nov-2018.pdf

206. Air Force, Department of Defense Fiscal Year (FY) 2019 Budget Estimates for Research, Development, Test & Evaluation, Air Force Vol-II, February 2018, pp. 657, 662, accessed August 9, 2019, https://www.saffm.hq.af.mil/Portals/84/documents/FY19/RDT_E/Air%20Force%20Research%20Development%20Test%20and%20Evaluation%20Vol%20II%20FY19%20v2.pdf?ver=2018-02-20-125551-480

Air Force had requested $283M for Space Rapid Capability Office (RCO) Solar Power for Fiscal Year 2019, pp. 657, 662

207. Air Force, Department of Defense Fiscal Year (FY) 2020 Budget Estimates for Research, Development, Test & Evaluation, Air Force Vol-II, March 2019, 420, 580, accessed August 9, 2019, https://www.saffm.hq.af.mil/Portals/84/documents/FY20/RDTE/FY20_PB_RDTE_Vol-II.PDF?ver=2019-03-18-153506-683

208. Air Force, Department of Defense Fiscal Year (FY) 2020.

209. Paul Jaffe, Kaylin Borders, Colin Browne, Chris DePuma, Lewis Longbottom, Hassan Nisar, Vinay Simlot, Avram Bar-Cohen, James McSpadden, Ryan Brandt, Eric Conrad, Katherine Duncan, William Rowley, Tony Thampan, Peter Garretson, John C. Mankins, Seth Potter, Eric Sundberg, Alexander Walts, *Opportunities and Challenges for Space Solar for Remote Installations* (Washington: Naval Research Lab, October 21, 2019).

210. Office of the Secretary of Defense, Department of Defense Fiscal Year (FY) 2020 Budget Estimates for Research, Development, Test & Evaluation, Defense-Wide, March 2019, P 420, accessed August 9, 2019, https://comptroller.defense.gov/Portals/45/Documents/defbudget/fy2020/budget_justification/pdfs/03_RDT_and_E/RDTE_Vol3_OSD_RDTE_PB20_Justification_Book.pdf

211. Office of the Secretary of Defense, Department of Defense Fiscal Year (FY) 2020 Budget Estimates for Research, Development, Test & Evaluation, Defense-Wide, March 2019, P 420, accessed August 9, 2019, https://comptroller.defense.gov/Portals/45/Documents/defbudget/fy2020/budget_justification/pdfs/03_RDT_and_E/RDTE_Vol3_OSD_RDTE_PB20_Justification_Book.pdf

212. Office of the Secretary of Defense, Department of Defense Fiscal Year (FY) 2020 Budget Estimates for Research, Development, Test & Evaluation, Defense-Wide, March 2019, P 420, accessed August 9, 2019, https://comptroller.defense.gov/Portals/45/Documents/defbudget/fy2020/budget_justification/pdfs/03_RDT_and_E/RDTE_Vol3_OSD_RDTE_PB20_Justification_Book.pdf

213. Sandra Erwin, "Air Force, DoD research organizations call for space industrial policies to counter China," *Space News*, June 17, 2019, accessed August 8, 2019, https://spacenews.com/air-force-dod-research-organizations-call-for-space-industrial-policies-to-counter-china/

214. Theresa Hitchens, "Pentagon Eyes Military Space Station," *Breaking Defense*, July 2, 2019, accessed August 8, 2019, https://breakingdefense.com/2019/07/pentagon-eyes-military-space-station/

215. Keith Cowling, "Who is DIU and Do They Want Their Own Space Station?," *NASA Watch*, July 5, 2019, accessed February 16, 2020, http://nasawatch.com/archives/2019/07/who-is-diu-and.html

216. Theresa Hitchens, "SDA's Kennedy: Cislunar Space The Next Military Frontier," *Breaking Defense*, April 17, 2019, accessed August 8, 2019, https://breakingdefense.com/2019/04/sdas-kennedy-cislunar-space-the-next-military-frontier/

217. The workshop provided this problem statement, "In an ever-changing world, the U.S. must remain vigilant and transform to continue protecting space-based national security and commercial enterprises. Current and potential adversaries recognize space dominance will provide significant political, economic, and military advantages. As a result, adversaries have developed and are executing long-term, integrated strategies and plans to ensure space dominance. The U.S. must be able to plan for the future and be strategically agile in conceptualizing future capability needs and leveraging game-changing technologies, prototyping, and experimentation" from NDIA, "AFSPC Space Futures Planning Meeting," March 5, 2019, accessed August 8, 2019, https://ndiarmc.org/product/afspc-space-futures-planning-meeting-march-5-2019/

218. Air Force Space Command, "The Future of Space 2060 and Implications for U.S. Strategy", Report on the Space Futures Workshop, September 5, 2019, accessed September 16, https://www.afspc.af.mil/Portals/3/documents/Future%20of%20Space%202060%20v2%20(5%20Sep).pdf?ver=2019-09-06-184933-230

219. NDIA, "AFSPC Space Futures Planning Meeting."

220. Patrick Tucker, "China's Moon Missions Could Threaten US Satellites: Pentagon," *DefenseOne*, October 16, 2018, accessed February 13, 2020, https://www

.defenseone.com/technology/2018/10/chinas-moon-missions-could-threaten-us-satellites-pentagon/152084/; Vivienne Machi, "U.S. Space Sectors Must Work Together to Combat Emerging Threats, Official Says," *Defense Daily*, October 12, 2018, accessed February 13, 2020, https://www.defensedaily.com/u-s-space-sectors-must-work-together-combat-emerging-threats/space/

221. Leonard David, "US Military Eyes Strategic Value of Earth-Moon Space," *Space.com*, August 29, 2019accessed February 13, 2020, https://www.space.com/us-military-strategic-value-earth-moon-space.html; Sean McClain (Major, USAF), *Celestial Sentinels: A Framework for Cis-Lunar Space Domain Awareness In 2035* (Master's Thesis), AIR COMMAND AND STAFF COLLEGE, AIR UNIVERSITY, March 2020.

222. Air Force Space Command, "The Future of Space 2060: Implications for U.S. Strategy," September 5, 2019, available at: https://www.afspc.af.mil/Portals/3/The%20Future%20of%20Space%202060%20-%203Oct19.pdf

223. Maj Gen John Shaw, Deputy Commander of Air Force Space Command: "When the Space Force stands up, it's going to be around for a long time, and its ultimate destiny is going to be providing security and projecting power for increasingly vast distances—from geosynchronous (orbit) to cislunar to beyond," from Rachel S. Cohen, "AFSPC Mulls Intel, Personnel Questions of the New Space Age," *Air Force Magazine*, September 5, 2019, accessed February 13, 2020, https://www.airforcemag.com/afspc-mulls-intel-personnel-questions-of-the-new-space-age/

224. Theresa Hitchens, "SDA's Kennedy: Cislunar Space The Next Military Frontier," *Breaking Defense*, April 17, 2019, accessed February 13, 2020, https://breakingdefense.com/2019/04/sdas-kennedy-cislunar-space-the-next-military-frontier/; Nathan Strout , "The Pentagon wants to solve a deep space problem with three vehicles," *C4ISRNET*, August 27, 2019, accessed February 13, 2020, https://www.c4isrnet.com/battlefield-tech/space/2019/08/27/the-pentagon-wants-to-solve-a-deep-space-problem-with-three-vehicles/

225. Said the AFWERX pre-solicitation. "The Air Force is seeking commercial innovation in support of space domain awareness for future cislunar operations" and listed Specific items the Air Force wants: payloads for providing space domain awareness from the lunar surface, lightweight sensors for space-based space domain awareness, and methodologies for orbit determination and catalog maintenance in cislunar space as well as interest in concepts for providing position, navigation and timing solutions for cislunar space operations; visualization of cislunar orbits; and terrestrial-based concepts for achieving space domain awareness of cislunar space. From Sandra Erwin, "Air Force Seeking Commercial Technologies for Cislunar Space Operations," *Space News*, December 12, 2019, accessed February 13, 2020, https://spacenews.com/air-force-seeking-commercial-technologies-for-cislunar-space-operations/

226. On July 2, 2019, DIU held an open solicitation seeking solutions for a self-contained and free flying orbital outpost which could support "space assembly, microgravity experimentation, logistics and storage, manufacturing, training, test and evaluation, hosting payloads, and other functions" and future capabilities to dock with unmanned and manned spacecraft and human rating. From Theresa Hitchens, "Pentagon Eyes Military Space Station," *Breaking Defense*, July 02, 2019, accessed February 13, 2020, https://breakingdefense.com/2019/07/pentagon-eyes-military-s

pace-station/; On February 12, 2020, DIU website posted an open solicitation for "Multi-Orbit Logistics" looking for ", a multi-orbit logistics vehicle (m-OLV) should be capable of autonomously transferring payloads or spacecraft to, within and between LEO, GEO, cislunar and other exotic orbits. Desired solutions should provide commercial services in the form of deep-space capable orbit transfer vehicles, fuel depots or other in-orbit mission enhancement opportunities, a multi-orbit logistics vehicle (m-OLV) should be capable of autonomously transferring payloads or spacecraft to, within and between LEO, GEO, cislunar and other exotic orbits. Desired solutions should provide commercial services in the form of deep-space capable orbit transfer vehicles, fuel depots or other in-orbit mission enhancement opportunities." From "Multi-Orbit Logistics Services," *DIU*, February 12, 2020, accessed February 13, 2020, https://www.diu.mil/work-with-us/open-solicitations

227. Kirtland Public Affairs, "U.S. Air Force Research Laboratory Developing Space Solar Power Beaming," Air Force Research Laboratory, October 24, 2019, accessed February 13, 2020, https://www.robins.af.mil/News/Article-Display/Article/1998062/us-air-force-research-laboratory-developing-space-solar-power-beaming/; "SSPIDR will study and analyze the concept of a space-based power collection and transmission capability using light weight, high efficiency solar cells coupled with RF transmitters to meet the project objectives. The project will culminate by testing the components of a space-based power collection and transmission demonstration system, capable of capturing solar energy and converting to RF energy for beaming power to the ground." From Doug Messier, "Northrop Grumman, AFRL Partner to Provide Critical Advanced Technology in Space Solar Power," *Parabolic Arc*, December 11, 2019, accessed February 13, 2020, http://www.parabolicarc.com/2019/12/11/northrop-grumman-afrl-partner-to-provide-critical-advanced-technology-in-space-solar-power/#more-71617

228. Bullock Museum, "Cowboys In Space And Fantastic Worlds: A Journey Through the History of Westerns in Science Fiction," *The Story of Texas*, December 1, 2019, accessed February 21, 2020, https://www.thestoryoftexas.com/press/media-kits/cowboys-in-space-20190715; Richard Whittaker, "'Cowboys in Space' Exhibit Finds the Frontier in the Stars: Exploring the infinite plains at the Bullock Museum," *Austin Chronicle*, October 11, 2019, accessed February 21, 2020, https://www.austinchronicle.com/screens/2019-10-11/cowboys-in-space-exhibit-finds-the-frontier-in-the-stars/

229. Jedediah Britton-Purdy, "Infinite Frontier: The Eternal Return of American Expansionism," *The Nation*, April 1, 2019, accessed February 21, 2020, https://www.thenation.com/article/archive/greg-grandin-end-of-the-myth-frontier-border-wall-book-review/

230. Mark Boardman, "Go West, Young Man? Did anybody actually say that famous phrase?," *True West Magazine*, June 5, 2015, accessed February 21, 2020, https://truewestmagazine.com/go-west-young-man/; Stephen J. Taylor, "Go West, Young Man": The Mystery Behind The Famous Phrase, Hoosier State Chronicles, July 9, 2015, accessed February 21, 2020, https://blog.newspapers.library.in.gov/go-west-young-man-the-mystery-behind-the-famous-phrase/

231. "Manifest Destiny," November 15, 2019, *History.com*, accessed February 19, 2020, https://www.history.com/topics/westward-expansion/manifest-destiny

232. Thomas G. Mahnken, “United States Strategic Culture,” SAIC (Prepared for Defense Threat Reduction Agency), 2006, accessed September 16, 2019, http://www.au.af.mil/au/awc/awcgate/dtra/mahnken_strat_culture.pdf: This Mahnken states: “impulse to *transform the international system in the service of liberal democratic ideals forms a strand that runs throughout American history*.”

233. Atlantic Council, “One Hundred Years of American Grand Strategy,” n.d., accessed August 13, 2019, https://www.atlanticcouncil.org/blogs/new-atlanticist/one-hundred-years-of-american-grand-strategy : America’s idealism, such as President Wilson’s “Fourteen Points were not abstract ‘idealism’ or charity, but reflected canny assumptions that: Yankee ingenuity would flourish best in a rules-based, open world without closed economic empires; The United States’ interests would advance with democracy and the rule of law; The United States would prosper when other nations did as well, and thus; The United States could make the world a better place and get rich in the process . . . U.S. leadership is essential. The alternative isn’t pretty, as we discovered the last time we tried to withdraw from our responsibilities.

234. Atlantic Council, “One Hundred Years of American Grand Strategy.”: “The West and the world are *composed of sovereign nations*. Patriotism is a good thing. That thought was integral to the Wilsonian Fourteen Points. But *sovereignty and the nation are not absolute*. They are *qualified by responsibility before higher values*. The Declaration of Independence makes that clear. Values are no luxury or artifice. As Abraham Lincoln observed, the American nation exists through common values rather than common blood. Thus, *we support, in our better moments, a rules-based international order reflecting these common values. The alternative—indifference to values and deference to power—was tried and found wanting*. Happily, we also learned that a rules-based world can work for our bottom line. The foundation of the American Grand Strategy remains valid.”

235. Martin Luther King Jr., “Letter from the Birmingham Jail” (April 16, 1963), Charter for Compassion, n.d., accessed February 21, 2020, https://charterforcompassion.org/images/SocialJustice/birmingham.pdf : “Injustice anywhere is a threat to justice everywhere. We are caught in an inescapable network of mutuality, tied in a single garment of destiny. Whatever affects one directly, affects all indirectly.”

236. Dan Deudney develops at length a theory about the unique problems of republican security. His “Republican Security Theory,” explores how the material context generated unique problems for the survival of republics. Elaborated in Dan H. Deudney, *Bounding Power: Repubican Security Theory from the Polis to the Global Village* (Princeton: Princeton University Press, 2007).

237. Patrick Porter, “Why America’s Grand Strategy Has Not Changed,” *International Security* 42, no. 4 (Spring 2018), 9–46, doi:10.1162/ISEC_a_00311, MIT Press, accessed August 13, 2019, https://www.mitpressjournals.org/doi/pdf/10.1162/isec_a_00311

238. Patrick Porter, “Why America’s Grand Strategy Has Not Changed: “Long before the fall of the Soviet Union, the *United States formed a grand strategy of ‘primacy,’ often coined as ‘leadership.’* This strategy was interrupted only occasionally. By the 1960s, it had set the parameters for Washington’s foreign policy debate. The strategy has *four interlocking parts: to be militarily preponderant*; to *reassure and*

contain allies; to integrate other states into U.S.-designed institutions and markets; and to inhibit the spread of nuclear weapons. These fundamental security commitments have proven hard to change, even amid shocks."

239. Hal Brands, *American Grand Strategy and the Liberal Order: Continuity, Change, and Options for the Future*, RAND, 2016, accessed August 13, 2019, https://www.rand.org/content/dam/rand/pubs/perspectives/PE200/PE209/RAND_PE209.pdf : "The international order is the body of rules, norms, and institutions that govern relations among the key players in the international environment. Since World War II, the dominant international order has been the liberal order anchored by the United States. That order has been broadly characterized by an *emphasis on liberal norms and values—including economic liberalism* in the form of relatively free trade and open markets; political liberalism in the form of representative government and *human rights; and other liberal concepts, such as nonaggression, self-determination, and the peaceful settlement of disputes.* These norms and values have been *manifested through such international institutions* as the International Monetary Fund (IMF), World Bank, World Trade Organization, United Nations, and European Union (EU). Both liberal norms and liberal institutions, moreover, have been bolstered by geopolitical arrangements that have fostered the international climate in which a liberal order can flourish. Alliances that have deterred aggression and promoted stability, arms control agreements that have reduced the danger of great-power war, and efforts to contain or *roll back the influence of illiberal powers*—these and similar initiatives have provided the foundation upon which the liberal order rests. As Robert Kagan, Michael Mandelbaum, and other scholars have noted, *international norms and rules are not divorced from underlying geopolitical realities and power dynamics; they are often directly a function thereof.*"

240. For example, the Organization for Economic Cooperation and Development (OECD), World Bank, International Monetary Fund (IMF), International Civil Aviation Organization (ICAO) and the dollar as the world's reserve currency.

241. R.D. Hooker, Jr., *The Grand Strategy of the United States*, INSS Strategic Monograph, Institute for National Strategic Studies, National Defense University, Washington D.C., October 2014, accessed August 13, 2019, https://inss.ndu.edu/Portals/68/Documents/Books/grand-strategy-us.pdf; "American grand strategy at mid-century continued to rest on the foundations described above and could be summarized concisely as *monitoring and enforcing a stable international order and economic system that preserved American sovereignty, security, and prosperity; ensuring the security of the homeland through nuclear deterrence, alliances, forward-deployed ground forces, and airpower and seapower;* and *preventing the rise of peer competitors that might challenge its economic and military superiority.* The isolationism that had always existed as a strain in American foreign policy would not disappear altogether, but it would never again contend for primacy in grand strategy."

242. The Democratic Peace literature has much to say on how Democracies perceive illiberal regimes as threats and may militate for war. Bruce Russett and John R. Oneal, *Triangulating Peace: Democracy, Interdependence, and International Organizations* (New York: Norton, 2001); Bruce Russett, *Grasping the Democratic*

Peace: Principles for a Post–Cold War World (Princeton, NJ: Princeton University Press, 1993); Michael W Doyle, *Liberal Peace: Selected Essays* (New York: Routledge, 2011).

243. Andrew Lamber, *Seapower States* (New Haven, CT: Yale University Press, 2019), 311: "The Western liberal order, largely shaped by the political, economic and intellectual legacy of seapower states, has retained strategic seapower since 1945, under a US umbrella. This has enabled sea states, and other global economic actors, to function as seapowers, at relatively low cost and with little strategic risk . . . Contemporary sea states, states with the sea at the heart of their identity and economy, include Japan, the Netherlands, Denmark, Britain, Norway, and Singapore and they are not alone. Such states are disproportionately engaged in international oceanic trade, maritime identities occupy prominent places in their culture and they will be first to respond when the peaceful use of the sea in threatened."

244. Mahnken, "United States Strategic Culture: "As a nation, American strategic culture was shaped by free security and imbued with exceptionalism. American strategic culture emphasizes liberal idealism and views war as a discontinuation of policy. American military culture, the so called 'American way of war', emphasizes direct strategies, an industrial approach to war, and firepower- and technology intensive approaches to combat." America's early experiences and its idealism leads to, "American strategic culture explicitly rejects the European tradition of power politics. Rather, from the founding Americans have seen themselves as exceptional. This exceptionalism has influenced the way the United States deals with others."

245. "*Americans have Shown a Disinclination to Use Force*, especially when it is *to Maintain Parity in Balance of Power Politics*," from Russell A. Moore, "Strategic Culture—How It Affects Strategic "Outputs," CSC 1998, accessed August 1, 2019, https://apps.dtic.mil/dtic/tr/fulltext/u2/a525913.pdf

246. Mahnken, "United States Strategic Culture."

247. Mahnken, "United States Strategic Culture"; Mahnken quotes J.C. Wylie: "*Is War in Fact a Continuation of Policy? For Us, I Think Not.* War for a nonaggressor nation is actually a nearly *complete collapse of policy*. Once war comes, then nearly *all prewar policy is utterly invalid* because the setting in which it was designed to function no Walter Lippmann, *Public Opinion and Foreign Policy in the United States.*:"*an aggression is an armed rebellion against the universal and eternal principles of the world society*. No war can end rightly, therefore, *except by the unconditional surrender* of the aggressor nation and by the overthrow and transformation of its political regime." longer corresponds with the facts of reality. *When war comes, we at once move into a radically different world*."

248. Jack Snyder, "The Broken Bargain How Nationalism Came Back," *Foreign Affairs*, March/April 2019, accessed August 13, 2019, https://www.foreignaffairs.com/articles/world/2019-02-12/broken-bargain

249. J.E. Stiglitz, "Trump and Globalization", *Journal of Policy Modeling* (2018), https://doi.org/10.1016/j.jpolmod.2018.03.006, accessed August 13, 2019, https://www8.gsb.columbia.edu/faculty/jstiglitz/sites/jstiglitz/files/Trump%20and%20Globalization_0.pdf

250. Hal Brands, *American Grand Strategy and the Liberal Order: Continuity, Change, and Options for the Future* (Santa Monica: RAND, 2016), accessed August 13, 2019, https://www.rand.org/content/dam/rand/pubs/perspectives/PE200/PE209/RAND_PE209.pdf : *"Support for the liberal international order has been the single most consistent theme of U.S. grand strategy* since World War II, and over seven decades, U.S. grand strategy has contributed markedly to the success and advancement of that order. Yet the precise form and manifestations of U.S. support have varied from year to year, and from presidential administration to administration, and America's relationship *with the liberal order may again be reaching an inflection point.*"

251. Colin S. Gray, *The Geopolitics of Super Power* (Lexington, KY: The University Press of Kentucky, 1988): 65.

252. Lippmann, *Public Opinion and Foreign Policy*. No war can end rightly, therefore, *except by the unconditional surrender* of the aggressor nation and by the overthrow and transformation of its political regime."

253. Russell A Moore, "Strategic Culture": *"limited (political) war is generally distasteful to Americans. Use of force must be portrayed as a crusade for values.*"

254. George F. Kennan, *American Diplomacy: Expanded Edition* (Chicago: University of Chicago Press, 1984): 100. "a war fought in the name of high moral principle finds no early end short of some form of total domination."

255. Walter Lippmann, *Public Opinion and Foreign Policy in the United States* (London: Allen and Unwin, 1952), 25–26.: *"an aggression is an armed rebellion against the universal and eternal principles of the world society.* No war can end rightly, therefore, *except by the unconditional surrender* of the aggressor nation and by the overthrow and transformation of its political regime."

256. For example, the Lone Ranger, or the many characters played by John Wayne; James T. Campbell, "'Print the Legend': John Wayne and Postwar American Culture," *Reviews in American History* 28, no. 3 (2000): 465–477.; 05 Michael E. Telzrow, "John Wayne: American Icon," *The New America*, August 5, 2016, accessed February 22, 2020, https://www.thenewamerican.com/culture/history/item/23714-john-wayne-american-icon; John Powers, "John Wayne: Icon Of America's Booming Confidence," October 7, 2011, accessed February 22, 2020, https://www.npr.org/2011/10/07/140870373/john-wayne-icon-of-americas-booming-confidence

257. Samuel Bazzi, Martin Fiszbein, and Mesay Gebresilasse. "Frontier Culture: The Roots and Persistence of 'Rugged Individualism' in the United States," *National Bureau of Economic Research* No. w23997, 2017, accessed February 21, 2020, https://www.bu.edu/econ/files/2017/03/Frontier-Culture-The-Roots-and-Persistence-of-%E2%80%9CRugged-Individualism%E2%80%9D-in-the-United-States-2.pdf

258. Nayef R. F Al-Rodhan, "U.S. Space Policy and Strategic Culture," *Colombia Journal of International Affairs*, April 16, 2018, accessed August 1, 2019, https://jia.sipa.columbia.edu/online-articles/us-space-policy-and-strategic-culture : Nayef R. F. Al-Rodhan both asserts a continuity, and asks a question we can answer: "the United States has a consistent Strategic Culture, centered on a sense of *exceptionalism, values-driven foreign policy*, casualty aversion, as well as *strong leadership*. These features reflect in its space policy and ambitions, where the *US has cooperated with*

other nations but assumes a position of natural leader. It *remains to be seen how this role will evolve – although the sense of exceptionalism may endure – as new actors, including private companies*, make their way into space."

259. Mahnken, "United States Strategic Culture.": "American insularity and the existence of free security bred the view that America may be averse to take actions that in effect constitute low-level persistent conflict for the purposes of power politics. This can create an unfortunate split between foreign policy and military strategy, and who will be the carrier of and coordinator of strategy. Leading thinkers like Mahnken assert that the United States tends to 'think astrategically' and see politics and strategy as things fundamentally apart."

260. Lamont Colucci, "A Space Service in support of American grand strategy," *The Space Review*, February 25, 2019, accessed August 13, 2019, http://www.thespacereview.com/article/3664/1 Colucci specifies: "A Space Service in support of American grand strategy,": "Grand strategy, at its core, attempts to harness military, economic, and political power to advance the nation. It is created organically, over decades and centuries, and for it to be successful must be forward-looking, peering across the horizon into the centuries uncounted. Grand strategy is the most critical form of statecraft. It implies the use of force to promote these interests. Grand strategy is married to hard power and military force; unlike domestic policy, it creates the conditions for either total triumph or total destruction. Grand strategy is often ignored because it is inconvenient, hard to change, and subject to the tyranny of the status quo. Its development requires a formidable depth of knowledge... To ignore grand strategy is to engage in ad-hoc policy anchored by nothing, moving nowhere. Grand strategy is further burdensome since it requires constant adaptation."

261. Speaking specifically of space, Colucci connected U.S. Grand Strategy, space and military space power, asserting, "*American grand strategy is fundamentally based on military primacy, and space dominance will determine which nation is in that position*" and argued, "the *creation of the Space Service would signal to America's adversaries the seriousness* in which we take grand strategy beyond rhetoric. *The Space Service would therefore be the foundation for American grand strategy of the 21st century and beyond.*"

262. Lamber, *Seapower States.*

263. Charles D. Lutes, et al., *Towards a Theory of Spacepower Selected Essays* (Washington D.C: National Defense University, 2001), accessed September 16, 2019, https://ndupress.ndu.edu/Portals/68/Documents/Books/spacepower.pdf

264. Ian W. Toll, *Six Frigates: The Epic History of the Founding of the U.S. Navy* (New York: W.W. Norton & Company, 2008)

265. Max Boot, *The Savage Wars of Peace: Small Wars and the Rise of American Power* (New York: Basic Books, 2002), 9.

266. Toll, *Six Frigates: The Epic History.*

267. Max Boot, *The Savage Wars of Peace: Small Wars and the Rise of American Power,* 8.

268. Boot, *The Savage Wars of Peace*, 8.

269. Toll, *Six Frigates: The Epic History.*

270. Toll, *Six Frigates.*

271. Toll, *Six Frigates*, 25.
272. Boot, *The Savage Wars of Peace*, 10.
273. Toll, *Six Frigates: The Epic History*.
274. Toll, *Six Frigates: The Epic History*.
275. Toll, *Six Frigates: The Epic History*.
276. Ian W Toll, *Six Frigates: The Epic History of the Founding of the U.S. Navy*.
277. Ian W Toll, *Six Frigates: The Epic History of the Founding of the U.S. Navy*.
278. Boot, *The Savage Wars of Peace*, 11.
279. Boot, *The Savage Wars of Peace*, 11.
280. Toll, *Six Frigates: The Epic History*.
281. Boot, *The Savage Wars of Peace*.
282. Peter A. Garretson., "The Future of the Asian Order Will Be Decided in Space," *The Diplomat*, July 3, 2019, accessed July 27, 2019, https://thediplomat.com/2019/07/the-future-of-the-asian-order-will-be-decided-in-space/
283. Lippmann, *Public Opinion and Foreign Policy*.
284. "Quotes about PREPARATION," Leadership Now, n.d., accessed February 19, 2020, https://www.leadershipnow.com/preparationquotes.html
285. For an excellent examination of America's clever use of commercial aviation to extend and protect its strategic footprint, see Mark Cotta Vaz and John H. Hill, *Pan Am at War: How the Airline Secretly Helped America Fight World War II* (New York: Skyhorse, 2019).
286. The alternative would look more like U.S. Naval Expansion through the Caribbean, Panama, and Western Pacific as it sought to develop bases and U.S. Airpower expansion first using PanAm.
287. Wolfgang Wegener, *The Naval Strategy of the World War* (Anapolis: US Naval Institute Press, 1989): 109. "Suppose our old army likewise had adopted our 'risk theory.' The army might then have said: To guarantee peace, we will reduce our military strength and maintain a force just strong enough to defend our position at the front…so we will not be so strong as to threaten France with a strategic offensive… This situation is precisely what our fleet faced"; and from 108: "Lacking the will to sea power, we consequently posed no peril for the enemy."
288. Wegener, *The Naval Strategy of the World War*, 86.
289. Wegener, *The Naval Strategy of the World War,* 95–117.
290. Wegener, *The Naval Strategy of the World War*, 91.
291. Wegener, *The Naval Strategy of the World War*. Notes in contrast a quote by Beadon that "The British fleet did not exist for the purpose of fighting the German Fleet, it existed for the purpose of obtaining and maintaining command of the seas."
292. U.S.-China Economic and Security Review Commission, Hearing on "China in Space: A Strategic Competition?," April 25, 2019, accessed July 27, 2019, https://www.uscc.gov/Hearings/china-space-strategic-competition

Chapter 5

China's Strategy and Space Resource Ambitions

China[1] is emerging as one of the most powerful nations with demonstrated capability to sustain space presence. At the time of the writing, the *Chang'e 4* rover *Yutu 2* entered its twentieth lunar day, having sustained frigid lunar temperatures for a year, since it soft-landed on the far side of the Moon on January 3, 2019. It continues to map and explore the lunar far side, one of the least known to humankind. With ambitions to scale asteroids, space mining, and establish capacity for space-based solar power (SBSP) power beaming, China funds one of the most ambitious space programs, among the spacefaring nations. This chapter offers an in-depth insight into that program, as well as identifies some pillars of China's strategic culture, its strategic traumas that influence its preferences, and offers guidance on expected behavior.

CHINA: CURRENT AND FORECAST POWER

At the time of writing, the People's Republic of China (PRC) finds itself in the first ranks of nations in the world order. Ranked #3 among the most powerful countries by U.S. News and World Report.[2] Global Firepower ranks it as #3 in military power.[3] The Stockholm International Peace Research Institute (SIPRI) estimates its military spending at $228,231Bn.[4] The International Institute for Strategic Studies (IISS) estimates it at $150.5Bn and places it as #2 in the military balance.[5] China has a population of 1,420,062,022 growing at 0.35 percent,[6] with a Human Development Index score of 0.752 (ranked #86/189 countries).[7] Its total energy consumption was 3,105 million tons of oil equivalent (MTOE[8]; 15.62 barrels of oil (BOE) per capita), growing at 2.9 percent, with a carbon output of 9,297 MtCO2.[9] Its electrical consumption is 5,683 TWh growing at 5.9 percent[10], with a per capita electrical usage of 3,927 Kw-hr/annum.[11] According to the IMF figures, its GDP (real) was $12,014.61Bn, and $23,159.11Bn at Purchasing Power Parity (PPP), with a per capita income of $8,643.11. It spent 31.5 percent of its GDP in government expenditures, its estimated gross national savings was 45.84 percent of GDP and had a current account balance of $164.887Bn (1.372 percent of GDP). According to the World Economic Forum, it spends $6.111Bn on its space program (0.07 percent of GDP)[12], whereas Euroconsult estimated its space spending at $4.909 billion.[13] The most recent (and somewhat dated) Futron Space competitiveness report, gave it an overall space competitive score of 19.44 (ranked #5),[14] with an overall military space capability score of 25.27 (ranked #4).[15] Latest media reports indicate that China's space budget in 2019 was about $8billion.[16]

According to PricewaterhouseCoopers, China's economy is expected to grow from $16 trillion (T) in 2020 to $49.9T in 2050 at market exchange rates (MER or "real" GDP) and from $26.9T to 58.5T in PPP, suggesting it will be the #1 economy in 2050 at MER and the #1 economy measured by PPP, with 20 percent share of the total global economy (up from 18 percent in 2016).[17] The Carnegie Endowment for International Peace (CEIP) similarly estimated it to be the #1 economy in 2050 at $46.265T.[18] The *Economist Intelligence Unit* (EIU) placed the PRC at #2 at MER, with a projected GDP rising from $10.335T in 2014 to $105.916T in 2050.

CHINA: SPACE CAPACITY

With the landing on the far side of the Moon on January 3, 2019, China demonstrated to the world its growing space capabilities. Never has another country landed any object, manned or unmanned on the Moon's far side. Consequently, when *Chang'e 4* landed on the Van Karmen crater, it registered humanity's first foray into the Moon's far side, an area believed to be rich in resources. China has gone ahead and announced follow-on ambitions for its lunar program, to include the *Chang'e 5* probe to the Moon later in 2020, that will collect lunar samples and bring them back to Earth for research and analysis.[19] As part of China's Lunar Exploration Program (CLEP),[20] China will continue to send probes to the Moon, to build capacity, and expertise on establishing a lunar research base by 2036. Missions that will build that capacity incrementally include the *Chang'e 6* that will bring back samples from the South Pole of the Moon, the *Chang'e 7* that aims to carry out detailed survey of the South Pole, to include its physical composition (terrain and landform), and the *Chang'e 8* that will test technologies in service of establishing a human settlement, to include 3D printing technology that can be used to construct objects from lunar resources.[21] China views the Moon as a critical step to enhance its capacity to live and work in space. It offers the nearest base from Earth to develop research, capacity, and resources; knowledge that can then be used to explore and exploit other bodies such as the asteroids in the asteroid belt as well as Mars and other planets.

China is investing in its space capacities as an overall strategy of national development and national rejuvenation. President Xi Jinping views China's space activities as part of his concept of the China dream.[22] The China dream is focused on the national rejuvenation of the Chinese nation, to include a modern People's Liberation Army (PLA). Xi's China dream has now been written into the Chinese Constitution, by a Constitutional amendment in 2017.[23] In achieving Xi's China dream, developing China's space capacities, especially in the cislunar space, is viewed as a critical component, and directly connected to maintain China's access and dominance in space-based assets that can be used to strengthen its earth-based military. This aspect was emphasized by Xi in the 18th and 19th National Congress of the Communist Party of China (CPC).[24] The first time the China dream was articulated was in November 2012 when Xi visited the "The Road towards Renewal" exhibition at the National Museum of China. He went on to further elaborate his concept in 2013.[25] Interestingly, in a speech given to heads of military academies and training schools in November 2019, Xi stressed the urgent requirement to create a new type of military personnel, proficient in new types of warfare, including space, and for whom loyalty to the CPC trumps any other loyalties.[26]

Within China's space dream is its long-term focus on space as interconnected to the national rejuvenation of the Chinese nation. As per its White Paper on China's Space Activities in 2016, "the Chinese government takes the space industry as an important part of the nation's overall development strategy."[27] To foster a national spirit for China's space ambitions and goals and build societal support, China declared April 24, 2016 (the same date in 1970 when its first satellite, the *Dong Fang Hong* 1-East is Red) reached across the skies to space, as National Space day.[28]

China's space capacities include an ability to send manned missions to Low Earth Orbit (LEO), orbit a space station, and sustain human presence for nearly a month there, land on the far side of the Moon, besides landings on the near side, flyby asteroids, space launches to include the launch of its indigenously built space cargo-ship. Besides this, China launched a quantum satellite called *Micius* (named after the famous fifth-century BC Chinese scientist and philosopher believed to have carried out the first optical experiments), in August 2016.[29] A quantum satellite is viewed as "hack-proof" for communications security, as it has a special kind of laser, that reveals any kind of interception known as the "observer effect." It changes the key thereby making any key obsolete once it is intercepted. This is because the photons are superimposed. This experiment will ensure how encrypted communications and keys can be distributed without the fear of hacks. If a quantum communication is intercepted or measured, the quantum state of the key will change, and the information that is being intercepted will self-destruct.[30]

This is the first time that such a quantum experiment is being tested from space thereby increasing the distance of such communications.[31] In August 2017, the first such quantum key distribution from space to Earth was realized. Pan Jianwei, the lead scientist of the Quantum Experiments at Space Scale (QUESS), who is with the Chinese Academy of Sciences (CAS) stated that the satellite sent out quantum keys to ground stations in Xinglong, in north China's Hebei Province, and Nanshan, in northwest China's Xinjiang Uygur Autonomous Region.[32] Bai Chunli, the President of CAS stated that "Micius has ushered in the construction of global quantum communication, the study of space quantum physics and experimental verification of quantum gravity theories. It helps China's race to control the command point of quantum science and technology and enables China to become a leader in the field."[33] QUESS has international collaborations with Austria's University of Vienna and the Austrian Academy of Sciences (AAS), since 2011.

China's capacities in terms of long-term space ambitions such as a lunar settlement, and asteroid mining, as well as SBSP is already entering stages of technology demonstration backed by clearly articulated future timelines. Regarding the Moon, the aim is to not simply carry out space science experiments but to develop long-term presence and human settlement. In its 2016 White paper on space activities, China identified asteroid exploration as one of its fundamental future space goals. In his presentation at the 2016 International Astronautical Congress in Mexico, vice president of CAST, Li Ming, specifically identified mining resources from the Moon and asteroids as a priority. China is focusing on exploiting resources such as titanium, Helium-3, and water from the far side of the Moon. Its *Chang'e* Lunar Exploration program, launched on *Long March* rockets, is an ongoing robotic mission to the Moon led by the China National Space Administration (CNSA). Besides discovering titanium and Helium-3, discovering water on the Moon surface is going to be vital for any ambitions for a lunar human settlement. Water is vital for creating propulsion, necessary for spacecrafts. Wu Weiren, the head designer of China's lunar missions, in an interview to the BBC, revealed that China aims for long-term exploration and a research base on the lunar surface.[34]

Toward this end, China is investing money in research to study the long-term effects of lunar presence on humans. In 2017, Beihang University in Beijing created a lunar lab, *Lunar Palace 1* or *Yuegong-1*,[35] simulating the Moon's surface here on Earth. Eight students lived in lunar-like conditions for about 365 days, the first by humanity again. The chief designer of

Yuegong-1, Liu Hong specified that this "test marked the longest stay in a bio-regenerative life support system (BLSS), in which humans, animals, plants and microorganisms co-exist in a closed environment, simulating a lunar base. Oxygen, water and food are recycled within the BLSS, creating an Earth-like environment."[36] The students grew potatoes, wheat, carrots, string beans, and onions.[37] According to Wang Ju, from the China Academy of Engineering, this test had critical implications for human ambitions to achieve long-term stays outside Earth, especially for the establishment of a lunar base. Similar but smaller bio-regenerative support systems will be taking rides on China's lunar and Mars probes to test the robustness of these experiments in the actual conditions of space. For a start, the *Chang'e 4* carried an 18-cm tall, 3 kg cylindrical tin, made of aluminum alloy—a lunar mini biosphere experiment that was designed by twenty-eight universities, led by Chongqing University. This mini biosphere contained seeds of cotton, potatoes, Arabidopsis, as well as eggs of fruit flies and yeast. The temperature was controlled, by human interception, between 1 and 30 degrees centigrade as temperatures on the lunar surface is extreme, ranging from below minus 100 degrees centigrade during the lunar night (fourteen earth days) to about 100 degrees centigrade or more during the lunar day (fourteen earth days). The cotton seeds did germinate and sprout, with Xie Gengxin, the chief designer of the project stating that "This is the first time humans have done biological growth experiments on the lunar surface."[38] According to Liu Hanlong, Chief Director of the Experiment and Vice President of Chongqing University, potatoes and Arabidopsis was included "because the growth period of arabidopsis is short and convenient to observe. And potato could become a major source of food for future space travelers . . . Our experiment might help accumulate knowledge for building a lunar base and long-term residence on the Moon."[39] The biosphere experiment was declared as closed by Professor Xie Gengxin, of Chongqing University and chief designer of the experiment on January 13, 2019, as the *Chang'e 4* entered a sleep mode during the lunar night to preserve itself, as temperatures drop below minus 150–170 degrees centigrade. Xie specified that since life in the simulated biosphere would not survive the lunar night, the experiment ends here and the CNSA indicated that all organisms will decompose themselves in the airtight aluminum canister.[40]

Significantly, in 2016, China successfully demonstrated to the world that reproduction could be possible in outer space when its *S-J 10* recoverable satellite sent 6000 mouse embryos to space, in which some of the embryos developed into advanced blastocysts in four days.[41] Professor Aaron Hsueh, from Stanford University, specializing in reproductive biology stated, "This represents an important milestone in human space exploration . . . One small step for mouse embryos, one giant leap for human reproduction."[42] To fulfill ambitions of space manufacturing and habitation, scientists from CAS's

Chongqing Institute of Green and Intelligent Technology and the Technology and Engineering Center for Space Utilization tested a space 3D printer in 2016. The head of the 3D printing research center (under CAS's Chongqing Institute) specified that the printer has finished ninety-three zero gravity flying tests in France.[43]

The roadmaps released by China's state-owned and -funded space agencies, strictly monitored and ideologically committed to the CPC, to include the China Aerospace Science and Technology Corporation (CASC) and the CNSA, tasked with setting policy for space, indicates that between 2020 and 2045, China aims to achieve several significant milestones regarding space technology. This includes a Mars probe by 2020, asteroids probe by 2022, Jupiter mission by 2029, probes to the lunar poles by 2030, a reusable carrier rocket by 2035, a manned mission to the Moon by 2036, and a nuclear-powered space shuttle by 2040.[44] In a report published in the front page of the *People's Daily*, a newspaper tightly controlled by the CPC, the China Academy of Launch Vehicle Technology (CALT) specified that nuclear-powered spacecrafts would augment larger payloads, and enable China to commercially explore and exploit the natural resources available in space by 2040.[45] Prof. Wang Chunghui, associate professor of Aerospace Propulsion at the School of Astronautics at Beihang University, stated, "The nuclear vessels are built to colonise the solar system and beyond."[46] This perspective to achieve the capacity to achieve human colonization in space was supported by Liu Hong, chief designer of the *Yuegong-1*.

To launch its human mission, China utilizes its *Shenzhou* spacecraft, launched on the *Long March* 2F or the *Changzheng-2F*. "The 7.8-tonne (7,800-kg) Shenzhou spacecraft has three sections: an Orbital Module, a Descent Module, and a Propulsion Module."[47] In March 2019, China's *Long March* rockets launched for the 300th time, after sending nearly 500 spacecrafts into space. "It took 37 years for the Long March rockets to complete the first 100 launches, 7.5 years to complete the second 100 launches, and only about four years to accomplish the final 100, with the average number of launches per year increasing from 2.7 to 13.3 and then to 23.5."[48] In June 2019, China launched its *Long March 11* from a mobile sea-based platform, joining the United States and Russia as the two other countries that had demonstrated this capability. The only difference is that the technology was built by China alone, whereas the United States and Russia built their capacity in conjunction with Norway and Ukraine.[49] *The Long March 11* will achieve several capabilities for China:

> The *Long March 11* will bring down launch costs, as launches close to the equator require less energy (fuel) to achieve speed boost to attain orbit. Second, sea-based mobile launch platforms offer China portable and flexible launch capability that enable rapid responses, especially in times of conflict. This will

> provide a boost for its PLA Strategic Support Force (SSF). Such sea-based launches are not tied to the geographic fixation of land launches and connected environmental hazards. Third, and more specifically, it enables "stealthy launches" boosting China's space-based military capability, as it is difficult to locate a launch platform in the Sea, then it is to find a fixed launch site on land. Sea-based launch system would require adversary navies to survey the vast seas constantly, making it difficult to track.[50]

China has also launched its own space lab, the *Tiangong 1*, in 2011 and hosted Chinese *taikonauts* (astronauts).[51] This was followed by the launch of the *Tiangong 2* in September 2016. In October 2016, the *Shenzhou 11* docked with the *Tiangong* 2, carrying two Chinese *taikonauts.* The two *taikonauts*, Jing Haipeng and Chen Dong, stayed thirty days abroad the *Tiangong 2*, before safely making it back to Earth. In April 2017, China demonstrated another space capability when its *Tianzhou 1* cargo spacecraft docked with the *Tiangong 2*.[52] All these activities are in preparation for the launch of China's permanent space station, the *Tiangong* space station (core module: *Tianhe*: two experimental modules: *Mengtian* and *Wentian*) in 2022.[53]

Some of China's follow-on space missions were reliant on China's successful launch of its heavy-lift rocket, the *Long March 5*. Two earlier test launches of the rocket had partially succeeded or ended in failure.[54] On December 27, 2019, the *Long March 5* successfully launched into orbit on its third attempt, sending the 8-tonne *Shijian-20* technological experiment satellite, into its planned orbit.[55] With a carrying capacity of 25 tons into LEO, and 14 tons to geostationary transfer orbit, and 8 tons to Earth-Moon transfer orbit, the *Long March 5* upgrades China's rocket carrying capacity, twice over earlier rockets.[56] China is developing the *Long March 9* (launch date 2028), its super-heavy rocket, with a payload capacity of 140 tons into LEO, and 50 tons to Earth-Moon transfer orbit. The *Long March 9* will be utilized to launch missions such as *Chang'e 7* and *Chang'e 8*.[57]

Military Space Capacities

China's military space capacities have been modernized and upgraded in the last few years. In December 2015, China established the PLA Strategic Support Force (PLASSF). Xi specified that the establishment of the PLASSF was aimed at accomplishing his core concept of the China dream. The PLASSF is a new type of combat force aimed at enhancing national security and combat readiness across the spectrum of conflict. The critical significance of political ideology within the PLASSF is determined from the fact that the Political Commissar of the PLASSF, Liu Fulian, is as important as its current military commander, Lieutenant General Li Fengbiao.[58] In July 2017,

President Xi, the Chairman of the Central Military Commission (CMC) promoted five senior officers to the rank of General, the highest rank in China's military. Among the five promoted was Gao Jin, former commander of the PLASSF along with Han Weiguo, of the Central Theater Command of the PLA, Liu Lei, Political Commissar of the PLA army, Yu Zhongfu, Political Commissar of the PLA Air Force, and Wang Jiasheng, Political Commissar of the PLA Rocket Force.[59] According to Major General Du Wenlong, from the PLA Academy of Military Science, "As for the Strategic Support Force, it better coordinates the cooperation between forces on the battlefield and logistic support."[60] Interestingly, what this implies is that the PLA will have control over space, and not the Chinese Air Force,[61] unlike in the United States, where space is the domain of the United States Air Force. The newly established U.S. Space Force is institutionally under the Department of the Air Force, a model similar to the U.S. Marine Corp, a separate service, but under the Department of the Navy.[62] The PLASSF has two distinct space-related organizations; the Space System Department and the Military Space Force.[63] Critically, the PLASSF is equal in grade to the other services (army, navy, air force, rocket forces),[64] and is a separate service directed by Xi's space dream. One of the most innovative leaders is PLASSF deputy commander, Lieutenant General Shang Hong, known for futuristic thinking on what space means for power projection capabilities. China is calibrating at the right time, contextually adapting to the changing logic of space. In December 2019, Xi promoted the current Commander of the PLASSF Li Fengbiao to General, the highest rank for active duty officers in China, along with six others.[65]

Besides bringing about institutional reform, China demonstrated its space military capacities when in 2007; it kinetically struck and killed its own aging weather satellite, the *Feng Yun* 1C polar weather satellite, with the help of a ballistic missile that was launched from China's Xichang Space Center. The 2007 Chinese ASAT test reached 850 kms above Earth rendering U.S. satellites in LEO vulnerable to adversary attack.[66] While the Chinese ASAT resulted in enormous debris creation,[67] it achieved one of the intended effects aimed from such a test by the PLA; to render U.S. military space assets vulnerable particularly the U.S. military's heavy dependence on its space-based assets for reconnaissance, precision-guided missiles, GPS, intelligence, and communications. This strategy is part of China's emphasis on developing asymmetric capabilities, articulated by thinkers within China such as Wang HuCheng. In July 5, 2000, *Liawang* (*Outlook*) article titled "The US Military's 'Soft Ribs,' A Strategic Weakness," Wang specified that "For countries that can never win a war with the U.S. by using the method of tanks and planes, attacking the U.S. space system may be an irresistible and most tempting choice."[68] In September 2006, China tested a ground-based laser beaming technology that was aimed at blinding and disabling U.S.

reconnaissance satellites as they pass over China.[69] Such jamming and blinding capabilities can be utilized to block communications between military command and control. In 2013, China demonstrated its capacity to send a rocket to geosynchronous orbit (GEO), 36,000 kms high. While China termed the May 2013 launch as a test of its scientific sounding rocket mission, the U.S. Air Force asserted that the rocket's trajectory indicated that China was testing technology designed to destroy satellites in GEO.[70] In its 2015 report to Congress, the Pentagon specified its concern stating:

> Analysis of the launch determined that the booster was not on the appropriate trajectory to place objects in orbit and that no new satellites were released. The post-boost vehicle continued its ballistic trajectory and re-entered Earth orbit 9.5 hours after launch. The launch profile was not consistent with traditional space-launch vehicles, ballistic missiles or sounding rocket launches used for scientific research. It could, however, have been a test of technologies with a counterspace mission in geosynchronous orbit.[71]

In July 2013, China's *Long March 4C* Rocket launched three secretive satellites into space, to include the *SY-7*, plausibly fitted with the robotic arm, that then conducted close in rendezvous with other satellites. The *SY-7* conducted several intelligent maneuvers in space, to include dropping 93 miles, practicing docking with a simulated space station. Over several days, it suddenly changed course and rendezvoused as close as 100 meters from another satellite. The SY-7's robotic arm, while claimed by China as a science experiment for debris removal simulation, revealed movement and rendezvous maneuvers that could be put to dual use, to include grabbing and destroying an adversary satellite, if the need arises.[72] Interestingly, a July 2019 report by *Xinhua* specifies that the robotic arm, specifically developed by the China 's CALT "has been tested to remove space debris, including rocket emissions and fragments from disintegration, erosion and collision."[73]

> There are billions of pieces of millimeter-sized debris around Earth, which require precise capture and disposal that could be provided by the robotic arm. It turns out that the high precision achieved in space is also very useful for sorting household garbage on Earth. Scientists from the CALT, using space and artificial intelligence technologies, have upgraded the arm into a high-efficient garbage sorting tool.

In 2014, China carried out another test that was perceived as a nondestructive anti-satellite test as per the U.S. State Department. This was refuted by Geng Shuang, a spokesman for the Chinese embassy in Washington, D.C., who asserted that it was a missile interceptor test, and cannot be associated with an anti-satellite test.[74] Following this, on October 30, 2015, China tested the *Dong Neng*-3 (DN-3) exoatmospheric vehicle from the Korla Missile Test Complex, and which is capable of ramming into U.S. satellites and destroying them.[75] The significant aspect of this test is that Chinese officials claim that it is a missile defense interceptor, while close inspection of the test reveal that it is a direct ascent missile capable of taking out satellites. In February 2018, China utilized a long-range missile interceptor or the *DN-3* to shoot down a missile in space, a capability that could potentially shoot down a satellite as well.[76]

According to the December 2018 National Air & Space Intelligence Center (NASIC) report titled, "Competing in Space":

> China has military units that have begun training with anti-satellite missiles . . . These missiles can destroy U.S. and allied space systems in low Earth orbit, making intelligence, surveillance, reconnaissance, and communications satellites vulnerable.[77]

Since 2012, Xi has personally pushed for greater "civil-military" integration strategy, and since then has gone about institutionalizing this process. In 2018, the Central Commission for Integrated Military and Civilian Development listed the "first civil-military integration demonstration zones, and a road map for the development of these zones."[78] Outer space is an integral part of that civil-military integration with Xi calling on new Chinese private space startups to work within that integration strategy. In continuation of that policy, the PLA opened up one of its space launch centers, the *Jiuquan Satellite Launch Center* for private launches.[79] In September 2019, China launched the *Ziyuan* (ZY-1 02D), fitted with high-resolution cameras. While described by China as a remote sensing satellite, a NASA spaceflight .com article argued that:

> Western military intelligence sources point out that while China states that the Ziyuan-2 are civilian application satellites, these were the first high-resolution military vehicles under the code name Jian Bing-3 (JB-3) . . . One of the salient features of this statement is that the first Ziyuan-2 was placed in a much lower orbit than the first Ziyuan-1. US intelligence services believe the first Ziyuan-2 was a photo recognition vehicle used exclusively for military purposes, such as the location of US and Taiwan military forces.[80]

Earlier, in 2017, China put three PLA-owned reconnaissance satellites into orbit. The three *Yaogan-30* satellites will form a constellation to "monitor electromagnetic signals" and other unspecified missions. The *Yaogan-30* satellites are capable of intercepting Earth-based radio signals, and pick up nuclear explosion-originated electromagnetic pulse. Li Xiaoming, a researcher at the Institute of Remote Sensing and Digital Earth at the CAS specified that "these are military assets—they are not for civilian use, there's no access."[81] As of March 31, 2019, there were 2,062 satellites in orbit, the United States operates 901, followed by China with 299, Russia,153, and India with 60.[82]

CHINA'S PRIVATE SPACE SECTOR

Since President Xi became the president, there has been an increased push toward developing China's private space sector. In 2014, China's State Council proposed in *Document 60*, to open up its space launch sector to private industry.[83] This was followed by China's civil-military integration strategy.[84] As a result, since then, China's private space sector has witnessed an increase in funding to the tune of $2 billion.[85] Some of these private space companies have already launched to space, and are invested in developing critical technologies such as reusable space launch capacity and including satellite communications as new infrastructure. Amongst the known private space companies are Onespace, one of the first company to successfully launch to sub-orbit,[86] followed by attempts made by Linkspace and Landspace. In August 2019, Linkspace experimented with the first of China's reusable launch vehicle, when its rocket reached a height of 300 meters above ground and then landed back intact.[87] Beijing Interstellar Glory Space Technology Ltd or iSpace launched its rocket *Hyperbola 1* into orbit in July 2019, marking the first such successful orbital launch by a Chinese private company.[88] By 2021, iSpace is aspiring to launch China's first reusable rocket into orbit, the *Hyperbola-2*.[89]

ELITE DISCOURSE ON SPACE RESOURCES

One of the most important markers of analyzing support toward a policy choice is to assess the support of decision-makers, high-ranking space institutions funded by the state, as well as influential individuals throwing their weight behind a policy decision. Such a cluster of individuals (decision-makers, policy planners, and implementors; powerful individuals shaping perception and policy) are known as elites in society. The critical importance of elite support for a space resource perspective within China can be gauged from who is talking about those resources, which institutions are articulating such a position, and if there are significant policy papers that support such a stance. To add to this is to analyze the level of support that could be determined from the highest echelons of the CPC.

The critical significance of space-based resources for China is connected to its requirement to generate resources far into the future to maintain and sustain the "national rejuvenation of the Chinese nation." The significance of such a discourse can be assessed from the linkage of space to the overall national development goals of the Chinese nation.[90] In its 2016 White paper on space, China specifically linked its space exploration to long-term economic developmental goals. These goals included bringing back samples from Mars for research as well as asteroid exploration.[91] In January 2019, as part of that goal, the *Chang'e 4* lunar lander and rover soft-landed on the far side of the Moon, to survey and map the side not visible to Earth. This is a part of a larger Lunar Exploration program.

Historically, China's space program has been supported by the highest echelons of the CPC. Ironically, the man who kickstarted China's space program was Qian Xuesen, a product of the Massachusetts Institute of Technology, and one of the founders of the Jet Propulsion Laboratory along with Theodore von Kármán at Caltech. He, along with Kármán, played an instrumental role in getting Nazi rocket scientists such as Wernher von Braun, to the United States. In the 1950s, at the height of the Cold War and the Korean War, Xuesen, a Chinese citizen and U.S. permanent resident at that time, was suspected of communist leanings, stripped of his U.S. government security clearance, and placed under house arrest. His application for U.S. citizenship was denied. He was deported to China, wherein, Mao Tse Tung welcomed his expertise and tasked him to work on the fledgling Chinese space program.[92] While leaving for China in 1955, Xuesen told reporters, "I do not plan to come back . . . I have no reason to come back . . . I plan to do my best to help the Chinese people build up the nation to where they can live with dignity and happiness."[93] The CPC named him as a member to the Central Committee of the CPC, and in 1964, China tested its first nuclear weapon.[94] While allegations against him could never be proven in the U.S.[95] Xuesen contributed to China's space program and never set foot in the U.S. again.

The effort to establish China's space program was supported foremost by the 863 program, conceived by four Chinese scientists, namely, Wang Daheng, Wang Ganchang, Yang Jiachi, and Chen Fangyun.[96] The 863-program aimed at accelerating Chinese development in high technology geared toward a specific end goal. The program was approved by Deng Xiaoping, then Premier of China, and spanned several five-year plans.[97] To oversee and implement the 863 program, the Science and Technology Leading Small Group under the State Council was made in charge, led by Premier Zhao Ziyang (1980–1987).[98] It was Ziyang who convinced the CPC that space should be made of priority for China.[99] Critically, space is administered by the State Administration on Science, Technology, and Industry for National Defense (SASTIND) which functions under the direction of the Ministry of

Industry and Information Technology (MIIT), and overseers the vital link between space technology and nuclear power, and communicates this aspect with other countries and international organizations.[100] Under SASTIND is the CNSA, which was established in 1993. The CNSA is responsible for articulating China's space policies, direct its manned space mission, the lunar mission, the Tiangong space station, and the Long March series of rockets.[101] CNSA secretary-general, and vice administrator of SASTIND, Yulong Tian, while speaking at the 33rd space symposium in Colorado Springs on April 5, 2017, stated that China's major space goals in the next five years is to launch robotic missions to the Moon, outline a policy for commercial space activities, an automated Mars sample return mission by 2030, and launch deep space exploration of Jupiter, Venus, and asteroids.[102]

CHINA: EVIDENCE OF PROGRAMS

Lunar Ambitions

If one analyzes China's space strategy and goals, the country is following an incremental development strategy of first focusing on developing the capacity to live and work on the Moon, to be followed by further deep space exploration and exploitation. The focus on the Moon was evident by the Earth-based experiments conducted to explore the impact of long-term stays on the lunar surface. As mentioned earlier, between May 2017 and May 2018, eight students lived for 370 days (taking shifts) on *Yuegong-1*, or Lunar Palace 1, a BLSS at Beihang University. According to Liu Hong, chief designer of Yuegong-1, "the BLSS is a critical piece of technology required for long-term human stays on the moon or other extraterrestrial bodies."[103] Wang Jun, an academician from the Chinese Academy of Engineering specified that "The test has important implications for human endeavors to achieve long-term stays outside the Earth. The experience, technology and findings will be conducive for future space exploration efforts."[104] A mini version of such a BLSS was sent up on the *Chang'e 4* to test its sustenance on the extreme temperatures of the lunar surface as mentioned earlier in the chapter. More such experiments will continue on China's future space probes.

On April 24, 2018, the CNSA released a video detailing China's ambitions for the Moon to include the establishment of a lunar palace. Wu Weiren, the chief designer of CLEP believes that such an end goal will become a reality by 2030, most likely on the South Pole of the Moon, due to the availability of water and sunshine.[105] Weiren asserts that "our long term goal is to explore, land, and settle" on the lunar surface.[106] Ye Peijian, an aerospace engineer and head of the Chinese Lunar Exploration program stated in an annual plenary session of the CPC in 2017 in Beijing that:

> The universe is an ocean, the moon is the Diaoyu Islands, Mars is Huangyan Island. If we don't go there now even though we're capable of doing so, then we will be blamed by our descendants. If others go there, then they will take over, and you won't be able to go even if you want to. This is reason enough.[107]

Ye Peijian, who leads deep space exploration at CAST stated that China is investing in research in both Mars and asteroid exploration. Asteroid exploration, specifically, will be carried out between 2020 and 2025.[108] He stated that "the detailed schedule and the target asteroid have yet to be determined, but we are working on them. We want to explore asteroids because their resources will be important to mankind's development in the future."[109] While some argue that statements such as those by Ye appear to be "part explainer, part propaganda, all dog whistle,"[110] our metric of measurement of success (elite discourse, followed by policy and demonstrated mission capabilities), to include the landing on the lunar far side, China's advancements in quantum satellites, the successful location of its relay satellite, *Queqiao* (Magpie Bridge) in L2, its capability to change orbits, and conduct multiple space missions in a single launch (*Chang'e 2* lunar mission left the lunar orbit, flew to Earth-Sun L2 and then flew past an asteroid, *Toutatis*),[111] should make us seriously ponder about some of these fact-less opiniated assertions that these statements by China's top scientists and policymakers are mere propaganda. We view them as building elite consensus and societal backing for China's space programs from the highest levels of the CPC, of which Ye is a high-ranking member. As part of the seventieth-year celebration of the establishment of the PRC, Ye Peijan, called the "father of *Chang'e* lunar probes," and a senior academician at CAS, was bestowed China's highest state honor for his contributions to CLEP.[112]

The seriousness of China's lunar efforts was visible from statements made by the lead in the *Chang'e* 4 probe. Sun Zezhou, chief designer of the *Chang'e-4* probe, from CAST stated that

> If we want to build a scientific research station on the moon, we will need to land multiple probes within the same area so that they can be assembled easily into a complex, which requires even greater landing accuracy . . . So solving the challenges of the Chang'e-4 mission can lay the foundation for the following lunar exploration and future landing on other planets.[113]

For China, the Moon is a means to an end to build industrial capacity on the lunar surface to then accomplish its goals of asteroid mining, deep space exploration, and exploitation.[114] A research base on the Moon, with an industrial capacity to build spacecrafts by using resources such as iron and water for rocket propellant, will bring down costs of interplanetary travel. Launches from Earth requires twenty-two times more energy, than it does from the Moon, due to Earth's gravity.[115]

Space Solar

Interconnected to the goals of establishing a research base on the Moon is the goal of harvesting the Sun's energy in outer space where access is twenty-four hours and does not suffer from the changes in weather. In a report published in 2010, CAST identified the following roadmap for SBSP as:

> In 2010, CAST will finish the concept design; in 2020, we will finish the industrial level testing of in-orbit construction and wireless transmissions. In 2025, we will complete the first 100kW SPS demonstration at LEO; and in 2035, the 100mW SPS will have electric generating capacity. Finally in 2050, the first commercial level SPS system will be in operation at GEO.[116]

In 2015, China expressed its intention to build a space solar station, 36, 000 kms above the Earth.[117] This power station will be placed in GEO and equipped with huge solar panels, and the solar electricity that will be generated, will be sent via microwaves or lasers to Earth. One of the biggest advocates of SBSP in China is Wang Xiji, the chief designer of China's first rocket, the *Long March 1*.[118] Wang believes that "The world will panic when the fossil fuels can no longer sustain human development. We must acquire space solar power technology before then . . . Whoever obtains the technology first could occupy the future energy market. So it's of great strategic significance . . . Construction of a space solar power station will be a milestone for human utilization of space resources. And it will promote technological progress in the fields of energy, electricity, materials and aerospace."[119] According to Duan Baoyan, from the Chinese Academy of Engineering (CAE), "If we have space solar power technology, hopefull [*sic*] we could solve the energy crisis on Earth."[120] SBSP is perceived as green energy that will address China's concerns about its environment, as well as its need for energy imports, especially fossil fuels.[121] Wang Li, Research Fellow with CAST, indicated that China will be leading in this field given the head start it has with regard to investments in SBSP. "Compared with traditional fossil energy, which has been increasingly exhausted and is responsible for severe environmental issues, space-based solar power is more efficient, sustainable, and clean. It can generate as much energy as a ground-based nuclear power plant."[122]

These views are supported by senior vice president of CAST, Li Ming, who believes that "China will build a space station in around 2020, which will open an opportunity to develop space solar power technology."[123] Ming indicates that once the space station is in place, China would then carry out experiments on developing an SBSP station.[124] In an interesting presentation for the 2016 International Astronautical Congress in Mexico, Ming further elaborated on the SBSP concept by suggesting that in-situ resource utilization and on-orbit 3-D printing could be applied using resources from asteroids to build SBSP satellites on a lunar base instead of having to lift them from Earth to space. This will bring down manufacturing costs from $536T ($50, 000 per kg) to $0.17T ($250 per kg). Materials present on the lunar surface and asteroids include silicon, aluminum required for solar panel production.[125] Ming pointed out that the low gravity of some asteroids or near-Earth objects (NEOs) makes it easier for spacecrafts to dock, park, or separate requiring less propulsion. NEOs are attractive as they are rich in resources required for SBSP purposes. More recently, Lieutenant General Zhang Yulin, former Deputy Chief of the General Armament Division (GAD), and China's Manned Mission, now with the PLASSF of the CMC stated that solar power generation in space was more efficient than Earth solar indicating that China would start developing technology for an industrial scale solar power station once it completes work on its permanent space station by 2022.[126] China has invested in developing a blueprint within a timeline of 2050 for its SBSP program. The CAST design by Hou Xinbin for SBSP satellites took the first position at the 2015 *SunSat* Design Competition.[127] Given CAST's timelines of completing the first 100kW SPS demonstration at LEO by 2025 and the first commercial level SPS system to be in operation at GEO by 2050, these could turn out to be China's own Sputnik moments.

In continuation of such statements, China became the first country in the world to establish a state-funded SBSP base plant in Chongqing's Bishan district in 2019.[128] The base plant is being constructed under the guidance of the Chongqing Collaborative Innovation Research Institute for Civil-Military Integration (CCIRICMI) in Southwestern China in partnerships with researchers from Chongqing University, CAST's Xi'an Branch in Shaanxi province, and Xidian University. The initial investment for the SBSP plant of $15 million has been made by the Bishan district government. Another $15 million has been pledged by CNSA, totaling $30 million. Technologies being tested include the construction of SBSP satellites in GEO using automated assembly and the wireless transmission of power.[129] Ming asserted that China will lead the world in this critical renewable energy source.[130] The key challenges that the plant will be testing are:

a. Microwave transmission of electricity,
b. In-space manufacturing of SBSP satellites.

Xie Gengxin, who is the Deputy Head of CCIRICMI stated that:

> We plan to launch four to six tethered balloons from the testing base and connect them with each other to set up a network at an altitude of around 1,000 meters . . . these balloons will collect sunlight and convert solar energy to microwave before beaming it back to Earth. Receiving stations on the ground will convert such microwaves to electricity and distribute it to a grid . . . if everything goes well, a Chinese solar power station will be put into orbit about 36,000 kilometers above Earth and start generating power before 2040.[131]

Pang Zhihou, from CAST, laid out the rationale for why SBSP is a critical investment focus for China's space program. First, it is available twenty-four hours; it is not susceptible to weather variations like ground solar; it can be utilized to power China's lunar base as well as spacecrafts, and increase efficiency "by freeing spacecraft from huge solar cell wings and greatly increasing power levels and control accuracy. It can also be used as a candidate for deep space exploration energy systems, and it can also be used for space fuel production and space processing manufacturing in the future to realize space industry development."[132]

Asteroid Mining

Chinese space scientists are working on plans to capture an NEA and bring it back to Earth to inspect and extract its resources. Researcher Li Mingtao, a researcher at the National Space Science Center under CAS, along with his team details that plan. The idea is for a spacecraft to bag an asteroid and push it over Earth, followed by a heat shield that unfolds reducing the velocity of the asteroid as it enters Earth's atmosphere. The landing must be minutely controlled so that it lands in an area far from human habitation. Li is working in collaboration with scientists from the Qian Xuesen Laboratory of Space Technology, under the CASC to locate satellites in the heliocentric Venus orbit, to search and analyze NEOs, with a diameter of 10 meters. The challenge after capture would be to drop the speed of the asteroid from 12.5 km per second to 140 meters per second before it touches down on its Earth-landing spot. The timeline for such a launch to capture an asteroid is 2029 and the aim is to bring it back to Earth around 2034. The logic behind such an experiment is the lucrative future of space mining. Li asserts that:

> Space mining might become a new engine for the global economy . . . Unlike missions to bring samples back, we aim to bring back a whole asteroid weighing several hundred tonnes, which could turn asteroids with a potential threat to Earth into usable resources . . . Our analysis shows that maneuvering a small asteroid is feasible in principle, and could bring enormous economic and social benefits.[133]

Li's perspective is supported by Huang Wei, Chief Engineer within CAST. Wei believes that ideas such as Li's would result in the promotion and development of futuristic space technologies.[134] Space mining is perceived to be China's future advantage given the rising demands for energy within China and projections that forecast China as the world's leading economy by 2050. Toward achieving the goals of harvesting space-based resources, a CASC report outlined in 2017 the focus of China on developing nuclear-powered space shuttles by 2040. Such capabilities make it feasible for China to invest and plan realistically for space-based mining as well as developing solar-powered satellites. Such a long-term strategy is geared toward reusability and developing a space post, that affects the development of space transportation and capabilities. By 2025, reusable suborbital carriers should come to fruition enabling the path toward a long-term vision.[135] China is developing the *Long March 9*, its super heavy lifter, designed to carry a payload of 140 metric tons to LEO, a 50-ton spacecraft to a lunar transfer orbit and a 44-ton payload to Mars transfer orbit.[136] Importantly, in its justifications for the *Long March 9*, China listed the following four goals: (1) launch a Mars robotic exploration mission that requires 41-ton payloads, (2) manned Mars missions, (3) deep space missions, and (4) "Constructing orbital solar power plant with 10,000 MW capacity, massing some 50,000 tons, requiring 620+ launches."[137]

Long-Term Strategy

To meet its first goal of permanent space presence, reducing launch costs is of priority. Consequently, China is developing abilities such as vertical takeoff and landing, as well as reusability. Long Lehao, an academician of the Chinese Academy of Engineering and a chief designer at the China CALT, that builds the *Long March* rockets, specified what the goals of launch for China are, by 2030:

> The capacity of Chinese rockets would reach 140 tonnes for low-Earth orbit, 44 tonnes for Earth-Mars transfer orbit, 50 tonnes for Earth-Moon transfer orbit and 66 tonnes for geosynchronous transfer orbit in 2030.[138]

The critical significance of China's investments in space is articulated at the top level by President Xi Jinping, who locates China's space dream within the contours of his China dream. In a ceremony on June 11, 2013, at the Jiuquan Satellite Launch Center to send off the astronauts on the *Shenzhou 10*, Xi articulated his vision for China in space. He believed that one of the priorities of China's space dream was to explore space to the advantage of the Chinese nations. "The mission's crew members carry a space dream of the Chinese nation, and represent the lofty aspirations of the Chinese people to explore space . . . Developing the space program and turning the country into a space power is the space dream that we have continuously pursued."[139]

Xi's vision about space was articulated again during the launch of the *Chang'e* 3 lunar probe that carried the rover "Yutu," which soft-landed on the Moon on December 14, 2013. Speaking to scientists from the mission in January 2014, Xi highlighted the fact that China's lunar probe demonstrated indigenous innovation. Xi stated, "As we made 'Yutu' lay its prints on the moon, we also imprinted the extraordinary creativity of the Chinese nation onto the history of the human civilization."[140] For Xi, the end goal of China's space ambitions were clearly articulated on the occasion of China's first space day, April 24, 2016, when he stated, "Exploring the vast universe, developing space programs and becoming an aerospace power have always been the dream we've been striving for."[141] In a separate report by CASC, China aims to become a leading space power by 2045, with timelines set for a 2020 launch of the *Long March 8* (a medium-lift carrier rocket), followed in 2025 by a suborbital spaceflight, a 2030 launch of a heavy-lift 100-ton rocket, a 2035 launch of a reusable rocket, and a 2040 launch of nuclear-powered space shuttle.[142] By 2045, China will be the leading power in this domain, at which point it would enjoy the capacity of manufacturing SBSP in orbit.

POLICIES AND LAW

For China, its space program has been supported since its inception in the 1950s by directed state policy and programs. Foremost among the policies was the development of China's rocket program after the return of America-based and then deported Chinese scientist, Qian Xuesen, in 1955 to China. In 1956, China established its space program. This was supported by the 1955 agreement between China and the Soviet Union, for transfer of both rocket and nuclear technology to China as well as the education of Chinese scientists in Moscow. The Russian transfer of its *R-2* rocket blueprints helped China advance its own understanding of rocketry. Soviet delegation led by E. A. Negin arrived in Beijing in 1958 to teach Chinese scientists how a nuclear weapon can be constructed. Though such transfer of technology was halted in 1960, with Qian Xuesen or Tsien's help, China launched the *DongFeng-1*, a licensed copy of the Soviet's R-2, a short-range ballistic missile.[143] The project that supported the development of an atomic bomb program was Project 596.[144] In 1970, China launched its first satellite *Dongfanghong 1* (the east is red) into space, registering its entry into the domain of outer space. Over the next few decades, progress was slow, but after the death of Mao Tse Tung, Deng Xiaoping took over the reins of the Chinese government, and its space program benefited from focused resources and long-term planning. As mentioned earlier, the 863 program, proposed by four Chinese scientists, was personally approved by Deng, and it focused China's policy on developing innovation and capacity in the field of high technology. The aim was to "leap frog" and connect any such activities to China's national economic development and national security especially toward the generation of resources for long-term benefit.[145]

Project 921-1 was issued by Shanghai Astronautics Bureau in October 1993, and while not all its proposed plans for six carrier rockets and eight spacecrafts were approved, the 921-1 spacecraft based on solid motor was approved based on storable propellant, the CZ-2E.[146] While China advanced and developed its capacity for launches, it also developed state-funded institutions that could manage that task. These included SASTIND, CNSA, as lead organizations for policy planning purposes. Other institutions tasked with space technology are CASC[147] (which includes under it CALT), the China Aerospace Science and Industry Corporation and CAS. Interestingly, the China Satellite Launch and Tracking Control General (CLTC) is run by the CMC.[148] The CLTC serves as a command and control center for PLA's space-related operations. This direct interlinkage resulted in very close supervision by the CMC, specifically its GAD, now merged into the PLASSF, which influenced space policy, as well as how goals were set in the long term. In collaboration with SASTIND, GAD (now PLASSF) issued regulations for defense industry procurements, as well as identified institutions that would deal with space technology. It was on December 31, 2015, China set up the PLASSF, now in charge of both cyber and space assets. The PLASSF is responsible for managing the human spaceflight program.[149]

As per policy statements from these leading space institutions, particularly CAST, CNSA, and SASTIND, three unique Chinese space goals highlighted earlier in this chapter, come to light. These are SBSP, lunar and asteroid mining, and establishing its own space station.[150] Since 2000, China has been issuing clear directives for its space activities as per its White Papers. The 2000 white paper on space clearly linked space, its exploration, and use to China's national security as well as building a sustainable, green model for China's comprehensive development. In its 2006 White paper on Space Activities, China stated that in the next twenty years, as part of their future development of outer space:

> To achieve industrialization and marketization of space technology and space applications. The exploration and *utilization of space resources* [emphasis added] shall meet a wide range of demands of economic construction, state security, science and technology development and social progress, and contribute to the strengthening of the comprehensive national strength.[151]

In its 2006 White Paper on Space Activities, China stated that "maintaining and serving the country's overall development strategy, and meeting the needs of the state and reflecting its will. China considers the development of its space industry as a strategic way to enhance its economic, scientific, technological and national defense strength, as well as a cohesive force for the unity of the Chinese people, in order to rejuvenate China."[152] In its 2011 White Paper on Space, China stated that "Strengthening legislative work. To actively carry out research on a national space law, gradually formulate and improve related laws, regulations and space industrial policies guiding and regulating space activities, and create a legislative environment favorable to the development of space activities."[153]

Given the growing interest in space-based resources, it is but natural that there will be the need for both international regulatory mechanisms as well as national legislation. In this regard, Director of CNSA, Xu Dazhe, stated that China hopes to establish its national legislation for space activities by 2020. At present, China does not have a national legislation for regulating space resources, though there are several institutions that oversee granting licenses for civil-space launches, space-debris management, and regulations.[154] It will be interesting to see what China's national space legislation will include, given its vast ambitions in space, to include the goal of establishing permanent presence, and a research base on the Moon. How will that be legally stipulated? While the Outer Space Treaty stipulates that "outer space shall be free for exploration and use by all states . . . outer space is not subject to national appropriation by claim of sovereignty, by means of use or occupation, or by any other means."[155] What will happen if one state (x) establishes a lunar base and engages in space mining, and then another state (y) comes along, and wants to mine in the same area? As we know, there are only certain areas on the lunar surface that are rich in resources such as thorium, titanium, water-ice, and iron ore. What if state x's lunar base depends on the limited extraction of those minerals for sustaining itself and extending their outer space capacities. What then? Will they agree to share those resources, and the profit that will entail from such extractions in the long run? If engaging in space-based activities are connected to a country's national rejuvenation, as is the case with China, will that lead to more nationalist behavior given resources on the Moon are precious? Does China's behavior with regard to its territorial claims here on Earth offer insights into how it will behave, once such resources are developed on the Moon and beyond.

This leads us to the next section, China's strategic culture.

CHINA'S STRATEGIC CULTURE

Toward a deeper understanding of how countries behave or may behave regarding strategic issues, knowledge of their strategic culture could throw some light. Strategic culture is "an ideational milieu by which the members of the national strategic community form their strategic preferences with regard to the use and efficacy of military power in response to the threat environment. Each country has its own way to interpret, analyze and react to external opportunities and threats."[156] In this regard, it is pertinent to understand the cultural, political, contextual, and philosophical influences that are unique to a country's identity. Historically informed strategic preferences tend to influence a country's response to the present strategic environment. For instance, there are some who argue that India, given its history of nonalignment, usually tends to stay away from military alliances in response to a threat situation even in the current context.[157] While it may not be the best choice, it is the preferred choice, based on history and culture. Alastair Johnston argues that a country may have more than one strategic culture that may emerge from a multiplicity of factors, and diversity of historical experiences, political culture, ideology, and geography within a society.[158] Significantly, while there are several strategic options available at a given time, a country's strategic preference that prioritizes one strategic option among many others is determined by the most dominant among its strategic cultures.[159] Johnston highlights the instrumental nature of strategic culture where it could be utilized by decision-makers to explain strategic behavior and may not always function as a conceptual tool that offers guidance to policy.[160]

Chinese strategic experts assert that China's dominant strategic preference is to prioritize harmonious existence despite differences.[161] For instance, the idea of *he* (harmony), written in ancient Chinese texts inspired by Daoism and Confucianism, and the ideal aspiration of "harmonization" attest to this assertion.[162] However, a divergent perspective on the concept of harmony is offered by Wei Xiaohong and Li Qingyuan, from Sichuan Agricultural University, China. They argue that while harmony is a fundamental guiding principle for social interactions in China, there are two types of harmony: genuine and surface harmony. Genuine harmony is sincere and holistic, while surface harmony is strategic tolerance, hiding conflicts under the surface.[163] For better or worse, surface harmony is preferred to direct confrontations. This idea of surface harmony creates suspicions regarding China's overall intent regarding its strategic behavior.

Given China's assertive behavior regarding "lost territories" intertwined with the idea of national resources, it is likely China's quest for space-based resources would be informed by a similar strategic logic.[164] Critics indicate that this harmony-based narrative is a CPC manufactured narrative aimed at hiding the dominant realist orientation of the Chinese state. Those in power in China manipulate concepts of harmony drawn from Confucius to deliberately project a pacifist bend whereas Chinese strategic culture is actually dominated by a firm belief in the efficacy of the use of force.[165] The famous

Deng Xiaoping dictum of "Hide your strength, bide your time," is often cited as proof of Chinese strategic deception.[166] It is in China's advantage to appear non-threatening while it is in the process of consolidating its power. Hence, in a fundamental sense, China's strategic culture stems from the realist's tradition and has used violence whenever the opportunity has presented itself. The insistence of the CPC that China is a pacifist state is mere propaganda.[167] Security analysts state that China is simply bidding its time and as reflected in its recent activities in the South China Sea (SCS) and East China Sea (ECS), where it has unilaterally staked its territorial claim by deception and the use of force. This is a strategic trait carried over from the warring state period (475–221 BCE).[168] More importantly, the moralizing tendency of the Chinese state may render it sensitive to criticism, especially from external forces.

While the CPC narrative on Chinese history is that China was never an expansionist power and remained at peace with its neighbors, Warren Cohen and Alastair Johnston argue that Chinese emperors consistently attacked smaller neighbors such as Vietnam, Tibet, Mongolia, Korea, for thousands of years.[169] Johnston, by analyzing the *Seven Military Classics* discovers that there are two compelling strategic culture traditions in China; the Mencian-Confucian, which gives primacy to diplomacy, especially when one is weaker than the enemy to buy time. Johnston asserts that this pacifist posture is largely cultural to package the narrative in acceptable terms. However, when it comes to operational doctrine, the second strategic culture based on *realpolitik* dominates and informs strategic preferences. Ancient Chinese texts on strategy prioritize the use of force in response to a threat.[170] Warren Cohen, for instance, asserts that "ancient Chinese writers on strategy invariably concluded that the best way to respond to a threat was to eliminate it by force."[171] His assertion is largely derived from his reading of Johnston's book.[172] A defensive posture would be viewed as a sign of Chinese weakness and ancient Chinese texts on strategy support offensive use of force against adversaries. However, the "use of force" option was always determined by the strength of the adversary; if the adversary was viewed as superior, then diplomacy was favored. And if the adversary was viewed as inferior, then the "use of force" was determined as the viable option. Some of the ideas by ancient Chinese philosopher, Han Fei Tzi or Tzu of the legalist tradition, known as the Realist Tradition or *Fa-chia*, specified that morality had no place in the realm of ruling, especially when the sole goal of ruling is the preservation of the state.[173] Borrowing from a Taoist tradition of transcendence and contemplation, Han advises the legalist ruler to be an absolutist ruler, such as the Taoist sage who assumes absolute understanding, based on withdrawal from the world, and rising above conventional ideas of right and wrong. Han advises the ruler to set up his administration but maintain a classic distance from his subordinates. This produces reverence from the subordinates, based on an aura of mystery and inscrutability. The legalist thought has a morbid view of human nature as evil and that human beings are irredeemable requiring strong laws to keep them in check.[174] Hence, while the ruler does have an enlightened view, the best way to ensure self-interest is using force, deceit, and demonstration of power.[175] This is not different from what Johnston surmises as preference for the use of force in Chinese strategic culture.

Chinese scholars are quick to point out that Alastair Johnston limited his analysis on Chinese strategic culture to the Ming dynasty and interpreted a defensive use of force by the Ming against the Mongol invaders into Chinese territory as representing a proclivity for offensive violence.[176] Chinese scholars on strategy, therefore, perceive Johnston's interpretation of Chinese strategic culture as overgeneralization or overdetermination from just one period; namely, the Ming period. They assert that the Ming period cannot be taken as a typical case to profess about the entirety of Chinese historical experience spanning thousands of years.[177] Others point out that while Deng Xiaoping's words are quoted repeatedly till date especially by Western scholars to represent China's foreign policy behavior, most miss the context of the 1980s within which Deng made those remarks. Moreover, according to Huang Youyi, secretary-general of the Translators Association of China, the phrase *tao quang yang hui* meaning "to keep a low profile," cannot be translated to mean "strategic trickery."[178] It could mean "self-effacing" but cannot be translated as an advice by Deng to hide one's capacity and pretend to be weak, while actually strong.

In contrast to these Chinese perspectives is Sun Tzu's *Art of War* offering several insights on ancient Chinese strategic thinking. For one, according to Sun Tzu, "all warfare is based on deception," in the tactical sense. For instance, one must appear weak when strong, and strong when weak.[179] While Sun Tzu offered several advisory insights into how victory can be won, he was generally in favor of winning by stratagem without fighting at all. He, however, provided detailed advisory on how to wage war to one's advantage conferring paramount importance to the moral high ground that a commander should possess to infuse his troops with a fiery spirit. Sun Tzu believed in taking advantage of superior intelligence and secrecy with regard to war plans, and to strike like a thunderbolt with no possibility of offering advantage to the enemy. Sun Tzu's advice to the commander during the Warring States Period was to adopt the spirit of a comprehensive grand strategy for success. These include an understanding of the power of norms (moral legitimacy), heaven, Earth (physical conditions), leadership, and finally, method and discipline (assessment of military capability, context, relative power potential/difference, logistics, resources). Once all elements come together, a state can benefit from a grand strategy for success.[180] Sun Tzu counseled that when circumstances were favorable, plans must be modified, with an element of deception maintained as the core guiding tactical principle. The critical core of his philosophy and strategy was that: behave in such a way that the adversary/competitor is unprepared for who you are. This is accomplished by feeding into adversary perceptions of who they believe you are; and not who you actually are (as the authors were told in Beijing, "USA thinks we only copy their space technology, cannot innovate. They are right. We are just followers"). Depending on who your "target audience" was, a king, according to Sun Tzu, could accomplish a particular desired end goal, by *stratagem*, and superior influence operations backed by targeted resources.

China's strategy to shape the path for its renaissance and emergence as a lead actor in outer space by 2045 can be understood through Sun Tzu inspired lens. For one, Xi has offered a compelling space vision. For another, China has demonstrated its space capability, to include killing its own satellite, sending humans to LEO, soft landing on the lunar surface, and sustained presence of its rover through several lunar nights and days.

Consequently, as is advised by Sun Tzu, to build a comprehensive context for the moral legitimacy of your power, President Xi and the CPC has gone ahead and build alliance structures, signed Memorandums of Understanding (MoUs) and offered to collaborate on lunar missions with other countries.[181] This is part of Xi's vision of creating a world order where China, not only has capacity but also legitimacy as the country that champions a peaceful and harmonious world order. Xi takes his insights of leadership from Chinese strategic culture that truly believes that China has the superior civilizational intelligence to lead others.[182] One such initiative is the Spatial Information Corridor,[183] where China is offering its Beidou Navigation System to the world, especially to the seventy-member countries of its Belt and Road Initiative (BRI). Categorically pitched within the United Nations agenda for making the world a better place, China offers its advanced space capacity as a force multiplier for a world free of poverty, backed by peace, justice, freedoms, and strong institutions.[184] This perspective[185] has been vindicated by none other than the United Nations Office for Outer Space Affairs (UNOOSA) Director, Simonetta Di Pippo.[186] China and UNOOSA signed an MoU that has called for applications from UN member states to be part of China's permanent space station.[187] China's BRI has been hailed by the African Union and dubbed the Marshall Plan, without a war.[188] Recently, China, in its quest for outer space resources, signed an MoU with Luxembourg and established its deep space exploration unit in the Grand Duchy, primarily to take advantage of Luxembourg's legislation on space resources.[189] Consequently, China hailed Luxembourg's entry into its BRI initiative in March 2019,[190] following which the Bank of China chose Luxembourg to list its $500 million BRI bond.[191]

What China has achieved under Xi are clear demonstrations of international legitimacy and the construction of a narrative that its space activities will lead to global freedom and economic development. This is a rather remarkable achievement, coming from a CPC regime that is brutal to internal dissidents,[192] and tolerates no dissent. President Xi, in a speech to the PLASSF categorically stated that loyalty to the CPC comes above everything else.[193] This loyalty was reiterated by other Chinese state officials[194] as well as Chinese media.[195] And there are dire consequences for those who fail in demonstrating that loyalty.[196] Sun Tzu's advice to build into the five elements of power, imbibed with *stratagem*, is clearly a guiding principle for Xi. As the author of the book, *China Dream*, Liu Mingfu, a retired PLA Colonel maintains; as per Sun Tzu's guidance, the breakout of war is the breakdown of strategy and demonstrates civilizational demise. To win a war by *stratagem*, without bloodshed, is the way of a superior grand strategy and civilization.[197]

Ancient Chinese texts such as the *Art of War*, the *Seven Military Classics,*[198] the *36 Stratagems*,[199] or the political realist legalist tradition that emerged strong during the warring state period,[200] offered detailed mental maps and policy advice to wage war, build a powerful state and army. Hence, there could be the predisposition to a calculated, clever use of force, which is swift and overwhelming in the Chinese strategic psyche.

In this context, it is important to realize that some of that ancient strategic thinking may influence contemporary Chinese leaders especially regarding gaining the advantage in a potential conflict in outer space. We witnessed the idea of "striking like a thunderbolt" in China's conflict with India in 1962, where it exercised a quick attack-withdrawal strategy at India's eastern border areas, leaving India with no time to galvanize its major forces.[201] This is conceptualized as an "active defense strategy," which includes both a deterrence and a retaliation capability. It is operationalized within the philosophy of launching an attack suddenly to gain the first initiative.[202] This is not to suggest that the only strategic culture dominant in China is to use deceptive force to gain the advantage. Yet its ancient classics do point to the advantages inherent in such a strategic approach to the use of force.

CHINA'S STRATEGIC TRAUMA AND EXPECTED BEHAVIOR

First Strategic Trauma: The Treasure Ships

China as a state and society has two significant strategic traumas that inform its present behavior regarding international relations. One of such traumas and deep-seated historical regret is the culling of the famous treasure ships of Zheng He, Chinese admiral and explorer, in the fifteenth century. Zheng

He's treasure ships were a marvel in engineering and is perhaps unmatched by any other during that period. From 1403 to 1433, *He* led seven expeditions to Southeast Asia, India, to the Persian Gulf and as far away as the east coast of Africa. These treasure ships were an engineering marvel and continues to inspire awe to this day.

> Over sixty of the three hundred seventeen ships on the first voyage were enormous "Treasure Ships," sailing vessels over 400 hundred (sic) feet long, 160 feet wide, with several stories, nine masts and twelve sails, and luxurious staterooms complete with balconies. The likes of these ships had never before been seen in the world, and it would not be until World War I that such an armada would be assembled again.[203]

The idea behind these expeditions was to showcase the power of Ming dynasty emperor, *Zhu Di* or the Yongle Emperor, who revived the tributary system, followed by the Han and Song dynasties. The tributary system worked on the principle that border states recognized the Chinese emperor of "all under heaven" and gave tribute to the emperor in exchange for trade treaties and military posts. Zhu Di chose to demonstrate Chinese power beyond China's borders via the ocean and he encouraged and funded his admiral and friend, *Zheng He*, to build those ships and undertake such voyages. Zheng He hailed from a Muslim family in Yunnan and was the ideal candidate to take up the job as the voyages included cultural and religious meetings with countries that followed Islam, to include major parts of north India, ruled at that time by the Delhi Sultanate (thirteenth to the sixteenth century). At that time, the context of the sea was significant as trade was vibrant between China and other Asian nations, to include trade in spices such as cardamom, cinnamon, ginger, turmeric, and pepper, clove, nutmeg, as well as ivory, gold, silver, lacquerware, woolen carpets, and so on. These voyages utilized advanced naval technologies such as propulsion, and magnetic compasses from the ninth century onward, and charts based on stars for precise navigation. China developed the double-hull technology that saved ships from sinking, as well as provided water-tight compartments for carrying food for passengers. China also developed as early as the first century, the sternpost rudder that helped navigate a ship through shallow waters near the shores. These inventions reached Europe hundreds of years after China. China invested in ships propelled by sailing, based on wind efficiency.[204] Eight centuries before Columbus sailed on similar ships to America, China had developed ships 200 ft long with a capacity to carry 500 men. *Zheng He*'s ship, a replica of which you can see in the Maritime Museum near Shanghai,[205] was 400 ft long, with twelve sails, four decks and with the capacity to carry 2,500 tons of cargo. These treasure ships would be accompanied by smaller ships thereby representing the reach power of the Ming emperor. Zheng He commanded the third expedition of forty-eight treasure ships between 1409 and 1411 that carried 30,000 people with it and traveled to Vietnam and Thailand. While carrying out seven voyages, with flotillas as large as sixty-three treasure ships, that carried 30,000 men between 1405 and 1433, Zheng He accomplished several wins in battles with pirates and local kings, bringing some of them back to China for punishment.[206] Significantly, the fifth voyage constituted a return trip for seventeen heads of state for South Asia, in which Zheng He visited the east coast of Africa to include Mogadishu, Malindi (Kenya), and Brawa (Somalia). However, after the death of the emperor and factions in court arguing against the enormous costs that the treasure ships entailed, any further voyages were ruled out. Much to the future detriment of China as a sea faring nation, all 3,500 ships were destroyed in 1525.[207] China's elites deeply regret the burning of those majestic ships. Today, while China glorifies the Zheng He expeditions in the China Maritime Museum near Shanghai,[208] the fact remains that had those ships not been destroyed and China continued to invest in naval technology, China might have avoided the experience of what is known as the "century of humiliation," its second major strategic trauma.

Second Strategic Trauma: The Century of Humiliation

For China, the century of humiliation plays an important part in its conception of who it is today, as a modern China. The British had sought Chinese tea, porcelain, and silk but China demanded silver as a mode of payment. The British then illegally smuggled opium into China, demanded payments in silver that was then used to payoff for three Chinese products they wanted.[209] When the Chinese empire stopped the opium trade, and the 1839 negotiations broke down between the British and the Qing dynasty, British troops occupied Canton (now Guangzhou), and then marched up the Yangtze River delta, to enter Central China. Within two years (1839–1842), the British troops led by the East India Company defeated the Qing dynasty. The British contingent, led by General Henry Pottinger, demanded China to pay for the costs of the war as well as loss in opium trade and open more ports for international trade and establish diplomatic relations. When China refused, the British contingent went on to occupy Shanghai and Nanking, whereby China pled for peace and signed the Treaty of Nanking (1842). As per the terms of that treaty, China ceded Hong Kong, and opened the ports of Guangzhou, Amoy, Foochow, Shanghai, and Ningpo to British trade.[210] This easy defeat at the hands of the British lowered the prestige of the Qing dynasty. This was followed by further humiliation during the Second Opium War, when the British wanted further access into China, a move supported by the French, the Americans, the Germans, and the Russians. The British demanded China to legalize the sale of opium. When the Qing dynasty refused, the British responded and quickly occupied Peking (Beijing), which included the infamous burning down of the Imperial summer palace.[211] In 1860, the Peking agreement was signed, which legalized the opium trade, opened trade on the Yangtze, as well as opened up more Chinese ports to international trade. The treaty enabled the free movement of Christian missionaries as well.[212] It was only after Mao Tse Tung delivered China from Japanese occupation in October 1949 and established the PRC that the "century of humiliation" was believed to be over. The discourse on humiliation is a powerful tool utilized to build legitimacy for the CPC, as well as recollected in maps where China's "lost territories" are viewed as part of such a humiliation.[213] And the loss of territory is seen as strategically disadvantageous as it amounts to the loss of ownership of the resources that territory has to offer. China has established museums,[214] dedicated to the memories of the century of humiliation, and have included the humiliation narrative in school textbooks.[215]

Based on such strategic traumas, what can we expect from China regarding space-based resources.

EXPECTED BEHAVIOR

While space resources are viewed as a more specific case of peacetime competition, when powerful states vie for influence, power, and standing in the larger community of nations, it has direct implications for national defense and security policy. China connects its space industry and development to the overall national rejuvenation of the Chinese nation. Consequently, first presence will be viewed as offering China special rights when it comes to ownership of precious resources that may be mined from the lunar surface or an asteroid surface. As mentioned earlier, the head of China's lunar mission, Ye Peijian stated that:

> The universe is an ocean, the moon is the Diaoyu Islands, Mars is Huangyan Island. If we don't go there now even though we're capable of doing so, then we will be blamed by our descendants. If others go there, then they will take over, and you won't be able to go even if you want to. This is reason enough.[216]

Based on how China has behaved in resource-rich areas on Earth to include Tibet, SCS, ECS, and Antarctica, we can expect China to behave in a nationalist manner when it comes to resources on the Moon and asteroids. The first plan in this goal is to build industrial capacity, especially with the use of robotic probes and robotic manufacturing, by mapping and surveying the

Moon for the best place to build a base. This is to be followed by a lunar sample return to comprehend how lunar resources can be extracted using robots and AI. Following such a phase, China aims to start industrial construction on the lunar surface with the help of 3D printing, following which the first Chinese settler will land on the Moon by 2036. Once the first presence is established, this will be followed by renaming areas where China establishes presence with Chinese names as China has done in Antarctica,[217] as well as other disputed areas with India.[218] Naming areas with Chinese names creates societal connection and memory and adds legitimacy to staking claim on the resources that will be mined. Following the establishment of the lunar base, China's national legislation will make it abundantly clear that those who establish presence first have the first right to mine resources. This inference is drawn from China's assertion of "first presence" claim to the resource-rich islands and the seas in SCS, and Tibet. A similar pattern appears to be emerging in Antarctica where China has unilaterally increased the size of its krill catch despite signing onto the Convention for the Conservation of Antarctic Marine Living Resources (CCAMLR, 1982) that restricts such catches to 680,000 tons. China unilaterally increased it to 2 million tons. More significantly, China's five Antarctic bases are in areas that are strategically rich in resources. Several other threats stem from Chinese involvement in Antarctica that will affect international security such as commercial tourism; bioprospecting; and the undeclared military actions. Several of the Antarctic concerns can be directly transposed into the space environment where China expresses similar intent of establishing a lunar research base for scientific experiments. Moreover, when we compare Antarctica to the SCS, we draw further similarities with Chinese mapping of the entire geography, the renaming of landscapes and significant landforms in Chinese, and a gradual, calculated, and intentional buildup of military equipment.[219]

While others offer insights that PRC space efforts are mostly aimed at anti-access, or information dominance strategies (informatized warfare), in our perspective, what should be of concern are the elements that are focused on material and economic strength and securing of space resources. Chinese space activities should be seen through the lens of territorial and resource competitive offensives such as the BRI and 5G internet. China's history of signing bilateral agreements with countries it has land disputes with, committing to de-escalation and peaceful resolution of disputes, while simultaneously engaging in behavior that is contrary to its signed commitments, do not give us much assurance it will behave any differently in the space domain.[220] This aspect was visible in the 2017 China-Bhutan border dispute. The dispute started when the Royal Bhutan Army (RBA) observed Chinese road-building activities in the Doklam area. The RBA tried to dissuade the PLA engineers from constructing the road but failed. This led to Indian military intervention on behalf of Bhutan, and de-escalation of the conflict after a seventy-three-day standoff between two nuclear-armed nations. Critically, China and Bhutan had signed agreements to resolve this dispute peacefully and have held twenty-four rounds of negotiations on the same, since 1984. The Bhutanese government's take is that China's road-building activities is against the 1988 and 1998 bilateral China-Bhutan agreements by which all boundary disputes will be resolved by negotiations. Bhutan states that "we have written agreements of 1988 and 1998 stating that the two sides agree to maintain peace and tranquility in their border areas pending a final settlement on the boundary question, and to maintain status quo on the boundary as before March 1959. The agreements also state that the two sides will refrain from taking unilateral action, or use of force, to change the status quo of the boundary."[221] On the other hand, the Chinese perspective is that its road-building activities do not infringe upon Bhutanese territory as Doklam has always been a part of China. Foreign Ministry Spokesperson, Lu Kang, asserted "Doklam has been a part of China since ancient times. That is an indisputable fact supported by historical and jurisprudential evidence, and the ground situation. China's activities in Doklam are acts of sovereignty on its own territory. It is completely justified and lawful."[222]

A similar pattern of Chinese behavior emerges with regard to the SCS as well that may inform expectations of future Chinese space behavior. China and ASEAN agreed on the framework on a Code of Conduct (CoC) in the SCS in May 2017.[223] The draft CoC commits the parties to resolve the crisis peacefully and avoid placing offensive weapons in the SCS islands. In 2002, a "Declaration on the Conduct of Parties in the South China Sea" was adopted by China and ASEAN.[224] Interestingly, art.5 of the declaration states:

> The Parties undertake to exercise self-restraint in the conduct of activities that would complicate or escalate disputes and affect peace and stability including, among others, refraining from action of inhabiting on the presently uninhabited islands, reefs, shoals, cays, and other features and to handle their differences in a constructive manner.

Yet, China is using early presence and the historical facts on the ground argument, to alter territorial claims despite its adoption of the 2002 declaration and has established exclusion zones and zones of military coercion in the SCS. In January 2014, it was discovered that Chinese vessels were dredging white sand and placing them onto corals at seven points in the disputed Spratlys, namely, Fiery Cross Reef, Mischief Reef, Gaven Reef, Cuarteron Reef, Subi Reef, South Johnson Reef, and Hughes Reef.[225] Once the artificial islands have been built, China has followed it up with erecting buildings, harbors, and airstrips, deploying radar and surveillance, as well as stationing its troops: all activities geared toward establishing ownership and sovereign control over disputed territory.

What China aims to achieve from its space ambitions is to establish alternative institutions, investment mechanisms, and capacities that not only challenges U.S. dominance in outer space but establishes a China-led space order that it projects as benefiting the world. We can already see this manifesting with its BRI. In 2018, China established the China International Commercial Court, in Xian, Shaanxi province, and Shenzhen under its Supreme People's Court in Beijing,[226] to adjudicate international commercial disputes, especially stemming from Chinese companies' engagement in BRI countries.[227]

The deliberate nature of Chinese actions both in the SCS and Antarctica should serve as an indicator to space powers that without regulations and ongoing dialogue the "first come, first served" mentality might prevail.

NOTES

1. This is an updated version of an article that was first published under the title, Namrata Goswami, "China in Space: Ambitions and Possible Conflicts," *Strategic Studies Quarterly*, 12, no. 1 (Spring 2018): 74–97, https://www.jstor.org/stable/26333878?seq=1#metadata_info_tab_contents

2. "The Power subranking is based on an equally weighted average of scores from five country attributes that related to a country's power: a leader, 00economically influential, politically influential, strong international alliances and strong military alliances."

"Power Rankings," *US News & World Report*, accessed January 10, 2020, https://www.usnews.com/news/best-countries/power-full-list

3. "2018 Military Strength Ranking," *Global Firepower*, accessed January 19, 2019. https://www.globalfirepower.com/countries-listing.asp

4. "World Military Spending in 2017," *SIPRI*, accessed January 19, 2019, https://www.sipri.org/sites/default/files/2018-04/sipri_fs_1805_milex_2017.pdf

5. "The Military Balance 2018," *IISS*, accessed September 18, 2019, report available at https://www.iiss.org/publications/the-military-balance/the-military-balance-2018

6. "Total Population by Country 2018," *World Population Review*, accessed January 19, 2019, http://worldpopulationreview.com/countries/

7. "Human Development Index (HDI)," *United Nations Development Program (UNDP)*, accessed January 19, 2019, http://hdr.undp.org/en/data

8. Energy is expressed in a number of different ways. 3105 MTOE = 22.18 Billion Barrel of Oils Equivalent (BBOE) = 130 Exajoules (EJ) = 123.22 Quadrillion

BTU (QUAD or PBtu) = 31,070.81 Megatons TNT equivalent = 36,111.15Terawatt Hours (TWh) = 4.12 Terawatts (continuous); "Global Energy Statistical Yearbook 2018," *Enerdata*, accessed January 19, 2019, https://yearbook.enerdata.net/total-energy/world-consumption-statistics.html

9. "Global Energy Statistical Yearbook 2018," *Enerdata*, accessed January 19, 2019, https://yearbook.enerdata.net/total-energy/world-consumption-statistics.html

10. "Global Energy Statistical Yearbook 2018," *Enerdata*, accessed January 19, 2019, https://yearbook.enerdata.net/total-energy/world-consumption-statistics.html

11. "Global Energy Statistical Yearbook 2018," *Enerdata*; "Electric Power Consumption (kWh per capita)," *World Bank*, accessed January 20, 2019, https://data.worldbank.org/indicator/EG.USE.ELEC.KH.PC

12. "Which Countries Spend the Most on Space Exploration?," *World Economic Forum*, accessed January 19, 2019, https://www.weforum.org/agenda/2016/01/which-countries-spend-the-most-on-space-exploration/

13. "Government Spending in Space Programs Reaches $62 Billion in 2016," *Euroconsult*, accessed January 19, 2019, http://www.euroconsult-ec.com/30_May_2017

14. "InSight: Futron The Space Competitiveness Index—Update, by Futron," *SatMagazine*, accessed January 19, 2019, http://www.satmagazine.com/story.php?number=1856883389

15. "INTEL: Global Military Space by Futron," *MilSat Magazine* (Sept 2009 Edition), accessed January 19, 2019, http://www.milsatmagazine.com/story.php?number=875113067

16. Charlie Campbell, "From Satellites to the Moon and Mars, China is Quickly Becoming a Space Superpower," *Time*, July 17, 2019, accessed January 13, 2020, https://time.com/5623537/china-space/

17. "The World in 2050," *Price Waterhouse Cooper (PwC)*, accessed January 19, 2019, https://www.pwc.com/gx/en/issues/economy/the-world-in-2050.html

18. Uri Dadush and Bennett Stancil, "The World Order in 2050," *Carnegie Endowment for International Peace*, April 2010, accessed January 19, 2019, https://carnegieendowment.org/files/World_Order_in_2050.pdf

19. "China Unveils Follow-up Lunar Exploration Missions," *Xinhua*, January 14, accessed on January 19, 2019, http://xinhuanet.com/english/2019-01/14/c_137743306.htm

20. "China's Lunar Exploration Program," *People's Daily Online*, accessed on January 19, 2019, http://en.people.cn/90002/91752/index.html

21. Xinhua, "China Unveils Follow-up Lunar Exploration Missions."

22. Namrata Goswami, "Waking up to China's Space Dream," *The Diplomat*, October 15, 2018, accessed September 18, 2019, https://thediplomat.com/2018/10/waking-up-to-chinas-space-dream/

23. "Full Text of Resolution on Amendment to CPC Constitution," *Xinhua*, October 24, 2017, accessed January 10, 2020, https://www.chinadaily.com.cn/china/19thcpcnationalcongress/2017-10/24/content_33656521.htm

24. Meng Xiangqing, "Chinese Dream Include Strong PLA," *China Daily*, October 10, 2013, accessed January 19, 2019, http://www.chinadaily.com.cn/opinion/2013-10/08/content_17012886.htm

25. "2014 Two Sessions NPC & CPPCC: Annual Legislative and Political Advisory Sessions," *China Daily*, 2014, accessed January 19, 2019, http://www.chinadaily.com.cn/china/2014npcandcppcc/2014-03/05/content_17324203.htm

26. "Xi Stresses Cultivating New Type of Military Personnel," *Xinhuanet*, November 27, 2019, accessed January 13, 2020, http://www.xinhuanet.com/english/2019-11/27/c_138587576.htm

27. "China's White Paper on Space Activities 2016," accessed on January 19, 2019, http://english.gov.cn/archive/white_paper/2016/12/28/content_281475527159496.htm

28. "China's White Paper on Space Activities 2016."

29. QUESS (Quantum Experiments at Space Scale), E-Portal Directory, accessed October 15, 2019, https://directory.eoportal.org/web/eoportal/satellite-missions/q/quess

30. Yu Fei, "China's Satellite Sends Unbreakable Cipher from Space," *Xinhua*, August 10, 2017, accessed January 20, 2019, http://www.xinhuanet.com//english/2017-08/10/c_136514705.htm

31. "Details of China's Long Range, Quantum Secured "unhackable" Messaging Revealed," *Xinhua*, January 20, 2018, accessed January 20, 2019, http://www.xinhuanet.com/english/2018-01/20/c_136909246.htm

32. Fei, "China's Satellite Sends Unbreakable Cipher."

33. Fei, "China's Satellite Sends Unbreakable Cipher." Elizabeth Gibney, "Chinese Satellite is One Giant Step for The Quantum Internet," *Nature*, July 27, 2016, accessed January 20, 2019, https://www.nature.com/news/chinese-satellite-is-one-giant-step-for-the-quantum-internet-1.20329

34. "China Plans for the Moon, Mars and Beyond," *BBC*, April 19, 2016, accessed September 18, 2019, http://www.bbc.com/news/world-asia-36085659

35. "Lunar Palace 365 Experiment Sets Record of 200 Days and Volunteers Change Shifts Again," *Beihang University*, January 30, 2018, accessed January 20, 2019, https://ev.buaa.edu.cn/info/1013/1534.htm

36. "China Focus: China Sets Record for Longest Stay in Self-Contained Moon Lab," *Xinhuanet*, May 15, 2018, accessed January 20, 2019, http://www.xinhuanet.com/english/2018-05/15/c_137180976.htm

37. Echo Huang, "Chinese Researchers are Figuring Out How to make Dumplings, Noodles and Stir-Fries in Space," *Quartz*, July 20, 2017, accessed January 20, 2019, https://qz.com/1034097/a-chinese-university-is-putting-students-on-a-grueling-space-diet-of-food-grown-with-their-own-poop/

38. Hannah Devlin, "Giant Leaf for Mankind? China Germinates First Seed on Moon," *The Guardian*, January 15, 2019, accessed January 20, 2019, https://www.theguardian.com/science/2019/jan/15/china-germinates-first-seed-on-moon-cotton-shoot-change4?CMP=share_btn_fb&fbclid=IwAR2SUPboveO7T06r7SakHqJ3eFUeDSbwouDJDMAM0-fprXiLAMuMdE458nk

39. "Flowers on the Moon? China's Chang'e 4 to Launch Lunar Spring," *Xinhuanet,* April 12, 2018, accessed January 20, 2019, http://www.xinhuanet.com/english/2018-04/12/c_137106440.htm

40. "China Focus: Moons Sees First Cotton Seed Sprout," *Xinhua*, January 15, 2019, accessed on January 21, 2019, http://www.xinhuanet.com/english/2019-01/15/c_137745505.htm,. Sasha Ingber, "Cotton Seeds Sprout in China's Lunar Lander," *NPR*, January 15, 2019, accessed on January 21, 2019, https://www.npr.org/2019/01/15/685449027/cotton-seeds-have-sprouted-in-chinas-lunar-lander.

41. Cheng Yingqi, "Embryos Growing in Space a 'Giant Leap,'" *China Daily*, April 18, 2014, accessed January 21, 2019, http://www.chinadaily.com.cn/china/2016-04/18/content_24616857.htm.

42. Yingqi, "Embryos Growing in Space."

43. "China Scientists Develop space 3D printer," *China Daily*, April 19, 2016, accessed September 18, 2019, http://www.chinadaily.com.cn/china/2016-04/19/content_24676024.htm

44. Ma Chi, "China Aims to be World-Leading Space Power by 2045," *China Daily*, November 17, 2017, accessed September 18, 2019, *http://www.chinadaily.com.cn/china/2017-11/17/content_34653486.htm*

45. Stephen Chen, "China's Nuclear Spaceships will be 'Mining Asteroids and flying tourists' as it aims to overtake US in Space Race," *South China Morning Post*, November 17, 2017, accessed January 13, 2020, https://www.scmp.com/news/china/policies-politics/article/2120425/chinas-nuclear-spaceships-will-be-mining-asteroids

46. Chen, "China's Nuclear Spaceships."

47. Wei Long, "New Details of Shenzhou and its Launcher Revealed," *Space Daily*, April 1, 2002, accessed September 18, 2019, *http://www.spacedaily.com/news/china-02zf.html*

48. "China's Long March Rockets Complete 300 Launches," *Xinhua*, March 3, 2019, accessed September 18, 2019, *http://www.xinhuanet.com/english/2019-03/10/c_137883443.htm*

49. Darrel Etherington, "China launches a Rocket to Space from a Shup at Sea for the First time," *TechCrunch*, June 5, 2019, accessed September 18, 2019, *https://techcrunch.com/2019/06/05/china-launches-a-rocket-to-space-from-a-ship-at-sea-for-the-first-time/*

50. Namrata Goswami, "Why China's Long March 11 Launch Matters," *The Diplomat*, June 10, 2019, accessed September 18, 2019, *https://thediplomat.com/2019/06/why-chinas-long-march-11-launch-matters/*

51. Elizabeth Howell, "Tiangong-1: China's First Space Station," *Space.com*, March 26, 2018, accessed September 19, 2019, *https://www.space.com/27320-tiangong-1.html*

52. "China Launches First Cargo Spacecraft *Tianzhou* 1," *Xinhua*, April 20, 2017, accessed September 19, 2019, *http://www.xinhuanet.com//english/2017-04/20/c_136223932.htm*

53. Jamie Carter, "China Hints at Surprise Space Station Launch, Says its New Rocket Can Challenge SpaceX," *Forbes*, March 5, 2019, accessed September 19,

2019, https://www.forbes.com/sites/jamiecartereurope/2019/03/05/china-hints-at-surprise-space-station-launch-says-its-new-rocket-can-challenge-spacex/#32a313f43e80; "China's Planned Space Station to Support Hundreds of Experiments," *Xinhua*, April 24, 2019, accessed January 13, 2020, *http://www.xinhuanet.com/english/2019-04/24/c_138005075.htm*

54. Matt Burgess, "China's Latest Rocket Launch Fails to Get to Orbit after "Anomaly," *Wired*, July 3, 2017, accessed January 13, 2020, https://www.wired.co.uk/article/china-rocket-long-march-5-fail

55. "China's Largest Carrier Rocket Long March-5 Makes New Flight," *Xinhua*, December 27, 2019, accessed January 13, 2020, http://www.xinhuanet.com/english/2019-12/27/c_138662139.htm

56. "Xinhua Headlines: The Return of Long March-5 Inspires China to Reach for the Stars," *Xinhuanet*, December 29, 2019, accessed January 13, 2020, http://www.xinhuanet.com/english/2019-12/28/c_138663650.htm

57. "China to Launch Long March-9 Rocket in 2028," *Xinhuanet*, September 19, 2018, accessed January 13, 2020, http://www.xinhuanet.com/english/2018-09/19/c_137477227.htm

58. "China Establishes Rocket Force and Strategic Support Force," *Ministry of National Defense, The People's Republic of China*, accessed January 22, 2019, *http://eng.mod.gov.cn/ArmedForces/ssf.htm*

59. "China Promotes 5 Officers to General," *Xinhuanet*, July 28, 2017, accessed January 22, 2019, *http://www.xinhuanet.com/english/2017-07/28/c_136480993.htm*

60. Megha Rajagopalan, "Chinese Military Force to Take Lead on Cyber, Space Defense," *Reuters*, January 29, 2016, accessed September 19, 2019, http://www.reuters.com/article/us-china-military-idUSKCN0V714B;

61. Deng Chang, "Look Out, America: China's New Military Force are Awakening," *The National Interest*, 11 January 2016, accessed September 19, 2019, *http://nationalinterest.org/blog/the-buzz/look-out-america-chinas-new-military-forces-are-awakening-14872*

62. "Barrett, Air Force Leaders Applauds Space Force Formal Creation," *Secretary of the Air Force Public Affairs, U.S. Air Force*, December 20, 2019, accessed January 14, 2020, https://www.af.mil/News/Article-Display/Article/2045939/barrett-air-force-leaders-applaud-space-forces-formal-creation/

63. "PLA Strategic Support Force," *SinoDefence*, January 1, 2017, accessed January 22, 2019, *https://sinodefence.com/organisation/pla-strategic-support-force/*

64. Rachael Burton and Mark Stokes, "The People's Liberation Army's Strategic Support Force Leadership and Structure," *Project 2049 Institute*, September 25, 2018, accessed September 19, 2019, *https://project2049.net/wp-content/uploads/2018/09/180925_PLA_SSF_Leadership-and-Structure_Stokes_Burton.pdf*

65. "7 Chinese Military Officers Promoted to Rank of General," *Xinhua*, December 12, 2019, accessed January 14, 2020, http://www.xinhuanet.com/english/2019-12/12/c_138626138.htm

66. "The U.S.-China Military Scorecard Forces, Geography, and the Evolving Balance of Power, 1996-2017," *RAND, Santa Monica*, 2015, accessed September 19, 2019, *https://www.rand.org/pubs/research_reports/RR392.html*

67. "Space Debris from Anti Satellite Weapons," *Union of Concerned Scientist*, Fact Sheet, accessed September 19, 2019, https://www.ucsusa.org/sites/default/files/legacy/assets/documents/nwgs/debris-in-brief-factsheet.pdf

68. As quoted in Mark Williams Pontin, "China's Anti-Satellite Missile Test: Why?," *MIT Technology Review*, March 8, 2007, accessed January 22, 2019, *https://www.technologyreview.com/s/407454/chinas-antisatellite-missile-test-why/*

69. Francis Harris, "Beijing Secretly Fires Lasers to Disable U.S. Satellites," *The Telegraph*, September 26, 2006, accessed January 22, 2019, https://www.telegraph.co.uk/news/worldnews/1529864/Beijing-secretly-fires-lasers-to-disable-US-satellites.html

70. Mike Gruss, "Pentagon Says 2013 Chinese Launch May Have Tested Antisatellite Technology," *Space News*, May 14, 2015, accessed on January 22, 2019, https://spacenews.com/pentagon-says-2013-chinese-launch-may-have-tested-antisatellite-technology/

71. Office of the Secretary of Defense, "Annual Report to Congress Military and Security Developments Involving the People's Republic of China," accessed January 22, 2019, https://dod.defense.gov/Portals/1/Documents/pubs/2015_China_Military_Power_Report.pdf

72. Robert Beck, "China's Mystery Satellite Could be a Dangerous New Weapon," *Medium*, August 22, 2013, accessed January 24, 2019, https://medium.com/war-is-boring/chinas-mystery-satellite-could-be-a-dangerous-new-weapon-630a858923ec; Defense Intelligence Agency, "China Military Power Modernizing a Force to Fight and Win," accessed January 14, 2020, http://www.dia.mil/Portals/27/Documents/News/Military%20Power%20Publications/China_Military_Power_FINAL_5MB_20190103.pdf?utm_source=Daily%20on%20Defense%20(Template)_01/16/2019&utm_medium=email&utm_campaign=WEX_Daily%20on%20Defense: 39—45.

73. "Space Robotic Arm Inspires Scientists for Sorting Garbage," *Xinhua*, July 26, 2019, accessed September 19, 2019, http://www.xinhuanet.com/english/2019-07/26/c_138260582.htm

74. Mike Gruss, "U.S. State Department: China Tested Anti-Satellite Weapon," *Space News*, July 28, 2014, accessed on January 22, 2019, https://spacenews.com/41413us-state-department-china-tested-anti-satellite-weapon/

75. Jayan Panthamakkada Acuthan "China's Outer Space program: Diplomacy of Competition of Cooperation?," *China's Perspectives* (2006), accessed September 19, 2019, https://chinaperspectives.revues.org/577

76. Jeffrey Lin and Peter W. Singer, "China Shot Down Another Missile in Space," *Popular Science*, February 13, 2018, accessed on January 24, 2019, https://www.popsci.com/china-space-missile-test .

77. National Air & Space Intelligence Center (NASIC), "Competing in Space," December 2018, accessed January 24, 2019, https://media.defense.gov/2019/Jan/16/2002080386/-1/-1/1/190115-F-NV711-0002.PDF: 20.

78. Zhao Lei, "Civil-Military Integration will Deepen," *China Daily*, March 3, 2018, accessed January 14, 2020, http://www.chinadaily.com.cn/a/201803/03/WS5a99d67ca3106e7dcc13f437.html

79. "PLA Welcomes Launches of Private Rockets," *China Daily*, February 18, 2019, accessed September 19, 2019, *http://eng.chinamil.com.cn/view/2019-02/18/content_9429066.htm*

80. Rui C. Barbosa, "China Returns Long March 4 to Service with Ziyuan Launch," *NASA Spaceflight.com*, September 12, 2019, accessed September 19, 2019, https://www.nasaspaceflight.com/2019/09/china-returns-long-march-4-ziyuan-2d/

81. Stephen Chen, "Months after Rocket failure, China Sends Spy Satellites into Space," *The South China Morning Post*, September 29, 2017, accessed September 19, 2019, *https://www.scmp.com/news/china/society/article/2113427/months-after-rocket-failure-china-sends-spy-satellites-space*

82. Union of Concerned Scientists, "USC Satellite Databasem," March 31, 2019, accessed September 24, 2019, https://www.ucsusa.org/nuclear-weapons/space-weapons/satellite-database

83. State Council Information Office of the People's Republic of China, "Guiding Opinions of the State Council on Encouraging Social Investment in Investment and Financing Mechanisms in Key Innovation Areas Document 60." November 16, 2014, accessed August 3, 2020, http://www.gov.cn/zhengce/content/2014-11/26/content_9260.htm

84. Zhao Lei, "Civil-Military Integration will Deepen," *China Daily*, March 3, 3018, accessed August 3, 2020, http://www.chinadaily.com.cn/a/201803/03/WS5a99d67ca3106e7dcc13f437.html

85. Michael Sheetz, "China Increases Investment in Emerging Private Space Industry," CNBC, October 10, 2018, accessed August 3, 2020, https://www.cnbc.com/2018/10/10/china-increases-investment-in-emerging-private-space-industry.html

86. Reuters, "China Launches First Rocket Designed by a Private Company," May 17, 2018, accessed August 3, 2020, https://www.reuters.com/article/us-space-launch-china-onespace/china-launches-first-rocket-designed-by-a-private-company-idUSKCN1II0FK

87. "China Successfully Launches Largest Reusable Rocket," *Global Times*, August 10, 2019, accessed August 3, 2020, https://www.globaltimes.cn/content/1160951.shtml

88. Reuters, "Chinese Rocket Startup Puts Satellites into Orbit for First Time," Reuters, July 25, 2019, accessed August 3, 2020, https://www.reuters.com/article/us-china-space/chinese-rocket-startup-puts-satellites-into-orbit-for-first-time-idUSKCN1UK0KY

89. "China's Private Reusable Rocket to be Launched in 2021," *Xinhua*, October 18, 2019, accessed on August 3, 2020, http://www.xinhuanet.com/english/2019-10/18/c_138482839.htm

90. "China's white paper on space, 2016."

91. "China's Space Activities in 2016."

92. Chris Impey, "How China Entered the Space Race," *The Wired*, April 13, 2015, accessed January 28, 2019, https://www.wired.com/2015/04/how-china-entered-the-space-race/

93. Evan Osnos, "The Two Lives of Qian Xuesen," *The New Yorker*, November 3, 2009, accessed January 28, 2019, https://www.newyorker.com/news/evan-osnos/the-two-lives-of-qian-xuesen

94. Osnos, "The Two Lives."
95. Jason Dean and Stephen Miller, "Trained in the U.S., Scientist Became China's Rocket King," *The Wall Street Journal*, November 4, 2009, accessed January 14, 2020, https://www.wsj.com/articles/SB125721495250424443
96. "National High-tech R&D Program (863 Program)," Ministry of Science and Technology of the People's Republic of China, accessed September 19, 2019, http://www.most.gov.cn/eng/programmes1/
97. "National High-tech R&D Program (863 Program)."
98. Stacey Solomone, *China's Strategy in Space* (Switzerland: Springer, 2013), 20.
99. Michael Sheehan, *The International Politics of Space* (London and New York: Routledge, 2007): 163.
100. "State Administration for Science, Technology, and Industry for National Defense," The State Council, People's Republic of China, October 6, 2014, accessed September 20, 2019, *http://english.gov.cn/state_council/2014/10/06/content_281474992893468.htm*; Marco Aliberti, *When China Goes to the Moon* (Switzerland: Springer, 2015): 10–11.
101. Elizabeth Howell, "China National Space Administration: Facts & Information," *Space.com*, May 25, 2016, accessed September 20, 2019, *http://www.space.com/22743-china-national-space-administration.html*
102. Leonard David, "CNSA Boss Outlines China's Space Exploration Agenda," Spacenews.com, April 5, 2017, accessed September 20, 2019, *http://spacenews.com/cnsa-boss-outlines-chinas-space-exploration-agenda/*
103. "China Focus: China Sets Record for Longest Stay in Self-Contained Moon Lab," *Xinhua*, May 15, 2018, accessed September 20, 2019, *http://www.xinhuanet.com/english/2018-05/15/c_137180976.htm*
104. "China Focus: China Sets Record for Longest Stay."
105. Echo Huang, "China lays Out its Ambitions to Colonize the Moon and Build a Lunar Palace," *Quartz*, April 26, 2018, accessed September 20, 2019, https://qz.com/1262581/china-lays-out-its-ambitions-to-colonize-the-moon-and-build-a-lunar-palace/
106. "When Will China Get to Mars," *BBC Interview with Wu Weiren, Chief Designer, Moon Missions*, April 20, 2017, accessed September 20, 2019, http://www.bbc.com/news/av/world-asia-36085659/when-will-china-get-to-mars
107. Brendon Hong, "China's Looming Land Grab in Outer Space," *Newsweek*, June 22, 2018, accessed January 14, 2020, https://www.thedailybeast.com/chinas-looming-land-grab-in-outer-space
108. "China at Technical Preparation Stage for Mars, Asteroid Mining," *Xinhua*, March 3, 2015, accessed January 14, 2020, http://www.globaltimes.cn/content/909948.shtml
109. Zhao Lei, "Asteroid Mission Planned by 2025," *China Daily*, March 4, 2017, accessed January 14, 2020, http://www.chinadaily.com.cn/china/2017-03/04/content_28430297.htm
110. Hong, "China's Looming Land Grab".
111. "Backgrounder: China's Journey to the Moon," *Xinhua*, December 8, 2012, accessed January 14, 2020, http://www.xinhuanet.com/english/2018-12/08/c_137658274.htm

112. Cao Siqi, "Heroes to Receive Highest Honor in China," *The Global Times*, August 08, 2019, accessed September 20, 2019, *http://www.globaltimes.cn/content/1162994.shtml*

113. "China Focus: Chang'e 4 Lands on Largest Crater in Solar System," *Xinhuanet*, January 3, 2019, accessed September 20, 2019, http://www.xinhuanet.com/english/2019-01/03/c_137717410.htm

114. "Riding an Asteroid: China's Next Goal in Space," *Chinese Academy of Sciences*, March 1, 2017, accessed January 28, 2019, http://english.cas.cn/newsroom/china_research/201703/t20170301_174455.shtml

115. "Theoretical Transportation Energy," accessed January 28, 2019, https://www.permanent.com/space-transportation-earth-moon.html

116. Gao Ji et al., "Solar Power Satellites Research in China," *Online Journal of Space Communication* 16 (Winter 2010), accessed September 20, 2019, https://spacejournal.ohio.edu/issue16/ji.html

117. Yuan Can, "China's Solar Power Station in Space is Under Research," *People's Daily Online*, March 08, 2016, accessed September 20, 2019, http://en.people.cn/n3/2016/0308/c90000-9027131.html

118. Wang Kong, "Xinhua Insight: China's Long March into Space," *Xinhua*, April 23, 2016, accessed September 20, 2019, http://news.xinhuanet.com/english/2016-04/23/c_135306099.htm; Also see "Chinese Scientists Mull Power Station in Space," *Xinhua*, March 30, 2015, accessed September 20, 2019, http://news.xinhuanet.com/english/2015-03/30/c_134109115.htm

119. Xinhua, "Chinese Scientists Mull Power."

120. Xinhua, "Chinese Scientists Mull Power."

121. Sun Wenyu, "China Holds Leading Position in Research of Space Based Solar Power," *People's Daily Online*, November 2, 2017, accessed September 20, 2019, *http://en.people.cn/n3/2017/1102/c90000-9288036.html*

122. Wenyu, "China Holds Leading Position."

123. Wenyu, "China Holds Leading Position."

124. Wenyu, "China Holds Leading Position."

125. Li Ming, "Using Resources on Asteroid for Manufacturing of SSPS-A New Attempt and its Potential," 2016 International Astronomical Congress, Guagalajara, Mexico, September 26, 2016, accessed September 23, 2019, https://www.youtube.com/watch?v=6kum9VbVmN8

126. "Exploiting Earth-Moon Space: China's Ambition After Space Station," *Xinhua*, March 7, 2016, accessed September 23, 2019, http://news.xinhuanet.com/english/2016-03/07/c_135164574.htm

127. "Multi-Rotary Joints SPS: 2015 SunSat Design Competition," *You Tube*, June 22, 2015, accessed September 23, 2019, https://www.youtube.com/watch?v=XhgJwnpYRGc

128. Zhao Lei, "Scientists Envision Solar Power Station in Space," *China Daily*, February 27, 2019, accessed April 13, 2019, at http://www.chinadaily.com.cn/a/201902/27/WS5c75c8b3a3106c65c34eb8e3.html. For more on SBSP, see Brain O'Leary, "Construction of Satellite Solar Power Stations from Non-Terrestrial Material," *Journal of Energy* 1, no. 3 (1977): 155–8.

129. Pan Zhaoyi, "China Starts to Build World's First Space Based Solar Power Plant," *CGTN*, February 18, 2019, accessed April 17, 2019, https://news.cgtn.com/news/3d3d514f34597a4e32457a6333566d54/index.html.

130. Zhaoyi, "China Starts to Build."

131. Zhaoyi, "China Starts to Build."

132. "China is Expected to take the Lead in Building a Space Solar Power Station," *Science and Technology Daily.com*, February 14, 2019, accessed on January 14, 2020, http://www.stdaily.com/index/kejixinwen/2019-02/14/content_750019.shtml

133. "China Focus: Capture an Asteroid, Bring it Back to Earth?," *Xinhua*, July 23, 2018, accessed January 29, 2019, http://www.xinhuanet.com/english/2018-07/23/c_137342866.htm

134. Xinhua, "China Focus: Capture an Asteroid."

135. "China to Achieve "Major Breakthroughs" in Nuclear Powered Space Shuttle around 2040: Report." *Xinhuanet*, November 16, 2017, accessed January 29, 2019, *http://www.xinhuanet.com/english/2017-11/16/c_136757737.htm*

136. Zhao Lei, "Super-powerful Long March 9 said to begin Missions Around 2030," *China Daily*, March 11, 2019, accessed February 17, 2020, http://www.chinadaily.com.cn/a/201903/11/WS5c859b62a3106c65c34edcc0.html

137. As quoted from Journal of Rocket Propulsion, January 2011, accessed September 23, 2109, https://www.globalsecurity.org/space/world/china/cz-x.htm

138. "China to Develop New Series of Carrier Rockets: Expert," *Xinhua*, July 2, 2018, accessed March 6, 2020, http://www.xinhuanet.com/english/2018-07/02/c_137295940.htm

139. "China's Space Dream Crystallized with Shenzhou-10 Launch," Economic and Commercial Counsellor's Office of the Embassy of the People's Republic of China in the Republic of Cameroon, June 13, 2013, accessed January 29, 2019, http://cm2.mofcom.gov.cn/article/headnews/201306/20130600160022.shtml

140. "Backgrounder: Xi Jinping's Vision for China's Space Development," April 24, 2017, accessed January 29, 2019, http://www.xinhuanet.com/english/2017-04/24/c_136232642.htm

141. Xinhua, "Backgrounder: Xi Jinping's Vision."

142. Ma Chi, "China Aims to be the World's Leading Space Power by 2045," *China Daily*, November 17, 2017, accessed January 29, 2019, http://www.chinadaily.com.cn/china/2017-11/17/content_34653486.htm

143. "DF-1," accessed September 23, 2019, http://www.astronautix.com/d/df-1.html

144. "Chinese Nuclear Program," *Atomic Heritage Foundation*, July 19, 2018, accessed January 30, 2019, https://www.atomicheritage.org/history/chinese-nuclear-program

145. "National High-tech R&D Program (863 Program)," *Ministry of Science and Technology of the People's Republic of China*, accessed January 30, 2019, http://www.most.gov.cn/eng/programmes1/

146. "Project 921-1," accessed January 30, 2019, http://www.astronautix.com/p/project921.html

147. "China Aerospace Science and Technology Corporation," accessed September 23, 2019, http://english.spacechina.com/n16421/index.html

148. Solomone, *China's Strategy in Space*, 13.

149. Solomone, *China's Strategy in Space*, 13.

150. Namrata Goswami, "China's Unique Space Ambitions," *The Diplomat*, August 3, 2016, accessed September 23, 2019, http://thediplomat.com/2016/08/chinas-unique-space-ambitions/

151. "China's Space Activities 2000," accessed September 23, 2019, *http://www.gov.cn/english/official/2005-07/27/content_17656.htm.*

152. "Chinese Space Activities in 2006," accessed January 30, 2019, http://www.china.org.cn/english/features/book/183672.htm.

153. "China Space Activities in 2011," accessed January 30, 2019, http://www.gov.cn/english/official/2011-12/29/content_2033200.htm.

154. "China Expects to Introduce Space Law by 2020," *China Daily*, November 17, 2014, accessed January 14, 2020, http://www.chinadaily.com.cn/china/2014-11/17/content_18930721.htm

155. United Nations Office for Outer Space Affairs, "Treaty on Principles Governing the Activities of States in the Exploration and Use of Outer Space, Including the Moon and Other Celestial Bodies," accessed February 11, 2019, http://www.unoosa.org/oosa/en/ourwork/spacelaw/treaties/introouterspacetreaty.html

156. Namrata Goswami, "India's Strategic Culture is Plain to See," *Asia Times*, April 5, 2013, accessed September 23, 2019, http://www.atimes.com/atimes/South_Asia/SOU-01-050413.html

157. "Can India become a Great Power?," *The Economist*, March 30, 2013, accessed September 23, 2019, http://www.economist.com/news/leaders/21574511-indias-lack-strategic-culture-hobbles-its-ambition-be-force-world-can-india

158. Alastair Iain Johnston, "Thinking about Strategic Culture," *International Security* 19, no. 4 (Spring, 1995): 32–64, accessed September 23, 2019, http://www.fb03.uni-frankfurt.de/45431264/Johnston-1995-Thinking-about-Strategic-Culture.pdf

159. Goswami, "India's Strategic Culture."

160. Johnston, "Thinking about Strategic Culture."

161. Author conversations and panel discussions at Tsinghua University, Peking University, China Institutes of Contemporary International Relations (CICIR), Pangoal Institution, Beijing Institute of Technology, Shanghai Institutes of International Studies (SIIS), Shanghai Tongji University and Fudan University, November 5–20, 2016.

162. Authors interactions at the SIIS, Shanghai, China, November 15, 2016. Also see Stacey Solomone, "China's Strategy in Space," 61; and for a detailed analysis, see Chenyang Li, "The Confucian Ideal of Harmony," *Philosophy East and West* 56, no. 4 (October 2006): 583–603, accessed September 23, 2019, http://doi.org/ftvnxn

163. Wei Xiaohong and Li Qingyuan, "The Confucian Value of Harmony and its Influence on Chinese Social Interactions," *Cross-Cultural Communication* 9, no. 1

(2013): 60–66, accessed September 23, 2019, http://citeseerx.ist.psu.edu/viewdoc/download?doi=10.1.1.958.6384&rep=rep1&type=pdf

164. For more on this topic, see William A. Callahan, "National Insecurities: Humiliation, Salvation, and Chinese Nationalism," *AlterNative* 29 (2004): 199–218, accessed September 23, 2019, http://doi.org/chkp

165. Mengjia Wan, "Behind the Official Narrative: China's Strategic Culture in Perspective," an interview with Christopher A. Ford," National Bureau of Asian Research, November 1, 2016, accessed September 23, 2019, http://www.nbr.org/research/activity.aspx?id=718

166. Richard Javed Heydarian, "Hide your Strength, Bid your Time," *al Jazeera*, November 21, 2014, accessed September 23, 2019, http://www.aljazeera.com/indepth/opinion/2014/11/china-hide-your-strength-bide-y-2014111198028498329.html

167. Heydarian, "Hide your Strength, Bid your Time."

168. Miles Maochun Yu, "Understanding China's Strategic Culture through its South China Sea Gambit," *Hoover Institution*, May 9, 2016, accessed January 16, 2020, https://www.hoover.org/research/understanding-chinas-strategic-culture-through-its-south-china-sea-gambit

169. Warren I. Cohen, "China's Strategic Culture," *The Atlantic*, March 1997, accessed September 23, 2019, http://www.theatlantic.com/magazine/archive/1997/03/chinas-strategic-culture/376815/

170. Cohen, "China's Strategic Culture." Also see Alastair Iain Johnston, *Cultural Realism: Strategic Culture and Grand Strategy in Chinese History* (New Jersey: Princeton University Press, 1998).

171. Cohen, "China's Strategic Culture."

172. Johnston, *Cultural Realism: Strategic Culture.*

173. 'Han Feizi," accessed February 11, 2019, http://www2.hawaii.edu/~freeman/courses/phil301/13.%20Han%20Feizi.pdf, 3. Also see *Han Fei Tzu: Basic Writings*, translated by Burton Watson (New York: Columbia University Press, 1964).

174. "Han Feizi," 5.

175. "Han Feizi," 1 and 11.

176. Authors interactions with security studies experts, Shanghai Institutes of International Studies (SIIS), Shanghai, China, November 15, 2015. Also see Pin-cheung Lo, "The 'Art of War' Corpus and Chinese Just War Ethics Past and Present," *The Journal of Religious Ethics* 40, No.3 (September 2012): 404–46.

177. Authors interaction with security studies experts, Shanghai Institutes.

178. Huang Youyi, "Context, Not History, Matters for Deng's Famous Phrase," *The Global Times*, June 15, 2011, accessed January 15, 2020, http://www.globaltimes.cn/content/661734.shtml

179. Sun Tzu, *Art of War*, translated by Lionel Giles, http://classics.mit.edu/Tzu/artwar.html

180. Parts of this section have been published earlier in *The Space Review*, the copyright of which is with the author, Namrata Goswami. As per *The Space Review* Submission guideline, "all authors retain copyright on their articles and are free to publish them elsewhere either before or after submitting them to *The Space Review*," please see submission guidelines, https://www.thespacereview.com/submissions

.html; Namrata Goswami, "China's Grand Strategy in Outer Space: To Establish Compelling Standards of Behavior," *The Space Review*, August 5, 2019, accessed January 15, 2020, https://www.thespacereview.com/article/3773/1

181. "China to Build Scientific Research Station on Moon's South Pole," *Xinhua*, April 24, 2019, accessed October 15, 2019, http://www.xinhuanet.com/english/2019-04/24/c_138004666.htm

182. Pin-cheung Lo, "The 'Art of War' Corpus and Chinese Just War Ethics." Liu Mingfu, *The China Dream: Great Power Thinking and Strategic Posture in the Post-American Era* (Beijing: N Times Beijing Media Time United Publishing Company Limited, 2015), 97–111.

183. Permanent Mission of the People's Republic of China to the United Nations and Other International Organizations in Vienna, "Statements by Chinese Delegation at the 55th Session of the Scientific and Technical Subcommittee of the Committee on the Peaceful Uses of Outer Space," February 14, 2018, accessed October 15, 2019, https://www.fmprc.gov.cn/ce/cgvienna/eng/hyyfy/t1535268.htm

184. Jiang Hui, "The Spatial Information Corridor Contributes to UNISPACE + 50," China National Space Administration, accessed October 15, 2019, http://www.unoosa.org/documents/pdf/copuos/stsc/2018/tech-08E.pdf

185. "United Nations/China Forum on Space Solutions," Co-Organized by United Nations Office for Outer Space Affairs and China National Space Administration, Changsha, China, April 24-25, 2019, accessed October 15, 2019, http://www.unoosa.org/oosa/en/ourwork/psa/schedule/2019/2019-un-china-forum-on-spacesolutions.html

186. United Nations Office for Outer Space Affairs, "United Nations and China Invite Applications to Conduct Experiments on-board China's Space Station," May 28, 2018, accessed October 15, 2019, http://www.unoosa.org/oosa/en/informationfor/media/2018-unis-os-496.html

187. United Nations Office for Outer Space Affairs, "United Nations and China."

188. Peter Sabine, "Belt and Road is 'Marshall Plan without a War' Analysts Say, as Beijing and Banks Woo Private Sector Investors," *South China Morning Post*, March 28, 2017, accessed October 15, 2019, https://www.scmp.com/special-reports/business/topics/one-belt-one-road/article/2082733/belt-and-road-marshall-plan

189. "Luxembourg Space Agency, Space Resources," accessed October 15, 2019, https://space-agency.public.lu/en/space-resources.html

190. Zhang Hongpei and Yang Kunyi, "Luxembourg Joins the Ranks of BRI Participants After Italy to be a Part of Inclusive Initiative," *Global Times*, March 28, 2019, accessed October 15, 2019, http://www.globaltimes.cn/content/1143864.shtml

191. "Bank of China Lists 500-mln-USD Bond in Luxembourg," *Xinhua*, April 25, 2019, accessed October 15, 2019, http://www.xinhuanet.com/english/2019-04/25/c_138006422.htm: also see Goswami, "China's Grand Strategy in Outer Space."

192. Amanda Erickson, "Here They Come Again': Chinese Police Arrest Dissident Professor During On-Air Interview," *The Washington Post*, August 3, 2018, accessed October 15, 2019, https://www.washingtonpost.com/news/worldviews/wp/2018/08

/03/here-they-come-again-chinese-police-arrest-dissident-professor-during-on-air-interview/

193. Ministry of National Defense The People's Republic of China, "China Establishes Rocket Force and Strategic Support Force," accessed October 15, 2019, http://eng.mod.gov.cn/ArmedForces/ssf.htm

194. Choi Chi-Yuk and Laura Zhou, "China's Leaders Reaffirm Loyalty to party Under Xi Jinping at top Level Gathering," *South China Morning Post*, December 28, 2018, accessed October 15, 2019, https://www.scmp.com/news/china/politics/article/2179722/chinas-leaders-reaffirm-loyalty-party-under-xi-jinping-top-level

195. "State Media Should Play Due Role in Properly Guiding Public Opinion," *China Daily*, February 22, 2016, accessed October 15, 2019, http://www.chinadaily.com.cn/opinion/2016-02/22/content_23580181.htm

196. Alexandra Ma, "Barging into Your Home, Threatening Your Family, or Making You Disappear: Here's What China does to People Who Speak Out Against Them," *Business Insider*, August 19, 2018, accessed October 15, 2019, https://www.businessinsider.com/how-china-deals-with-dissent-threats-family-arrests-2018-8

197. Liu Mingfu, *The China Dream: Great Power Thinking and Strategic Posture*: 97–100. Also see Goswami, "China's Grand Strategy in Outer Space."

198. *Seven Military Classics of Ancient China*, translated by Ralph D. Sawyer (New York: Basic Books, 2007).

199. Stefan H. Verstappen, The *Thirty-Six Strategies of Ancient China* (San Francisco: China Books and Periodicals, 1999).

200. "Legalism in Chinese Philosophy," Stanford Encyclopedia of Philosophy, December 10, 2014, accessed September 23, 2019, *https://plato.stanford.edu/entries/chinese-legalism/*

201. Bertil Litner, *China's India War: Collison Course on the Roof of the World* (New Delhi: Oxford University Press, 2018); Neville Maxwell, *India's China War* (New York: Random House, 2000).

202. Authors interactions at Tongji University, Shanghai, November 16, 2016.

203. "The Ming Voyages," accessed September 23, 2019, *http://afe.easia.columbia.edu/special/china_1000ce_mingvoyages.htm*; Edward L. Dreyer, *Zheng He: China and the Oceans in the Early Ming Dynasty, 1403-1433* (New York: Pearson Longman, 2007).

204. For more on this, see Louise Levathes, *When China Ruled the Seas: The Treasure Fleet of the Dragon Throne, 1405–33* (New York: Oxford University Press, 1997).

205. Authors visited the maritime museum, on the outskirts of Shanghai in November 2016.

206. Wan Ming, "Reflections on the Study of Zheng He's Expeditions," *Ming Studies*, 1, 2004, accessed September 23, 2019, https://www.tandfonline.com/doi/abs/10.1179/014703704788762899

207. Jim Edwards, "500 Years Ago, China Destroyed its World-Dominating Navy Because its Political Elite was Afraid of Free Trade," *Business Insider*, February 26, 2017, accessed January 17, 2020, https://www.businessinsider.com/china-zhenge-he-treasure-fleet-elite-free-trade-2017-2

208. "China Maritime Museum," accessed September 23, 2019, http://www.timeoutshanghai.com/venue/Things_to_Do-Museums/7542/China-Maritime-Museum.html

209. Callahan, "National Insecurities: Humiliation."

210. "The Opium War," National Army Museum, accessed September 23, 2019, *https://www.nam.ac.uk/explore/opium-war-1839-1842*

211. Orville Schell and John Delury, *Wealth and Power: China's Long March to the Twenty-first Century* (New York: Random House, 2013).

212. Terry Stewart, "The Second Opium War," Historic UK, accessed February 12, 2019, https://www.historic-uk.com/HistoryUK/HistoryofBritain/Second-Opium-War/

213. William A. Callahan, "*The Cartography of National Humiliation and the Emergence of China's Geobody*," *Public Culture*, 2009 21, No.1 (2009): 141–73.

214. Callahan, "National Insecurities: Humiliation".

215. Zheng Wang, *Never Forget National Humiliation: Historical Memory in Chinese Politics and Foreign Relations* (New York Columbia University Press, 2012).

216. Frank Sieren, "Sieren's China: Speaking Mandarin on Mars," *DW*, October 19, 2018, accessed February 12, 2019, https://www.dw.com/en/sierens-china-speaking-mandarin-on-mars/a-45964091

217. Alice Slevison, "Considering China's Strategic Interests in Antarctica," *The Strategist*, February 5, 2016, accessed January 17, 2020, https://www.aspistrategist.org.au/considering-chinas-strategic-interests-in-antarctica/

218. Shaurya Karanbir Gurung, "All About the Names of Six Places in Arunachal that China Gave," *The Economic Times*, July 12, 2018, accessed January 17, 2020, https://economictimes.indiatimes.com/news/defence/all-about-the-names-of-six-places-in-arunachal-that-china-gave/articleshow/58259491.cms

219. L. M. Foster and Namrata Goswami, "What China's Antarctic Behavior Tells us About the Future of Space," *The Diplomat*, January 11, 2019, accessed January 17, 2020, https://thediplomat.com/2019/01/what-chinas-antarctic-behavior-tells-us-about-the-future-of-space/

220. Namrata Goswami, "Can China be Taken Seriously on its 'Word' to Negotiate Disputed Territory?," *The Diplomat*, August 18, 2017, accessed April 17, 2019, https://thediplomat.com/2017/08/can-china-be-taken-seriously-on-its-word-to-negotiate-disputed-territory/

221. Ministry of Foreign Affairs, Government of Bhutan, "*Press Release*," June 29, 2017, accessed September 19, 2017 at http://www.mfa.gov.bt/?p=4799.

222. Ministry of Foreign Affairs of the People's Republic of China, "Foreign Ministry Spokesperson Lu Kang's Regular Press Conference on June 29, 2017," accessed September 19, 2017, http://www.fmprc.gov.cn/mfa_eng/xwfw_665399/s2510_665401/t1474166.shtml

223. "China, ASEAN Agree on Framework for South China Sea Code of Conduct," *Reuters*, May 18, 2017, accessed April 17, 2019, https://www.reuters.com/article/us-southchinasea-china-philippines-idUSKCN18E1FS

224. "Declaration on the Conduct of Parties in the South China Sea," *Association of Southeast Asian Nations*, October 17, 2012, accessed April 17, 2019https://asean.org/?static_post=declaration-on-the-conduct-of-parties-in-the-south-china-sea-2.

225. Derek Watkins, "What China has been Building in the South China Sea," *The New York Times*, February 29, 2016, accessed April 17, 2019, https://www.nytimes.com/interactive/2015/07/30/world/asia/what-china-has-been-building-in-the-south-china-sea-2016.html

226. "China International Commercial Court," accessed April 16, 2019, http://cicc.court.gov.cn/html/1/219/index.html.

227. Freshfields Bruckhaus Deringer, "China Establishes International Commercial Courts to handle Belt and Road Initiative Disputes," University of Oxford, Faculty of Law, August 17, 2018, accessed October 15, 2019, https://www.law.ox.ac.uk/business-law-blog/blog/2018/08/china-establishes-international-commercial-courts-handle-belt-and

Chapter 6

India's Strategy and Space Resource Ambitions

India is one of the major spacefaring nations in Asia and is rapidly becoming a global space power with demonstrated capability to go to Mars and lunar orbits, as well as launch a variety of satellites on a regular basis. On September 7, 2019, India attempted to land near the lunar South Pole with its *Chandrayaan 2* Moon mission and came very close to accomplishing the feat. Its lander *Vikram* malfunctioned with minutes to spare, and the lander was later discovered by the *Chandrayaan 2* orbiter.[1] On December 2, 2019, NASA's Lunar Reconnaissance Orbiter Camera released details of the impact side (70.8810°S, 22.7840°E, 834 m elevation) and the associated debris field from the lander, just 2.1 km short of its intended target.[2] With future ambitions to send humans to Low Earth Orbit (LEO), a *Chandrayaan 3* lunar mission in 2022,[3] as well as missions to Venus and the Sun, India's space ambitions are showing signs of expansion. This chapter offers a detailed perspective of India's discourse on space resources, their instantiation in policy, law, and programs. It also investigates India's strategic culture, strategic trauma, and expected behavior with regard to space resources.

INDIA: CURRENT AND FORECAST POWER

At the time of writing (2019), India finds itself significantly behind the United States and China but projected to gain in the next two decades. India is ranked #15 on the list of the most powerful countries by U.S. News and World Report.[4] Global Firepower ranks India as #4 in military power.[5] The Stockholm International Peace Research Institute estimates its military spending at $63.924Bn.[6] The International Institute for Strategic Studies estimates it at 52.5$Bn and places it as #5 in the military balance.[7]

India has a population of 1,368,737,513 growing at 1.08 percent,[8] with a Human Development Index score of 0.640 (ranked #130/189 countries).[9] Its total energy consumption was 934 million tons of oil equivalent[10] (4.87 barrels of oil per capita), growing at 4.4 percent, with a carbon output of 2,234 MtCO2.[11] Its electrical consumption is 1,156 TWh growing at 5.3 percent,[12] with a per capita electrical usage of 806 Kw-hr/annum.[13]

According to International Monetary Fund figures, its GDP (real) was $2,611.01Bn ($2.6 T), and $9,459.00Bn at purchasing power parity (PPP), with a per capita income of $1,982.70. It spent 27.7 percent of its GDP in government expenditures, its estimated gross national savings was 29.7 percent of GDP, and had a current account balance of $51.214Bn (1.961 percent of GDP).

According to PricewaterhouseCoopers, its economy is expected to grow from $3.6T in 2020 to $28.0T in 2050 at market exchange rates (MER or "real" GDP) and from $11.8T to $44.1T in PPP, suggesting it will be the #3 economy in 2050 at MER and the #2 economy (after China) measured by PPP, with 15 percent share of the total global economy (up from 7 percent in 2016).[14] The Carnegie Endowment for International Peace similarly estimated it to be the #3 economy in 2050 at $15.384T.[15] The Economist Intelligence Unit also placed India at #2 at MER, with a projected GDP rising from $2.055T in 2014 to $63.842T in 2050.[16]

INDIA: SPACE CAPACITY

According to the World Economic Forum, India spends $1.159 billion on its space program (0.06 percent of GDP),[17] whereas Euroconsult estimated its space spending at $1.092 billion.[18] The most recent (and somewhat dated) Futron Space competitiveness report, gave it an overall space competitive score of 15.33 (ranked #7),[19] with an overall military space capability score of 10 (ranked #7).[20]

India's space activities began in 1962 when the Department of Atomic Energy (DAE) formed the Indian National Committee for Space Research under the leadership of Dr. Vikram Sarabhai and Dr.K.R. Ramanathan. The Indian Space Research Organization (ISRO) was formed in August 1969. India became capable of producing its own satellites in 1975 and launching its own satellites in 1980.[21] As of March 31, 2019, India is listed as the owner/operator of fifty-seven satellites in the Union of Concerned Scientists (UCS) database.[22] In 2018, India accounted for 6 percent of global launches (7 of 111 global successful orbital launches behind China: 38, United States: 29, Russia: 19, and equal to all of Europe: 7).[23]

India's space program is undergoing a major expansion and reorientation. So much is happening at such a fast pace since our field visit to India in

July–August 2017 that it can prove difficult to stay abreast of all the changes. In fact, at the time of writing, India was in the midst of a major policy shift with regard to the domain of outer space.[24]

Civilian Space Activities

India's space activity remains dominated by its civilian space program run by ISRO. A full exploration of India's civilian space program is beyond the scope of this work.[25]

ISRO develops and operates a family of launch vehicles including the Polar Satellite Launch Vehicle (PSLV), and the Geosynchronous Satellite Launch Vehicle (GSLV). ISRO has proven its reliability, low cost,[26] and expertise in launching multiple payloads, including a record-breaking 104 satellites on a single launch in 2017,[27] and another record of 29 satellites to three different orbits in 2019.[28]

ISRO's major function has been to build space infrastructure to aid in national development. Toward that end, ISRO operates a broad constellation of satellites that accomplish telecommunications (including for telemedicine, tele-education, satellite television), remote sensing, navigation, and timing. ISRO has produced and launched satellites with dual use and military capabilities.

The ISRO *Vision 2025 plan* put out by the Vikram Sarabhai Space Centre, identifies the following future space goals:

1. To develop satellite-based communication and navigation systems for rural connectivity, security needs, and mobile services.
2. To enhance imaging capability for the management of natural resources, weather, and assist in climate change studies.
3. To further develop "space science" missions to enhance understanding of the solar system and the universe.
4. To develop a heavy-lift launcher capability.
5. To develop reusable launch vehicles.
6. To develop a human flight program.[29]

While space has been and will be connected to national development goals, ISRO's achievements in recent years project an ambition to demonstrate technological capability and success globally.[30] Fairly recently, ISRO has become engaged in robotic planetary exploration. Starting in 2008, India launched its first Moon orbiter, *Chandrayaan 1*, which detected water on the Moon's surface. Subsequently, India successfully sent a Mars orbiter, *Mangalyaan*, in its first attempt in 2014, with a low cost of US$74 million.[31] This was a significant geopolitical achievement given it was the first Asian nation to do so.[32] Its *Chandrayaan 2* lunar mission successfully entered lunar orbit in August 2019 and placed an orbiter there to map the lunar poles.[33] However, India's attempt to soft-land on the lunar South Pole was not successful. In 2017, India announced a second Mars mission, expected to put a lander on Mars by 2021–2022.[34] ISRO's former chairperson A. S. Kiran Kumar specified in September 2016 that India will be investing in launching vehicles to Venus and to asteroids.[35] *Shukrayaan-1*, the announced mission to Venus,[36] is expected to launch in 2023.[37]

India has recently announced its entry into human space. In August 2018, Prime Minister Narendra Modi announced that India would send a man or woman into space by 2022 in a mission called *Gaganyaan*.[38] Four Indian astronauts have started their year-long training in the Gagarin Cosmonaut Training Center (GCTV). According to Pavel Vlasov, head of GCTV, "this is a program specifically designed to [suit the needs of the Indian astronauts] and coordinated with the Indian side."[39] The course focuses on enhanced and advanced engineering, general space training, and physical conditioning to survive the space environment.

Then, in 2019, India announced its intention to launch its own space station by 2030, specifically stating there will be no cooperation with any other country on this project.[40]

Military Space Program

India is a latecomer to the military use of space but is rapidly undergoing significant changes. As with nuclear weapons, India was reticent to encourage a global arms race and hoped for a more peaceful international regime.

Early advocacy of military space came from the former Chief of Air Staff, Air Chief Marshall S. K. Sareen's address to the Air Force (AF) Day parade on October 8, 1998: "I have often emphasized that in the years ahead, the exploitation of space-based resources for the conduct of air operations will assume increasing importance. . . . The necessity to progress from an Air Force to an 'Air and Space Force' is growing in importance everyday."[41]

The Defense Research and Development Organisation (DRDO) had long-held ambitions of developing space technology,[42] including hypersonic air-breathing launch vehicles (AVATAR, Hyperplane), small Inter Continental Ballistic Missiles-derived mobile smallsat launchers,[43] a ready staple of operationally responsive smallsats,[44] and anti-satellite missiles (kinetic ASATS) capable of striking both LEO and geosynchronous orbit (GEO).[45] But such ambitions had apparently been kept in check by the political influence of the civilian ISRO, and the reticence of the Congress party. Similarly, there were calls for a dedicated military space branch or command, but seemed to both lack a political constituency and failed to find consensus among competing military branches.

An early vision "Military Dimensions in the Future of the Indian Presence in Space," published in 2000 by V. Siddhartha, a former officer on the personal staff of the DRDO chief, laid out a vision that included AVATAR, a reusable single stage to orbit space plane, KALI, a kinetic attack loitering interceptor, and DURGA and SHIVA which were directed energy weapons systems assumed to be technically fieldable by 2010.[46]

India's foray into military space applications began circa 2000, when, "India began planning a constellation of high-resolution satellites for tactical military applications after an inquiry into the country's failure to detect the May 1999 Pakistani intrusion in the Kargil area of Kashmir."[47] The Technology Experiment Satellite (TES) was India's first satellite with military applications. *SpaceNews* reported that "officially, ISRO is a 100-percent civilian agency. But scientists, who asked not to be named, said ISRO's Technology Experiment Satellite, launched in 2001, was meant to validate technologies for future dedicated military satellites. The 1-meter-resolution imagery collected by the satellite is used exclusively by the defense services, they say, adding that once the aerospace command gets government sanction, ISRO may be asked to build and launch a dedicated satellite for the military."[48] By 2004, in response to shortcomings in the Kargil conflict, India had established the "National Technical Research Organization (NTRO)—modeled on the U.S. National Security Agency—which would be the repository of the nation's technical intelligence-spy satellites, unmanned aerial vehicles (UAVs), and spy planes."[49]

Advocates developed a more ambitious organizational construct, "the Aerospace Command concept was conceived in 2000, and India's Air Force officials envision it as the headquarters of space technologies that will help link radar and communications networks and be used for ballistic missile defense and intelligence-gathering."[50]

Subsequent Indian Air Force (IAF) chiefs expanded on the vision. In 2003,[51] IAF chief S. Krishnaswamy specified that "the work has started on an aerospace command to have weapon platforms in space," and "any country on the fringe of space technology like India has to work towards such a command as advanced countries are already moving towards laser weapon platforms in space and killer satellites." In 2005,[52] "Indian Air Force (IAF) Chief Air Marshall S. P. Tyagi said an Aerospace Command has to be established to lay the groundwork to develop capability to degrade space weapons. He said such a command for futuristic warfare has to be developed with the help and guidance of the country's indigenous space agency [ISRO]."[53]

Following the 2007 Chinese ASAT test, Indian thinkers took note. A monograph titled "Militarization of Space," by Kaza Lalitendra for the Center for Airpower Studies (CAPS) captured the new thinking,

> Orbital space debris, anti-satellite weapons (ASAT), jamming, hacking and ground station vulnerabilities are posing a real danger to civil, military, and commercial satellites. Therefore there exists a strong urgency for global governance . . . With existing legal frameworks and unofficial UN guideline considered to be ambiguous in nature, lacking verification or enforcement provisions and unable to expressing [sic] prohibit conventional weapons in outer space or ground-based ASATs. There exists today a dormant threat of space weaponization, which if allowed would radically change the way countries operate in space.[54]

On June 10, 2008,

> The Defence Minister Shri AK Antony today announced the formation of an Integrated Space Cell under the aegis of the Integrated Defence Services Head Quarters to counter "the growing threat to our space assets." Addressing the Unified Commanders' Conference here, he said, although we want to utilize space for peaceful purposes and remain committed to our policy of non-weaponisation of space, "offensive counter space systems like anti-satellite weaponry, new classes of heavy-lift and small boosters and an improved array of Military Space Systems have emerged in our neighbourhood." Shri Antony said the new Cell will act as a single window for integration among the Armed Forces, the Department of Space and the Indian Space Research Organisation (ISRO).[55]

In 2010, India set up a Space Security Coordination Group chaired by its national security adviser with representatives from DRDO, IAF, and NTRO, which "besides laying down the Government's space policy, this body will also coordinate response on an international code of conduct in space."[56] India's military technical arm, the DRDO also responded. In June 2010, DRDO's Technology Perspective and Capability Roadmap (TPCR) stated an ambition to "development of ASAT for electronic or physical destruction of satellites in both lower earth orbit (LEO) and Geo-synchronous orbits" by 2015.[57]

In April of 2012, scientific adviser to the Defense Minister Dr. Vijay Kumar Saraswat told *India Today* that "India has all the building blocks for an anti-satellite capability."[58] Circa 2013, futurists at the Institute for Defense Studies and Analyses (IDSA) predicted that "Militarization of Space" would be a driver of strategic change; that "geopolitical rivalries and national security considerations drive military use of space in 2050 with more than 50 countries routinely applying space capabilities to their military operations."[59]

The stability of Indian military satellites has progressively grown and is presently in the midst of a major growth spurt. Following its first high-resolution spy satellite TES, the first dedicated military satellite would be RISAT-2 (a radar satellite) in 2008,[60] and GSAT-7 (also designated INSAT-4F) to support Indian Navy communications in 2013.[61] As per the 2020 U.S. National Defense Authorization Act (NDAA), India's NAVIC global navigation satellite system (GNSS) is to be identified as an "allied system," under section 1601, subtitled, Space Activities. This implies that "the GNSS system will be on a par with Japan's GNSS Quasi-Zenith Satellite System (QZSS) and Europe's Galileo GNSS. This will allow U.S. military and government users to utilise NAVIC along with the U.S. Global Positioning System (GPS), QZSS, and Galileo."[62] The 2020 NDAA specifies:

> Waiver Authority for Trusted Signals Capabilities—Section 1609(a)(2)(B) of the John S. McCain National Defense Authorization Act for Fiscal Year 2019 (Public Law 115-232; 10 U.S.C. 2281 note) is amended by striking "such capability" and inserting "the capability to add multi-Global Navigation Satellite System signals to provide substantive military utility." (g) Definitions—In this section: (1) The term "allied systems" means—(A) the Galileo system of the European Union; (B) the QZSS system of Japan; and (C) upon designation by the Secretary of Defense, in consultation with the Director of National Intelligence—(i) the NAVIC system of India . . .[63]

Already the ISRO has launched payloads on Microsat-R (an ASAT target), RISAT-2B (radar imaging), and EMISAT[64] (electronic surveillance),[65] and it was announced that beginning in May 2019, the ISRO was expected to launch a string of defense satellites at the rate of nearly one a month for the next year (2020), including high-resolution imaging for Earth Observation (Cartosat-3), ground forces communication (GSAT-32), and three additional radar imaging satellites (RISAT-2BR1, 2BR2,1-A), and GISAT 1 & 2.[66] Over and above these announced launches, DRDO officials suggest they will launch at least three additional electronic surveillance satellites to compliment EMISAT.[67]

In 2019, after over a decade of forbearance, the Modi government executed a direct ascent kinetic impact anti-satellite (KE-ASAT) test, "mission Shakti" against the target Microsat-R. The test was executed in a manner to minimize long-lasting debris. Immediately afterward, the DRDO released a detailed video explaining the test.[68]

Following the ASAT test, the *Press Trust of India* (PTI) reported that India's government has finalized the broad contours of a new defense space agency; "The Defence Space Agency will evolve strategy to protect India's interests in outer space including dealing with threat of space wars," and "is being tasked with developing a range of platforms and co-orbital weapons to protect Indian assets in space and to have deterrence."[69] PTI reported "the government is also planning to set up a defense space research organization to carry out research and development relating to military dimension of use of outer space."[70] According to Rajeswari Pillai Rajagopalan, the new tri-service Defence Space Agency (DSA), based in Bangalore, is an important institutional initiative that combines two key functions performed by the Defense Imagery Processing and Analysis Centre in Delhi and the Defense Satellite Control Centre in Bhopal and represents the growing integration of India's space capabilities.[71] The DSA will be commanded by an Air Vice-Marshall from the IAF.[72] To gain a better understanding of the current and emerging challenges in the space domain, the Integrated Defense Staff of the Ministry of Defense held the first-ever space wargame, "IndSpaceEx" in July 2019.[73]

Commercial Space Activities

While dominated by its civilian sector, India's commercial sector is undergoing significant growth. Until very recently, private companies only supplied components or subcomponents to ISRO, which served as the designer, and operator of space assets. ISRO would sell services such as launch or satellite imagery internationally through a state-owned enterprise, *Antrix*. Unlike in other countries, no domestic private space companies serviced the Indian domestic market or international markets.

However, both because of pressure from private industry and increasing demand, ISRO is in the process of transferring both satellite-building and launch vehicle technology to the Indian private sector. Already a consortium led by Alpha Design Technologies delivered the first consortium-manufactured navigation satellite in 2017,[74] the first successful launch in 2018, and ISRO expects to procure nine consortium-built satellites from 2018 to 2021.[75] ISRO is also privatizing its proven low-cost, high-reliability launch platform, and it is expected that the first consortium-built PSLV will fly in 2020.[76] The aim of this endeavor is to enable India to increase its satellite launches. Says ISRO former chairperson, A.S. Kiran Kumar: "Now, we are trying to go up to 8-9 PSLV per year, two GSLV-Mk II and one GSLV-Mk III. Total about 12 per year."[77] India's government-sponsored commercialization efforts also received a recent upgrade:

> A public sector enterprise, New Space India Limited, has been incorporated as a new commercial arm of the Department of Space to tap the benefits of research and development carried out by ISRO. The company will spearhead commercialisation of various space products, including the production of launch vehicles, transfer of technologies, and marketing of space products.[78]

While the high-tech nature of ISRO is credited with having created the Indian "silicon valley" in Bangalore, it was India's Information Technology

(IT) sector with its experience in fast product turn around and international markets that has given birth to the Indian NewSpace sector. Different from traditional contractors that seek to service the Indian Government market, NewSpace hopes to create a commercial space sector that can service a global market directly, and supply launch, satellite components, satellite manufacture, and space-based services such as remote sensing and communications. The Indian NewSpace sector began with "Team Indus" that joined the Google Lunar X-prize competition and emboldened other entrepreneurs to step out from the shadow of the state-run space enterprise. When we began writing this book, you could count every Indian NewSpace company on one hand. But the number of truly commercial companies has grown steadily[79] and now includes ARDL,[80] Agnikul Cosmos,[81] Astrome,[82] Bellatrix,[83] Blue Sky Analytics,[84] Dhruva Space,[85] Earth2Orbit,[86] Exseed Space,[87] Kawa Space,[88] Pixxel,[89] ReBeam,[90] Satsure,[91] Skyroot Aerospace,[92] SmartEnovations,[93] TeamIndus,[94] and XOVIAN,[95] as well as Rocketeers,[96] a youth-focused rocketry club to encourage future workforce development. Already it is seeing successes. Bellatrix Aerospace[97] that focuses on propulsion and hopes to build launch vehicles is the first company to "spin-in" a technology into ISRO. It also broke ground by being the first Indian space start-up to raise a significant amount of funding—$3M.[98] *Exseed Space* is the first entirely private Indian space start-up to succeed in building and launching its own satellite on an international commercial launch (a *SpaceX* Falcon 9).[99] TeamIndus (*Axiom Research Labs Private Limited), the Indian space company that competed in the Google Lunar X-Prize and kickstarted the Indian NewSpace ecosystem, successfully competed for*[100],[101] *(and then withdrew from*[102],[103]) *a NASA* Commercial Lunar Payload Service (CLPS)[104] contract, but remains committed to providing cislunar delivery services.[105]

INDIA: ELITE DISCOURSE

Since its inception, the Indian space program has been connected to national development. The father of the Indian space program, Dr. Vikram Sarabhai (chairman from 1963 to 1971) set the course, and ISRO has followed, with the following quote:

> There are some who question the relevance of space activities in a developing nation. To us, there is no ambiguity of purpose. We do not have the fantasy of competing with the economically advanced nations in the exploration of the moon or the planets or manned space-flight . . . But we are convinced that if we are to play a meaningful role nationally, and in the community of nations, we must be second to none in the application of advanced technologies to the real problems of man and society.[106]

While Mars or Venus probes seem to diverge from this mission, it would seem that the tapping of resources (development vs. exploration) specifically to aid in national economic development would fall squarely within the Sarabhai vision. Yet this has not been the case. The startling exception was Dr. A. P. J. Abdul Kalam, the eleventh president of India.[107] No head of state before or since came with his credentials—an aerospace engineer having developed rockets at both DRDO and ISRO—or had such a comprehensive vision.[108]

Dr. Kalam's grand vision, articulated at Boston University in 2007, encompassed the mining of asteroids and the use of the Moon as an industrial base and source of energy. He saw an extra-terrestrial habitat on Mars as a fail-safe mechanism for problems on Earth. He hoped to see these activities blossom into a mature regime of space security, with Space Traffic Control, Space Debris Management, and an International Space Force for defense against asteroids.[109] He sought to catalyze United States-India[110] and global efforts to pursue space solar power.

He elaborated his vision for Space Solar Power—Key to a Livable Planet in May of 2013[111] and his broader "World Space Vision 2050" to the Indian Institute of Space Technology (IIST) in June of 2013.[112] At these events, Dr. Kalam spoke of a "Space Industrial Revolution,"[113] and of:

> Evolving a Global Strategic Plan for space industrialization so as to create large scale markets and advanced space systems and technologies, for clean energy, drinking water, tele-education, tele-medicine, communications, resource management and science; and for undertaking planetary exploration mission.[114]

Dr. Kalam spoke of "Large Scale Societal missions (including Space Solar Power mission) enabled by low cost access to space," stating:

> Such a mission in space industrialization could start up with an intensely human international mission in societal-critical areas of energy, water, environment and security which would expand over the decades and centuries to all other human needs and activities on planet earth through space exploration and a new order in space through space security. I think there cannot be a greater vision for all nations other than transforming our terrestrial habitat into a liveable planet earth through international collaboration in space.[115]

Dr. Kalam's vision was not limited to space solar power. He stated, "mining in planets or asteroids would need innovative methods for exploring, processing and transporting large quantities of rare materials to earth,"[116] and "Helium 3 from Moon is seen as a valuable fuel for thermonuclear reactors,"[117] and

> The Moon also has other advantages as a source of construction materials for near Earth orbit. Its weak surface gravity is only one-sixth as strong as Earth's. As a result, in combination with its small diameter, it takes less than five percent as much energy to boost materials from the lunar surface into orbit compared with the launch energy needed from Earth's surface into orbit. Electromagnetic mass drivers powered by solar energy could provide low-cost transportation of lunar materials to space construction sites. Low gravity manufacturing holds tremendous promise for mankind in new materials and medicines.[118]

Dr. Kalam's vision was motivated by humanitarian and environmental concerns:

(1) Massive shortages, escalating costs of energy, and rapidly depleting fossil fuel reserves world over; (2) Global environment degradation and climate change; (3) Depletion of Mineral Resources.[119] In fact, it was this vision that motivated one of the authors to come to India to examine the possibility of a U.S.-India collaboration in Space Solar Power.[120]

It was a puzzle thereafter, that the vision of a man so well regarded by the citizens of India found no significant echo within ISRO. His vision for long was shared by only one other advocate, a close friend of one of the authors, Air Commodore (Retired) Raghavan Gopalaswami[121] who was a tireless advocate of Space Solar Power[122] and reusable space access.[123] Though Dr. Kalam made significant efforts—the NSS-Kalam Initiative, including attempts under President Barack Obama to kick-start U.S.-India collaboration on SBSP—it remains a tragedy for both nations that these efforts have yet to come to any fruition.

It seems that ever-so-slowly, the themes so presciently identified by Dr. Kalam are beginning to permeate Indian discourses.

An exception to this is the discussion on Lunar Helium-3. Perhaps this is because of the historic close association between ISRO and the DAE from which it was birthed. Consideration of Helium-3 appears to have been a motivation or consideration of India's lunar program from the start. The dialogue showcases Indian strategic concerns about the continuity of the Outer Space Treaty (OST) and the need to hedge by establishing a justification for a future territorial claim should the need arise.

As early as September 2006, the *Hindustan Times* carried a headline, "Chandrayaan will search Moon's Surface for Deposits of Helium-3 that can be used to power future nuclear reactors."[124] The article reported a top scientific official, then ISRO chairman Madhavan Nair, stating that India's ambitious lunar mission Chandrayaan will search the Moon's surface for deposits of Helium-3, which can be used to power future nuclear reactors. Speaking at the Bhabha Atomic Research Centre in Mumbai, Nair stated, "The quantity of helium-3 is also very important as it will determine the economics before we exploit it." Noting that Helium-3 would be useful as a nuclear fuel, Atomic Energy Commission (AEC) former chairman Anil Kakodkar noted that there are miniscule deposits of Helium-3 on the Earth but the Moon had an abundance and was quoted as stating, "Our energy requirements are increasing by the day and for how long will we be dependent on terrestrial fuels? We have to look at external sources."[125]

In 2008, *OneIndia* reported that former ISRO chairman Nair said that *Chandrayaan I* will plant the Indian national flag on the Moon, thus helping India establish claim on the lunar surface. Moreover, his remarks clearly showed consideration of the future: "Today, as per the international charter, the Moon belongs to the global community. Nobody can make special claim on the surface. In due course, we don't know how things will change. But, our presence will be established through this mission."[126] *OneIndia* reported,

> India will be the fourth country after USA, Russia and Japan to put its Flag on the Moon. And, so like Antarctica, if ever the resources on the Moon are shared, India will be a party to it. Dr Nair said the claim on the Moon would also help India do mining for precious metals and look for Helium-3, a tonne of which can sustain the entire country's energy for one year. The mission will also look for water on the Moon, which is essential for establishing a lunar colony, the ISRO chief added.[127]

Circa 2012, a team of futurists at India's IDSA forecast the following vision for 2050:

> Several futuristic space products such as space tourism, on-orbit services, microgravity materials processing, lunar resource applications, asteroid mining and solar power satellites become viable. The United States leads the space innovation industry. In 2050 several American private corporations are engaged in operating bases on the moon and extracting resources from asteroids located between the orbits of Mars and Jupiter.[128]

The conversation remained largely dormant until in February of 2017, ISRO Professor, Dr. Sivathanu Pillai ignited significant attention when it was reported that ISRO "plans to mine Helium-3 rich lunar dust, generate energy and transport it back to Earth . . . former chief of BrahMos Aerospace, said that mining lunar dust was a priority programme for his organization,"[129] further stating that India may be able to meet its energy requirements from Helium-3 mined from the Moon.[130]

On the eve of the success of *Chandrayaan-1*, author of the TeamIndus blog (India's entrant into the Google Lunar XPrize) stated,

> I am often asked why a country like India should mount a Moon mission. I have thought deeply about it and one of the reasons that I feel strongly about is that many a time countries who have access to resources, tend to form a club and keep everyone else away from it. Moon is a fantastic source for many valuable resources including Helium 3, which can take care of the energy needs of all mankind. The new space race promises big returns for investments in space exploration, evident by the moon rush kick started by the private space companies. Moon mining for the 3 crucial elements: Water, Helium-3, and Rare Earth Metals has captured the attention of the entire space industry, inspiring commercial ventures. Being space infrastructure ready is critical to become a part of the elite club of countries that are future ready.[131]

Most recently, *Bloomberg* reported that resource justifications are significant with regard to India's second lunar mission:

> India's space program wants to go where no nation has gone before—to the south side of the moon. And once it gets there, it will study the potential for mining a source of waste-free nuclear energy that could be worth trillions of dollars. The nation's equivalent of NASA will launch a rover in October to explore virgin territory on the lunar surface and analyze crust samples for signs of water and helium-3. That isotope is limited on Earth yet so abundant on the moon that it theoretically could meet global energy demands for 250 years if harnessed. "The countries which have the capacity to bring that source from the moon to Earth will dictate the process," said K. Sivan, chairman of the Indian Space Research Organisation. "I don't want to be just a part of them, I want to lead them."[132]

With respect to space solar power, it appears that the elite discourse is slowly catching up to Dr. Kalam. In earlier research prior to 2010, one of the authors was aware of exploratory interest by Tata[133] in the concept.

In late 2010, as a result of earlier work,[134] space solar power was considered by the Policy Coordinating Committee as a top-level "deliverable" for President Barack Obama's first visit to India in November of 2010, but narrowly failed to reach sufficient consensus among the interagency staff due to its perception at the time as a "controversial idea."[135] In 2011, the DRDO formally raised the idea of space solar power cooperation as a topic for the Joint Technical Group, but at the time, there was no parallel interest within the U.S. Department of Defense.

Similarly, it had been reported to the authors that there had been internal studies within ISRO.[136] There had also been formal but peripheral interest at DRDO. In 2011, two Indian strategists noted that, "China Walks the US-India Space Solar Power Dream," stating that:

> If China wins the race in developing Space Based Solar Power as a feasible source of energy, which would meet the world's growing energy demand, it will result in huge economic and strategic gains for China. While India and the US make pledges about potential collaboration on space, others walk those promises and potentials. In the India-US context, space has remained a potential area of cooperation for the last decade or so whereas China, which has studied the Indo-US joint communications carefully, has made fast progress on space-based solar power (SBSP), in terms of devoting financial and human resources into the project. The need of the hour is for democracies like India, US and may be even Japan to come together, structure large collaborations around space and capture the political space in this regard. The political leadership in both India and the US should recognise the importance of it and act accordingly before it is too late . . . Under such a scenario, there lies a strong imperative for India, the US and other like-minded democracies to come together and realise the SBSP utilisation dream. This will not only provide economic gains but also give a strategic advantage in the changing security environment in Asia. For India, such collaboration would meet its growing energy demand and provide other spinoff benefits like job creation and access to advanced technology, much-needed for sustaining India's growth story. It is estimated that India's energy requirements would double by 2030, making it imperative for it to explore other feasible options. Also, if India becomes a part of the process of realising the SBSP dream, it will augment India's position in Asia as well as the world as a responsible leader.[137]

The next year, in 2012, China proposed Space Solar Power with India during a visit by Dr. A. P. J. Abdul Kalam,[138] but perhaps because of the strategic logic by the above authors, it failed to fructify. Formal government or policy interest appeared lacking, but academic and commercial sectors showed occasional interest. For example, a paper by Rugved Bidkar in 2012 stated,

> The advantage of placing solar cells in space is the 24 hour availability of sunlight. Also, the urgency of finding an alternative energy source due to the depleting energy resources on Earth calls for Space Based Solar Power. Here we study the concept of Solar Power Satellites (SPS), investigate the feasibility of implementation, the overall architecture & the underlying components. The results highlight the effectiveness of this system as an environment friendly, low-loss and large-scale method of energy transfer.[139]

In 2013, N.R. Sonkavday, an electrical engineer who had extensive experience in the power industry, authored the book on space solar power, published in India. He made the case that, "Continuing economic progress will require a four-fold increase in annual energy use by the end of the century . . .

by 2100, 90% off all energy used must be from renewable energy sources," to address CO2, therefore "In order to keep pace with ever increasing population of this earth and meeting its energy demand, there is need to take this concept to a development stage," and "These new Sunsats, will not only breathe new life into space industry, they will be a key to the future economic strength and improve environmental health globally and benefit to all nations."[140] In 2015, two Indian NewSpace entrepreneurs authored a paper examining how

> usage of SWCNT [Single Wall Carbon Nano Tube] based solar cells and electron emitters would substantially reduce the mass of energy harnessing systems by eliminating radiation shields, protection barriers and reducing weight of support structures & microwave transmitters. This is possible due to the inherent properties of the material . . . We make a case for space based solar power with a 50MW satellite that would weigh about 115tons if its core components were made of SWCNT. A modular design allows for this system to be built to full capacity over time using existing launch options at a cost of US$340 Million. The system would be profitable within 2 years of launch.[141]

They subsequently created a start-up called ReBeam.

In 2017, the think tank CSTEP, Center for Study of Science, Technology and Policy https://www.cstep.in/ evidenced interest in space solar power. Till date, advocates have not succeeded in securing any government programmatic interest. However, a potentially important change occurred in 2018. Former chairman of ISRO and then-secretary of the Department of Space, K. Sivan "made a strong pitch for creating a solar power satellite," detailing a satellite 10 by 30 kilometers beaming microwaves to Earth, and serviced by a logistics system of reusable launch vehicles capable of making a thousand sorties per day to transport thousands of tons of material.[142] He noted the importance of Helium-3 on the Moon and stated, "we need to tap the asteroids and extract all their benefits when they come near earth."[143] It appears the essential elements of the space resources industrial chain have been linked at the highest level of Indian civil space policy. Yet another author, S.S. Verma, writing in *Scientific India* in 2019 similarly brought all the elements of the space industrial vision together:

> In the longer term, with sufficient investments in space infrastructure, space solar power can be built from materials from space. The full environmental benefits of space solar power derive from doing most of the work outside of Earth's biosphere. With materials extraction from the Moon or near-Earth asteroids, and space-based manufacturer of components, space solar power would have essentially zero terrestrial environmental impact. Only the energy receivers need be built on Earth. Space solar power can completely solve our energy problems

> long term. The sooner we start and the harder we work, the shorter "long term" will be.[144]

It appears that Dr. A. P. J. Abdul Kalam's vision for space industrialization to address large-scale societal missions has at last found important resonance. We even see the occasional glint of a vision for colonization or space settlement. In February 2017, former ISRO chairman, K. Kasturirangan specified that "India should be part of this global adventure and exploring Venus and Mars is very worthwhile since humans definitely need another habitation beyond Earth."[145]

The confluence of a resource and economic focus on nationalism can be seen in the discourses of India's NewSpace entrepreneurs. Serial entrepreneur Narayan Prasad, cofounder of Rebeam, Dhruva Space, and Chief Operating Officer (COO) of SatSearch authored an article whose title alone says much: "As the race for Moon heats up, Chandrayaan-2 could give ISRO an edge, and spoils"[146] in which he summarizes the new landscape,

> The race to space is heating up around the world, but differently from how it played out five decades ago. Instead of trying to achieve goals for national pride and prestige, the current competition, driven by commercial companies, seeks to monetise space-based services, access to space and space-based resources.[147]

A fellow entrepreneur, Jatin Singh, Founder & Managing Director at Skymet Weather Services Pvt Ltd[148] provides perhaps the clearest articulation of a nationalist space discourse, connected to previous colonial endeavors and future colonization and economic returns. He stated that while space has delivered amazing dividends in telecommunications, agriculture, weather, and knowledge of climate change, India should manage its resources so as not be dependent on anyone:

> But the next 50 years is going to be very different. It seems that space travel is going to become cheaper. Private companies and individuals will explore space. It is private enterprise that is going to lead. For instance the National Aeronautical and Space Administration (NASA) is planning to use private companies like Elon Musk's Space X and Jeff Bezos' Blue Origin for satellite and manned space launches in the future. Blue Origin is already planning lunar modules. Private sector is working on reducing costs by making reusable rockets. Richard Branson's Virgin Galactic will take you on a trip to a low earth orbit for USD 250,000.
>
> Maybe in the next 20 years are also looking at some kind colonization of space. And the dawn of a new space race. A race that will activate most basic of human

> concepts, nationalism. It will be like the race for colonies. In the 17th/ 18thcentury, one that conquered the sea, conquered the world, in the next 50 years, the battle ground will be space and countries will compete.
>
> This where India needs to make an institutional adjustment. India's future in space is dependent on how we conceive of it. So far we see space as a tool for nation building that is purely the preserve of government. Almost like a Soviet style programme.
>
> Our space programme needs to be closely integrated with our commerce. We need to have a mandate for spin-offs. Let me ask you a question, the next 10 years is going to be about 5G networks and self driving electric cars; where is India in all this ? We don't have a single 5G network manufacturer and are unable to build self driving cars or even good reliable electric cars. China and US are leading by miles, and it seems that China is ahead of the US. We have been reduced to franchises choosing a platform. Just the way we have adopted to Google, Netflix, Facebook, Airbus and Boeing. In the future we will be dealing with a new type of colonialism, where end up paying Google or Facebook tax or may be Blue Origin or a Space X tax in some form.

He concludes:

> Our space, both civil and defense need to be connected with our commerce . . . Unless private capital, more importantly private initiative are allowed to lead in space. India will never have equivalents of Space X, Blue Origin, Virgin Galactic, Intel, Boeing, Airbus, or Lockheed Martin.
>
> If this is truly an Asian century, then it should be our endeavor that first person to set foot in Mars is an Indian, the first flag is the tricolor, and the first manned mission is completely Indian. And this can only be done if the companies work in partnership with government.
>
> And if we don't, we will forever be franchises, customers, markets but not leaders. India needs to own space in the way Britannia used to own the seas. Nothing less. Space is India's manifest destiny. Lets take it.[149]

Vasudevan Mukunth[150] summarizes the pressures of a perceived race, as "The development race."

> At the moment, going to space has two purposes: research and development. Research precedes development, but development triggers the race. Scientists have built and launched satellites to understand the Solar System in great detail. But if someone is rushing to go to the Moon or Mars in the name of exploiting resources there to benefit humankind, it is because someone else is doing the same thing.

INDIA: POLICIES AND LAW

In 2017, our assessment was that "Indian space policy is not postured at present to enable India to take leadership regarding space-based resources. Unlike China who has established the fundamental research for SBSP as well as identified asteroid exploration as an area of priority in their 2016 white paper, and have identified the settlement of the Earth-Moon space as sustaining the great rejuvenation of the Chinese nation, such a cogent articulation of space ambitions is missing from the Indian national discourse" and that "with regard to law, our assessment is that India is significantly behind UAE and Luxembourg in their national legislation regarding exploitation of space resources."[151]

Jatin Singh, founder and managing director at Skymet Weather Services Pvt Ltd, summed up the current state of affairs:

> I want to put a dedicated network of satellite over the Indian ocean to improve weather forecasting as well agricultural remote sensing. The market for this data is big enough, and there are paying customers. But there is no legal framework under which this could be done (although draft space bill is ready, but it has been on the anvil for a long time). ISRO has actually worked with universities; SRM University has put a satellite into space. But nothing really of scale has happened. And if I want to do this now, I will deploy our hard earned capital in a foreign country, invest in foreign technology to get Indian data. What a waste![152]

Serial space entrepreneur Narayan Prasad says, "With no particular, dedicated, independent space commerce assessment body for space-related activities or a dedicated road map within ISRO for commercial space in India, there is a need to establish an Office of Space Commerce that is independent of ISRO"[153] stating that,

> Given that the varied nature of activities and the several institutions within the government that need to be involved in providing regulatory clarity to the nascent space industry in India, the space bill should consider setting up of an independent coordinating body which can act as a single point of contact for all types of activities. *By establishing a single coordinating body, India has a chance to leapfrog established space powers such as the US, where space entrepreneurs are struggling to deal with multiple agencies and institutions* [emphasis added].[154]

While "India's government is reacting to the influx of commercial firms in space by drafting legislation to regulate satellite launches, company registrations and liability, said G. V. Anand Bhushan, a Chennai-based partner at the Shardul Amarchand Mangaldas & Co. law firm. *It doesn't cover moon mining*."[155]

The draft space utilities bill crafted by the Indian government has been made public.[156] As of December 12, 2018, "The Government has invited suggestions from the public or stakeholders regarding the draft *Space Activities Bill, 2017*."[157]

The stated purposes of the proposed Bill include,

> to promote and regulate the space activities of India . . . encourages the participation of non-governmental/private sector agencies in space activities in India under the guidance and authorisation of the government through the Department of Space . . . formulate the appropriate mechanism for licensing, eligibility criteria, and fees for licence . . . maintain a register of all space objects . . . provide professional and technical support for commercial space activity and regulate the procedures for conduct and operation of space activity . . . ensure safety requirements and supervise the conduct of every space activity of India and investigate any incident or accident in connection with the operation of a space activity.[158]

The Space Activities Bill, 2019 has been praised for reflecting India's international obligations and commits to the establishment of a mechanism that enforces such obligations by offering a clear definition of transgressions and the punishments thereof. Most see this bill as creating the right climate to boost investor sentiment in India's space sector. However, some of the shortfalls are its lack of clarity with regard to offering criteria in choosing who will be the licensed operators, or "what barriers of entry such space aspirants will face, as the question of timelines for granting of licenses and the licensing fee."[159]

G.V. Ashok asks, "Does the space activities bill provide to the state and to the industry, the missing links necessary to mitigate international liability and improve investor sentiment respectively?" He concludes, it does not:

> The Space Activities Bill gets many things right—it reflects India's international obligations so far as space activities are concerned, it provides a mechanism for enforcing those obligations by defining offences to address transgressions in space and provides for appropriate punishment for the same" and that "The Space Activities Bill can augment that support and encouragement and greatly improve investor sentiment in the Indian space sector. If the setting up of the Special Purpose Vehicle is anything to go by, the Indian space tech businesses have much to celebrate and to hope for, for the Government is clearly inclined to unleash the private sector initiatives in space." Though he cautions that, *it*

does "not provide the clarity on the critical question of who will be eligible to be the licensed operator of space activities and what barriers of entry such space aspirants will face, as the question of timelines for granting of licenses and the licensing fee."[160]

The "provisions of the Bill" also raise some concerns:

> The term "Central Government" has not been defined and hence the authority responsible for administering the Bill and the sector is unclear. Some provisions also have potential for harassment such as allowing Central Government to seek such information from a licensee "relating to its affairs" as the Central Government requires, or permitting Central Government to take copies of documents from the licensee. *The Bill is also silent in respect of new opportunities such as space mining and tourism.*[161]

INDIA: EVIDENCE OF PROGRAMS

The case for any evidence of extant or planned exploitation of space resources by India is weak. India's current commercial focus is on servicing domestic and international markets for space services and launch, and still largely views its planetary missions in the context of science and prestige.

The motivations for *Chandrayaan-2* likely preceded the global dialogue on space resources, however, though India certainly contextualized the mission within the discourse of space resources.

The South Pole appears to be among the most important areas for industrial exploitation, and India made an attempt to land on an ancient high plane just 600 km from the Lunar South Pole.[162] *Chandrayan-2* consisted of an orbiter, a lander, and a rover.[163] The ISRO had stated before the launch that "the payloads will collect scientific information on lunar topography, mineralogy, elemental abundance, lunar exosphere and signatures of hydrogel and water-ice."[164] While the Helium-3 isotope is limited on Earth, it is abundant on the lunar surface, and could theoretically meet global energy demands for the next 250 years if harnessed.[165] "'The countries which have the capacity to bring that source from the moon to Earth will dictate the process,' said K. Sivan, former chairman of the ISRO. 'I don't want to be just a part of them, I want to lead them.'"[166]

Unfortunately, *Chandrayaan 2* would not achieve these aims. Initially planned for launch in April 2018, and *Chandrayaan 2* experienced some delays.[167] Scheduled to launch on July 14, a technical snag caused a week's delay[168] with a successful launch occurring on July 22, 2019.[169] The orbiter arrived on the lunar orbit on August 20 and the lander attempted (unsuccessfully) to soft-land close to the lunar South Pole around September 7, 2019.[170] The chairman of the Department of Space, Dr. K. Sivan briefed the Parliamentary Standing Committee on Science & Technology that a large initial error from its expected altitude and velocity combined with an error in on-board software resulted in a larger demand for angle correction that could not be met, resulting in rotation of the lander and by the time the control system had stabilized, the lander lost altitude and hard landed.[171] Undeterred, India plans a follow-up landing attempt in 2022.[172]

With regard to asteroid exploration, while no clear policy insight is available, certain statements by former ISRO chairperson, A.S. Kiran Kumar suggested in both 2016[173] and 2017 that ISRO had teams working on an asteroid mission and would at some point seek approval.[174]

The company that started India's NewSpace sector, Axiom Research Labs, better known as "Team Indus" is still in the game. Axion fell short of sufficient funding and flight readiness to win the Google Lunar X-Prize by the March 31, 2018 deadline, and terminated its PSLV launch contract with ISRO.[175] Subsequently they stated, "India's TeamIndus is working on a series of lunar lander missions as well for sending payloads to the moon inexpensively." "The key to creating this market is to bring the cost down by an order of magnitude. You have to think about exponential reductions in cost for this market to really come online," stated Rahul Narayan, founder and chief executive of TeamIndus.

Narayan went on to specify that "TeamIndus is planning an annual series of lunar lander missions, starting in mid-2019. The company has completed a set of tests on a qualification model of the lander . . . they were ready to develop flight hardware for the lander."[176] Subsequently, *TeamIndus* has teamed with Honeybee Robotics, Advanced Space, Ceres Robotics, and *Apollo Fusion* to form a consortium called *OrbitBeyond*.[177] *OrbitBeyond*[178] aspired to "deliver your scientific and commercial payloads to the lunar surface and in lunar orbit" as a commercial service. The consortium was one of nine companies NASA selected for its CLPS program in November 2018[179], and one of only three selected on June 3, 2019, for the first series of contracts to deliver payloads to the Moon, with *OrbitBeyond* stating its Z-01 lander will be ready to land on the Moon by September 2020,[180] leading to the headline, "America's first private moon lander will be engineered in India."[181] However, shortly after award, *OrbitBeyond* notified NASA of "internal corporate challenges that will prevent the timely completion of its awarded task order," resulting in termination of its $97 million contract to send science instruments to the lunar surface in 2020.[182],[183] *Axiom Research Labs Private Limited* (*TeamIndus*) appears to have terminated its relationship with *OrbitBeyond* and is now collaborating with Agile Planetary Exploration, inc (APEX) to enable "affordable payload delivery services to the cis lunar ecosystem."[184]

A difficult-to-categorize, but nevertheless significant, factor is a broader technological and epistemic community available to India in its space industry expats and diaspora. This community remains connected to the Indian NewSpace sector, and hopeful that India's business climate will improve.

For example, Naveen Jain, founder and Executive chairman of *Moon Express*, asserted,

> What's interesting about the moon is that it has been collecting asteroidal material for the last billions of years. With naked eyes you can see the crater sites. That means asteroids have been coming and crashing there and all the asteroidal riches are right on the surface of the moon, solidified. All you do is collect those things. What are those? Platinum grade material, that is, platinum, gold, silver, all the rare earth elements, but the most important part of the thing really comes from solar radiation. It's called helium-3. A small quantity of helium-3 could power this planet . . . when you have lower gravity things crystallise very differently in low gravity than in high gravity . . . We chose to go to the moon not because it's easy, but because it's good business . . . . Our long-term goal is to be able to save humanity from extinction because it's only a matter of when, not if, that we're going to become dinosaurs, sooner or later. We're on this spacecraft called Earth that is flying around space naked and it's a dangerous place out there. If we get hit by an asteroid, we will all essentially get wiped out. And if you don't believe me, ask any dinosaur.[185]

Jain believes that if humanity learns to live on the Moon, they could live anywhere, similar to what Paul Spudis advocated for, in his book, *The Value of the Moon*.[186] Jain asserts that private space companies would offer consumer and industrial services, to include tourism, games and mining over the next couple of decades. "Entrepreneurs are now doing things that used to be done by the state," he says. "In the next 20 years, people will be going to the moon just like they go to Sydney. . . . not very often, but they do go."[187]

We now turn to examine the impact of India's strategic culture and strategic trauma on its ambitions for outer space and space-based resources.

INDIA: STRATEGIC CULTURE

Indian strategic culture is currently undergoing a significant shift, evidenced by a renaissance in realist thinking. For the first half-century of its existence as a modern nation-state, Indian strategic thought took place in a context of economic developmental backwardness and military weakness. An emergent consciousness among the dominant elites hoped to interpret a broader cultural and geographic identity within the fashion of the modern nation-state and to craft an ideology to create a national identity upon enlightenment ideals while avoiding foreign entanglements. The ideological preferences of the elites encouraged participation in United Nation's frameworks, and dissuaded (though never forsook) emphasis on hard power and nationalism, and emphasized either western humanist ideals or pacifist traditions of Indian thought or a combination of both.[188]

India largely succeeded in steering clear of entangling alliances and oversees adventures and proved surprisingly successful in crafting and maintaining a nation-state with an amazing diversity of internal identities and a challenging neighborhood.

However, with increased economic success have come a changed sense of self and a more confident and assertive India is emerging. A new global context is emerging where the key features are no longer an ideological struggle between the United States and USSR but rather a general rising of developing nations, an increasing importance of Asia, and the rising economic and military power of China. The expectation of both external[189] and internal audiences[190] that India can also see an equivalent rise, and that India is, and should be a balance to China has fueled new aspirations and a new nationalism.

Strategic culture concerns the implicit patterns by which elites perceive and interpret threats and opportunities within a particular nation-state.[191] More generally, cultures are those positive and negative feedback loops that encourage certain thinking and behavior in a society, group, or organization and dissuade or make taboo, other thinking and behavior.[192] Of interest then, is what positive and negative feedback loops exist which may channel India's response to space resources.

The new nationalism seeks to draw from India's unique pre-colonial history and past[193] and has resulted in a renaissance of classical Indian realist and geostrategic thought. Beginning circa 2012, a vibrant conversation has emerged (centered at the IDSA to examine and reinterpret classical Indian thinking on statecraft.[194]

While India's elites generally are familiar with and use western terms in discussing statecraft, they are also inheritors to a different set of epics/myths (Vyasa's *Mahabharata*,[195] Valmiki's *Ramayana*)[196] Kautilya's *Arthashastra*,[197] Vishnu Sharma's *Panchatantra* (mirrors for princes),[198] Narayana's *Hitopadesa*,[199] and Tiruvalluvar's *Kural*,[200] which offer a distinct experience and perspective on conflict, statecraft, and righteous conduct. These interpretations remain salient in popular culture through books, television, comic books (e.g., *Armarchitrakatha*),[201] and Bollywood movies. They take place in, and repeatedly reference a broader context of religious and philosophical thinking.

Generally, familiar concepts have specific cultural nuances and connotations that are worth introducing for the purposes of a richer discussion[202] of India's strategic culture[203]:

Policy (*Niti*). *Niti* is the choice among difficult alternatives, often involving choosing unworthy means to achieve a worthy end, especially perpetuation or extension of the state's prosperity.

Prosperity and Flourishing (*Artha*). This includes wealth and power, and is the proximate goal of policy (*niti*).

Moral Duty (*Dharma*). The force or obligation, custom, social and religious duty, culture. This is seen both as an end of statecraft to maintain, and a constraint on statecraft. A key duty of statecraft is to enable and enforce the other web of mutual duties. Adherence to *Dharma* is the source of legitimacy. Actions taken out of duty do not incur the same moral consequences (*Karma*) as actions arising from emotion or personal ambition.

Societal Order & Cohesion (*Dharma*). "Holding together." A different connotation of the same word. In this case, an end that statecraft is to work toward.

Geopolitical configuration (*Mandala*). More generally a geographic figure that represents the universe. In statecraft, it is a spatial ordering of important considerations. A key feature in a geopolitical *Mandala* are which states are border states or immediately proximate.

Use of force (*Danda*). Literally, "the rod." Notice that the choice of a rod over a sword implies a disciplining punishment where the default position is to preserve life. *Danda* is employed to maintain *Dharma* (cohesion and order) internal to a state, and among states in the *Mandala*.

In this context, it is likely to perceive that space resources are a key ingredient in *Artha* (prosperity) and therefore among the *Dharma* (duty) of the state to pursue. It is likely to locate such resources and the actions of other states within a broadened geopolitical *Mandala*. It will adapt its *Niti* (policies) toward this end, including procuring whatever *Danda* (Rod of Discipline) would be necessary to protect its interests.

A strategic culture where the dominant narrative is nationalism is likely to perceive space as an important vanguard of national expression, and welcome and encourage formulations based upon destiny, security, and expansion. For example,

> If this is truly an Asian century, then it should be our endeavor that first person to set foot in Mars is an Indian, the first flag is the tricolor, and the first manned mission is completely Indian. And this can only be done if the companies work in partnership with government . . . And if we don't, we will forever be franchises, customers, markets but not leaders. India needs to own space in the way

> Britannia used to own the seas. Nothing less. Space is India's manifest destiny. Let's take it.[204]

When confronted in evaluating a future articulated by India's defense strategists as:

> With commercial competitiveness, military dominance and resource security at stake, outer space, in 2050, has also become a playground for vested interests and geopolitical rivalries and national security considerations continue to drive military use of space. The number of states with orbital capabilities has almost doubled since 2010 when only 11 states had this capability and each decade since 2010 has seen the average launch of 1200 satellites. Newer types of ASAT capabilities such as pellet cloud attacks and low-orbit satellites, micro-satellite technology, EMP devices, high-energy lasers, and particle beam weapons are available with many countries. Airborne lasers and super-maneuverable satellites are routinely deployed.[205]

Then prescriptions such as Ajey Lele's is likely to hold sway: "India needs its own space force" arguing that it needs to:

> For the foreseeable future, there is no need for India to conduct any debris-creating ASAT tests to further establish its space agenda in the strategic realm. However, from a space security perspective, it is important for India to evolve a mechanism for the generation of space situational awareness (SSA) . . . develop various counter-space capabilities like electromagnetic pulse systems, lasers, jamming techniques and cyber options. In addition, satellite-hardening technologies and space debris removal techniques are required to be mastered, too. Spaceplanes, satellite swarms and launch-on-demand services are required for network-centric warfare. India should also develop the ability for the human spacecraft to move from one orbit to another. New quantum-based communications systems and cells for studying space weather forecasting are the requirements of the present and the future.[206]

Cultural Drag

However, India is likely to encounter significant internal drag that will compromise its ability to compete on a level playing field.

First is a discomfort or rejection of an Indian identity as a great power. For example, Jean Drèze, a famous development economist observing that the Mars mission was "part of the Indian elite's delusional quest for superpower status."[207]

A second is a tendency toward insularity. This includes a discomfort, or questionable legitimacy of behaviors that might appear "colonial" or which fall outside the mandala of India's local regional concerns. An example was a self-imposed prohibition on the learned class (Brahmins) crossing oceans chronicled in Pradeep Gautam's "A Longue Durée Perspective on Military Science in India" where he examines how a tendency to be inward looking was one ingredient in a failure to culture military science.[208]

A third is a cultural, intellectual, and bargaining proclivity to agree on top-down idealistic first principles[209] before any endeavor can be undertaken, and to be less concerned about time.

A fourth is a desire for multi-lateral solutions over aggressive national action. This can perhaps be attributed to a moral/ethical view where Dharma imposes such a duty to first seek a multi-lateral and binding solution. An example of this tendency on the specific topic of space resources can be seen in Senjuti Mallick and Rajeswari Pillai Rajagopalan's "If space is 'the province of mankind', who owns its resources?"[210] where they argue that "this paper appreciates the economic benefits of space mining but argues against the national legislations legalising extraterrestrial appropriation of resources due to inconsistency with international treaties and customary international law."

These features operate even among Indian realists, who appear constrained by idealistic multi-lateral ethical concerns to a greater degree than seems asked of them by the broader world. These conceptions of duty often seem to trump ambitious agenda and outcome-based considerations.[211]

Collectively, these mean that India may be slow to perceive space resources as a source of *Artha* (prosperity) and within its Mandala (geopolitical map) of concerns, that it might not see spacefaring as legitimate *Dharma* (duty) or consistent with an imagined insular identity. The lack of temporal urgency and orientation toward the future as unknowable or fated further disincline timely and purposeful action. Most likely, these features mean that India may delay taking effective initiative while it seeks an unrealistic, multi-lateral, top-down, idealistic, and legally binding rule-set before proceeding.

INDIA: STRATEGIC TRAUMA

Observers of Indian culture (e.g., George Tanham[212]) have noted the importance of status, symbolism, and opportunity based (vs. requirement-based thinking) with respect to technology associated with power status.[213] One component of Indian strategic culture that transcends the left-right divide is a concern to avoid any possibility of having its strategic options constrained by an international normative regime creating a club of powers from which India is excluded. Of the many important factors of India's strategic culture likely to exert an influence on its behavior in outer space, the one likely to drive the most predictable response is a determination to ensure India can meet whatever bar is necessary to be in the "in club."

The roots of this preoccupation can be traced to the learning that took place in the context of the Nuclear Nonproliferation Treaty (NPT).[214] Because India had not yet tested a nuclear capability at the time of the treaty (1967–1968), it was locked out from the established nuclear weapons states (the "P5"). This placed it outside the "haves" of the nuclear weapons states, and significantly constrained it from nuclear trade or knowhow. It delegitimized any move by India to become a recognized nuclear state, and India had to pay a significant price to become one. A similar experience occurred in the context of the Missile Technology Control Regime (MTCR).[215] This experience of having been "left out" from what became the rule-making powers constituted a significant strategic trauma Indian policymaker are determined not to repeat.

Such thinking was apparent as early as 2010, when former DRDO scientist V. Siddhartha in a presentation titled "Military Dimensions in the Future of the Indian Presence in Space," argued: "ASAT NPT on the horizon?," "Only those nations which demonstrate an ASAT capability before a certain fiducial date will be 'allowed' mil-Space assets." And "all 'rouge mil-satellites' will be destroyed by (his fictional) UNCLES—United Network Command for Law Enforcement in Space"[216]

The influence of the NPT trauma can be clearly seen in the justifications of the 2019 ASAT test by leading Indian thinkers. Says Rajeswari Pillai Rajagopalan,

> This [ASAT TEST] was required in order to avoid some of the mistakes that India had done in other areas, such as the nuclear domain. India refused to test its nuclear capability in the 1960s, thus being left out of the Non Proliferation Treaty's category of "nuclear weapon state." Clearly, India does not want an NPT for space to be developed and then be banned from developing and demonstrating its ASAT capability.[217]

Said V. Siddhartha,

> 1. Given experience of Italy blocking India's membership in MTCR (India became full member of MTCR only in 2016).) for a completely unrelated issue of release of Italian sailors arrested in India (for homicide of Indian fishermen), did not want this (too) to become casus beli for a whole lot of EU members of MTCR, who were/are: 2 . . . pushing off-line in Geneva for a ban on ASAT testing Once bitten by NPT, twice shy.[218]

This concern of "once bitten, twice shy" is likely to exert a powerful influence on the direction of the Indian space program to ensure that India has demonstrated whatever technological threshold capability will enable it to remain within the club of influential nations.

INDIA: EXPECTED BEHAVIOR

While the discourse on space-based resources in India has not reached the level of national-level articulation as we see in the United States or China, it is not unrealistic to forecast that it will become an integral part of India's space policy given its growing capacity for space access and power projection. As seen by India's recent ASAT test, India's fears of being shut out from a governance regime (as happened with the nuclear Non-Proliferation Treaty) are likely to force sufficient steps before the emergence of a technology governance regime to ensure New Delhi will at least have a seat at the table. Most conversations stressed the contingent view of the topic—that it would deserve serious attention if and when it was demonstrated, but that it did not at present require a competitive or proactive response. In the event it did, our conversations suggested that India would have a solid desire to compete should such activities become clear and present.[219] Our expectation is that to the extent that lunar and asteroid mining proves possible and salient, India will bandwagon to maintain status, access, and profit.

But ensuring one is not too late or too far behind to be in the in-club is a different criterion of success than leadership. India is certainly capable of playing a leadership role. It is among the most capable spacefaring states. Its budget, capabilities, and access to talent greatly exceed, for example, Luxembourg or UAE. Its strategic elites appear to understand the potential of space resources, and it has a nascent though vibrant commercial space sector that would respond to policies and incentives. However, various internal structural and cultural forces are likely to inhibit proactive law and policy—even in the face of a rebirth of nationalism and political realism—resulting in an India that—with respect to space resources—elects to be a laggard rather than a pace-setter.

NOTES

1. "Chandrayaan 2: Vikram Lander in Tilted Position after Hard Landing, Says ISRO," *The Hindu Business Line*, September 09, 2019, accessed September 17, 2019, https://www.thehindubusinessline.com/news/chandrayaan-2-vikram-lander-tilted-after-hard-hit-says-isro/article29374671.ece

2. NASA, "Vikram Lander Found," December 2, 2019, accessed January 22, 2020, https://www.nasa.gov/image-feature/goddard/2019/vikram-lander-found

3. Mike Wall, "India Is Officially Going Back to the Moon with Chandrayaan-3 Lunar Lander," *Space.com*, January 02, 2020, accessed February 16, 2020, https://www.space.com/india-confirms-moon-landing-mission-chandrayaan-3.html

4. "The Power sub-ranking is based on an equally weighted average of scores from five country attributes that related to a country's power: A leader, economically influential, politically influential, strong international alliances and strong military alliances."

"Power Rankings," *US News & World Report*, accessed September 17, 2019, https://www.usnews.com/news/best-countries/power-full-list

5. "2018 Military Strength Ranking," *Global Firepower*, November 22, 2018, accessed January 19, 2019. https://www.globalfirepower.com/countries-listing.asp

6. Nan Tian, Aude Fleurant, Alexandra Kuimova, Pieter D. Wezeman, and Siemon T. Wezeman, *Trends in World Military Expenditure, 2017* (Stockholm International Peace Research Institute, 2018); "Trends in World Military Expenditure 2017," *SIPRI*, accessed January 19, 2019, https://www.sipri.org/publications/2018/sipri-fact-sheets/trends-world-military-expenditure-2017

7. "The Military Balance 2018," *IISS*, accessed January 19, 2019, https://www.iiss.org/publications/the-military-balance/the-military-balance-2018

8. "Total Population by Country 2018," World Population Review, accessed January 19, 2019, http://worldpopulationreview.com/countries/

9. "Human Development Index (HDI)," United Nations Development Program (UNDP), accessed January 19, 2019, http://hdr.undp.org/en/data

10. Energy is expressed a number of different ways. 934 MTOE = 6.67 Billion Barrel of Oils Equivalent (BBOE) = 39.10 Exajoules (EJ) = 37.06 Quadrillion BTU (QUAD or PBtu) = 9,346.26

Megatons TNT equivalent = 10,862.42 Terawatt Hours (TWh) = 1.24 Terawatts (continuous).

11. "Global Energy Statistical Yearbook 2018," Enerdata, July 9, 2019, accessed January 19, 2019, https://yearbook.enerdata.net/total-energy/world-consumption-statistics.html

12. "Global Energy Statistical Yearbook 2018."

13. "Electric power consumption (kWh per capita)," World Bank, March 22, 2010, accessed January 20, 2019, https://data.worldbank.org/indicator/EG.USE.ELEC.KH.PC

14. "The World in 2050," *Price Waterhouse Cooper (PwC)*, December 3, 2015, accessed January 19, 2019, https://www.pwc.com/gx/en/issues/economy/the-world-in-2050.html

15. Uri Dadush and Bennett Stancil, "The World Order in 2050," *Carnegie Endowment for International Peace*, April 2010, accessed January 19, 2019, https://carnegieendowment.org/files/World_Order_in_2050.pdf

16. Brian Wang, "EIU GDP Forecasts to 2050," *Next Big Future*, June 24, 2015, accessed October 11, 2019, https://www.nextbigfuture.com/2015/06/eiu-gdp-forecasts-to-2050.html

17. "Which countries spend the most on space exploration?," World Economic Forum, January 11, 2016, accessed January 19, 2019, https://www.weforum.org/agenda/2016/01/which-countries-spend-the-most-on-space-exploration/

18. "Government Spending in Space Programs Reaches $62 Billion in 2016," Euroconsult, May 30, 2017, accessed January 19, 2019, http://www.euroconsult-ec.com/30_May_2017

19. "InSight: Futron The Space Competitiveness Index—Update, by Futron," SatMagazine, December 9, 2009, accessed January 19, 2019, http://www.satmagazine.com/story.php?number=1856883389

20. "INTEL: Global Military Space by Futron," MilSat Magazine (September 2009 Edition), September 9, 2009, accessed January 19, 2019, http://www.milsatmagazine.com/story.php?number=875113067

21. Vikaspedia, "Space Programmes of India," accessed June 12, 2019, http://vikaspedia.in/education/childrens-corner/science-section/isro-created-history-by-launching-104-satellites-together

22. Union of Concerned Scientists, "UCS Satellite Database," March 31, 2019, accessed May 20, 2019, https://www.ucsusa.org/nuclear-weapons/space-weapons/satellite-database

23. Ed Kyle, "Space Launch Report," December 29, 2018, accessed May 20, 2019, http://www.spacelaunchreport.com/log2018.html

24. Namrata Goswami and Peter Garretson, "Critical Shifts in India's Outer Space Policy," *The Diplomat*, April 16, 2019, accessed May 19, 2019, https://thediplomat.com/2019/04/critical-shifts-in-indias-outer-space-policy/

25. For those interested, please see Gurbir Singh, *India's Space Programme: India's Incredible Journey from the Third World Towards the First* (Manchester: Astrotalkuk Publications, 2017); Indian Space Research Organisation (ISRO), *From Fishing Hamlet to Red Planet* (Bengaluru: ISRO, 2015).

26. Samanth Subramanian, "India's Frugal Mission to Mars," *The New Yorker*, November 7, 2013, accessed September 15, 2017, https://www.newyorker.com/tech/elements/indias-frugal-mission-to-mars

27. Barry Ellen, "India Launches 104 Satellites From a Single Rocket, Ramping Up a Space Race," *The New York Times*, accessed June 13, 2019, https://www.nytimes.com/2017/02/15/world/asia/india-satellites-rocket.html

28. Indian Express, "Explained: What makes PSLV-C45 special," April 2, 2019, accessed June 13, 2019, https://indianexpress.com/article/explained/pslv-c45-rocket-launch-indian-space-research-organisation-isroc-5653610/

29. Indian Express, "Explained: What Makes PSLV-C45 Special."

30. Michael Safi, "India Launches Record Breaking 104 Satellites from Single Rocket," *The Guardian*, February 15, 2017, accessed September 17, 2019, https://www.theguardian.com/science/2017/feb/15/india-launches-record-breaking-104-satellites-from-single-rocket

31. Sanjay Kumar, "India Joins Elite MARS Club," *NATURE*, September 24, 2014, accessed September 17, 2019, https://www.nature.com/news/india-joins-elite-mars-club-1.15997

32. "Mangalyaan India's Mission to Mars," *National Geographic*, November 24, 2016, accessed September 15, 2017, https://www.youtube.com/watch?v=Lv-4fiVadVk; Also see Ajey Lele, *Mission Mars India's Quest for the Red Planet* (New Delhi: Springer Briefs in Applied Sciences and Technology, 2014).

33. Jason Davis, "Chandrayaan-2 Enters Lunar Orbit," *The Planetary Society*, August 20, 2019, accessed September 17, 2019, http://www.planetary.org/blogs/jason-davis/chandrayaan-2-successful-loi.html

34. Doug Messier, "ISRO Budget to Increase 23 Percent, Venus & Mars Missions Planned," *Parabolic*, February 26, 2017, accessed February 18, 2020, http://www.parabolicarc.com/2017/02/26/isro-budget-increase-23-percent-venus-mars-missions-planned/

35. Vijay Karthik, "ISRO Working on Launching Spacecraft to Venus, Asteroids: A.S. Kiran Kumar," *LiveMint*, September 08, 2016, accessed September 17, 2019, http://www.livemint.com/Science/1Qn4uyqO0pe8iwIbqKX9AN/Isro-working-on-launching-spacecraft-to-Venus-asteroids-A.html

36. Pallava Bagla, "India's Eyes a Return to Mars and a First Run at Venus," *Science*, February 17, 2017, accessed September 15, 2017, *http://www.sciencemag.org/news/2017/02/india-eyes-return-mars-and-first-run-venus*

37. "ISRO 'Makes Announcement of Opportunity' for its Venus Mission," *The Hindustan Times*, April 24, 2017, accessed September 17, 2019, http://www.hindustantimes.com/india-news/isro-makes-announcement-of-opportunity-for-its-venus-mission/story-radr7VgPoTM40Li7yoVaZL.html

38. Press Trust of India, "Gaganyaan, India's first Manned Mission, Set to be a Reality by 2022," *Live Mint*, August 15, 2018, accessed June 13, 2019, https://www.livemint.com/Politics/glFe56EP0V2EkF25mQshXO/Narendra-Modi-announces-Gaganyaan-in-Independence-Day-speech.html

39. "Star City adventure: Take a Sneak Peek into India's First Space Crew Training in Russia," *Russia Today*, February 17, 2020, accessed February 18, 2020, https://www.rt.com/news/480892-india-first-space-crew-russia/

40. Pallava Bagla, "India Planning To Launch Own Space Station By 2030, Says ISRO Chief K Sivan," *NDTV*, June 13, 2019, accessed June 13, 2019, https://www.ndtv.com/india-news/indian-space-station-india-planning-to-launch-own-space-station-says-isro-chief-2052686

41. V. Sidhartha, "Military Dimensions in the Future of the Indian Presence in Space," Presentation to Centre for Airpower Studies (CAPS), September 7, 2010.

42. A full exploration of India's pre-ASAT military thought is beyond the scope of this work. However, authors interested in exploring this area should seek to interview past DRDO directors, including Dr. V.K. Saraswat, Dr. Aatre, Dr. V. Sidhartha and Dr. Yavendra who were prominent advocates.

43. Authors interactions with anonymous DRDO personnel, September 2008-April 2009.

44. Authors interactions with anonymous DRDO personnel, September 2008-April 2009.

45. HQ Integrated Defense Staff PP&FD/WSOI Division Maritime DTE, Technology Perspective and Capability Roadmap (TPCR 01/10), May 20, 2010.

46. Sidhartha, "Military Dimensions in the Future"; V. Siddhartha. "Military Dimensions in the Future of the Indian Presence in Space." *USI of India* (2000): 243–58.

47. Vivek Raghuvanshi, "Indian Aerospace Command To Operate Military Satellites," *SpaceNews*, April 4, 2005, accessed June 11, 2019, https://spacenews.com/indian-aerospace-command-operate-military-satellites/

48. K. S. Jayaraman, "Indian Government Set To Create Aerospace Command," *SpaceNews*, October 31, 2005, accessed June 11, 2019, https://spacenews.com/indian-government-set-create-aerospace-command/

49. Jayshree Bajoria, "RAW: India's External Intelligence Agency," *CFR*, November 7, 2008, accessed June 12, 2019, https://www.cfr.org/backgrounder/raw-indias-external-intelligence-agency

50. Raghuvanshi, "Indian Aerospace Command."

51. "India Working on Space Weapons: IAF Chief," *Rediff India Abroad*, October 07, 2003, accessed June 11, 2019, https://www.rediff.com/news/2003/oct/06iaf1.htm

52. "India to Set Up Strategic Aerospace Command," *India Daily*, April 27, 2005, accessed June 11, 2019, https://web.archive.org/web/20060902214721/http://www.indiadaily.com/breaking_news/33107.asp

53. Jeffrey Lewis, "An Aerospace Command for India?," *Arms Control Wonk*, May 1, 2005, accessed June 11, 2019, https://www.armscontrolwonk.com/archive/200570/an-aerospace-command-for-india/

54. Kaza Lalitendra, *Militarization of Space*, Center for Air Power Studies, New Delhi, 2010, accessed September 17, 2019, https://books.google.com/books/about/Militarisation_of_Space.html?id=R2SPSQAACAAJ

55. Ministry of Defense Press Release (Sitanshu Kar/RAJ (Release ID :39503)), "Special cell set up to counter growing threat to space assets," Press Information Bureau, Government of India, June 10, 2008, accessed June 12, 2019, http://pib.nic.in/newsite/erelease.aspx?relid=39503

56. Sandeep Unnithan, "India Attains the Capability to Target, Destroy Space Satellites in Orbit," *India Today*, April 28, 2012, accessed June 11, 2019, https://www.indiatoday.in/magazine/nation/story/20120507-agni-v-launch-india-takes-on-china-drdo-vijay-saraswat-758208-2012-04-28

57. Matt Hoey, "Space Alert 21: India Developing Space Weapons," *Military Space Transparency Project*, June 2010, accessed June 11, 2019, http://www.space4peace.org/articles/india_developing_space_weapons.htm

58. Sandeep Unnithan, "India has all the Building Blocks for an Anti-Satellite Capability," *India Today*, April 27, 2012, accessed March 6, 2020, https://www.indiatoday.in/india/story/agni-v-drdo-chief-dr-vijay-kumar-saraswat-interview-100405-2012-04-27

59. Institute for Defence Studies & Analyses, *Strategic Trends 2050 Project*, 2013, accessed September 17, 2019, https://www.facebook.com/indiaon15august2047/posts/idsa-drdo-strategic-trends-2050projectthe-strategic-trends-2050-projectst2050-jo/1590640924502650/

60. EO Portal Directory, "RISAT-2 (Radar Imaging Satellite-2)," accessed June 12, 2019, https://earth.esa.int/web/eoportal/satellite-missions/r/risat-2

61. Rajat Pandit, "India's first military satellite successfully launched," *Times of India*, August 30, 2013, accessed June 12, 2019, https://timesofindia.indiatimes.com/india/Indias-first-military-satellite-successfully-launched/articleshow/22168268.cms;

62. "India's Navic GNSS Capability Declared an Allied System by U.S. Congress," *Spaceflight Asia Pacific*, n.d, accessed February 18, 2020, https://spacewatch.global/2019/12/indias-navic-gnss-capability-declared-an-allied-system-by-u-s-congress/

63. National Defense Authorization Act for Fiscal Year 2020, *Congress.Gov*, accessed February 19, 2020, https://www.congress.gov/bill/116th-congress/senate-bill/1790/text?format=txt

64. D.S Madhumathi, "Starting May, ISRO to launch a string of 'defence' satellites," *The Hindu*, April 2, 2019, accessed June 13, 2019, https://www.thehindu.com/news/national/starting-may-isro-to-launch-a-string-of-defence-satellites/article26714439.ece

65. Press Trust of India, "India successfully launches EMISAT satellite," *The Hindu Business Line*, accessed June 13, 2019, https://www.thehindubusinessline.com/news/science/india-successfully-launches-emisat-satellite/article26699384.ece

66. D.S. Madhumathi, "Starting May, ISRO to launch a string of 'defence' satellites,".

67. D. S. Madhumathi, "India Needs at Least Three More Spy Satellites," *The Hindu*, April 1, 2019, accessed June 13, 2019, https://www.thehindu.com/news/national/india-needs-at-least-three-more-spy-satellites/article26705497.ece

68. DRDO, Official video on Mission Shakti (Project XSV1), *You Tube*, Apr 6, 2019, accessed June 13, 2019, https://www.youtube.com/watch?v=1-J4jVlx5Do

69. Press Trust of India, "Centre Finalises Broad Contours of Defence Space Agency," *NDTV*, June 11, 2019, accessed June 12, 2019, https://www.ndtv.com/india-news/centre-finalises-broad-contours-of-defence-space-agency-2051699?pfrom=home-topstories

70. Press Trust of India, "Centre Finalises Broad Contours."

71. Rajeswari Pillai Rajagopalan, "A First: India to Launch First Simulated Space Warfare Exercise," *The Diplomat*, June 12, 2019, accessed June 12, 2019, https://thediplomat.com/2019/06/a-first-india-to-launch-first-simulated-space-warfare-exercise/

72. Rajat Pandit, "Agencies Take Shape for Special Operations, Space, Cyber War," *Times of India*, May 15, 2019, accessed May 19, 2019, https://m.timesofindia.com/india/india-begins-setting-up-new-tri-service-agencies-to-handle-special-operations-space-and-cyberspace/articleshow/69346012.cms

73. Rajeswari Pillai Rajagopalan, "A First: India to Launch."

74. Pallava Bagla, "India's First Private Sector-Built Satellite Launch Unsuccessful, Says ISRO 10 Facts," "NDTV, August 31, 2017, accessed June 14, 2019, https://www.ndtv.com/india-news/indias-first-private-sector-built-satellite-launched-by-isro-from-sriharikota-1744575

75. Pallava Bagla, "India's First Private Sector-Built Satellite Launch Unsuccessful, Says ISRO 10 Facts," "*NDTV*, August 31, 2017, accessed June 14, 2019, https://www.ndtv.com/india-news/indias-first-private-sector-built-satellite-launched-by-isro-from-sriharikota-1744575; IW2 Research Team, "Bangalore Startup Alpha Design Signs Pact with ISRO to Assemble Satellites," Indian Web2, June 18, 2018, accessed June 14, 2019, https://www.indianweb2.com/2018/07/18/bangalore-startup-alpha-design-signs-pact-with-isro-to-assemble-satellite/

76. "First Industry Built PSLV by 2020, says VSSC Director Sivan," *The Hindu*, November 17, 2017, accessed June 14, 2019, https://www.thehindu.com/sci-tech/science/first-industry-built-pslv-by-2020/article20535383.ece

77. "ISRO to Increase Frequency of Launches to 12 per Year," *The Times of India*, April 23, 2017, accessed September 17, 2019, http://timesofindia.indiatimes.com/india/isro-to-increase-frequency-of-launches-to-12-per-year/articleshow/58326033.cms.

78. News18, "Sitharaman Launches New Space India Limited, ISRO's Commercial Arm to Market, Sell Products," July 5, 2019, accessed July 7, 2019, https://www.news18.com/news/business/sitharaman-launches-new-space-india-limited-isros-commercial-arm-to-market-sell-products-2219263.html

79. Astropreneurs, "New Space In India," August 19, 2018, accessed October 13, 2019, http://astropreneurs.space/2018/07/19/new-space-in-india/; M. Ramesh, "India's start-ups find opening in space," *The Hindu Busines Line*, July 14, 2019, accessed October 13, 2019, https://www.thehindubusinessline.com/news/science/indias-start-ups-find-opening-in-space/article28427597.ece; Disha, Missal, "9 Indian Startups That Are Taking India's Space Industry To The Next Level," Curious Dose, March 11, 2019, accessed October 13, 2019, https://curiousdose.com/2019/03/9-indian-startups-that-are-taking-indias-space-industry-to-the-next-level/

80. ARDL Website, accessed October 13, 2019, https://www.ardl.co.in/

81. Agnikul Cosmos Website, accessed October 13, 2019, http://www.agnikul.in/#

82. Astrome Website, accessed June 14, 2019, http://www.astrome.co/

83. Bellatrix Aerospace Website, accessed June 14, 2019, http://www.bellatrix-aerospace.com/

84. Blue Sky Analytics Website, accessed October 13, 2019, https://blueskyhq.in/about

85. Dhruva Space Website, accessed February 18, 2020, https://www.dhruvaspace.com/

86. Earth2Orbit Website, accessed June 14, 2019, http://www.earth2orbit.com/

87. Exseed Space Website, accessed June 14, 2019, https://www.exseed-space.com/

88. Kawa Space Website, accessed October 13, 2019, https://kawa.space/about.html

89. Pixxel Website, accessed October 13, 2019, https://pixxel.co.in/

90. ReBeam Website, accessed June 14, 2019, https://nopo.in/web/rebeam/

91. Satsure Website, accessed June 14, 2019, http://www.satsure.in/

92. Skyroot Aerospace Website, accessed October 13, 2019, https://skyroot.in/

93. Smart Enovations Website, accessed June 14, 2019, http://www.smartenovations.com/

94. TeamIndus Website, accessed June 14, 2019, https://medium.com/teamindus

95. XOVIAN Website, accessed February, http://www.xovian.co.in/

96. Rocketeers Website, accessed June 14, 2019, https://www.rocketeers.in/

97. "Bellatrix Aerospace Website."

98. Caleb Henry, "Indian Startup Bellatrix Aerospace Raises $3 Million," *Space News*, June 21, 2019, accessed July 9, 2019, https://spacenews.com/indian-startup-bellatrix-aerospace-raises-3-million/

99. Surendra Singh, & Srinivas Laxman, "India's First Private Satellite Launched," *Times of India*, December 4, 2018, accessed June 14, 2019, https://timesofindia.indiatimes.com/india/space-x-launches-indias-first-privately-built-satellite-exceedsat-1/articleshow/66937238.cms

100. Arc Parabolic, "OrbitBeyond Teams TeamIndus Honeybee Robotics NASA Lunar Program," November 29, 2018, accessed June 14, 2019, http://www.parabolicarc.com/2018/11/29/orbitbeyond-teams-team-indus-honeybee-robotics-nasa-lunar-program/

101. Stephen Clark, "NASA Picks Three Companies to Send Commercial Landers to the Moon," *Spaceflight Now*, June 4, 2019, accessed June 5, 2019, https://spaceflightnow.com/2019/06/04/nasa-picks-three-companies-to-send-commercial-landers-to-the-moon/

102. Stephen Clark, "NASA Terminates Lunar Lander Contract with OrbitBeyond," *Spaceflightnow*, July 30, 2019, accessed February 12, 2020, https://spaceflightnow.com/2019/07/30/nasa-terminates-lunar-lander-contract-with-orbitbeyond/

103. Mike Wall, "Private Company Orbit Beyond Drops Out of 2020 NASA Moon-Landing Deal," *Space.Com*, July 30, 2019, accessed February 12, 2020, https://www.space.com/nasa-drops-orbit-beyond-moon-landing-contract.html

104. NASA, "Commercial Lunar Payload Services," accessed June 14, 2019, https://www.nasa.gov/content/commercial-lunar-payload-services-1

105. Interactions with Indian space entrepreneurs to author Peter Garretson, August 26, 2019.

106. "Dr. Vikram Ambalal Sarabhai (1963–1971)," *ISRO*, accessed June 4, 2019, https://www.isro.gov.in/about-isro/dr-vikram-ambalal-sarabhai-1963-1971

107. See the tribute to his life Peter Garretson authored on behalf of the NSS: "The National Space Society Pays Tribute to Dr. Kalam—One of Our Leading Lights Has Joined the Stars," NSS, July 31, 2015, accessed June 4, 2019, https://space.nss.org/the-national-space-society-pays-tribute-to-dr-kalam-one-of-our-leading-lights-has-joined-the-stars/

108. Dr. A. P. J. Abdul Kalam, "Space Future in the 21st Century," Address to the Engineers, Scientists, Technologists of VSSC and also Students, faculties, and professors of Indian Institute of Space Technology (IIST), Thiruvanathapuram, June 2, 2013, accessed September 17, 2019, https://space.nss.org/media/Kalam-2013-Space-Future-in-the-21st-Century.pdf

109. A.P.J Kalam, Address through multi-media tele-conference to The Symposium on "The Future of Space Exploration: Solutions to Earthly Problems," Boston University, U.S, April 12, 2007, accessed June 4, 2019, http://www.bu.edu/csp/Conferences/Space_Exploration/Day1/Presentations/Kalam_Space%20Exploration%20and%20Human%20Life.pdf see video of talk here: https://www.bu.edu/pardee/space-abdul-kalam/

A.P.J Kalam, "The Future of Space Exploration and Human Development," *The Pardee Papers*, No. 1, Boston University, August 2008, June 4, 2019, accessed

September 17, 2019, https://www.bu.edu/csp/Conferences/Space_Exploration/Pardee_ThinkBooklet_andCover.pdf

110. Ajey Lele, "Kalam-NSS Indian-American Energy Initiative," IDSA, November 9, 2010, accessed June 4, 2019, https://idsa.in/idsacomments/KalamNSSIndianAmericanEnergyInitiative_alele_091110

111. A.P.J Kalam, "Space Solar Power—Key to a Livable Planet," Address at the 32nd International Space Development Conference (ISDC) and Acceptance speech on receiving Wernher Von Braun Memorial Award.

San Diego, California, May 24, 2013, accessed June 4, 2019, https://space.nss.org/media/Kalam-Address-ISDC-2013.pdf video here: https://space.nss.org/2013-isdc-video-dr-abdul-kalam-space-solar-power-key-to-a-livable-planet-earth/

112. A.P.J Kalam, "Space Future in the 21st Century: World Space Vision 2050," Address to the engineers, scientists, technologists of VSSC and also Students, faculties and professors of IIST, Thiruvanathapuram, June 2, 2013, accessed June 4, 2019, https://space.nss.org/media/Kalam-2013-Space-Future-in-the-21st-Century.pdf

113. A. P. J. Kalam, Address through multi-media tele-conference to The Symposium on "The Future of Space Exploration: Solutions to Earthly Problems," *Boston University*, U.S, April 12, 2007, accessed June 4, 2019, http://www.bu.edu/csp/Conferences/Space_Exploration/Day1/Presentations/Kalam_Space%20Exploration%20and%20Human%20Life.pdf see video of talk here: https://www.bu.edu/pardee/space-abdul-kalam/

A. P. J. Kalam, "The Future of Space Exploration and Human Development," *The Pardee Papers*, No. 1, *Boston University*, August 2008, June 4, 2019, accessed September 17, 2019, https://www.bu.edu/csp/Conferences/Space_Exploration/Pardee_ThinkBooklet_andCover.pdf

114. A. P. J. Kalam, Address through multi-media tele-conference to The Symposium on "The Future of Space Exploration: Solutions to Earthly Problems," *Boston University*, U.S, April 12, 2007, accessed June 4, 2019, http://www.bu.edu/csp/Conferences/Space_Exploration/Day1/Presentations/Kalam_Space%20Exploration%20and%20Human%20Life.pdf see video of talk here: https://www.bu.edu/pardee/space-abdul-kalam/

A. P. J. Kalam, "The Future of Space Exploration and Human Development," *The Pardee Papers*, No. 1, *Boston University*, August 2008, June 4, 2019, accessed September 17, 2019, https://www.bu.edu/csp/Conferences/Space_Exploration/Pardee_ThinkBooklet_andCover.pdf

115. A. P. J. Kalam, "Space Solar Power."

116. A. P. J Kalam, Address through multi-media tele-conference to The Symposium on "The Future of Space Exploration: Solutions to Earthly Problems," Boston University, USA, April 12, 2007, accessed June 4, 2019, http://www.bu.edu/csp/Conferences/Space_Exploration/Day1/Presentations/Kalam_Space%20Exploration%20and%20Human%20Life.pdf

117. "The Future of Space Exploration: Solutions to Earthly Problems,"

118. "The Future of Space Exploration: Solutions to Earthly Problems."

119. A. P. J. Abdul Kalam, "Space Solar Power—Key to a Livable Planet."

120. Peter A. Garretson, "Sky's No Limit: Space-Based Solar Power, The Next Major Step in the Indo-US Strategic Partnership?," IDSA Occasional Paper No.9, *Institute for Defence Studies and Analyses (IDSA)*, New Delhi, August 2010, accessed June 4, 2019, https://idsa.in/sites/default/files/OP_SkysNoLimit.pdf also available at https://spacejournal.ohio.edu/issue16/papers/OP_SkysNoLimit.pdf and at https://space.nss.org/media/Skys-No-Limit-space-based-solar-power-IDSA-Garretson-2010.pdf

121. Air Commodore Raghavan Gopalaswami is a Rocket technologist, pioneered liquid rocket propulsion technology in India during early 1960s. He retired in 1988 from the Air Force and DRDO as Air Commodore and later as Chairman and Managing Director of Bharat Dynamics Ltd in 1994. He was a close professional associate of the late former President Dr APJ Abdul Kalam for over four decades. The work of author Peter Garretson would not have been possible without the tremendous help and mentorship of Air Cmde Gopalaswami.

See additional biographical details at *https://spacejournal.ohio.edu/issue16/contributors.html#gopal and http://www.spacetechasia.com/author/raghavan-gopalaswami/ and http://www.bharat-rakshak.com/IAF/Database/5826*

122. For example, see: Raghavan Gopalaswami, "Sustaining India's Economic Growth," *Online Journal of Space Communication*, Winter 2010, accessed June 4, 2019, https://spacejournal.ohio.edu/issue16/gopal.html and "An International Preliminary Feasibility Study on Space Based Solar Power," presented for the Kalam-National Space Society Energy Technology Universal Initiative, accessed June 4, 2019, https://space.nss.org/media/KALAM-NSS-Initiative.pdf

123. Numerous articles, for example, Raghavan Gopalaswami, "The Spaceplane Equation," *Online Journal of Space Communication*, Winter 2010, accessed June 4, 2019, https://spacejournal.ohio.edu/issue16/gopal2.html

124. Press Trust of India, "India to Explore Moon's Surface for Helium 3," *The Hindustan Times*, September 1, 2006, accessed June 9, 2019, Arcive.org https://web.archive.org/web/20170528033036/https://www.hindustantimes.com/india/india-to-explore-moon-s-surface-for-helium-3/story-K4HAfpqIhRCJm6zeKVsGxM.html

125. "India to Explore Moon's Surface for Helium-3," *Hindustan Times*, September 1, 2006, accessed June 9, 2019, Arcive.org, https://web.archive.org/web/20170528033036/https://www.hindustantimes.com/india/india-to-explore-moon-s-surface-for-helium-3/story-K4HAfpqIhRCJm6zeKVsGxM.html

126. "Planting Tricolour May Help India's Claim on Moon: Nair," *One India*, October 20, 2008, accessed June 9, 2019, https://www.oneindia.com/2008/10/20/planting-tricolour-may-help-indias-claim-on-moon-nair-1224511741.html

127. "Planting Tricolour May Help."

128. Institute for Defence Studies & Analyses, *Strategic Trends 2050 Project.*

129. Utpal Bhaskar, "Isro Plans to Mine Energy from Moon by 2030 to help meet India needs," *LiveMint*, April 21, 2017, accessed June 9, 2019, https://www.livemint.com/Science/W5WjJCdqqxXYpHvrB2TTHP/Isro-plans-to-mine-energy-from-Moon-by-2030-to-help-meet-Ind.html

130. Dipanjan Roy Choudhury, "India Needs Dedicated Military Space Program: Lt. Gen. PM Bali," *The Economic Times*, February 18, 2017, accessed April 11, 2017,

http://economictimes.indiatimes.com/news/defence/india-needs-dedicated-military-space-programme-lt-gen-pm-bali/articleshow/57225348.cms

131. "When India set a Date with the Moon," *Team Indus Blog*, November 13, 2017, accessed June 9, 2019, https://medium.com/teamindus/when-india-set-a-date-with-the-moon-8c095a7b5f37

132. Anurag Kotoky, "The Quest to Find a Trillion-Dollar Nuclear Fuel on the Moon," *Bloomberg*, June 26, 2018, accessed June 9, 2019, https://www.bloomberg.com/news/articles/2018-06-26/the-quest-to-find-a-trillion-dollar-nuclear-fuel-on-the-moon

133. Rajeswari Pillai Rajagopalan, "Space Based Solar Power: Time to Put it on the New US-India S&T Endowment Fund?," *ORF Online*, April 2, 2011, accessed July 25, 2019, https://www.orfonline.org/research/space-based-solar-power-time-to-put-it-on-the-new-us-india-st-endowment-fund/

134. Peter A. Garretson, "Sky's No Limit: Space-Based Solar Power".

135. Notes by author Peter Garretson from October 1, 2010, Interagency Policy Coordination (IPC) meeting to coordinate ideas for US-India Space Cooperation for President Obama's November 2010 (first) trip to India.

136. Authors interactions with anonymous ISRO personnel, ISRO Headquarters, Bengaluru, July 2017.

137. Rahul Prakash and Rajeswari Pillai Rajagopalan, "China Walks the US-India Space Solar Power Dream," *ORF*, September 21, 2011, accessed June 11, 2019, https://www.orfonline.org/research/china-walks-the-us-india-space-solar-power-dream/

138. Palash Ghosh, "China Proposes Space Solar Power Collaboration With India," *International Business Times*, November 2, 2012, accessed June 11, 2019, https://www.ibtimes.com/china-proposes-space-solar-power-collaboration-india-858330

139. Rugved Bidkar, "Space Based Solar Power (SBSP): An Emerging Technology," December 8, 2012, accessed May 20, 2019, https://ieeexplore.ieee.org/abstract/document/6450440/authors#authors

140. N. R. Sonkavday, *Space Based Solar Power: An Eternal Source of Power from Space Volume* 1 (Pune: First Impression Print Technology, 2013).

141. Gadhadar Reddy, Narayan Prasad, and Divyashree Iyer, "Carbon-Nanotube Based Solar Power (CASSP)," 4th Space Solar Power International Student and Young Professional Design Competition, ESA, October 2015, accessed May 20, 2019, https://www.esa.int/gsp/ACT/doc/nrg/SPS%202015_winner%20paper.pdf

142. Express News Service, "India Needs to Create Solar Power Satellite: Indian Space Research Organization," *The New Indian Express*, July 8, 2018, accessed May 20, 2019, http://www.newindianexpress.com/states/karnataka/2018/jul/08/india-needs-to-create-solar-power-satellite-indian-space-research-organisation-1840117.html

143. Express News Service, "India Needs to Create Solar Power Satellite."

144. S.S Verma, "Space Solar Power," *Scientific India*, November 19, 2014, accessed May 20, 2019, http://www.scind.org/41/Space/space-solar-power.html

145. Bagla, "India's Eyes a Return."

146. Narayan Prasad, "As the Race for Moon Heats Up, Chandrayaan-2 Could Give ISRO an Edge, and Spoils," *ThePrint*, June 28, 2019, accessed July 7, 2019,

https://theprint.in/science/as-the-race-for-moon-heats-up-chandrayaan-2-could-give-isro-an-edge-and-spoils/254145/

147. Narayan Prasad, "As the Race for Moon Heats Up, Chandrayaan-2 Could Give ISRO an Edge, and Spoils," *ThePrint*, June 28, 2019, accessed July 7, 2019, https://theprint.in/science/as-the-race-for-moon-heats-up-chandrayaan-2-could-give-isro-an-edge-and-spoils/254145/

148. Skymet Weather Services Pvt Ltd Website, accessed February 18, 2020, https://www.skymetweather.com/

149. Jatin Singh, "India, the Next 50 Years in Space; Will She Grab her Manifest Destiny ?," *LinkedIn*, June 26, 2019, accessed July 7, 2019, https://www.linkedin.com/pulse/india-next-50-years-space-she-grab-her-manifest-destiny-jatin-singh/

150. Vasudevan Mukunth, "Don't Let the Excitement Over Chandrayaan 2 Distract From the Bigger Picture," *The Wire*, July 4, 2019, accessed July 9, 2019, https://thewire.in/space/isro-chandrayaan-2-gslv-mk-iii-gaganyaan

151. Peter Garretson and Namrata Goswami, "Is India Looking Towards Space-Based Resources?," *The Space Review*, October 2, 2017, accessed June 4, 2019, http://www.thespacereview.com/article/3338/1

152. Singh, "India, the Next 50 Years in Space."

153. Kendall Russell, "What Will It Take to Unleash NewSpace in India?," *Satellite Today*, February 12, 2018, accessed October 13, 2019, http://interactive.satellitetoday.com/what-will-it-take-to-unleash-newspace-in-india/

154. Narayan Prasad, "India is a Maturing Space Power, But Can Rival the US With an Independent Regulator," *The Print*, August 8, 2019, accessed October 13, 2019, https://theprint.in/opinion/india-is-a-maturing-space-power-but-can-rival-the-us-with-an-independent-regulator/273886/

155. Anurag Kotoky, "ISRO in Quest to Find a Trillion-Dollar Nuclear Fuel on the Moon," *Hindustan Times*, June 27, 2018, accessed July 7, 2019, https://www.hindustantimes.com/science/isro-s-quest-to-find-a-trillion-dollar-nuclear-fuel-on-the-moon/story-2JXQrUGaUbhfUDe6Wu8NKO.html

156. The Draft bill can be seen at: https://www.prsindia.org/billtrack/draft-space-activities-bill-2017 and its progress tracked at https://www.insightsonindia.com/2019/01/01/space-activities-bill-2017/

157. Press Information Bureau, Government of India, Department of Space, "Space Activities Bill," December 12, 2018, accessed July 7, 2019, http://pib.nic.in/newsite/printrelease.aspx?relid=186315

158. InsightsIAS, "Space Activities Bill, 2017," January 1, 2019, Accessed July 7, 2019, https://www.insightsonindia.com/2019/01/01/space-activities-bill-2017/

159. GV Ashok, "The Space Activities bill- Does it deliver?," *ORF Online*, APR 19 2019, accessed July 7, 2019, https://www.orfonline.org/expert-speak/the-space-activities-bill-does-it-deliver-50019/

160. GV Ashok, "The Space Activities bill- Does it Deliver?"

161. George Alice, Alice and Sheikh Kaynat, "Race for Space."

162. Pallava Bagla, "India Plans Tricky and Unprecedented Landing Near Moon's South Pole," *Science Magazine*, January 31, 2018, accessed May 20, 2019, https://www.sciencemag.org/news/2018/01/india-plans-tricky-and-unprecedented

-landing-near-moon-s-south-pole; also see Namrata Goswami, "Chang'e 4 and Chandrayaan 2: To the Moon and Beyond," *The Diplomat*, September 9, 2019, accessed September 17, 2019, https://thediplomat.com/2019/09/change-4-and-chandrayaan-2-to-the-moon-and-beyond/

163. Department of Space, Indian Space Research Organisation, "Chandrayaan-2," accessed September 15, 2017, http://www.isro.gov.in/chandrayaan-2.

164. TK Rohit, "Second Lunar Mission Chandrayaan-2 to be Launched in July," *The Hindu*, May 2, 2019, accessed June 3, 2019, https://www.thehindu.com/sci-tech/science/second-lunar-mission-chandrayaan-2-to-be-launched-in-july/article27003022.ece

165. Kotoky, "The Quest to Find a Trillion-Dollar."

166. Kotoky, "The Quest to Find a Trillion-Dollar."

167. Surendra Singh, "Chandrayaan-2 Mission Deferred Again, Scheduled for May launch," *Times of India*, April 1, 2019, accessed June 3, 2019, https://timesofindia.indiatimes.com/india/chandrayaan-2-mission-deferred-again-scheduled-for-may-launch/articleshow/68675904.cms

168. Tariq Malik, "India Delays Chandrayaan-2 Moon Lander Launch Over 'Technical Snag'," *Space.com*, July 14, 2019, accessed July 25, 2019, https://www.space.com/india-delays-moon-lander-chandrayaan-2-launch.html

169. "Chandrayaan-2: India launches second Moon mission," BBC, July 22, 2019, accessed July 25, 2019, https://www.bbc.com/news/world-asia-india-49032603

170. "Chandrayaan-2 to Reach Moon by August 20: ISRO," *Times of India*, July 24, 2019, accessed July 25, 2019, http://timesofindia.indiatimes.com/articleshow/70366062.cms

171. K. Sivan, "Presentation to Parliamentary Standing Committee on Science & Technology, Environment and Forests," *Government of India, Department of Space*, September 25, 2019, slide 13.

172. Mike Wall, "India Is Officially Going Back."

173. Vijay Karthick, "Isro working on launching spacecraft to Venus, asteroids: A.S. Kiran Kumar," *LiveMint*, September 8, 2016, accessed July 25, 2019, https://www.livemint.com/Science/1Qn4uyqO0pe8iwIbqKX9AN/Isro-working-on-launching-spacecraft-to-Venus-asteroids-A.html

174. "ISRO to Increase Frequency of Launches to 12 per year," *The Times of India*, April 23, 2017, accessed September 17, 2019, http://timesofindia.indiatimes.com/india/isro-to-increase-frequency-of-launches-to-12-per-year/articleshow/58326033.cms

175. Manu P. Toms, "Team Indus Ends Pursuit of Google Lunar XPrize," *VC Circle*, January 24, 2018, accessed June 4, 2019, https://www.vccircle.com/team-indus-ends-its-pursuit-of-google-lunar-xprize/

176. Jeff Foust, "Former Google Lunar X-Prize Teams Focused on New Commercial and Government Opportunities," *Space.com*, May 25, 2018, accessed September 17, 2019, *https://spacenews.com/former-google-lunar-x-prize-teams-focused-on-new-commercial-and-government-opportunities/*

177. Tim Fernholz, "America's First Private Moon Lander Will be Engineered in India," *Quartz*, June 3, 2019, accessed June 4, 2019, https://qz.com/1633918/americas-first-private-moon-lander-will-be-made-in-india/

178. Orbit Beyond, accessed June 4, 2019, https://www.orbitbeyond.com/

179. Jeff Foust, "NASA Selects Nine Companies for Commercial Lunar Lander Program," *Space News*, November 29, 2018, accessed June 4, 2019, https://spacenews.com/nasa-selects-nine-companies-for-commercial-lunar-lander-program/

180. Stephen Clark, "NASA picks three companies to send commercial landers to the moon," *Spaceflight Now*, June 4, 2019, accessed June 5, 2019, https://spaceflightnow.com/2019/06/04/nasa-picks-three-companies-to-send-commercial-landers-to-the-moon/

181. Tim Fernholz, "America's first private moon lander will be engineered in India,".

182. Stephen Clark, "NASA terminates lunar lander contract with OrbitBeyond," Spaceflightnow, July 30, 2019, accessed February 12, 2020, https://spaceflightnow.com/2019/07/30/nasa-terminates-lunar-lander-contract-with-orbitbeyond/

183. "Mike Wall, "Private Company Orbit Beyond Drops Out of 2020 NASA Moon-Landing Deal," Space.Com,

July 30, 2019, accessed February 12, 2020, https://www.space.com/nasa-drops-orbit-beyond-moon-landing-contract.html

184. According to Daudi Barne's link in site https://www.linkedin.com/in/daudi-barnes-0b2737132/, Spun out of Agile Space Industries, APEX has procured flight hardware and licensed IP to deliver its first Lunar landing in Q2-2021. "Agile Planetary Exploration (APEX) is a commercial space exploration company, committed to providing global customers with reliable, repeated and affordable payload delivery services to the cis lunar ecosystem."

185. Naveen Jain, "If We can Learn to Live on the Moon We Can Live Anywhere in Space," *LSE Business Review*, May 11, 2017, accessed June 5, 2019, https://blogs.lse.ac.uk/businessreview/2017/05/11/naveen-jain-if-we-can-learn-to-live-on-the-moon-we-can-live-anywhere-in-space/

186. Paul D. Spudis, *The Value of the Moon: How to Explore, Live and Prosper in Space Using the Moon's Resources* (Washington, DC: Smithsonian Books, 2016).

187. Jain, "If We Can Learn to Live on the Moon."

188. Namrata Goswami, "India's Strategic Culture is Plain to see," *Asia Times*, April 06, 2013.

189. Peter Valente, "The Tiger and the Dragon: India as a Counterbalance to China in the Indo-Pacific," Modern War Institute, West Point, September 4, 2018, accessed September 18, 2019, *https://mwi.usma.edu/tiger-dragon-india-counterbalance-china-indo-pacific/*; Lisa Curtis, "U.S.-India Relations: The China Factor," *The Heritage Foundation*, November 25, 2008, accessed September 18, 2019, *https://www.heritage.org/asia/report/us-india-relations-the-china-factor*

190. Manoj Joshi, "India- A Counterweight to the Rise of China," *Observer Research Foundation*, September 17, 2014, accessed September 18, 2019, https://www.orfonline.org/research/india-a-counterweight-to-the-rise-of-china/

191. Goswami, "India's Strategic Culture is Plain to See."

192. Peter Garretson, "Space Force's Jupiter-Sized Culture Problem," *War on the Rocks*, July 11, 2019, accessed July 27, 2019, https://warontherocks.com/2019/07/space-forces-jupiter-sized-culture-problem/

193. Rupam Jain and Tom Lasseter, "By Rewriting History, Hindu Nationalists Aim to Assert their Dominance over India," *Reuters*, March 6, 2018, accessed September 18, 2019, https://www.reuters.com/investigates/special-report/india-modi-culture/; Tarun Vijay, *Saffron Surge: India's Re-emergence on the Global Scene and Hindu Ethos* (New Delhi: Har Anand, 2008).

194. Exploring the Roots of India's Strategic Culture, Institute for Defence and Analyses Conference, October 5, 2017, accessed July 22, 2019, https://idsa.in/event/exploring-the-roots-of-indias-strategic-culture

195. Krishna Dwaipayana Vyasa, *The Mahabharat,* Translated by Bibek Debroy (New Delhi: Penguin, 2010).

196. Valmiki, *Ramayana* (New Delhi: Penguin, 2010).

197. Kautilya, *Arthasashtra* (New Delhi: Penguin, 1992).

198. Vishnu Sharma, *The Pancantantra* (New Delhi: Penguin, 2007).

199. Hitopadesha, translated by Sir Edwin Arnold, London, 1861, accessed September 18, 2019, *http://www.columbia.edu/itc/mealac/pritchett/00litlinks/hitopadesha_arnold/index.html*

200. Tiruvalluvar's *Kural,* translated by P.S. Sundaram (New Delhi: Penguin, 2005).

201. See https://www.amarchitrakatha.com/us/

202. For an even richer discussion Indian strategic terms and their connotations, see Medha Bisht, *Kautilya's Arthashastra: Philosophy of Strategy* (New York and London: Routledge, 2020): 30. Also see P.K. Gautam, *One Hundred Years of Kautilya's Arthasashtra*, IDSA Monograph, Series No. 20, 2013, accessed September 18, 2019, *https://idsa.in/monograph/OneHundredYearsofKautilyasArthasastra*

203. The specific importance of such works on Indian strategic thought is debated. However, the Thomas Theorem states: "If men define situations as real, they are real in their consequences". In other words, the interpretation of a situation causes the action." Thus, if modern rediscovery and reinterpretation of ancient Indian strategic thinkers such as *Arthashastra* are believed to inform modern Indian thinking, such a belief will be real in its consequences. See Thomas Theorem, *Oxford Index*, accessed September 18, 2019, *https://oxfordindex.oup.com/view/10.1093/oi/authority.20110803104247382*

204. Jatin Singh, "India, the next 50 years in space; will she grab her manifest destiny ?"

205. Institute of Defence Studies & Analyses, *Strategic Trends 2050 Project.*

206. Ajey Lele, "Op-ed | India Needs Its Own Space Force," *Space News*, May 28, 2019, accessed June 16, 2019, https://spacenews.com/op-ed-india-needs-its-own-space-force/

207. "Space Vision 2025," Vikram Sarabhai Space Centre, accessed August 22, 2017, http://www.vssc.gov.in/VSSC_V4/index.php/space-vision-2025/117-space-vision-2025.

208. Pradeep Kumar Gautam, "A Longue Durée Perspective on Military Science in India," *Journal of Defence Studies* 12, no. 1 (January-March 2018): 15–38, accessed July 22, 2019, https://idsa.in/system/files/jds/jds-12-1-2018-military-science.pdf

209. Rajesh Kumar, "Negotiating With The Complex, Imaginative Indian," Ivey Business Journal, March 2005, accessed July 27, 2019, https://iveybusinessjournal.com/publication/negotiating-with-the-complex-imaginative-indian/

210. Senjuti Mallick and Rajeswari Pillai Rajagopalan, "If Space is 'the Province of Mankind', Who Owns its Resources? The Potential of Space Mining and its Legal Implications," *ORF Occasional Paper No. 182*, January 2019, accessed September 18, 2019, https://www.orfonline.org/research/if-space-is-the-province-of-mankind-who-owns-its-resources-47561/

211. See a broader discussion of Indian Strategic culture here: Peter Garretson, "Tanham in Retrospect: 18 Years of Evolution in Indian Strategic Culture," *South Asia Journal*, January 22, 2013, accessed May 19, 2019, http://southasiajournal.net/tanham-in-retrospect-18-years-of-evolution-in-indian-strategic-culture/

212. George K. Tanham, "Indian Strategic Thought: An Interpretive Essay," RAND, 1992, accessed July 27, 2019, https://www.rand.org/pubs/reports/R4207.html

213. Peter Garretson, "Tanham in Retrospect: 18 Years of Evolution in Indian Strategic Culture,".

214. "Treaty on the Non-Proliferation of Nuclear Weapons (NPT)," United Nations Office of Disarmament Affairs, accessed September 18, 2019, https://www.un.org/disarmament/wmd/nuclear/npt/

215. "Missile Technology Control Regime," accessed September 18, 2019, https://mtcr.info/

216. Sidhartha, "Military Dimensions in the Future," Presentation to Centre for Airpower Studies (CAPS), September 7, 2010.

217. Rajeswari Pillai Rajagopalan, "Having Tested Its ASAT Capability, India Should Help Shape Global Space Norms," *The Wire*, March 28, 2019, accessed May 19, 2019, https://thewire.in/space/having-tested-its-asat-capability-india-should-help-shape-global-space-norms

218. V. Siddhartha, Email Message to Author Peter Garretson, April 1, 2019.

219. Garretson and Goswami, "Is India Looking Towards Space-Based Resources?"

Chapter 7

Middle Power Strategy and Ambitions for Space Resources

Luxembourg and the UAE

This chapter provides an overview of the most active space middle powers, the United Arab Emirates (UAE) and Luxembourg. First, it examines the role of the middle power generally and in the context of space resources. It then discusses the space ambitions of UAE and Luxembourg, respectively. The chapter makes the case that Luxembourg and UAE have invested their resources in building space agencies that view space resources as a significant future component of their economy, and an expanded role in space governance.

WHAT IT IS TO BE A MIDDLE POWER?

We define "Middle Powers" as those states in the international system whose position is below the great powers, but with enough power, capacity, and influence at their disposal to shape international regimes and events. These powers utilize their constitutive capacities, especially through their foreign policy preferences and actions, to not only create global norms and standards of behavior but also add legitimacy to international regimes through their support.[1] Both Luxembourg and UAE are middle powers given their material resources, institutional strength, and visible demonstration of willingness to invest in outer space. Both have established space agencies and articulated long-term vision plans for outer space resource extraction. Both want to utilize space to create influence, for economic development, as well as for regime construction. Both are well placed, with crafted legislation, and the resource capacity to build such a future.

THE CONTEXT FOR THE MIDDLE POWERS

As discussed in the earlier chapters, the great powers have begun to take an active interest in space resources and to frame the meaning of space as a competitive commercial endeavor. This marks a significant change from the framing of the geopolitical meaning of space during the Cold War. At that time, the "flags and footprints" model was about showcasing technological progress to the audience of the developing world. Space was not viewed as a domain of commerce but rather as an arena for short-duration exploration sorties to showcase national vibrancy. The Soviet Union and the United States were competing for great power influence on Earth, and utilized space as a medium for achieving that goal; that is, who had the more attractive ideological paradigm, that would provide the basis for other countries to align with and develop their own societies.[2] But the new context features interest by the great powers (United States, China, India) in space industrialization and resource utilization.[3] As discussed in chapter 1, space resources are expected to be lucrative, worth trillions of dollars.[4] For example, some asteroids may be made of pure metal, such as platinum, and worth trillions of dollars.[5] An example is 3554 Amun (2 kilometers) in size, which is estimated to be worth $20 trillion.[6] Well over a million asteroids have been cataloged till date, with about 17,000–18,000 considered near-Earth asteroids (NEA).

Within this new context, lesser but capable states may wish to benefit from this vast wealth. To ensure that they can benefit, middle powers, Luxembourg and UAE, can be seen to be taking action in three arenas: first, each is pursing domestic legislation to attract global investment and international companies to incorporate their efforts within their tax structures; second, efforts to shape the global governance regime through example-setting legislation and policy initiatives; and third, efforts to shape themselves as the space hub for their respective geographic regions. If successful, these states will help enable a broader opportunity for states and countries to participate in space resource extraction and receive economic benefit in the context of a stable international governance regime adhered to by the great powers.

THE ANALOGY OF A COSMIC STAG HUNT

Middle powers face a classic collective action problem. We can frame this problem in terms of a well-known parable. In *A Discourse on Inequality*, Jean Jacques Rousseau's introduced the concept of a "Stag Hunt."[7] In this classic collective action game, a group of hunters would do better if they can coordinate to bring down a large stag rather than individually pursue a hare.[8]

Today's middle powers face similar coordination problems. We can imagine that asteroids or other desirable resources are the stags we want to hunt. Unlike in the classic stag hunt scenario, in our modern scenario, we have just a few large hunters able to bring down a stag, but a diversity of weaker powers who are unable to bring down the big game themselves, but could certainly provide lookout, carry the spear, pass information, or help carry the big game home.

The problem for them is that the big hunters in the tribe don't always get along. On the one hand, competition between the big hunters is likely to generate more hunting activity, benefiting the entire tribe. But if the big hunters spend too much time fighting, then everybody in the tribe goes hungry. If the big hunters create rival factions, that means the less capable members must choose one to the exclusion of the other.

The ideal state for the less-capable members (middle powers) is for the big hunters (major powers) to be rivalrous enough that they have lots of opportunities to go hunting, but amicable enough that property is respected and fighting doesn't disrupt the hunting enterprise. Therefore, we expect that the less-capable members of the tribe (middle powers) will take the agency to moderate the conflict, offer compromises, and bandwagon with powers (signing MoUs) who are taking action to pursue collective gains or establish precedents for broad sharing and distribution.

LUXEMBOURG

Luxembourg is prominent in encouraging the exploitation of space resources. In June 2016, Luxembourg established a $227 million fund to attract companies focused on asteroid mining.[9] It became the first European country to develop a legal framework (similar to the 2015 U.S. CSLCA) to adjudicate the commercial exploitation of space resources. In 2016, its Ministry of Economy announced the establishment of the Space Resources Initiative (SRI), whose role "will be the development of a legal and regulatory framework confirming certainty about the future ownership of minerals extracted in space from Near Earth Objects such as asteroids."[10] The SRI website states that Luxembourg aims to establish itself as a European hub (similar to UAE's goal for the Middle-East), in the "exploration and use of space-based resources."[11] Luxembourg has a legislation that states that companies can keep the resources they have mined. And unlike the CSLCA that requires a 50 percent U.S.-backed equity, the Luxembourg Space Resources Act has no such limits. What it achieves though is to offer a legal framework for those who would like to set up shop and utilize Luxembourg's legal framework to own the space resources that they mine, in the process. This positive atmosphere has drawn asteroid mining companies such as U.S.-based *Deep Space Industries* (now Bradford Space Group), *Planetary Resources* (now acquired by Consensys Inc),[12] and a Japanese company, *iSpace*, developing lunar landers, to open shop in Luxembourg. In May 2019, the United States and Luxembourg signed an MoU to deepen cooperation in space, to include space situational awareness.[13] The U.S. Secretary of Commerce, Wilbur Ross specified:

> The United States and Luxembourg share a rich history and a common vision of a future in space that is overwhelmingly commercial. This Memorandum forms the basis for increased collaboration across a wide range of space activities, including research, exploration, defense, and space commerce.[14]

Luxembourg was already cooperating with *Deep Space Industries* on its *Prospector X* mission that aimed to use a nano-spacecraft to test its asteroid technologies. To add to its institutional capacity, Luxembourg established its Space Agency on September 12, 2018, especially aimed at space resources. Deputy Prime Minister and Minister of the Economy, Étienne Schneider, announced that "The agency will be well-equipped to support industry in their daily challenges, and it leads to the most favorable environment for this sector to continue to grow."[15] He announced the creation of $116 million Luxembourg Space Fund.

Luxembourg's penchant for understanding the lucrative nature of outer space was vindicated by its early intervention in communication satellites. In the 1980s, when the space commercial satellite industry was still in its infancy, Luxembourg passed legislations and invested heavily in its home-grown space company, *Société Européenne des Satellites* (SES), which enabled it to thrive and enjoy revenues to the tune of 2 billion (Bn) Euros in 2015.[16] Consequently, Luxembourg is taking deliberate constitutive actions, contextually situated within the future-oriented space resources discourse, taking advantage of its state capacity rooted in high GDP per capita growth rates ($115.203).[17]

LUXEMBOURG: SPACE CAPACITY

Luxembourg's space capacity is built around its willingness to enable international partners to access space resources. A document titled "Luxembourg Space Capabilities," released in 2017 basically offered a legal and normative framework that showcased the attractiveness of Luxembourg as a hub for future space technologies. Luxembourg aims to establish itself as "a pioneer in the development of the space resources economy."[18] Within

Luxembourg, the Ministry of the Economy oversees space affairs, and crafts its National Space policy and maintains contact with the European Space Agency (ESA). Luxinnovation is the unit within the Ministry that is in charge of Luxembourg space activities and manages the Luxembourg Space Cluster. Included into the body of space institutions is Groupement Luxembourgeois de l'Aéronautique et de l'Espace (GLAE).[19] GLAE was established following Luxembourg's accession to the ESA in 2005. GLAE's objectives are to "provide a permanent link between its members, advise them and defend their shared professional, economic and social interests at national and international level."[20] GLAE partners with the Luxembourg Government to address all initiatives taken by them as well as plays a key role in the establishment of a Luxembourg space policy. In 2017, there were about thirty companies and two public research organizations involved in space. Among the core companies are Creation International, Deep Space Industries (now acquired by Bradford Space), EarthLab Luxembourg, EmTronix, EC Group, FTA Communications Technologies SARL, and GovSat, a partnership between the government and SES (one of the leading satellite operators globally). GovSat supports NATO Alliance Ground Surveillance, as well as HITEC Luxembourg. Among the public space institutions are the Environmental Research and Innovation (ERIN) Department of the Luxembourg Institute of Science and Technology (LIST), as well as IT for Innovative Services Department, Material Research & Technology Department under LIST. Through its university, Luxembourg is actively shaping thinking about the global governance regime: the University of Luxembourg is a member of the Hague Space Resources Working Group (HSRWG),[21] and Luxembourg hosted one of the HSRWG face-to-face meetings in 2018.[22]

Luxembourg has built a reputation for efficient legislative and economic support for space services. This can be inferred from the MoU signed between China and Luxembourg in January 2019 in which both sides agreed to work together toward certain common goals: to include scientific, economic, technical, as well as build capacity in the utilization of space resources. During this signing, another agreement was signed by the National Space Science Center (NSSC) of the Chinese Academy of Sciences (CAS) and the Ministry of the Economy, Luxembourg, to establish a Research laboratory for Deep Space Exploration.[23] As per Xinhua, "the lab will focus on the coordinated design and analysis of deep-space probes, as well as the development of key technologies in exploring and utilizing space resources in the solar system, according to a press release from the National Space Science Center (NSSC) with the Chinese Academy of Sciences."[24]

As per the website of Luxembourg Trade and Invest,

> The Ministry of the Economy and the China National Space Administration (CNSA) signed today a Memorandum of Understanding (MoU) that provides a framework for the development and implementation of scientific, technical, economic and political cooperation between Luxembourg and China in the exploration and use of outer space for peaceful purposes. The areas of potential cooperation include, but are not limited to economic, legal, regulatory and technological aspects of the utilization of space resources.[25]

In March 2019, Luxembourg became a member of the Belt and Road Initiative (BRI) during Prime Minister Xavier Bettel's visit to China.[26] Since then, Chinese banks have increased investments in Luxembourg, with the Bank of China choosing Luxembourg to list its $500 million bond, the first BRI themed bond to be listed in the Luxembourg Stock Exchange.[27] Mathias Link, deputy director of Space Affairs at the Luxembourg Ministry of the Economy specified that "the long-term objective here is to create a new industry, to really seize the opportunities that space resources bring to create a new space economy." Link believes that the push for space resources requires collaboration with China as in the end he thinks it will be an international effort.[28] Significantly, as mentioned earlier, Luxembourg has signed an MoU with the United States as well. Russia, another serious spacefaring nation expressed an interest in signing an MoU with Luxembourg in January 2019, for collaborations in space mining. This was expressed by Russian Deputy Prime Minister Tatyana Golikova.[29]

The power of legal framework established by Luxembourg to enable space mining and ownership of such resources reveals how attractive that can be for a middle power such as Luxembourg, to then offer those services to major countries that possess serious spacefaring capacities. In this, both China and Russia's interest in signing MoUs with Luxembourg as well as establish institutions that can take advantage of its laws and tax cuts, signify how a smaller nation such as Luxembourg, can play a major role in incentivizing and creating governance mechanisms for the future trillion dollars' worth space economy.

LUXEMBOURG: STRATEGIC CULTURE

Luxembourg's strategic culture is influenced by its geographic location, at the heart of Western Europe, and its history. Historically, Luxembourg played a key strategic role, in connecting the Germanic and the Frankish territories. It was known as the Gibraltar of the North. Its role and location in Europe mean that it has played a key role in forming several of European institutions, to include the European Economic Community, now the European Union (EU).

Luxembourg was among the original countries that signed the Treaty of Rome on March 25, 1957, which was instrumental in the integration of Europe into a common economic zone. Significantly, Luxembourg in partnership with Belgium, formed an economic union in 1921, and has played a key role in European geopolitics.[30] While Luxembourg is a small nation, its ability to invest in a common European project has offered it a platform where it has been able to influence institutional policy-making in a way that has benefited it. With the EU in search of a way to foment a common strategic and defense culture, Luxembourg's influence in the domain has grown significantly.[31] Being a constitutional monarchy with a representative parliamentary system, based on consensus formation, ideas such as deeper European integration plays a key role in Luxembourg's strategic culture. Sharing borders with Belgium, Germany, and France, Luxembourg's political culture has been one of accommodation and partnership and taking advantage of inter-governmental institutions, that offers it a level playing field. Jean Claude Juncker, a leading politician of Luxembourg served as president of the European Commission from 2014 to 2019. Luxembourg hosts the European Court of Justice, the European Court of Auditors, the Secretariat of the European Parliament, the European Investment Bank, and the European Stability Mechanism. Hosting these institutions ensures that Luxembourg plays a central role in EU geopolitics.

A similar strain is observed with its space policy. Luxembourg is a key member of United Nations Office for Outer Space Affairs (UNOOSA) activities. In November 2019, UNOOSA and Luxembourg signed a funding agreement, to support UNOOSA's "New Space Law for New Space Actors" project.

> The "Space Law for New Space Actors" project will offer UN Member States tailored capacity building to facilitate their drafting of national space legislation and/or national space policies in line with international space law, promoting the long-term sustainability of outer space activities. Such capacity building will support in particular new and emerging space-faring nations to conduct space activities in a responsible and sustainable manner.[32]

Adherence to Outer Space Treaty (OST) and other international regulations for space activities has become of priority in a situation where states are increasingly looking toward the exploitation of space-based resources. Luxembourg is a member of the United Nations Committee on the Peaceful Uses of Outer Space (COPUOS) established in 1959. As per Luxembourg's Minister for Development Cooperation and Humanitarian Affairs, Ms. Paulette Lenert,

> Space-based activities are becoming more important in our daily lives and research has shown that around 40% of the 169 targets behind the 17

> Sustainable Development Goals (SDGs) benefit from the use of space. This innovative and forward-looking project specifically aims to include developing and least developed countries, to give them a voice and a chance to participate in the discussions around space law and to become "new space actors" in the framework of international space law.[33]

It is in that context that Luxembourg's space laws and its involvement in the global space regulatory framework becomes significant. It showcases a strategic culture that is future oriented and willing to take risks to ensure that Luxembourg plays a key role once that future becomes a reality. Luxembourg's strategic culture has been to diversify investment, and legislation to attract new actors in space. Luxembourg's focus on the space sector is to enable private space companies set up shop, and succeed in ventures such as space mining, and claim ownership of space-based resources, propelled by its laws. From its support to a private satellite-based communication in the 1980s, to its push for a space resource industry, it is clear that Luxembourg perceives itself as an important player in this emerging field.

UAE: SPACE CAPACITY

The GDP of the UAE was $414.18Bn in 2018. The GDP per capita of the UAE was $40,782 in 2018 with an annual growth rate of 2.2 percent.[34] The UAE's space budget is about $5.2Bn in government, private, and semi-private funding.[35] According to Stockholm International Peace Research Institute (SIPRI), UAE's military expenditure was about $22Bn in 2018; a figure the report speculated was similar to its 2014 expenditure.[36] The UAE does not publicly report on its military expenditure.

In 2014, the UAE announced the establishment of its space agency with the explicit aim to develop UAE as the regional hub for outer space activities in the Middle-East.[37] The UAE Space Agency seeks to help resolve global issues of natural disasters and share space expertise to mitigate the problems arising from shrinking resources and climate change. The Mohammad Bin Rashid Space Center (MBRSC), the commercial satellite communication companies, *Thuraya* and *Al Yah Sat*, are taking the lead in this domain. The UAE Space Agency and Exolaunch aims to jointly launch the *MeznSat*, a satellite developed by students from the American University of Ras Al Khaimah and Khalifa University. MeznSat will monitor and measure the methane and carbon dioxide levels in the UAE's atmosphere.[38] According to Dr Mohammed Al Ahbabi, the director-general of the UAE Space Agency "The MeznSat project falls within the framework of the UAE Space Agency's strategy, which aims to develop Emirati capacities and expertise and support scientific research."[39] The UAE Space Agency and the Abu Dhabi-based Krypto Labs jointly launched the New Space Innovation Programme, which aims to encourage UAE's new space sector, specifically targeted at the private spaceflight industry.[40] This initiative is within UAE's National Space Investment Promotion Plan. According to Ahbabi, "This initiative comes in line with the UAE Space Strategy 2030 and National Space Investment Promotion Plan, which seeks to encourage investment in the UAE's emerging space sector. We always encourage the development and implementation of space science and technology in the UAE, as part of the UAE Space Agency's objectives."[41] The UAE launched its *National Space Strategy* in March 2019, in which the UAE Space Agency has been directed to craft a national space structure that supports space science and industry, space policy as well as develop UAE into a major hub for space development.[42]

One of UAE's major space projects is its Mars Mission, named "Hope Probe," an indigenously built spacecraft that will orbit MARS and study its climate and atmosphere by 2021.[43] 2021 is significant for the UAE as it marks the 50th year of its establishment, and the Hope Probe is planned to utilize space in that celebration. On April 22, 2019, the UAE Space Agency and the MBRSC issued a statement that the Hope Probe was 85 percent complete. director-general of the MBRSC, Yousuf Hamad Al Shaibani, asserted that "Completing 85 per cent of the Hope Probe in this short period was a great challenge that we overcame through the guidance of our wise leadership and the efforts of our youth. The UAE has reached an advanced stage in achieving our wise leadership's vision to reach the Mars orbit by December 2021."[44]

Interestingly, the UAE space program is advertised as the "first Arab, Islamic probe to reach MARS by encouraging the peaceful application of space research."[45] On January 5, 2020, the Vice President and Ruler of Dubai, Sheikh Mohammed bin Rashid, and Crown Prince of Abu Dhabi and Deputy Supreme Commander of the Armed Forces, Sheikh Mohamed bin Zayed, inscribed a message on the outer casting of the Hope Probe in Arabic, translated as "the power of hope shortens the distance between the Earth and sky."[46] Sheikh Mohammed later tweeted, "Today me and my brother Mohamed bin Zayed signed the last piece of the outer structure of the Hope probe, the first Arab-Islamic probe to reach Mars." The Hope Probe successfully launched to Mars on July 19, 2020 on a H2-A rocket from the Tanegashima Space Centre in Japan, and will take seven to nine months to reach Martian orbit. The UAE strategic narrative is clear: this is the first Mars mission to be launched by a Muslim majority Arab nation, a matter of great national pride. Hope probe will analyze how the upper and lower Martian atmosphere interact, to provide us with a comprehensive picture. UAE space scientists hope to collect scientific data on Mars, in pursuance of their aspirational goal of a city in Mars by 2117.

The significance of the *Hope Probe* was specified by Ahmad Al Falasi, minister of state for higher education and advanced skills and chairman of the UAE Space Agency:

> The UAE is on the verge of making history, after turning its dream of becoming the first Arabic and Islamic country to send a spacecraft to Mars into reality. This monumental endeavour is the culmination of the efforts of a skilled and experienced team of young Emiratis who, with the support of the nation and its visionary leadership, will secure the UAE's position at the forefront of space exploration.[47]

According to the UAE's Ambassador to the United States, Yousef Al Otaiba, "This is the Arab world's version of President Kennedy's Moon shot—it's a vision for the future that can engage and excite a new generation of Emirati and Arab youth."[48]

In October 2015, the UAE Space Agency became a member of the International Space Exploration Coordination Group (ISECG).[49] Dr. Mohammed Al Ahbabi, director-general of the UAE Space Agency, specifies, "certain countries might have problems here on Earth, but you will see them cooperate in space."[50] UAE is starting to play a prominent role in the UN's COPUOS and by hosting several international meetings and conferences on space law and policy. Through its membership in ISECG and UN space forums, the UAE can bring about substantial influence to construct this future space regime relevant to a resource-based future of space, especially to ensure that the future is based on cooperation.[51] The UAE participated in the HSRWG.[52]

As part of an institutional building strategy, the UAE is an integral part of the UNOOSA. During the fiftieth-anniversary celebrations of UNOOSA, the UAE Space Agency adopted a resolution to develop space as a key driver for the sustainable development of the planet, in tune with the 2030 Agenda for Sustainable Development. Dr. Ahmad Abdullah Humaid Bel Houl Al Falasi, minister of state for higher education and advanced skills and chairman of the UAE Space Agency, stated during the celebrations that:

> This strategic resolution recognises the importance of capacity-building, education and training for space science and technology. It's critical that, with the support of the United Nations, member states work together to enhance capabilities and overall capacity across these areas. Space science and technology and their applications will continue to have a significant role in achieving the goals and targets of the 2030 Agenda for Sustainable Development, which is of great importance for humanity and the planet. The UAE Space Agency is honoured to be part of this long-term strategy and fully supports the objectives of the new resolution.[53]

In 2017, UNOOSA and the UAE Space Agency signed an MoU to work on the legal aspects of space exploration for social and economic benefits. Interestingly, the UNOOSA and UAE pledged to work for peaceful regional cooperation in space, especially the UAE's support for "UNISPACE+50 and activities under the Space2030 agenda, which will result from UNISPACE+50." UNOOSA director Simonetta Di Pippo, specified that:

> The UAE Space Agency's achievements in space science and technology are extensive and impressive. I am therefore thrilled that the Agency is extending its expertise and dedication to working with UNOOSA as we strive to ensure that the benefits of space reach everyone, everywhere. Together we will work on a number of initiatives to promote and facilitate the peaceful uses of outer space and the use of space as a tool for the achievement of the 2030 Agenda for Sustainable Development and its 17 Sustainable Development Goals. This will include a particular focus on the Middle East region, as well as science, technology, engineering and mathematics education (STEM), especially for women and girls.[54]

This narrative that UAE is the first Arab nation in space was clearly visible when the first Emirati astronaut, Ḥazzaa Al Mansoori made it to the International Space Station (ISS) on September 25, 2019, aboard a Russian Soyuz spacecraft. As Hazzaa broadcasted from the ISS, he stated, "I am so happy to hear your voices. I wish I could have shared what I saw with you. Beautiful scenery of sunsets and sunrises [*sic*] the earth. I say hello to our leaders, our people and all the Arabs." In a tweet by Sheikh Mohammed bin Rashid Al Maktoum, vice president and prime minister of the UAE and ruler of Dubai, he highlighted, "The arrival of the UAE's first astronaut Hazaa Al Mansouri in space is a message to all Arab youth that we can make a progress and move forward. We can catch up with the others. Our next stop is Mars via the Hope probe, which is designed by our young people with competency."[55] Besides the Hope probe, the UAE has an ambitious program to build a city on Mars by 2117, the "City of Wisdom" with 600,000 permanent residents. The simulated video of life on Mars was released by the UAE in the World Government Summit in February 2017.[56] The UAE University's College of Food and Agriculture sent up seeds of palm trees to the ISS, for their astronaut, Hazzaa to conduct experiments in the environment of space, and then bring the seeds back to the UAE and plant them, to study if they grow in a unique way. The idea behind sending date palm seeds to the ISS was backed by UAE's societal lore of the palm trees' iconic image of surviving very harsh conditions. Muhammad Nasser Al Ahbabi, director-general of the UAE Space Agency specified that "palm trees are deeply rooted in our nation's heritage and capable of thriving in harsh environments, which inspired us to launch the Palm in space project and attempt to germinate a palm seed in space."[57] The palm seeds were sent up to the ISS on July 27, 2019, on a Space X's Dragon capsule, in time for Hazzaa to carry out his experiments. As per Khalid Al Hashimi, the director of space missions, science, and technology, at the UAE Space Agency, "we chose to send palm tree seeds to the International Space Station both because of their importance to the UAE's culture and heritage, as well as the fact that they are able to withstand the harshest conditions and are often planted in circumstances that resemble aspects of the Martian environment."[58]

As mentioned before, in March 2019, the UAE adopted its National Space Strategy 2030, in a meeting chaired by Sheikh Mohammed bin Rashid Al Maktoum. The National Space Strategy 2030 is ambitious, and calls for investments on space manufacturing, in-space assembly, space science, and research, as well as commercial space. Critically, the UAE is committed to establish a space regulatory framework, which includes four components: National Space Policy, Space Sector Law, Space Regulations, and National Space Strategy.[59] From UAE's space strategy, it is clear that it wants to invest in its financial, and regulatory frameworks to develop UAE into a hub for outer space activities.[60] To bolster its space capacity, UAE signed an agreement with NASA, termed an Implementing Arrangement in October 2018 at the 69th International Astronautical Congress, held in Bremen, Germany. NASA Administrator, Jim Bridenstine specified that:

> As NASA builds cooperation for the return of humans to the Moon for long-term exploration and utilization, we welcome the opportunity to expand our partnership with the UAE Space Agency as it builds its significant capabilities on Earth, in low-Earth orbit, and beyond. UAE is currently working with U.S. universities to build an orbiter "Hope," to launch in 2020 and reach Mars in 2021. I'm delighted to sign this agreement signifying our deepening relationship as we move forward into the next phase of exploration.[61]

The UAE signed a Memorandum of Understanding (MoU) with Luxembourg on space resources in 2017. The MoU is for a period of five years and covers space resource exploration and utilization, as well as international space governance.

Minister of State for Higher Education and Chairman of the Board of Directors of the UAE Space Agency, Dr. Ahmad bin Abdulla Humaid Belhoul Al Falasi, said:

> Our collaboration with Luxembourg is aligned with the strategic visions of both the Space Agency and the UAE. This includes working towards closer

> international cooperation, establishing mutually beneficial international partnerships and exchanging scientific knowledge with the rest of the world. The UAE Space Agency recognizes the importance of international collaboration in the field of peaceful exploration of outer space, as this field is considered to be part of humankind's common heritage. It is extremely important in our view for all competent entities in this field to work together towards the common goal of enhancing the welfare of humanity.[62]

UAE's articulation of its ambitions from outer space falls under three distinctive categories. First, build capacity, to include expertise, of its space institutions. Second, invest in a long-term goal of what are the priority areas: space science; develop Emirati capability for indigenous space innovation; space resource investment, develop the legal regimes, and national regulation. Third, develop UAE as the Arab-Islamic representative for outer space. Toward that goal, UAE signed an agreement with the Bahrain National Space Science Agency. Under this agreement, the UAE Space Agency aims to train the Bahrain Space Team in satellite technology, design, construction as well as launch.[63]

UAE: ELITE DISCOURSES AND STRATEGIC CULTURE ON OUTER SPACE POLICY

There has been an emphasis, from the highest levels of policy-making within the UAE, to focus on developing UAE's space capacities. This is visible from the involvement of Sheikh Mohammed bin Rashid Al Maktoum, vice president of UAE as well as the prime minister, and the ruler of Dubai. He presided over the adoption of UAE's National Space Strategy. The UAE's space agency, MBRSC, was established by him in 2014. A decade ago, in 2006, Al Maktoum was instrumental in establishing the Emirates Institution for Advanced Science and Technology (EIAST). Al Maktoum issued a decree to appoint Crown Prince of Dubai, Hamdan bin Mohammed Hamdan Al Maktoum, as chairman and general supervisor of the space agency. The involvement of the royal family in developing the UAE's 2014 National Innovation Strategy is highly visible, and space falls within that gambit. The idea behind the innovation strategy is to create an environment that is friendly to innovation especially in science and technology, based on a supportive regulatory framework. Seven sectors have been prioritized for innovation, with space being of priority, along with renewable energy, transport, education, health, technology, and water.[64] In its *Vision 2021*, the UAE aspires to achieve an innovation-friendly environment that encourages "cutting edge" indigenous innovation and establishes UAE as a leader in this regard.

In that environment of enabling leadership focus, the committed focus on outer space is going to be long term. The UAE's strategic culture is to demonstrate the material, cultural, and innovation potential and investments of the country, to attract talent. Outer space is marketed within that sphere. Alastair Johnston defines strategic culture, as "Strategic culture is an integrated" system of symbols (e.g., argumentation structures, languages, analogies, metaphors) which acts to establish pervasive and longlasting strategic preferences by formulating concepts of the role and efficacy of military force in interstate political affairs, and by clothing these conceptions with such an aura of factuality that the strategic preferences seem uniquely realistic and efficacious."[65] The word "culture" therein, implies that there are certain assumptions, that are inherited, and which are then utilized to form attitudes about life. Critically, what this implies is that there are certain assumptions about the political environment that are inherent. Culture signifies a shared assumption and codes about the political environment by the political elite. This informs the understanding of causality as well. From one's strategic culture flows the preferred strategic choices. One key factor in analyzing strategic culture is to trace it to the sources of the socialization process, especially of the key decision-maker; in this context, Sheikh Mohammed bin Rashid Al Maktoum, vice president of UAE as well as the prime minister, and the ruler of Dubai. Rashid has been instrumental in turning the UAE into a competitive economy and is focused on developing innovation and technology. The UAE ranks fifth in the world in the IMD World Competitiveness Rankings. According to the report, "The United Arab Emirates—ranked 15th as recently as 2016—entered the top five for the first time. The UAE now ranks first globally for business efficiency, outshining other economies in areas such as productivity, digital transformation and entrepreneurship."[66] This kind of capabilities add credence to their investment on outer space as UAE has proven capacity to innovate and draw talent. The IMD World Competitiveness Rankings have four macro indicators, to include economic performance, infrastructure, government efficiency, and business efficiency, along with 235 other indicators.[67]

The historicity of the environment plays a role. The UAE emerged as a nation in 1971, when the seven emirates: Abu Dhabi, Ajman, Dubai, Fujayrah, Sharjah, and Umm al-Qaywayn formed the UAE. Immediately after, UAE joined the Arab League. Consequently, the environment of fellow Arab states, as well as membership in the Arab League and the Cooperation Council of the Arab States of the Gulf, indicated a central role for the UAE. The internal environment based on the leadership offered by Rashid and the external institutional memberships offered UAE a unique opportunity to lead the Arab states into space. Innovative business practices as well as focus on the future have enabled UAE to take up leadership roles.

CONCLUSION

This chapter clearly reveals the impact that middle powers have on the structure of governing space regimes for the future.

We can surmise that both Luxembourg and the UAE, with their wealth and service sector-based economy, are in a position to offer incentives for other spacefaring nations to set up shop within their territories from which both they and foreign private space companies can benefit. Moreover, Luxembourg and the UAE have taken leadership among the middle powers to influence a global space regime that will be beneficial to them in their long-term interest. Through the shaping of international norms, principles, and standards of behavior, Luxembourg and UAE are bringing about important changes in space resources policy. This approach, then, supports the needs of the broader global ecosystem of private commercial space ventures who are seeking legal stability and international capital.[68]

Middle powers routinely play the role of being the "stabilizers and legitimizers of the world order, whether in time of hegemony or not."[69] Knowing their limited ability to bring about a change in an established international regime, they seek to be proactive about securing their interests and manage the risk introduced by more powerful actors. For middle powers, it makes rational sense to invest resources in a global space regime, which can rein in the great powers. The strategic key for middle powers such as Luxembourg and UAE is to ensure their presence and influence in international partnerships in space; that they are part of the norm constituting, building, and enforcement countries. Then, ideally, they can construct a regime within the present order from within which their position is privileged by their economic robustness and institutional centrality.

NOTES

1. For an interesting perspective on middle powers, please see Eduard Jordaan, "The Concept of a Middle Power in International Relations: Distinguishing Between Emerging and Traditional Middle Powers," *Politikon: South African Journal of Political Studies* 30, no. 1 (2003): 165–81, accessed September 14, 2018, https://ink.library.smu.edu.sg/cgi/viewcontent.cgi?article=1393&context=soss_research

2. Mark Kramer, "Ideology and the Cold War," *Review of International Studies* 25, no.4 (October 1999): 539–76; Asif A. Siddiqi, *The Red Rockets' Glare Cambridge Centennial of Flight* (Cambridge: Cambridge University Press, 2014); Deborah Gadbury, *The Space Race: The Epic Battle between America and the Soviet Union for Dominion of Space* (New York: Harper Collins, 2006).

3. "Remarks by Vice President Pence at the Fifth Meeting of the National Space Council," Huntsville, AL, *The White House*, March 26, 2019, accessed September 30,

2019, https://www.whitehouse.gov/briefings-statements/remarks-vice-president-pence-fifth-meeting-national-space-council-huntsville-al/; Alexander Bowe, "China's Pursuit of Space Power Status and Implications for the United States," U.S-China Economic and Security Review Commission, April 11, 2019, accessed October 8, 2019, https://www.uscc.gov/sites/default/files/Research/USCC_China%27s%20Space%20Power%20Goals.pdf

4. Jeff Desjardins, "There's Big Money to be Made in Asteroid Mining," *Business Insider*, November 3, 2016, accessed June 3, 2019, https://www.businessinsider.com/the-value-of-asteroid-mining-2016-11

5. John Lewis, *Asteroid Mining 101: Wealth for the New Space Economy* (San Jose, CA: Deep Space Industries, 2014).

6. Marc M. Cohen, et al., "Robotic Asteroid Prospector (RAP)," Final Report, July 09, 2013, accessed September 30, 2019, https://www.nasa.gov/sites/default/files/files/Cohen_2012_PhI_RAP.pdf

7. Brian Skyrms and U. C. Irvine, "The Stag Hunt," Presidential Address to Pacific Division of the American Philosophical Association, March 2001, accessed March 4, 2019, http://www.socsci.uci.edu/~bskyrms/bio/papers/StagHunt.pdf

8. Jean Jacques Rousseau, "Discourse on Inequality," translated by G.D.H. Cole, accessed January 23, 2020, https://www.aub.edu.lb/fas/cvsp/Documents/DiscourseonInequality.pdf879500092.pdf,

9. David Z. Morris, "Luxembourg to Invest $227 Million in Asteroid Mining," *Fortune*, June 5, 2016, accessed September 14, 2018, http://fortune.com/2016/06/05/luxembourg-asteroid-mining/

10. "Space Resources.Lu," accessed September 14, 2018, https://spaceresources.public.lu/en.html

11. "Space Resources.Lu."

12. Jeff Froust, "Asteroid Mining Company Planetary Resources Acquired by Blockchain Firm," *SpaceNews*, October 31, 2018, accessed June 14, 2019, https://spacenews.com/asteroid-mining-company-planetary-resources-acquired-by-blockchain-firm/

13. "The United States and Luxembourg Sign Memorandum on Space Cooperation on May 10, 2019," *U.S. Embassy in Luxembourg*, May 10, 2019, accessed June 14, 2019, https://lu.usembassy.gov/the-united-states-and-luxembourg-sign-memorandum-on-space-co-operation-on-may-10-2019/

14. Ibid.

15. Jeff Foust, "Luxembourg Establishes Space Agency and New Fund," *SpaceNews*, September 13, 2018, accessed September 30, 2019, https://spacenews.com/luxembourg-establishes-space-agency-and-new-fund/

16. Atossa Araxia Abrahamian, "How a Tax Haven is Leading the Race to Privatise Space," *The Guardian*, September 15, 2017, accessed September 14, 2018, https://www.theguardian.com/news/2017/sep/15/luxembourg-tax-haven-privatise-space

17. The World Bank, "GDP Per Capita (Current US$), accessed September 14, 2018, https://data.worldbank.org/indicator/ny.gdp.pcap.cd

18. "Luxembourg Space Capabilities Turning Innovation into Business 2017," accessed September 30, 2019, https://clustermembers.luxinnovation.lu/space/wp

-content/uploads/sites/4/2016/10/07697_LUXINNOVATION_SpaceCapabilities_05-2017-Web.pdf, 6

19. GLAE, accessed July 16, 2019, https://glae.lu/

20. GLAE.

21. "The Hague International Space Resources Governance Working Group: The Consortium," Institute of Air and Space Law of Leiden University (The Netherlands), N.D., Accessed February 16, 2020, https://www.universiteitleiden.nl/en/law/institute-of-public-law/institute-of-air-space-law/the-hague-space-resources-governance-working-group

22. "Uni.lu hosts Hague Space Resources Group," University of Luxembourg, December 12, 2018, accessed February 16, 2020, https://wwwfr.uni.lu/fdef/actualites/uni_lu_hosts_hague_space_resources_group

23. "Luxembourg Cooperates with China in the Exploration and Use of Outer Space Resources for Peaceful Purpose," *Space Resources.LU*, accessed July 16, 2019https://spaceresources.public.lu/en/actualites/2018/Luxembourg-cooperates-with-China-in-the-exploration-and-use-of-outer-space-for-peaceful-purpose.html.

24. "China to Build Deep Space Lab in Luxemborug," *Xinhua*, January 16, 2018, accessed October 1, 2019, http://www.xinhuanet.com/english/2018-01/16/c_136900477.htm

25. Luxembourg Trade & Invest, "Luxembourg and China Agreed to Cooperate on Space Exploration," January 17, 2018, accessed September 30, 2019, https://www.tradeandinvest.lu/news/luxembourg-china-sign-agreement/

26. "Luxembourg, a Member of the BRI," *OBOReurope*, accessed on July 16, 2019, https://www.oboreurope.com/en/luxembourg-member-bri/

27. "Bank of China Lists 500-mln-dollars USD BRI Bond in Luxembourg," *Xinhua*, April 25, 2019, accessed July 16, 2019, http://www.xinhuanet.com/english/2019-04/25/c_138006422.htm

28. Sarah Scoles, "China Wants to Make a Mark in Space-But It will Need a Little Help," *Wired*, February 14, 2018, at https://www.wired.com/story/china-wants-to-make-a-mark-in-spacebut-itll-need-a-little-help/ (Accessed on July 16, 2019).

29. Vladimir Soldatkin, "Russia Wants to Join Luxembourg in Space Mining," *Reuters*, March 6, 2019, accessed July 16, 2019, https://www.reuters.com/article/us-luxembourg-russia-space/russia-wants-to-join-luxembourg-in-space-mining-idUSKCN1QN1OQ

30. Alex Gorka, "It Starts: EU Founders to Form Federal Union of European States," *Strategic Culture Foundation*, March 3, 2017, accessed February 7, 2020, https://www.strategic-culture.org/news/2017/03/03/it-starts-eu-founders-form-federal-union-european-states/

31. Sten Rynning, The European Union: Towards a Strategic Culture?," *Security Dialogue* 34, no.4 (December 2003): 479–96.

32. Luxembourg Space Agency, "UNOOSA and Luxembourg Launch New 'Space Law for New Space Actors" Project," November 13, 2019, accessed February 7, 2020, https://space-agency.public.lu/en/news-media/news/2019/UN_and_Luxembourg_sign_project.html

33. Luxembourg Space Agency, "UNOOSA and Luxembourg."

34. "United Arab Emirates GDP," accessed September 30, 2019, https://trading economics.com/united-arab-emirates/gdp; Satish Kumar, "UAE Central Bank Says GDP Growth in First Quarter of 2019 was 2.2% Year on Year: WHAM," *Reuters*, June 23, 2019, accessed September 30, 2019, https://www.reuters.com/article/us-emirates-economy-gdp/uae-central-bank-says-gdp-growth-in-first-quarter-of-2019-was-22-year-on-year-wam-idUSKCN1TO08D

35. Kate Greene, "Why the United Arab Emirates is Building a Space Program from Scratch," *Slate*, March 30, 2017, accessed September 30, 2019, https://slate.com/technology/2017/03/why-the-united-arab-emirates-is-building-a-space-program.html; Binsal Abdul Kader, "UAE Space Agency Unveils Ambitious Plans," *Gulfnews*, May 25, 2017, accessed February 16, 2020,

36. Pieter D. Wezeman and Alexandra Kuimova, "Military Spending and Arms Imports by Iran: Saudi Arabia, Qatar, and the UAE," *SIPRI Fact Sheet*, May 2019, accessed September 30, 2019, https://gulfnews.com/uae/government/uae-space-agency-unveils-ambitious-plans-1.1521609; https://www.sipri.org/sites/default/files/2019-05/fs_1905_gulf_milex_and_arms_transfers.pdf

37. Muhammad Bin Rashid Space Center, accessed September 14, 2018, https://mbrsc.ae/en/page/aboutus

38. Jack Dutton, "UAE Space Agency to Launch Satellite Developed by Students," *The National*, May 12, 2019, accessed May 13, 2019, https://www.thenational.ae/uae/science/uae-space-agency-to-launch-satellite-developed-by-students-1.860640

39. Dutton, "UAE Space Agency to Launch Satellite."

40. Staff Report, "Space Agency Launches NewSpace Innovation Programme," *Gulf News*, January 11, 2020, accessed January 23, 2020, https://gulfnews.com/uae/space-agency-launches-newspace-innovation-programme-1.68942607

41. Staff Report, "Space Agency Launches NewSpace."

42. "UAE Cabinet Approves National Space Strategy 2030," Government of Dubai, Media Office, March 11, 2019, accessed June 3, 2019, http://www.mediaoffice.ae/en/media-center/news/11/3/2019/smnationalstrategy.aspx.

43. "Emirates Mars Mission," UAE Space Agency, accessed January 24, 2020, https://www.mbrsc.ae/emirates-mars-mission.

44. "Video: UAE's Hope Probe Bound for Mars is 85 per cent complete," *The Khaleej Times*, April 24, 2019, accessed May 13, 2019, https://www.khaleejtimes.com/news/general/video-uaes-hope-probe-bound-for-mars-is-85-complete-

45. Muhammad Bin Rashid Space Center.

46. Staff, "UAE Leaders Sign Final Piece of Hope Probe Destined for Mars," *The National*, January 5, 2020, accessed January 24, 2020, https://www.thenational.ae/uae/science/uae-leaders-sign-final-piece-of-hope-probe-destined-for-mars-1.960118

47. Staff, "UAE Leaders Sign Final Piece."

48. Taimur Khan, "Mars Mission 'Arab World's Kennedy Moon Shot', Says UAE." *The National*, December 3, 2015, accessed September 14, 2018, https://www.thenational.ae/world/mars-mission-arab-world-s-kennedy-moon-shot-says-uae-ambassador-to-us-1.103660

49. "International Space Exploration Coordination Group, "Annual Report 2015," accessed September 14, 2018, https://www.dlr.de/Portaldata/28/Resources/dokumente/rm/ISECG_AnnualReport_2015.pdf

50. Thamer Al Subaihi, "UAE Space Programme a Conduit for Cooperation," *The National*, November 28, 2015, accessed September 14, 2018, https://www.thenational.ae/uae/uae-space-programme-a-conduit-for-cooperation-1.67069

51. "UAE Space Agency Adopts New Resolution at UN Summit," *Gulf News*, June 23, 2018, accessed June 5, 2019, https://gulfnews.com/going-out/society/uae-space-agency-adopts-new-resolution-at-un-summit-1.2240898

52. The Hague Space Resources Governance Working Group, *Final Report of The Hague Space Resources Governance Working Group* Annex I – Participants of The Hague Space Resources Governance Working Group, December 18, 2017, accessed February 16, 2020, 21, https://www.universiteitleiden.nl/binaries/content/assets/rechtsgeleerdheid/instituut-voor-publiekrecht/lucht--en-ruimterecht/space-resources/final-report_the-hague-space-resources-governance-working-group.pdf

53. "UAE Space Agency Adopts New Resolution."

54. "UNOOSA and UAE Space Agency Sign MoU for Increased Cooperation," November 08, 2017, accessed June 5, 2019, http://www.unoosa.org/oosa/en/informationfor/media/2017-unis-os-487.html; also see United Nations Office for Outer Space Affairs, "Capacity-Building in Space Science and Technology," United Nations, New York, 2008, accessed September 30, 2019, http://www.unoosa.org/pdf/publications/st_space_41E.pdf

55. Evangeline Elsa, "UAE in Space-as it Happened: First Emirati Astronaut Hazaa AlMansoori and Team Enter ISS: makes history for UAE," *Gulf News*, September 25, 2019, accessed September 30, 2019, https://gulfnews.com/uae/uae-in-space---as-it-happened-first-emirati-astronaut-hazzaa-almansoori-and-team-enter-iss-make-history-for-uae-1.1569396150288

56. "Mars 2117," You Tube, November 27, 2017, accessed September 30, 2019, https://www.youtube.com/watch?v=JboqeUqQiv4#action=share

57. Dominic Doodley, "Dates in Space: UAE Floats Idea of Planting Palm Trees on Mars," Forbes, August 5, 2019, accessed January 24, 2020, https://www.forbes.com/sites/dominicdudley/2019/08/05/dates-in-space/#52c83a734ef6

58. Doodley, "Dates in Space."

59. "UAE Cabinet Approves National Space Strategy 2030," *Khaleej Times*, March 12, 2019, accessed May 13, 2019, https://www.khaleejtimes.com/news/government/uae-cabinet-approves-national-space-strategy-2030--12. Also see National Space Strategy 2030, accessed May 13, 2019, https://www.government.ae/en/about-the-uae/strategies-initiatives-and-awards/federal-governments-strategies-and-plans/national-space-strategy-2030

60. UAE Voice, "National Space Strategy 2030 Gets the Approval of the UAE Cabinet," UAE Voice, March 12, 2019, accessed June 4, 2019, https://www.uae-voice.com/national-space-strategy-2030-gets-approval-uae-cabinet/

61. "NASA-UAE Space Agency Sign Historic Implementing Arrangement for Cooperation in Human Spaceflight," *NASA*, October 8, 2018, accessed June 4, 2019,

https://www.nasa.gov/press-release/nasa-uae-space-agency-sign-historic-implementing-arrangement-for-cooperation-in-human

62. "Luxembourg and the United Arab Emirates Sign MoU on Space Resources," *Space Resouces.LU*, October 10, 2017, accessed June 4, 2019, https://spaceresources.public.lu/en/actualites/2017/MoU-UAE.html

63. "Bahrain Boldly Going into Space, with the help of the UAE," *The National*, November 17, 2018, accessed on June 4, 2019, https://www.thenational.ae/uae/science/bahrain-boldly-going-into-space-with-the-help-of-the-uae-1.792920.

64. "National Innovation Strategy," *United Arab Emirates*, 2015, updated May 30, 2019, accessed June 4, 2019, https://government.ae/en/about-the-uae/strategies-initiatives-and-awards/federal-governments-strategies-and-plans/national-innovation-strategy

65. Alastair Johnston, "Thinking about Strategic Culture," *International Security* 19, no. 4 (Spring 1995): 32–64

66. "Singapore Topples United States as World's Most Competitive Economy," *IMD*, May 2019, accessed February 16, 2020, https://www.imd.org/news/updates/singapore-topples-united-states-as-worlds-most-competitive-economy/

67. "Singapore Topples United States."

68. Kristi J. Bradford, "A Model for Space Sector Growth: A Luxembourg Case Study," *The Aerospace Corporation*, October 2018, accessed February 7, 2020, https://aerospace.org/sites/default/files/2018-10/Bradford_Luxembourg_10292018.pdf

69. Jordaan, "The Concept of a Middle Power," 4.

Chapter 8

Are We Observing the Beginning of a Race or Scramble for Space Resources?

Scenarios and Concluding Thoughts

Our research began with a central question, "Are we observing the beginning of a race or scramble for space resources?" We hypothesized that we would find an observable shift in the orientation of national space programs and national space policy toward a greater emphasis on the exploitation of space resources.

Chapters on the United States, China, and India (the three major spacefaring countries) clearly indicate such a shift. As stated earlier, what is not yet observable at the time of writing (2019–2020) is actual commercial exploitation or conflict over the actual exploitation of space resources. Nevertheless, elite discourses are anticipating the development of space resources and the possibility of conflict or competition as a result. Official policy statements and documents indicate this fact.[1] By extension, we infer that a "scramble for space resources" is beginning and national space programs are beginning to reorient toward space development and exploitation.

Such behavior is clearly evident in the United States, China, India, as well as Luxembourg and the UAE. From our analyses, based on primary and secondary sources, supported by field work, it is clear that within the United States and China, there is a growing conversation at the level of policy on the salience on space-based resources. The U.S. space policies and advocacy literature are articulating the need to create an environment, where the utilization of space becomes a priority. This perspective is apparent from the Trump Administration's space policy directives,[2] speeches offered by Scott Pace, Executive Secretary, National Space Council,[3] and Vice President Mike Pence,[4] as well as Trump's establishment of a separate Space Force.[5] China is taking a lead role in articulating long-term space ambitions, in which utilization of space-based resources is critical for the national rejuvenation of the Chinese nation.[6] President Xi Jinping in his policy advocacy for the "China

Dream" has included China's space capacity as an integral part of developing comprehensive national power.[7] In pursuance of such goals, China has established a separate PLA Strategic Support Force (PLASSF),[8] and drafted a Civil-Military Integration Strategy that encompasses space as a key component.[9] India, another lead space player is beginning to explore the importance of space resources at the level of policy[10] and is making military institutional changes in anticipation of a more militarized future in outer space.[11]

Middle powers such as UAE and Luxembourg are at the forefront of asserting their policy position on space-based resources. UAE plans to build a city on Mars by 2117.[12] Luxembourg has passed legislation to enable private companies to own resources mined in space.[13] Both countries aim to become the hub for space activities in their regional setting, as well as influence great power behavior in setting global standards for space-based resource utilization.

The race, the scramble, is on. In the beginning of this book, we hoped to clearly answer three questions for the reader:

- Are states beginning to compete over space resources?
- What are their ambitions?
- What behavior should we anticipate?

The preceding chapters sought to answer those questions for individual nation-states. Now, in the final chapter, we hope to offer further insights regarding what behavior we might anticipate as state policies and programs interact. Below we discuss potential alternate future scenarios at the systemic level. We also explore scenarios for specific provocations that we might expect will arise out of the scramble for space resources. Each suggests important forks-in-the-road leading alternately to more cooperative, more competitive, or more adversarial futures. Scenarios are not expected to predict the future, but to anticipate causal pathways, implications, and interactions that inform policy choice and judgment in the present.

ALTERNATE FUTURES METHODOLOGY

One of the most utilized methods to analyze the future is the method of scenario building. Scenario building is a creative tool that offers policymakers and academics alike, an ability to examine various interdependent variables, that could shape the alternative futures.[14] Scenario planning helps us anticipate the future by exploring the interaction of key drivers of change on the subject of interest; in this case, outer space. Ultimately, scenario building is not really about the future *per se*, but rather it brings a clearer focus on today's trends, where they could lead us, and what we could do to affect them. Thus, scenario building is amongst the most powerful analytical tools to answer two fundamental questions of interest to policymakers, academia, think tanks, and their stakeholders:

1. How will the future be different from today?
2. What would states need to do in the present to arrive at, or avoid a potential future?

The answers to these questions inform and orient policymakers and policy implementers to position their systems proactively. By understanding which future states are desirable to approach or avoid, they can make informed decisions about the path they wish to take. Scenarios operate at two key levels. First, by providing a mental map of the future, they help orient us toward clues in our environment. Second, they operate at the level of strategy, enabling us to understand what is the art of the possible, and enable us to choose our ends and match our means and actions to accomplish those ends.[15]

To build scenarios, to construct these "maps of reality," we follow an alternate futures method:

Identify the Issue Area

This is the critical first step. It is here that the researcher should ask and answer the difficult question: what are the most important factors that will make the future different from today for that specific system, area, or region. In this case, we are interested in issues surrounding space policy with regard to space resources.

Pick a Specific Date

A specific date constrains how far each trend and the system as a whole can move or change in a given period. For instance, an extrapolation of U.S. GDP growth or China's Space Capacity and Military Budget will look quite different in 2030 as opposed to 2050. Similarly, a specific date and historical knowledge provides insight into how much of a window exists to change a policy or strategy and see it come to fruition. Any change in policy, for example, "A Lunar Industrial Base on the South Pole of the Moon" takes time to bear fruit, as projects and meetings take place over years, and there is a delay between sowing and reaping. For our purposes, we are primarily interested in the future between now and 2060.

Identify the Strategic Drivers of Change

This requires a certain degree of imagination. It can be difficult to look beyond our present lived experience. We don't want to echo the lack of foresight displayed by the *New York Times* when they wrote off the Wright Brothers[16] or Winston Churchill's inability to imagine India could succeed as a democracy.[17] It is important to consider how tomorrow may be different (and therefore requires different thinking and policies to remain adaptive) than today.

It is important to identify the high-impact drivers (major trends), playing themselves out that have considerable momentum, and can be forecast with relative confidence, barring some major catastrophe (wild cards).[18] These typically include demographic and macroeconomic forecasts. Some drivers of change reflect clear trends or roadmaps for which we have a high level of confidence. For example, China's success in meeting its announced timelines might give us relatively high confidence they would attempt to establish a base on the Lunar South Pole by 2036. In identifying drivers, the key question to ask is: will this driver of change, really result in a qualitative change in the system? An example for space is: will space mining activities have a significant effect on the global economy?

Identify the Key Uncertainties

While certain trends may be clear, other drivers of change exhibit a significant degree of uncertainty or disagreement on the direction of change. For instance, will India adopt a space resource policy? Will India grow faster or slower than China?

Subsequently, it is pertinent for a well-rounded picture to identify the drivers of change about which there is great uncertainty. Some factors can have a high element of embedded uncertainty in them, meaning that these are highly consequential factors that will cause the future to be different than today, but could also go either way, or the rate of change is deeply uncertain.

For a successful scenario-building exercise, it is generally helpful to limit the major drivers of uncertainty to three, allowing easy visualization along three axes. Scenario building that attempts to analyze together more than three drivers has historically proven cumbersome and difficult to digest, resulting in less useful analytic products. While there are likely to be many uncertainties about the future, it is the job of the researcher to consolidate and pare down to identify the top three drivers that are distinct from each other and which, in the researcher's opinion, are likely to have a greater strategic effect on the system than the others.

Generate the Scenario Narratives

Usually, at least two positive and two negative scenarios prove useful. For purposes of clarity, it is useful to construct a narrative of each alternative future. The purpose of constructing the narrative timeline is not to actually predict any course of events, but strengthen the researcher's understanding and insight into cause and effect, second-order effects, and to be able to talk concretely rather than vaguely about what such uncertainties might actually mean if played out in reality. The timeline has a second purpose: it allows the researcher to "live through" a simulated future of influential events. This puts the researcher in a strong position to describe and make judgments about that future—What is it like to live in that end state? Is it a safer or more dangerous world than today or the other scenarios that have been crafted?

Develop Preferences for the Desired Futures

Finally, once the scenarios are crafted, it is useful to make policy recommendations as to what should be the realistic long-term vision of a desirable future. That becomes the strategic end-state. An example of this method can be found in the Air Force Space Command product, *The Future of Space 2060 and Implications for U.S. Strategy*.[19] Both authors participated in the workshop and assisted as facilitators in constructing the scenarios. The report followed this methodology and asked three important questions about which we can have little certainty:

(1) How broadly and in what numbers do humans live and work in space?
(2) What is the economic level of new, persistent revenue from space activities?
(3) To what degree does the U.S. coalition lead in the creation of the civil, commercial, and military space capabilities and in establishing the norms, standards, and international regulations in space?

The report used these questions to explore eight radically different futures than today. We will not repeat that work here, but recommend it to anyone interested in exploring alternate futures for space.

STRATEGIC CONFIGURATION SCENARIOS

Because our focus is on the structure of the international system, we take it for granted that there will be new and persistent revenue from space activities and that humans will live and work in space. Technology might move faster or slower, but the key unknown in the period of interest (circa 2060) is: which of the three great powers (United States, China, India) will lead the global economy.[20]

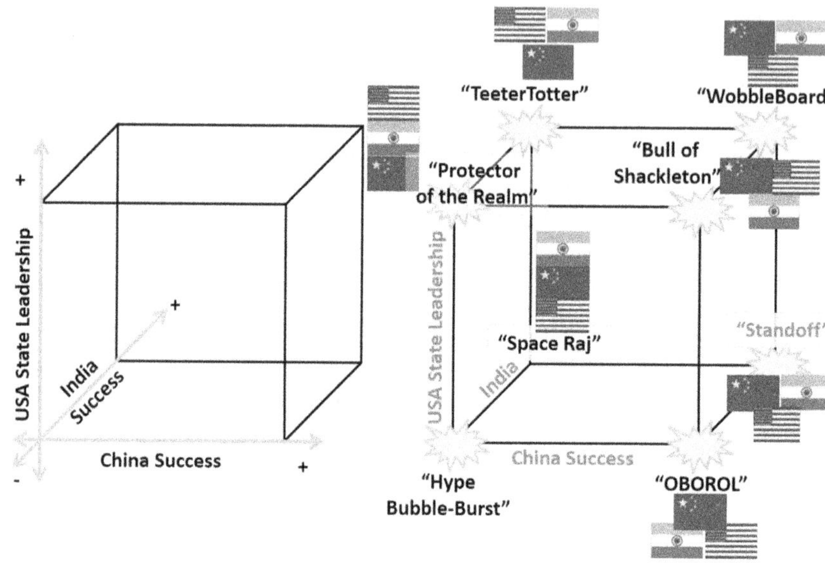

Figure 8.1 Future Strategic Cube Representation. *Source*: Compiled by authors, October 13, 2019.

One driver remains to what extent the United States maintains overall leadership, the next two uncertainties have to do with the relative economic and space performance of China and India. Each strategic configuration suggests a different dynamic with different tensions.[21] Below we offer the following macro-scenarios for 2060 to allow the reader to imagine how different the world of the future might be with regard to the broader geopolitical order, and its potential effect on the exploitation of space.

Monopolar Scenarios

Protector of the Realm

(United States#1, India#2, China#3) The United States provides exceptionally strong leadership, proactively extending a structure of a liberal economic order into space. While China and India both are doing well, the leadership of the U.S. commercial sector and the security brought by its Space Force enables a stable commerce that looks much like the maritime domain does today.

OBOROL *("Overall")*

(China#1, United States & India tied for #2) China's proactive space development plans are unstoppable. The United States continues to withdraw from a leadership role, and fails to provide a clear and proactive policy to enable its private sector, and as a result, the laissez-faire attitude results in a noncompetitive sector against a predatory and mercantilist China. India's economy grows slowly and is reticent and late to the game for space resources. China has latitude to set up an alternate governance structure along the lines of the Belt and Road Initiative and hierarchical patronage system.[22]

Space Raj

(India#1, China#2, United States#3) The United States-China competition continues to decay into mutually harmful military spending and trade wars resulting in both economies flagging and leaving fewer resources for space development. In contrast, India's economy surges and both its government and NewSpace industry prove remarkably successful. India is adjusting to its new role as the largest economy on Earth with a preponderance of power for rule-making in space. Seeing an attractive way forward, the middle powers bandwagon with India.

Bipolar/Balancing Scenarios:

Teeter Totter

(India & United States tied for #1, China#3) China's economy has flagged while a nationalist India has surged ahead. Competition between India and the United States provides space for China to seek advantageous deals and play one off against the other.

Bull of Shackleton

(United States & China tied for #1, India#3) In this bipolar world, China and the United States are engaged in an all-out scramble for space resources seeking an edge in the international system. China and the United States have agreed to divide the spoils and are largely constructing separate systems. India is a late-comer, courted by both the United States and China to be part of their alternate governance arrangements, but largely seeking autonomy.

Standoff

(China & India tied for #1, United States#3) The United States has retreated from economic and global leadership. A nearly evenly matched China and India vie for leadership in space and the international system. A weak United States has a limited role, supporting India in constructing a liberal order in a manner similar to how the U.K. or Canada supports the United States today.

Multipolar Scenarios:

Hype Bubble Burst

A series of spectacular failures by the United States, China, and India have burst the expectations and financial bubble of an in-space economy. The hype of development of space resources touted in the 2020s has proved just that.

Wobble Board

(China, India, and the United States are all tied for #1) In this tri-polar world, India has risen to be a peer economy of China. All three powers are exploiting space resources, and constantly jockeying for the top position. A rising and nationalist India makes it less conciliatory and a sunsetting China is more conciliatory to the United States, who as the third power is now in the position of a balancer.

The most familiar scenarios are those with a United States or China-centric world order or a bipolar U.S-China peer competition. The least familiar scenarios are those where India is the leading power or bipolar scenario where India is among the leading states, and the United States or China plays the role of the balancer. Such scenarios require additional study.

Within each of these scenarios, there is the additional uncertainty of how successful are the middle powers in moderating the conflict? Do the great powers feel threatened enough by competition that they welcome moderation? Do they care enough about the opinion of the middle powers to allow moderation?

Expected Primary Configuration 2030–2060: China#1, US#2, India#3

China will lead, India will Hedge, the United States will over-react (and succeed or fail), middle powers will moderate and prosper. The authors' expected scenario assumes that the United States fails to initially mount a comprehensive response, allowing China a strong lead toward space resources, followed by a United States over-reaction[23] to attempt to regain leadership resulting in success ("Protector of the Realm"), failure ("OBORAL"), or a division of spoils and continued and piqued competition ("Bull of Shackleton").[24] Within such evolving scenarios, we can imagine several specific provocations that have happened on Earth that appear plausible to occur in space. Each offers the potential to challenge stability in the system and would likely result in a crisis.

Situational Scenarios

Some scenarios can arise without malice from states but could nevertheless cause problems:

Claim Disputes

In the absence of a bilateral or multi-lateral agreement recognizing each other's claims, two entities could arrive at the same location (whether on the Moon or an asteroid) and have competing claims over who has the right to exclusive use. This could arise if states have different concepts of a legitimate area to claim for use or as a safety zone.

Trafficking

Rogue elements within a state might traffic persons (labor, sex trafficking) or illegal goods (dangerous materials (biohazards), illegal weapons,[25] drugs, other contraband), which require action by another state.

Unsafe Practices

Actors from one state (such as a mining company) might operate in a way that is directly harmful or dangerous to the activities of another state and require intervention. Such intervention might involve arrest and detention, seizure of assets, harmful interference, or stoppage of operations.

Piracy

Individuals from one state might choose to conduct illegal seizure for private financial gain, which would necessitate intervention by another state against the first state's citizens.

Defection of Citizens or Facilities

Once any level of facility is established, it is possible that for any number of reasons, the individuals of that facility might wish to renounce their citizenship and cede their facility to another nation. This might be on grounds of maltreatment (such as human rights abuses or religious freedom),[26] desire to escape criminal prosecution,[27] or desire for a more profitable future.

Contemporary territorial state behavior suggests several possible scenarios that might challenge easy cooperation and provoke a policy response if not a crisis. These "micro-aggressions" constitute grey-zone "elbowing"[28] or "salami-slicing"[29] to secure progressive advantage without provoking all-out war. While the United States is in theory capable of making every one of these aggressions, its commitments to a global liberal order based upon upholding contracts make it more likely to be the one challenged rather than the one challenging.

False Promises via Lawfare

The aggressor state might follow the example it has set of offering to negotiate on broad principles in an effort to restrain the activity of the other party, only to renege on such agreements claiming the other party has misinterpreted things.[30]

"Over-fishing"

An aggressor state might agree to communal limits on resource extraction, only to violate such restrictions later when to their advantage.[31]

One-Sided National Lawfare[32]:

The aggressor state might author national law to bolster a specific territorial claim to strengthen its case or force the hand of its executive to use force to back up a commercial claim.

Cartographic Aggression

The aggressor state might publish a map claiming exclusive use of territory as a means to constrain or intimidate another nation, its citizens, or businesses.[33]

Infrastructure Aggression

The aggressor state might build roads or infrastructure within what the other party understands to be their claim for exclusive use, operations, or safety.[34]

Limitations on Innocent Passage

The aggressor state might claim exclusive economic rights to an area and deny innocent passage to all or military vehicles and personnel. Examples include China's attempts to exclude U.S. Navy vessels from its Exclusive Economic Zone (EEZ).

Defense Identification Zones

The aggressor state might declare that anyone transiting within a certain radius of their facility must make certain advance declarations to be allowed safe passage.[35]

Constructing Deniable Dual-Use Military Infrastructure

The aggressor state might use legitimate civil or commercial purposes to build up fortifications, defenses, or offensive capabilities. For example, a laser used for mining, or a mass driver used for moving material to orbit could be repurposed to deny approach and landing.[36]

Maliciously Funding Competitors

An aggressor state might fund a competitor state or company in a way that is distracting or sapping.[37]

Hostile Financial Takeover[38]

An aggressor state might seek to purchase a controlling interest in a competitor's company or facility to take it over.

Claim Jumping

An aggressor state might aid and abet their own company to rush to exploit a site already claimed by another state's company.

Facility Blockage

An aggressor state might deny critical infrastructure such as an airlock or landing pad using construction equipment to shut down operations as a coercive tactic.[39]

Blockade

An aggressor state might threaten not to allow transit to or from a facility or heavenly body as a coercive tactic.

Impounding

An aggressor state might take possession of a rival state or rival state company's property or vessels under some claimed violation of law or safety, or for ransom. An example of this was the seizure of U.S. vessels by the Barbary states.[40]

Impressment

An aggressor state might coerce another states' citizens to provide service. Examples of this include Great Britain's impressment of American sailors after its war of independence.[41]

Hostage Taking

An aggressor state might arrest and detain the citizens of a rival nation as a coercive measure or for ransom. An example of this was the seizure of U.S. crew and passengers by the Barbary states.[42]

Tourism Aggression

An aggressor state might seek to use tourists as a way to circumvent limits on resource extraction or personnel by expanding tourists.[43]

Retributive Economic Punishment

An aggressor or aggrieved state may seek to harm a rival state by turning off business on which they are dependent (perhaps cutting back space tourism), or failing to provide vital resources (water, air, rocket fuel, access to launch pads, communication) as a means of coercion.[44]

Privateering

An aggressor state might issue licenses to harass or negate the operations of a rival state, and grant bounties for any property seized.

Deliberate Deniable Harmful Interference

An aggressor state might find continual ways to harass or slow operations, "gumming up the works" through various means of "accidental" harmful interference.

Sabotage

An aggressor state might find deniable ways of causing damage to a competitor state's infrastructure or facilities.

Issuing Private Law

An aggressor state might also provide a legal basis for a company to make its own law (as a company town or along the lines of European colonial Joint Stock Companies[45]) and act as its own armed force.[46] Allowing such autonomy might allow claims that they are the actions of rogue elements not under the control of the government.[47]

How might we place some of these in an imaginative scenario?

A Fork in the Road Circa 2047

On our road to 2060, we could imagine an important fork in the road circa 2047. By 2047, the in-space economy is quite mature, likely exceeding $3 trillion, with a reusable and affordable transportation system, lunar and asteroid mining, and solar power satellites under construction from space resources. China has had a sustained human presence on the Moon for over a decade, and the United States a sustained human presence for nearly twenty years. This is two years after China's established goal of being the most advanced nation in space (2045), and two years before its 100-year celebration of the People's Republic of China (2049). China has enjoyed being the largest economy since around 2030 but is now in decline. India still has a long way to go but is closing the gap fast with significant growth.

The Common Road to Crisis

Between 2020 and 2028, continued innovation in reusable launch and on-orbit logistics, and proactive state policies facilitate an expanded level of space access and commercial investment enabling regular access to the Moon, and a common ambition among the major players to industrialize the Moon and near-Earth asteroids to build in-space commercial industries such as propellant depots, private space stations for research and tourism, and construct solar power satellites. By 2028, fully reusable launch vehicles have proliferated from the United States to China and India; the cost of launch has dropped 10-fold, and larger rockets with higher flight rates result in access to space at least 10-fold above 2020 levels, with significantly larger payloads. This new level of access creates a flood of new investment and vastly increased salience for policymakers. Nobody wants to be left out of something that could increase their economy, and nobody wants another great power to gain in relative power and advantage.

Beginning in 2028, both United States and Chinese firms seek to exploit water resources on the lunar South Pole. However, the lack of a United States-China bilateral agreement means that both firms claim portions of each other's safety area. Without the support of the U.S. government, the U.S. company hoping to slow down its competitor, finds ways of continually harassing the Chinese company, jamming their wireless communication network whenever they trespass in the U.S. zone. The Chinese company responded in 2032 by blocking the United States landing pad with a robotic prospector. This forces the United States to forcibly remove and impound the vehicle, which becomes an international incident. By 2034, brokered by Luxembourg, China, and the United States agree to principles to govern the disputed areas, which include no infrastructure in the disputed areas. The next year, 2035, nationalist forces in China succeed in passing a national law asserting administrative control over its claimed facility and domestically defined safety zone, and asserting no innocent passage is allowed in the area it claims, as well as releasing a map asserting the area in dispute is undisputedly for Chinese operations. This angers U.S. companies, who pressure the U.S. Congress to have hearings, and heightens fears that China might deny access to the Moon. The United States conducts repeated freedom of navigation operations in the disputed zone from 2035 to 2037. China responds in 2038 by building roads and facilities in the disputed area, followed up by regular tourist visits within the disputed area, and then emplacement of "mining lasers." These steps provoke a *demarche* from the United States, and Congressional hearings to consider withdrawal from the Outer Space Treaty, but no further action. China participates in crafting a 2040 multi-lateral agreement with several states that specifies how much water ice can be extracted per year from the craters, and how they should be protected. However, China allows ecotourists to extract a certain amount of water ice that it does not count as commercial mining. A group of multi-national businessmen stage a protest of this practice in 2041. China withholds its regular supply of rocket fuel, resulting in a constriction of landing opportunities to resupply the broader community. Two multi-national companies go under, and China attempts—successfully—to purchase a controlling share in the failing companies in 2043. By 2046, the individuals from these companies find their liberties restricted, and seek to defect and give their facility to the United States. The United States publicly supports the protesters. China attempts to restrict all travel to and from the lunar South Pole until it agrees to let China manage its own affairs. This leads to the crisis of 2047, placing the United States in an untenable position—it cannot allow the precedent to be set of shutting down the celestial lines of commerce and restricting trade or access to its citizens.

Divergence

We can imagine that a common path leads to a fork in the road that might take us to very different futures, depending on what forces are operating. Here, the key uncertainties are:

- The relative political will of the United States and China to seek a decision or to delay and moderate the conflict further.
- The relative military strength of the United States vs. China within the space domain.
- The initiative and influence of the middle powers.

Below explore how differences in these variables could lead us in a few interesting directions.

The Road to the Bull of Shackleton

In this scenario, both powers are concerned with escalation, and the middle powers are proactive and influential. The relative strength of military power and outcome of an engagement is uncertain to both China and the United States

The middle powers step in and moderate, negotiating "the Bull of Shackleton" creating a clear dividing line on the Moon for development, and a scheme that divides up the administration of resources in the asteroid belt between the two powers. While the powers now keep their distance, wary of such an encounter again, they scramble within their own zones of control to maximize their industrial and military power for a future conflict that becomes the primary security concern of both countries.

The Road to OBOROL

In this scenario, the political will of China and the United States is asymmetric, and China's military capabilities and intelligence is superior to the United States. The middle powers are passive and fail to take initiative to construct the system, already concerned they may need to accommodate China due to perceived U.S. hollowness.

The United States begins to challenge China, but China unveils various "assassins mace" of dual-use capabilities. The sitting president and the U.S. Congress have no stomach to risk the safety of its citizens and lucrative facilities, and backs acceding to the Chinese demands for its safety zones. With the United States failing to stand up to the Chinese provocation, there is a general bandwagonning effect of other nations onto the Chinese system. United States companies also defect and are increasingly part of the Chinese patronage system. The Chinese aim at general prosperity, but take their cut and expect full compliance. Having once backed down, it becomes a habit for the United States and its companies.

The Road to Protector of the Realm

In this scenario, both the United States and China have the political will to seek a decision through military might, and the middle powers either lack initiative and influence or passively back the United States in its assertion. U.S. military and intelligence capabilities are superior to China.

The United States withdraws from the Outer Space Treaty and engages its Space Force to destroy the blockade. When China unveils its "assassin's mace," it finds that the U.S. intelligence and Space Domain Awareness knew about them in advance and had been countered through sabotage and asymmetric means. True to the American form of over-reaction and massive force, there was massive attrition of Chinese military space capabilities. The decisive show of force had left the United States as the unchallenged provider of security in the space domain and the liberal institutions it championed came to be favored by the other nations.

The Road to Space Raj:

We could also imagine that a United States-China conflict in space might not be decisive and instead prolonged and sapping, or might never turn to a hot war, but involve a series of tit-for-tat blockades and strangulation of commerce. If neutral parties were not affected, that could open a path for India. In this scenario, neither the United States nor China have the political will to seek an all-out contest and engage in prolonged attempts at coercion that sap their economies and cause investment to flee to other locations. Neither the United States nor China believe a decisive conflict would be to their benefit, and prefer coercive tools such as temporary blockades followed by concessions and negotiations. The role of the middle powers in this scenario would be to moderate crisis as well as to shift investment. Such a scenario might be more likely in a world where both the United States and Chinese economies were flagging, and both might be experiencing a relative decline in global economic share.

China blockades U.S. facilities. The United States responds with its own blockade. The middle powers negotiate the ability of neutral parties to enjoy freedom of navigation and provide humanitarian supplies to both sides. India as a neutral party is able to continue securing additional ground, as it is not a direct party to the dispute. While China had passed the United States in nominal GDP in 2030, India is on a steady trajectory to pass a stagnating China just before 2050. The prolonged and sapping conflict between the United States and China has meant that much of the world's private space investment has fled the United States and China, and moved to India, Luxembourg, and the UAE, which continue commercial economic development. India is also able to slowly improve its military space capabilities because both China and the United States are engaged with each other.

Familiar Playbooks

The playbooks available to nations are not new. Space merely provides a new canvas for geostrategic moves to be played. One can imagine similar events on an asteroid or on Mars or between in-space facilities. Such scenarios have played out many times in human history, and routinely play out on the surface of the Earth, and yet rarely result in war. War does not seem any more or less likely in space. But though war may or may not happen, aggressions and missteps are likely to happen. Such scenarios help us anticipate the future and be ready for policy responses for each occasion.

Early Middle Power Influence

We could also imagine that very early and proactive initiative (circa 2020–2030) on the part of the middle powers could succeed in establishing a cooperative regime for mutual recognition of mining claims, resolution mechanisms, and standards for safety zones. They might also prove to be a successful hub of capital, shifting the perception of space as a zone of great power competition, to a zone where international private capital funds private companies in the United States, China, and India. If a broader banking industry benefitted from multi-national companies housed in great powers, it might have significant moderating power to reduce threats to security by threatening to withdraw capital or depress the market. Moreover, in this world, the role of military space forces might more closely approximate Naval work to keep the sea lanes free.

While such scenarios are far from exhaustive, hopefully, they provide an analytic tool for both policymakers and students of the subject to anticipate events and possible policy failure.

CONCLUSION

In this book, we have sought to provide the reader with analyses based on fieldwork, with broad evidence from primary and secondary sources. We hope we have convinced the reader that the subject of space resources now enjoys a salience among policy elites of several important state actors.

Our conclusion is that we are witnessing a second space race. This time the primary components are not status-seeking for a global audience, but lasting economic power. It is the beginning stage of a competition to secure positional, logistical, and industrial advantages. It is a competition to secure the key astrostrategic positions to enable states to profit from the vast energy and material wealth of space resources, and to convert that wealth into power to shape the international system. This opening stage is comparatively muted as states make their preparations. Nevertheless, we have every expectation that it will become a rush, a scramble.

NOTES

1. The White House, "Remarks by Vice President Pence at the Fifth Meeting of the National Space Council, Huntsville, AL," March 26, 2019, August 7, 2019, accessed February 7, 2020, https://www.whitehouse.gov/briefings-statements/remarks-vice-president-pence-fifth-meeting-national-space-council-huntsville-al/; "China's Space Activities in 2016," accessed February 7, 2020, http://news.xinhuane

t.com/english/china/2016-12/27/c_135935416_5.htm; Dr. APJ Abdul Kalam, "Space Future in the 21st Century," Address to the Engineers, Scientists, Technologists of VSSC and also Students, faculties and professors of Indian Institute of Space Technology (IIST), Thiruvanathapuram, June 2, 2013, accessed September 17, 2019, https://space.nss.org/media/Kalam-2013-Space-Future-in-the-21st-Century.pdf

2. Presidential Memorandum on Reinvigorating America's Human Space Exploration Program, The White House, December 11, 2017, accessed August 1, 2019, https://www.whitehouse.gov/presidential-actions/presidential-memorandum-reinvigorating-americas-human-space-exploration-program/; Space Policy Directive-2, Streamlining Regulations on Commercial Use of Space, The White House, May 24, 2018, accessed August 1, 2019, https://www.whitehouse.gov/presidential-actions/space-policy-directive-2-streamlining-regulations-commercial-use-space/; Space Policy Directive-3, National Space Traffic Management Policy, The White House, June 18, 2018, August 1, 2019, https://www.whitehouse.gov/presidential-actions/space-policy-directive-3-national-space-traffic-management-policy/; White House, "Recommendations Approved by the National Space Council to President Trump," March 26, 2019, accessed August 8, 2019, http://www.spaceref.com/news/viewsr.html?pid=52358

3. Scott Pace, Executive Secretary, National Space Council, "Space 2.0: U.S. Competitiveness and Policy in the New Space Era," Hudson Institute, Keynote Address, May 17, 2018, accessed March 9, 2019, https://www.hudson.org/research/14341-full-transcript-space-2-0-u-s-competitiveness-and-policy-in-the-new-space-era;

4. The White House, "Remarks by Vice President Pence at the Fifth Meeting of the National Space Council | Huntsville, AL," March 26, 2019, August 7, 2019, accessed February 7, 2020, https://www.whitehouse.gov/briefings-statements/remarks-vice-president-pence-fifth-meeting-national-space-council-huntsville-al/

5. "Trump to Establish Space Force as Sixth Branch of Military," *CNBC*, June 18, 2018, accessed October 11, 2019, https://www.cnbc.com/video/2018/06/18/trump-space-force-sixth-military-branch.html

6. "2014 Two Sessions NPC & CPPCC: Annual Legislative and Political Advisory Sessions," *China Daily*, March 5, 2014, accessed January 19, 2019, http://www.chinadaily.com.cn/china/2014npcandcppcc/2014-03/05/content_17324203.htm

7. Namrata Goswami, "Waking up to China's Space Dream," *The Diplomat*, October 15, 2018, accessed September 18, 2019, https://thediplomat.com/2018/10/waking-up-to-chinas-space-dream/

8. No author, "China Establishes Rocket Force and Strategic Support Force," Ministry of National Defense, The People's Republic of China, accessed January 22, 2019, *http://eng.mod.gov.cn/ArmedForces/ssf.htm*

9. Zhao Lei, "Civil-Military Integration will Deepen," *China Daily*, March 3, 2018, accessed September 19, 2019, http://www.chinadaily.com.cn/a/201803/03/WS5a99d67ca3106e7dcc13f437.html

10. GV Ashok, "The Space Activities bill- Does it deliver?," *ORF Online*, April 19, 2019, accessed July 7, 2019, https://www.orfonline.org/expert-speak/the-space-activities-bill-does-it-deliver-50019/

11. Press Trust of India, "Centre Finalises Broad Contours Of Defence Space Agency," NDTV, June 11, 2019, accessed June 12, 2019, https://www.ndtv.com/india-news/centre-finalises-broad-contours-of-defence-space-agency-2051699?pfrom=home-topstories

12. "Mars 2117," You Tube, November 27, 2017, accessed September 30, 2019, https://www.youtube.com/watch?v=JboqeUqQiv4#action=share

13. "Space Resources.Lu," accessed September 14, 2018, https://spaceresources.public.lu/en.html

14. Air Force Space Command, "The Future of Space 2060 and Implications for U.S. Strategy: Report on Space Futures Workshop," September 5, 2019, accessed October 11, 2019, https://www.afspc.af.mil/Portals/3/documents/Future%20of%20Space%202060%20v2%20(5%20Sep).pdf?ver=2019-09-06-184933-230; also see Peter Schwartz, *The Art of the Long View: Planning for the Future in an Uncertain World* (New York: Penguin Random House, 1991).

15. Peter Garretson, "Scenario Building: A Suggested Methodology," *Institute for Defence Studies and Analyses (IDSA)*, November 25, 2009, accessed October 11, 2019, via internet archive: https://web.archive.org/web/20121122211758/http://www.idsa.in/asc/opinion11.html

16. Aviation Quotations, Great Aviation Quotes: Predictions, accessed August 15, 2019, http://www.aviationquotations.com/predictions.html; also see Glenn Firebaugh, *Seven Rules for Social Research* (Princeton and Oxford: Princeton University Press, 2008): 27.

17. Ishaan Tharoor, "The Dark Side of Winston Churchill's Legacy No One Should Forget," *The Washington Post*, February 3, 2015, accessed October 11, 2019, https://www.washingtonpost.com/news/worldviews/wp/2015/02/03/the-dark-side-of-winston-churchills-legacy-no-one-should-forget/?fbclid=IwAR2s_r7l3-YblgSJrLd5CkfzZsStknhVffvUquQIeJJev4vmHeZ6Lnlpf04

18. For more on scenario methods, please see "Shell Scenarios," accessed October 11, 2019, https://www.shell.com/energy-and-innovation/the-energy-future/scenarios.html

19. Air Force Space Command, "The Future of Space 2060."

20. Five distinct scenarios of how this might play out are discussed in an article on Quora which provides images from a Westpac Bank report very different trajectories of the top four economies from 2012 to 2100. Mrityunjay Nigam, "Many Population Analysts Predict India will Overtake China in 4 to 3 Decades Your Thoughts?," *Quora*, March 5, 2017, accessed October 13, 2019, https://www.quora.com/Many-population-analysts-predict-India-will-overtake-China-in-4-to-3-decades-your-thoughts

21. A. F. K Organski, *World Politics* (2d ed.) (New York: Knopf, 1968); R. Powell, "Stability and the Distribution of Power," *World Politics* 48, no. 2 (1996): 239–67; R. S. Ross and Z. Feng, *China's Ascent: Power, Security, and the Future of International Politics* (New York: Cornell University Press, 1998); R. Tammen, "The Impact of Asia on World Politics: China and India Options for the United States," *International Studies Review* 8, no. 4 (2006): 563–80.

22. The World Bank, "Belt and Road Initiative," March 29, 2018, accessed October 14, 2019, https://www.worldbank.org/en/topic/regional-integration/brief/

belt-and-road-initiative; The State Council, The People's Republic of China, "The Belt and Road Initiative," accessed October 14, 2019, http://english.www.gov.cn/beltAndRoad/

23. Robert Gilpin, *War and Change in World Politics* (Cambridge: Cambridge University Press, 1983); Robert O' Keohane, *After Hegemony: Cooperation and Discord in the World Political Economy* (Princeton, NJ: Princeton University Press, 1984); Michael C. Webb and Steven D. Krasner, "Hegemonic Stability Theory: An Empirical Assessment", *Review of International Studies* 15, No. 2 Special Issue on the Balance of Power (April, 1989):183-198

24. Jonathan Hart, *Contesting Empires: Opposition, Promotion and Slavery* (New York: Palgrave, MacMillan, 2005); Pope Alexander VI, "Inter Caetera Division of the Undiscovered World Between Spain and Portugal," 1493, Papal Encyclicals Online, accessed October 14, 2019, https://www.papalencyclicals.net/alex06/alex06inter.htm

25. Mary Beth Nikitin, "Proliferation Security Initiative," Congressional Research Service, August 9, 2018, accessed October 14, 2019, https://fas.org/sgp/crs/nuke/RL34327.pdf

26. Ewelina U. Ochab, "Is China Conducting a Crackdown on Religion?," *Forbes*, April 20, 2019, accessed October 14, 2019, https://www.forbes.com/sites/ewelinaochab/2019/04/20/is-china-conducting-a-crackdown-on-religion/#3dea378c19d3; Lily Kuo, "In China, They are Closing Churches, Jailing Pastors-and Even Rewriting Scriptures," *The Guardian*, January 13, 2019, accessed October 14, 2019, https://www.theguardian.com/world/2019/jan/13/china-christians-religious-persecution-translation-bible

27. Mary Hui, "These are all the Crimes Hong Kong is Considering Extraditing People for Under a New Law," *Quartz*, June 7, 2019, accessed October 14, 2019, https://qz.com/1636663/the-37-crimes-included-in-hong-kongs-proposed-extradition-law/

28. John Raine, "War or Peace? Understanding the Grey Zone," *IISS*, April 3, 2019, accessed October 14, 2019, https://www.iiss.org/blogs/analysis/2019/04/understanding-the-grey-zone

29. S D Pradhan, "Chinese Grand Strategy of Salami Slicing Continues in its Periphery," *The Times of India*, April 1, 2018, accessed October 14, 2019, https://timesofindia.indiatimes.com/blogs/ChanakyaCode/chinese-grand-strategy-of-salami-slicing-continues-in-its-periphery/

30. Examples of such behavior include China's behavior in the South China Sea, Nepal, and Bhutan. Orde F. Kittrie, "The Chinese Government Adopts and Implements a Lawfare Strategy," in *Lawfare: Law as a Weapon of War* Orde F. Kittrie ed. (Oxford: Oxford University Press, 2016).

31. An example would be China's overfishing behavior in the Antarctic. See: L M Foster and Namrata Goswami, "What China's Antarctic Behavior Tells us About the Future of Space," *The Diplomat*, January 11, 2019, accessed October 14, 2019, https://thediplomat.com/2019/01/what-chinas-antarctic-behavior-tells-us-about-the-future-of-space/; Anne Marie Brady, *China as a Polar Great Power* (Cambridge: Cambridge University Press, 2017).

32. Michael Listner, "The Art of Lawfare and the Real War in Outer Space," *The Space Review*, September 17, 2018, accessed October 13, 2019, http://www.thespacereview.com/article/3571/1

33. A terrestrial example might be China's publication of maps claiming large parts of India's Northeast or "nine-dash line" claiming the South China sea. See: Jonathan Kaiman, "Chinese Passport Maps Causes Diplomatic Row," *The Guardian*, November 27, 2012, accessed October 14, 2019, https://www.theguardian.com/world/2012/nov/27/chinese-passport-row-diplomatic-dispute

34. Examples of this include China's road-building activities within India's and Bhutan's borders. See: "Tiny Bhutan Protests to China Over Roadbuilding in Disputed Territory," *The Strait Times*, June 29, 2017, accessed October 14, 2019, https://www.straitstimes.com/asia/south-asia/tiny-bhutan-protests-to-china-over-road-building-in-disputed-territory

35. A terrestrial example would be China's Air Defense Identification Zone (ADIZ). See: "Special: China's Air Defense Identification Zone Triggers Mixed Response," *China Daily*, November 27, 2013, accessed October 14, 2019, http://www.chinadaily.com.cn/china/2013-11/27/content_17135438.htm

36. Examples include China's construction of "tourist" facilities on South China Sea islands which enable fighters, bombers, and anti-air defenses to be emplaced, see: Xue Gong, "Welcome to Conflict Tourism: How Chinese State Firms are Using the South China Sea," *The South China Morning Post*, September 19, 2016, accessed October 14, 2019, https://www.scmp.com/week-asia/geopolitics/article/2020096/welcome-conflict-tourism-how-chinese-state-firms-are-using

37. Examples of this include China's support of Pakistan's military and nuclear program, for example: Jonah Blank, "Pakistan and China's Almost Alliance," *Foreign Affairs*, October 16, 2015, accessed October 14, 2019, https://www.rand.org/blog/2015/10/pakistan-and-chinas-almost-alliance.html

38. Thomas Cooley, Eric Felt and Steven J. Butow, "State of the Space Industrial Base: Threats, Challenges and Actions," *AFRL & DIU*, May 30, 2019, accessed October 13, 2019, https://aerospace.csis.org/wp-content/uploads/2019/08/AFRL_DIU_Report_State_of_Space_Ind_Base_30May2019_Final.pdf

39. Examples of this might include China's use of "grey" actors such as fishermen to blockade access to disputed islands. See: Megha Rajagopalan, "China Trains 'Fishing Militia' to Sail into Disputed Waters," *Reuters,* April 30, 2016, accessed October 14, 2019, https://www.reuters.com/article/us-southchinasea-china-fishingboats-idUSKCN0XS0RS

40. Office of the Historian, "Barbary Wars, 1801–1805, and 1815-1816," accessed October 14, 2019, https://history.state.gov/milestones/1801-1829/barbary-wars

41. Amanda Foreman, "The British View the War of 1812 Quite Differently than Americans Do," *Smithsonian Magazine*, July 2104, accessed October 14, 2019, https://www.smithsonianmag.com/history/british-view-war-1812-quite-differently-americans-do-180951852/

42. Office of the Historian, "Barbary Wars, 1801–1805."

43. Examples of this might include China's use of 'eco-tourists' and 'citizen explorers' in Antarctica. See: Max Daniels, "The Need to Consider Chinese Tourism

in Antarctica," *The Polar Connection*, April 18, 2017, accessed October 14, 2019, http://polarconnection.org/chinese-tourism-antarctica/; "Protecting Antarctica from Ugly Tourists," *The Star Online*, February 4, 2018, accessed October 14, 2019, https://www.thestar.com.my/news/regional/2018/02/04/protecting-antarctica-from-ugly-tourists-the-growing-popularity-of-the-icy-continent-among-chinese-v/

44. Examples of this include China's restriction on rare earth metals. See, Panos Mourdoukoutas, "China Sends America Another Warning on Rare Earth – Seriously!," *Forbes*, July 18, 2019, accessed October 14, 2019, https://www.forbes.com/sites/panosmourdoukoutas/2019/07/18/china-sends-america-another-warning-on-rare-earths-seriously/#104dac07748c; Tomio Shida, "History Offers Lessons in Escaping China's Rare-Earth Dominance," *Nikkei Asian Review*, June 25, 2019, accessed October 14, 2019, https://asia.nikkei.com/Economy/Trade-war/History-offers-lessons-in-escaping-China-s-rare-earth-dominance2

45. A.E. Scrivener, "Private Company Rule in the Congo," *King Leopold's Rule in Africa* (London: William Heinemann, 1904):181–186, http://www.classzone.com/books/wh_05_shared/pdf/WHS05_027_774a_PS.pdf ; Ben Johnson, "The East India Company and its role in ruling India," *Historic UK*, September 2, 2012, accessed October 14, 2019, https://www.historic-uk.com/HistoryUK/HistoryofEngland/The-East-India-Company/

46. A. E. Scrivener, "Private Company Rule in the Congo," *King Leopold's Rule in Africa* (London: William Heinemann, 1904), 181–186, http://www.classzone.com/books/wh_05_shared/pdf/WHS05_027_774a_PS.pdf; Ben Johnson, "The East India Company and Its Role in Ruling India," *Historic UK*, September 2, 2012, accessed October 14, 2019, https://www.historic-uk.com/HistoryUK/HistoryofEngland/The-East-India-Company/; Larry Getlen, "Could Private Armies Fight the Next Major Conflict?," *New York Post*, December 28, 2014, accessed October 14, 2019, https://nypost.com/2014/12/28/could-private-armies-fight-the-next-major-conflict/

47. Examples might include China's Red Hacker Alliance. See ointhelibrarian, "Chinese Red Hacker Army//Red Hacker Alliance//Honkers//Honker's Union//HUC//中国红客队//红客," Steemit, June 29, 2017, accessed October 14, 2019, https://steemit.com/hacking/@odinthelibrarian/chinese-red-hacker-army-red-hacker-alliance-honkers-honker-s-union-huc

Appendix A

Theorizing Space Resources within International Relations Concepts and Positioning of the Authors

This afterword is provided for the academic community, and has three aims. First, we lay out a causal theory to explain state behavior and variation with respect to space resources. Second, we discuss our motivations and perspective. Third, we outline future avenues of research.

THEORY

We came to the work with certain implicit theories of causality about the world, which, through the process of writing we are now able to make more explicit. During the research and writing of this work, we observed certain regularities that suggest to us a causal mechanism. Here, we will lay out our post hoc causal theory, in hopes it will be useful to future researchers who wish to take up the topic.

Variables of Interest:

Dependent Variable (DV): State Behavior with respect to exploiting Space Resources.

Situated within the discipline of international relations, it is of interest to understand and anticipate state behavior. The units of analysis are individual states. We can expect variance in individual state behavior along a spectrum of non-interest to strong interest.[1]

States are theorized to demonstrate interest and investment along a spectrum that begins with elite discourse, becomes instantiated in policy documents, become funded programs leading to prospecting and actual exploitation.

Independent Variable (IV): Salience of perceived possibility of exploiting space resources.

Salience is the quality of being particularly noticeable or important; prominence[2] or the fact of being important to or connected with what is happening or being discussed.[3]

Hypothesis

- As space resources appear to be more exploitable, we will see an increase in competition to exploit those resources by the top tier of great powers.

The basic logic is that as salience for space resources increases, states will increase their investment toward space resource exploitation.

Causal Process Linking IV and DV:

Space resources are seen as a specific case of a more general process of peacetime competition between powerful states which vie for influence, power, and standing within a larger community. Top states seek to improve their relative position with respect to one another and pay attention to opportunities that would affect their relative power position.[4],[5] The distribution of power and technical capability determines which states have a reasonable chance to compete. Strategic culture is an *intervening factor* that accelerates or slows their initiatives and accounts for the greatest degree of state variation.[6]

Following Kingdon,[7] we conceptualize a general internal process where separate political, policy, and problem streams combine to put a topic on the agenda. We conceptualize a general process whereby *policy proposal entrepreneurs attempt to move an idea from elite discourse onto the national agenda to enact formal policy, legislation, and funded programs.*

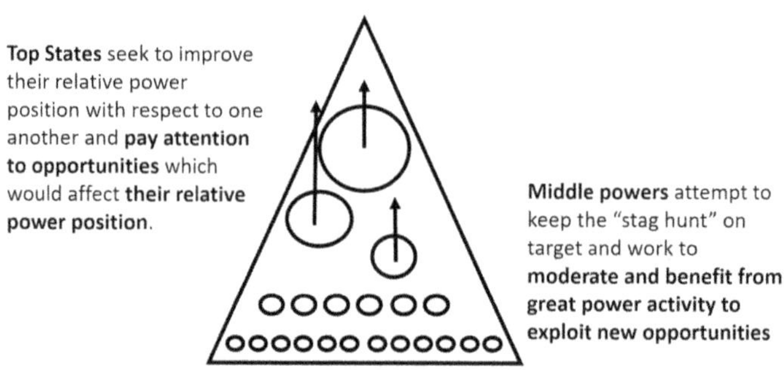

Figure A.1 Systemic Theory. *Source*: Compiled by authors, February 10, 2020.

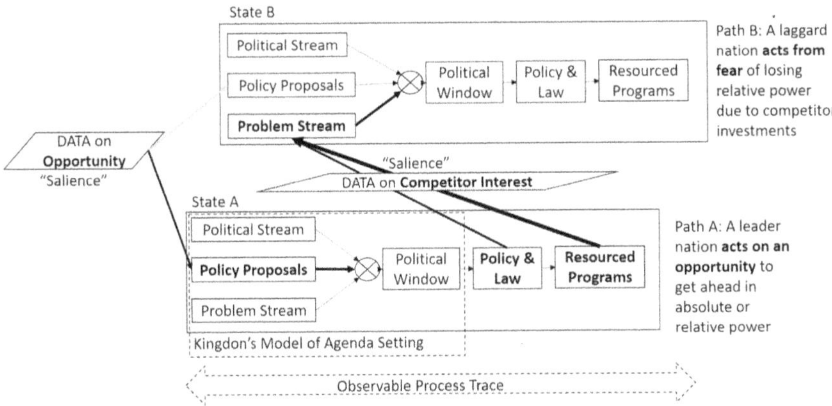

Figure A.2 Theory and Causal Process. *Source*: Compiled by authors, February 10, 2020, incorporates Kingdon's Multiple Streams Framework.

Data on Opportunity → Salience → Elite Discourse → Policies → Legislation → Programs

In this model, as data on the opportunity to exploit space resources become available, some elites will see this as an opportunity to get ahead, and experience success in moving it onto the national agenda (Path A in figure 3).

As rival states observe such actions, *elites in competitor nations will see this as a problem where they might get behind and will attempt to move it onto their own national agenda*. (Path B in figure A-2).[8]

Salience → Investment

Such a process is mutually reinforcing and creates a positive feedback loop.[9] Therefore, *as the opportunity to exploit space resources becomes more noticeable and prominent* (more discussed in important forum), *we expect to see states increase their investment toward exploiting space resources*.

While states follow this causal logic, their alacrity of response will be conditioned by a variety of factors, including technical and economic capacity, regime type, and strategic culture. The elites of states understand threats and opportunities differently.[10]

A mutually reinforcing feedback loop of competition constitutes the primary systemic influence. However, it may be moderated by two factors. First, where there are uncertain gains but very large risks (one side might lose catastrophically), states may agree to limit the areas or scope of competition. Second, less powerful states also play a moderating role. States that cannot independently compete for this opportunity and who have no reasonable opportunity for hegemonic leadership nevertheless desire to benefit from this opportunity. They may, therefore, join the coalition of one of the competing states in a specialized role, or attempt to build institutional structures that moderate the competition with a level of cooperation that allows them to benefit.[11] Other less competitive states may attempt to use law to restrain those in the lead to maintain position or extract benefits (a "weak power strategy").[12] Since the great powers have a broad range of interests,[13] they must pay attention to the desires of these lesser powers and may moderate their own behavior to co-opt or seek legitimacy from them.

Observable implications of the theory:

- Great powers will reference the activity of other great powers as an example or problem.
- Great powers will "race" or "scramble" for space resources.
- Increases in spacefaring capability, the salience of opportunity, or threat of competitor behavior should accelerate activity.
- Intent to race or scramble should be observable in enabling actions: policy, legislation, programs.

Operationalization of Variables:

Dependent Variable (DV): State Behavior with respect to exploiting Space Resources.

We can operationalize this variable through objectively observable indicators:

- Elite Discourse: What statements do thought leaders make about their ambitions to exploit space resources?
- Policy Statements: What policy statements address space resources?
- Legislation: Is there legislation that addresses space resources?
- Programs: Are there formal programs seeking to prospect or exploit space resources? What is their level of funding?

Independent Variable (IV): Salience of perceived possibility of exploiting space resources.

Salience can best be measured by whether or not there is a prominent discourse on the subject. Of greatest interest is when the elite discourse of one state references discourse or actions by other states.[14]

- What statements on space resources mention competitor interest?

Intervening Variable: Strategic culture. While we conceptualize the primary drivers at the systemic[15] (salience of opportunity to any capable actor due to technology readiness) and interstate level (actions of competitor state),[16] we see strategic culture as a significant intervening variable. States exhibit bounded rationality, conditioned by historical or path-dependent learning that biases them to interpret stimuli in particular ways and bias them toward or away from actions (scripts)[17] based on previous reinforcement.

States are information processing systems that—like all organizations—use schema and scripts[18] to minimize energy costs and satisfice.[19] While the general logic at the systemic level is true for all states, states differ in their biases in information processing leading to different rates of recognition, urgency for action, and action. While we start with a realist lens, believing that "states respond to material needs incentives and power"[20] and will self-help in the absence of an overarching legal authority, we accept insights from organizational theorists of the cognitive school[21], that those responses are conditioned by social cognition and history, including history of interaction, successes and failures.

How the theory addresses Kellstedt and Whitten's four hurdles of a "good" causal theory:

1. "Is there a credible causal mechanism that connects X to Y?"[22]: We have provided a causal mechanism above.
2. "Can we rule out the possibility that Y could cause X?"[23]: Logically one must perceive an opportunity before one invests in attempts to exploit it.
3. "Is there covariation between X & Y?"[24]: Our study has provided qualitative evidence of covariation.
4. "Have we controlled for all confounding variables Z that might make the association between X and Y spurious?"[25]: Given the diversity of variables that impinge upon state behavior and the impossibility of a true experiment to eliminate all possible confounding variables, we cannot—as in most observational studies—meet this bar.

A Potential Coding Scheme

What was an implicit coding scheme for us, might be formally specified as follows (see table A.1):

NOTES ON AUTHORS' MOTIVATIONS FOR RESEARCH DESIGN

As mentioned in the introductory chapter, a secondary audience for this book are academics themselves, who may have an interest to understand and develop models for more general purposes rather than any specific applied policy situation. For them, this may constitute a brick on which they seek

Table A.1 Codebook Key

Code	*Description*	*Example*	*Close but "No"*
Epistemic community discourse	Discourse coming from within the epistemic community of scientists, technologists, or advocates who are not prominent cultural or political figures or involved in the policymaking community	"NASA's goal should be to establish a Moon Base"—National Space Society/ Space Author	"NASA's goal should be to establish a Moon Base"—Billionaire Space Industrialist or Former Speaker of the U.S. House of Representative (this would be elite discourse)
Elite discourse	Discourse coming from individuals who have power or influence (including cultural power due to recognition) in the national policymaking process. This includes current and former politicians, public intellectuals, cultural icons and prominent businessmen	"NASA's goal should be to establish a Moon Base"—Billionaire Space Industrialist or Former Speaker of the House	"NASA's goal should be to establish a Moon Base"—National Space Society/Space Author [this would be epistemic community discourse] "NASA will go to the south pole of the Moon to exploit space resources"—Vice President Mike Pence
Law or Policy	On-record statements by officials or extant legislation which establishes policy	"The President will use the ambit of US Power to protect Space Commerce"—U.S. Law	"America needs the capability for Cis-Lunar Power Projection"—US Congressman (This would be elite discourse)
Evidence of Programs	Resourced programs to create material capabilities or missions	A program directed and justified to discover and exploit space resources	A program justified as a science or exploration program that could reveal space resources

Source: Compiled by authors, October 13, 2019.

to build or counter some other model or debate. They may have additional questions, in particular, how to evaluate our research in terms of relevance and theoretical and methodical rigor. In this section, we attempt to be as transparent as possible.

POSITIONING THE AUTHORS

Important in any interpretive work is providing the reader with transparency regarding the position of the authors, and how their own identity and philosophies may have influenced not only their telling of the story but their viewing of the subjects as well.

Criswell and Poth identify three important lenses that color a researcher's interpretation[26]:

Axiology Lens of the Authors/Our Normative Stance (What Do We Value? What Motivates Us?)

The research is motivated by a desire to aid policymakers struggling to create policies that advance their national gains but minimize or moderate conflict. The authors are sympathetic to the needs of policymakers to make policy, and value academic work that serves those ends. The authors sympathize with the aims of the space development advocates who want to improve the long-term survival and prosperity of humanity by expanding into space.

Ontological Lens of the Researcher (What Do We Believe about Reality?)

The authors are "realist" in the ontological sense.[27] There is a real world, but we can only know part of it based upon our perspective. We may mistake the map for the territory, but there is a territory out there we can all be mistaken about. Perspectives and world views matter.[28] While there are likely to be many different perspectives on a scramble for space resources, and we may never have a complete picture, there is a single reality that is occurring.

Epistemological Lens of the Authors (On What Basis Do We Claim We Can Gain Knowledge of What Is Real?)

We can't know with certainty but our best indicators are behaviors and their match/mismatch with statements of motive. Motives can be discerned through observation over time. Both researchers are "participant-observers" in this research"; actively participating in debates, and taking positions of advocacy and stated opinions, on specific policy prescriptions in testimony,[29] and Op-eds[30] to position free societies to lead and structure this future area of state competition. Given that, most of the observations and data collected by the authors on spacefaring countries such as the United States, China, and India are based on extensive research including fieldwork, elite interviews, and societal observations.

This research was animated by the following question, research hypothesis, and inferences.

Central Question: Are we observing the beginning of a race or scramble for space resources?

- Are states beginning to compete over space resources?
- What are their ambitions?
- What behavior should we anticipate?

Central Hypothesis

We are observing a shift in the orientation of national space programs and national space policy toward a greater emphasis on the exploitation of space resources.

Inferences: If we are correct that a scramble is beginning and national space programs are beginning to reorient toward space development and exploitation, then we ought to observe this shift among the great powers who are also spacefaring powers.

What Should Be Observable to the Authors?

In seeking to answer our central question and provide evidence for our hypothesis, we must consider what is observable. What is not yet observable at the time of writing (2019-2020) is actual commercial exploitation or conflict over the actual exploitation of space resources. What we should be able to observe are precursors to this reality. We should be able to observe elite discourse on the subject. We should be able to observe official policy statements and documents. We should be able to observe material resources dedicated toward prospecting or exploiting resources, or justified on those grounds. The preceding chapters in the book reflect such observations.

Unit of Analysis

The units of interest are individual nation-states expressed as behavior as in policy and programs.

Level of Analysis

We believe a significant amount of state behavior can be explained by forces present at the systemic level[31] (see theory section above). However, the primary level of analysis within our chapters is domestic—at the level of policymaking internal to a state—where individuals, organizations, and the influence of strategic culture matter.[32] However, the salience of competition for space resources at the systemic level directly influences domestic factors.

Expected Research Answer

We expected to find strong evidence of an emerging competition for space resources. We found evidence of extant ambitions within states to exploit space resources, and that in at least some states this has moved beyond elite discourse to formal policies and legislation, and even funded programs and activities. We expected and found evidence that major differences in the nuanced responses can be explained by strategic culture.

RESEARCH DESIGN

Choice of Approach

As a plausibility probe, we undertook a series of process-tracing case studies of the United States, China, and India, and UAE and Luxembourg to illustrate the underlying causal processes. We have chosen to approach the subject through a series of case studies of individual states who are contenders or have an interest in the subject of space resources. The cases follow a basic approach of process tracing,[33] how elite discourse leads to policy and law that lead to specific space programs. The individual country-focused chapters (and the fieldwork they are based on) resemble a series of ethnographies. For the United States, China, and India, we conducted extensive travel, fieldwork, and interviews, which pointed us toward a volume of literature that we have chosen to emphasize because of its prominence and availability to other researchers.

Cases of What?

We see each chapter as a case of the broader phenomenon of how individual states react to the salient possibility of exploiting space resources by themselves or others. These can be seen as cases of a broader phenomenon of competition over resources, and interstate competition most generally. Each chapter provided an individual case study of how ideas move from discourse to policy to resourcing. There are also case studies of individual state policy formation specific to their strategic culture.

Case Selection

Our topic admits only of cases of contemporary states in a very narrow temporal window. The novelty of the topic neither allows us to look at historical cases or to do longitudinal cases examining the same country in different periods.

Ideally, we would have liked to examine the entire population of states, or at least the entire population of spacefaring great powers and space-interested middle powers. However, this is beyond the scope of what is possible. Because the primary purpose of this study is to inform policymakers rather than rigorously test a causal theory, we have selected a sample of convenience based on criteria of states known to be most active, and accessible to the authors. We are of course aware of the selection bias of choosing cases on which we based our theory, and of selection based on the dependent variable. It cannot be helped, and we hope that the insights can still be of value.

The omission of Russia and Japan was purely a matter of scope and accessibility.

Instrument Design

We used semi-structured, snowball interviews. We would typically provide a ten-minute presentation introducing our topic and providing a number of images and statements by global thought leaders that provided an open-ended framing to solicit reaction. Sometimes this was one-on-one, but more often it was in small groups at policy-focused institutions within or advising (think tanks, academia) governments, or groups with technology or business interests. We would generally seek additional information related to discourse and attitudes, extant or expected policies, and extent or expected programs. Typically, we would seek insight into attitudes toward other actors, competition, and preferences toward unilateral or multilateral (institutional) actions. Wherever possible we would seek specific references to publications that established official positions. Data (specific statements) was collected on:

- Elite discourse related to space resources, in particular, statements that portray ambitions, threats and opportunities, and justifications for policies, legislation, and proposals.
- Policy statements and legislation that specifically address space resources.
- Funded programs that advance the ability to exploit space resources.

Data was collected via:

- Publicly available policy statements and legislation;
- Publicly available statements by policymakers and thought leaders in the media;
- Publicly available announcements of space programs;
- Interviews with policymakers, businessmen, and thought leaders.

Coding: Coding was informal and implicit. Above we provide a more formal coding scheme that reflects our judgment in structuring the data in the chapters into epistemic community discourse, elite discourse, law and policy, evidence of programs.

POSITIONING THE RESEARCH

This research is conceptualized to be qualitative research of policy relevance within the field of international relations. The appropriate type of research to observe the causal logic is the case study using process tracing.[34] The unit of analysis are individual states. While the primary causal theory is at the level of the international system (systemic level), the level of analysis of the study itself is conducted at the domestic level where the causal processes can be observed in action. This study is conceived to consist of case studies of the three major economies in the 2040–2060 timeframe, who are also major spacefaring states: China, India, United States. A minor treatment is also provided for less powerful states (middle powers) that are active in the field of study (Luxembourg, United Arab Emirates (UAE)). Our plan, and execution was to conduct in-country snowball semi-structured interviews.[35]

Relevance (Why Is This Relevant Research?)

Critical to policy formation is an accurate understanding of the ambitions and intentions of other actors and an expectation of how they will behave in the future. It is not yet established in the broader academic or policy community if space resources will become an area for great power competition. Nor is there a clear picture of the emphasis or scale of such ambitions.

Policymakers are keenly interested in opportunities that can alter the global balance of relative power, especially if it might alter the trajectory of the international system toward or away from the status quo. Space resources are so abundant that in principle they exceed the wealth available on Earth manifold, and could therefore significantly alter the relative balance of wealth and power.

LIMITATIONS OF OUR RESEARCH DESIGN

Generalizability and Transferability

As a primarily interpretive qualitative study and theory plausibility probe, it is inappropriate to speak in terms of "validity" sought by quantitative methods. We cannot know that a different researcher observing the same system at a different time, with a different lens of identity would not observe different things. However, our method should be suitably transparent that other researchers could follow our method. While our conclusions may not be "generalizable" in the broadest sense, they should be transferrable to other fields.

Hypothesis Testability

A test of our hypothesis is only sufficient to allow our theory to continue to stand (a "hoop" test) and to show evidence that supports it (a "smoking gun" test).[36] This is exploratory research for theory building. As an observational study and not an experiment, it cannot definitively rule out other explanations for state behavior, and we make no such claim.

However, our central assertion is testable, and our hypothesis is falsifiable. If our hypotheses were false, we would expect to find no evidence of an emerging international competition for space resources. We would not observe states moving along the continuum from discourse to investment. We would not observe justifications of such investments to reference the discourse and investment of other states.

If the hypothesis is correct, and there is an emerging struggle for space resources among the major spacefaring powers, then we would expect to:

- Observe space resources as a subject in elite discourse
- Observe the process as discourse moves to policy proposals into formal policies and legislation
- Observe the process as states move from policies and legislation to programs and activities

If our causal process is correct, then we should see states citing the ambitions and investments of rival states as a major reason for action. It appears technologically feasible to begin to access space resources now. Actual exploitation is likely to require significant timelines to mobilize technology and capital at scale. If states are contemplating such activities in the future, the first steps should be observable now.

These processes should be observable in the major spacefaring nations: United States, China, India, Russia. Since this process is expected to come to full maturity circa 2050,[37] the largest three economies will be China, the United States, and India.

Benefits and Limitations of Ethnography-Inspired Approach

The ethnographic approach allows us to look at the unique strategic culture and policymaking conversation within a given country's context. It allows close association with the subject, and a chance to approach perceiving the world through the lens of the subjects. It provides a rich description where emic (perspective of the subject) and etic (perspective of the observer) insights interact.[38] The approach allows us to tell the story of "what's going on here?" (with space resources in X country) in their own words or "in Vivo."[39] However, an ethnographic approach can at best help educate the reader's *Verstehen* (understanding) or *Erklären* (explaining),[40] and to provide an interpretation.[41] Interpretivism is inherently limited in its generalizability because its primary approach is to examine each case as unique in and of itself, rather than to discover general laws. Moreover, as researchers and authors, we bring our own identity and motivations that provide the initial lens of subjectivity for interpretive works.[42] For readers to properly understand and interpret these works, they require detailed knowledge of where the researcher is positioned to understand what reflexivity effects (how our role as a researcher can influence a study) may be in play[43]—which we have provided in this chapter.

Limitations of the Approach and Ethical Concerns

Ethnographies and participant–researcher study bring with them important limitations and concerns. While ethnographies allow the researcher to get close to the subject being studied, both the actors and their natural environment, ethnographies are not without their limitations and concerns. Chief among these are a concern of a lack of objectivity and interactive effects that happen in participant-observation. Creswell and Poth list reflexivity, reactivity, reciprocality, "going native," deception and divulging private information as typical problems encountered with the ethnology approach.[44] Policy-focused journalism with an ethnographical leaning provide minimal ethical concerns because interactions were with policymakers and thought leaders in the subject of their public roles, where privacy is not a concern. Nevertheless, our research was reviewed through the Department of Defense's Human Subject Institutional Review Board.[45]

Nevertheless, there are ethical and validity concerns arising due to reflexivity and reactivity of the system. This is because it is impossible to study the system without interacting with it. The very act of study itself may have meaningfully changed the system, and perhaps not for the better. This is a concern because, as the Thomas Theorem States, "if men define situations as real, they are real in their consequences."[46] An academic work may cause practitioners to define a situation as real that previously had escaped such a definition. Conversation to achieve shared cognitive consensuality[47] may alter the schema,[48] organizing paradigms[49] or generative metaphors of policy frames.[50] Ideas such as Halford Mackinder's Heartland Theory[51] may have contributed to a paradigm that led to two world wars.[52]

It is inescapable that in asking the questions—and in providing our results—this provides a feedback loop that was not present on its own in the system under study. It thus creates a confirming and amplifying signal that helps co-construct a perception of reality that there is a race (where perhaps there wasn't one).[53] This concern is particularly acute because the policy-attentive community is so small, as is the subject matter literature they are likely to reference. We are mindful that the mere perception of scarcity or competition can feed a security dilemma,[54] or accelerate securitization of an issue.[55] However, the alternatives are to refrain from the study of reality, to keep the results to oneself, to falsely report a rosier picture—none of which seem to offer a superior ethical choice.

As stated by Mars Society founder Robert Zubrin,

> Human civilization currently faces many serious dangers. The most immediate catastrophic threat, however, does not come from environmental degradation, resource depletion, or even asteroidal impact. *It comes from bad ideas. Ideas have consequences. Bad ideas can have really bad consequences. The worst idea that has ever been is that the total amount of potential resources is fixed. It is a catastrophic idea, because it sets all against all.* Currently, such limited-resource views are quite fashionable among not only futurists but much of the body politic. But if they prevail, then human freedoms must be curtailed. Furthermore, world war and genocide would be inevitable, for if the belief persists that there is only so much to go around, then the haves and the want-to-haves are going to have to duke it out, the only question being when. This is not an academic question. *The twentieth century was one of unprecedented plenty. Yet it saw tens of millions of people slaughtered in the name of a struggle for existence that was entirely fictitious. The results of similar thinking in the twenty-first could be far worse.*[56]

The most elegant response we can make to this charge is a paraphrase of Ray Bradbury: the function of social science is not only to anticipate the future conflict but sometimes to prevent it.[57]

EVALUATION

As stated at the outset, we think we have filled a gap in the literature at the intersection of international relations, policy studies, and space resources. We think we have also identified an important emerging security challenge and provided information to assist policymakers in formulating policy.

There are numerous works that explore the "militarization of space," which interpret state actions in space through the familiar lens of Cold War nuclear deterrence and arms control.[58] Few, if any, works within the international relations literature have explored how space resources may fundamentally alter the ordering of power in the international system. This work is meant to fill that gap in the literature. Increasingly, major states view space as a major component of their identity[59] and a component of comprehensive national power. They encourage commercial development and see space as part of the broader eco-system of great power competition. The great powers have begun to articulate and pursue policies of *astropolitik*[60]—extending the art of geopolitical maneuver into the domain of space. Given the potential of space resources to profoundly affect the balance of power and leadership in the international system, we hope our exploration of the strategic aspects of space resources will interest our fellow academics.

Academics lack an orientation to what appears to be a new area of great power competition. Policymakers lack both a coherent theory by which to frame policy debates as well as clarity regarding the ambitions of other states. Therefore, the objectives of our study were to:

- Frame a theory to explain and anticipate a class of previously unobserved behavior by spacefaring states.
- Provide illustrative cases that act as a plausibility probe for the underlying theory.
- Provide in-depth case studies providing evidence of the animating ambitions of contemporary spacefaring powers.
- Make the case for an emerging area for great power competition to motivate a broader research agenda.

As a descriptive work, this study fills a gap in the literature on international relations and public policy. For the first time, students of these disciplines as well as area studies students focusing on individual nations have available a temporal snapshot and historical collection of elite discourse, policies, and funded programs on the subject of space resources. This is also the first time that the broader topic of space resource claims has been discussed in the context of international relations academic work, or considered from the perspective of public policy theory. It is likely the first time that an interested reader can be made aware of the breadth of elite discourse and policies in a single place, and should provide a valuable reference for future students.

As a series of case studies,[61] the book provides amplifying evidence for the broader phenomena of state competition, resource competition, policymaking, or consistent strategic behavior or culture of individual nations. As a work of policy insight, it claims to predict an emerging area of great power competition. A claim that is falsifiable, but only by the history of the forthcoming century, not by the study itself.[62] As a work of theory, we propose a causal mechanism to explain why states behave as they do with respect to space resources. This theory is generally adaptable to the broader class of state competition for resources.[63] The study can provide only a plausibility test, but cannot exclude alternate causal theories.

We elaborated (above) what we think is a novel interpretation of the theory of state behavior or international relations applied to space resources.

We have only attempted here to provide a plausibility probe, to show confirmatory evidence consistent with our theory (survive the hoop test), as well as to show direct statements of ambition (smoking gun).[64] For further research probes, a null hypothesis might be that there is no significant shift in emphasis toward exploitation of space resources, and that it remains a marginal focus, with a continued emphasis on prestige-seeking or exploration focus, and an even stronger case that no scramble for resources will emerge. Such a test of emphasis we leave to a future researcher who might wish to develop a coding scheme and track word count consistently across important outlets. We await the test of history as to whether our central claim is correct.

Alternate explanations—or alternate emphasis of causal factors—might also be admitted. It could be that the role of domestic business interests is stronger, or that states are more motivated by relative gains than absolute gains or vice versa. It could be that strategic culture is more or less mediating that we propose, or that ideology plays a greater or lesser role. It might be that structural conditions of anarchy or institutions have a stronger or lesser effect than we propose. But our purpose was not to reduce the complexity of human nature to a clock-like mechanism, but rather to provide causal mechanisms that allow policymakers to anticipate, and students of international relations and public policy an explanatory framework. The model is far from mathematical precision but should provide explanatory power.

RETROSPECTIVE

While the trajectory we describe in the book may now appear quite obvious, when we began, we were definitely articulating a minority position, detecting and amplifying a "weak signal" among a lot of noise. How much the world has changed in the intervening years to produce this book. We began the research that would become this book on November 27, 2015. That was just two days after the U.S. Congress had passed its asteroid mining law[65] and three days after Blue Origin's first successful suborbital flight and landing.[66] To many, the evidence for a new space race based on resource motivations appeared very thin. SpaceX would not succeed in its first successful landing until mid-December.[67] China had yet to create its Strategic Support Force (that December). No other state had passed similar legislation and would not for another two years. Even some of our reviewers were skeptical. When we submitted our proposal, framing space resources as an emerging area of great power "grey zone" competition, our proposal drew skepticism, with one reviewer stating "this is not what great power competition is about." Fortunately, we were funded. Formal recognition of the patterns we perceived only began to be expressed in formal policy after our first draft was submitted, with key recognitions coming from the *Space Futures 2060* report and the *U.S.-China Economic and Security Review Commission Report of 2019.* The reader likely understands the U.S. Space Force and the U.S. Space command as permanent features. But neither they nor their emerging analogs in France,[68] India,[69] Japan,[70] Australia,[71] Canada,[72] or NATO[73] were in existence when we submitted our initial draft in mid-October 2019.

FUTURE AVENUES OF RESEARCH:

In this work, we have argued that there is an emerging competition over space resources and offered a causal theory as to why and how it unfolds. We have provided rich evidence of the foundations of the dialogue, and advanced a (strategic) culture explanation of why states vary. We hope it will stimulate further research on the topic. A rich variety of potential topics exist. A broad analytic agenda to better explore the topic might include:

- Greater area studies focus on the specific plans and ambitions of the space-faring powers;
- Exploring the topic through alternate methodological frameworks;
- Exploring the topic through alternate causal frameworks and levels of analysis;
- Better quantitative indices related to budget allocation for space resources;
- Better indices for relevant relative capability;
- Systematic word-count tracking of key terms in specific journals and media to measure salience;
- Increased attention to the development of cooperative or moderating institution design;
- Game Theory simulations to explore competitive behavior and claims.

Area and Country Studies on Space Resource Ambitions

Policy-relevant research demands data that is currently up to the moment and specific with regard to friendly or competitor states of interest. While we explored the state of things in 2020 for the United States, China, India, UAE, and Luxembourg, left unexplored were several current spacefaring powers (Russia, Japan, France, Israel) and emerging spacefaring states (such as North Korea, Iran). There is a need for well-articulated studies of individual states' space resource ambitions, timelines, and investments.

Different Policy Theory Frameworks

There are numerous theories of the policy process.[74] Our own conceptualization was inspired by John Kingdon's Multiple Streams Framework (MSF). However, our emphasis was primarily on activity within policy proposal and problem streams that put the broad topic of space resources on the agenda. Left unexplored was a greater attention to policy entrepreneurs and political windows. Future research on the topic using MSF might focus on a single major piece of law or policy (such as the passage of the 2015 asteroid act, the U.S. Space Force, the Luxembourg Space Law, the PRC's Space Solar Power Program) and explore the role of the specific policy and political entrepreneurs involved.

The use of alternate frameworks might result in equally rich insights. For example, the same major changes in legislation could be usefully approached using Punctuated Equilibrium Theory (PET),[75] bringing into focus the change in the dominant policy image, the venues of importance, and the factors that enabled new actors to contest the dominant policy image.

What we have explained using the IR concept of a security dilemma, scholars who use Innovation and Diffusion Models (IDM)[76] might characterize as competitive pressures that enable diffusion of innovations among states.

Our chapters provided rich evidence of policy beliefs held by space resource policy advocates and their attempt to use narrative to convince others. To the extent scholars may wish to categorize these in a disciplined manner; the Advocacy Coalition Framework (ACF)[77] and the Narrative Policy Framework (NPF)[78] provide useful tools. The ACF provides a structured way to understand the advocacy community and differentiate policy core beliefs from secondary beliefs. The NPF[79] provides a method to deconstruct the structure and content of the narrative. Moreover, both ACF and NPF lend themselves well to contested issues. There is already a rich literature arguing against space expansionism.[80] ACF could be used to explore the structure and beliefs of both communities while NPF could compare the narrative structure and content employed by the expansionists and anti-expansionists.

Alternate Causal Perspectives

Clearly social phenomena are affected by extant systemic factors (anarchy, political economy), system history and path dependence, social structure, culture and identity, ideas, and individuals. Our understanding can be enriched by examining the same subject from different perspectives. With regard to causal theories and level of analysis, we made a number of analytic choices that might not satisfy researchers with different interests or perspectives. Several inversions suggest themselves:

Rather than focus as we have on drivers and pathways to conflict that concern realists, liberal institutionalists might instead look for opportunities for absolute gains and focus on how previous scrambles have resulted in accommodation or moderating institutions.

Rather than treat strategic culture as a kind of learned bias that conditions attention and rational choice as we have, a more constructivist scholar might emphasize choice based on the logic of appropriateness[81] based upon identity.

Rather than focus on states as the primary actors with initiative, social conflict scholars and structuralists might explore how structural economic forces use states as vehicles to ensure continued economic growth for empowered classes of actors.

Rather than focus primarily at the domestic level as we have, either structural realists or world-system theorists[82] might focus at the system level. Alternately, scholars interested in exploring the causal impact of individuals or ideas might choose to focus at the sub-state level.

A remaining puzzle is why only some (and not all) middle powers choose to address space resources. This could be conceptualized as a broader question of why particular middle powers choose particular policy-economic niches and ignore others. This begs for an explanatory theory.

Better Indices

The topic could be advanced with better quantitative indices related to budget allocation for space resources. At present, no nation or NGO tracks spending on space resources specifically, and separate from other space programs.

Study of space resources will also benefit from better indices of space industrial capability. While various assessments have attempted to gauge relative space capability or competitiveness, none are focused on space industrial capability. Such an index might track current indicators such as annual metric tons to orbit and total metric tons of material on orbit. It might anticipate future industrial activity, and structure its database to track such indicators as: total metric tons processed on extraterrestrial bodies, total metric tons launched into space (from any planetary body), total logistical flow (in tons, ton-kilometers, ton-seconds-deltaV). Such indices should attempt to gauge the scale and value of the in-space economy and the relative contribution of important actors.

Quantitative Opportunities

Our research question focused on whether we are observing the beginning of a race or scramble for space resources using qualitative methods. But if there is such a phenomenon as a scramble for space resources, it would be valuable to have objective numerical data that would allow us to notice and characterize trends.

How could we objectively and numerically compare states on their comparative investment or activity? How do we know if the salience of the topic (our independent variable) is increasing or decreasing? How do we know if state activity to exploit space resources is increasing or decreasing? If we say the topic is salient, then it is salient compared to what? If we say state activity or investment is increasing, increasing compared to what? These "questions of degree" may interest quantitative scholars, who may have ideas for how to operationalize these variables.

Measurement of salience would seem approachable through quantitative content analysis[83] on Qualitative Data Analysis (QDA) software that could do automated word-counts to track issue-attention cycles[84] such as "asteroid mining" or "space solar power" in policy documents, speeches, journals news, and social media to measure acceleration—the degree to which it is heating up or cooling down.

We thought that strategic culture was a persuasive explanation of why states differ in their responses. But quantitative scholars have also attempted to measure culture. For example, multiple measures have been advanced to measure national culture,[85] national personality, national traits, or national negotiating style. It would be interesting to know if any of these correlates predict that states display interest in space expansionism and space resources. For example, there are likely several states of comparative development and power to the UAE and Luxembourg—is there some combination of measured cultural traits that predicts their entrepreneurial interest in space resources?

Application of Qualitative Data Analysis Software to Social Network Analysis

Our understanding of how ideas become policy could be significantly advanced if we better understood the structure of the involved communities. It would be of value for a researcher to use qualitative data analysis software[86] to conduct social network analysis (SNA)[87] of the various epistemic communities and advocacy as has been done for the Geoengineering community by Dr. Ina Moller.[88] This would be valuable both within each state (and on both sides of the issue), as well as at the global level.

Likewise, our ability to trace how ideas move from their beginnings in the epistemic community to elite discourse and into policy and law could be advanced in its precision through software tools that can find and visually display the "epidemiology of ideas" helping us understand the timing and sequence as ideas and phrases move from agent to agent. A basic application is just a visualization of citation network mapping. More advanced methods would attempt to apply textual analysis to determine the heritage/parentage of derivative works similar to how biologists are able to place species within a tree-of-life from similarities and differences in DNA.

Opportunities to Advance Research for Praxis

If states are choosing to compete for space resources, then there is an urgent need to develop ways to limit the dangers and costs of conflict to the community of nations through governance and institutions. Scholars can draw on the rich literature of institutional design to suggest institutional frameworks with the best chance to work in space. Elinor Ostrom's Social Ecological Systems and Institutional Analysis and Development Frameworks[89] that have been used to examine the management of Common Pool Resources in fisheries appear to be a framework that could be usefully applied. There are a range of potential institutions to be explored: bilateral agreements recognizing claims,[90] maritime analogs,[91] economic development institutions (Organization for Economic Cooperation and Development, World Bank, International Monetary Fund, International Civil Aviation Organization).

There is an equal need to explore when actors are likely to be open to such frameworks, and how they are likely to react to such frameworks. Part of this can be accomplished through behavioral economics and game theory work to theorize costs and payoffs and what sort of equilibria is likely to emerge.[92] Useful work likely needs to consider the insights of prospect theory to properly take into account loss aversion.[93] A second line of effort is to explore the human element through role-playing simulations ("wargames")[94] that explore scenarios where actors might come into conflict over resource extraction activity or claims to see how they resolve or escalate.

NOTES

1. Interactions between states might produce an international environment ranging from non-competition to fierce competition; from peaceful market-based competition under multilateral rules to hostility and conflict; from unilateral action to

coordinated action among like-minded states in spheres of influence to system-wide multi-lateral cooperative frameworks.

2. Oxford Dictionary, "Salience," accessed August 26, 2019, https://en.oxforddictionaries.com/definition/salience

3. Oxford Dictionary, "Salience."

4. Kenneth Waltz: "Contending states imitate the military innovations contrived by the country of greatest capability and ingenuity," from Kenneth Waltz, *Theory of International Politics* (Reading, MA: Addison-Wesley, 1979); see esp. pp. 123–128.

5. Barry Posen, for instance, "stresses the causal role of the international system; structural realism deduces from the anarchical condition of international politics that states that wish to remain autonomous will compete for security. Military capabilities are a key means to such security, and thus states will pay close attention to them. States will be concerned about the size and effectiveness of their military organizations relative to their neighbors. As in any competitive system, successful practices will be imitated. Those who fail to imitate are unlikely to survive." Barry R. Posen, "Nationalism, the mass army, and military power," *International security* 18, no. 2 (1993): 80–124.

6. This is analogous to the extensive literature on how states differ in converting material endowments into military effectiveness.

7. John W. Kingdon, *Agenda's, Alternatives, and Public Policies* (New York: Longman, 2003).

8. An alternate logic for imitation to the "security dilemma" logic was termed the "leadership principle" by Dr. Tong Zhao of Carnegie Tsinghua whereby states are inspired to imitation by observing others doing well and believe they can too. Author conversations.

9. Though not directly military, this dynamic is strongly reminiscent of the classic security dilemma which drive arms races, where attempts by one state to improve is security undermines the security of other states.

10. There is literature that has studied these aspects. Please see Christian Cantir and Juliet Karrbo, "Contested Roles and Domestic Politics: Reflections on Role Theory in Foreign Policy Analysis and IR Theory," *Foreign Policy Analysis* 8, no. 1 (January 2012): 5–24; Peter Katzenstein, "International Relations and Domestic Structures: Foreign Economic Policies of Advanced Industrial States," *International Organization* 30, no. 1 (Winter 1976): 1–45.

11. Randall Schweller, "Bandwagoning for Profit: Bringing the Revisionist State In," *International Security* 19, no. 1 (Summer 1994): 72–107.

12. Literature on weak states include Christina J. Schneider, "Weak States and Institutionalized Bargaining Power in International Organizations," *International Studies Quarterly* 55, no. 2 (June 2011): 331–55; Hanna Samir Kassab, *Grand Strategies of Weak States and Great Powers* (Basingstoke: Palgrave Macmillan, 2018).

13. Joshua B. Spero, "Great Power Security Dilemmas for Pivotal Middle Power Bridging," *Contemporary Security Policy* 30, no. 1 (2009): 147–71, accessed on September 12, 2019, https://doi.org/10.1080/13523260902760355

14. Vice President Mike Pence clearly referenced China and Russia's achievements in space as a motivator for the U.S. interest on the Moon. Please see "Remarks by Vice President Pence at Fifth Meeting of the National Space Council Huntsville, AL," *The White House*, March 26, 2019, accessed September 12, 2019, https://www.whitehouse.gov/briefings-statements/remarks-vice-president-pence-fifth-meeting-national-space-council-huntsville-al/

15. Morton A Kaplan, *System and Process in International Politics* (New York: John Wiley, 1957); Robert Gilpin, *War and Change in World Politics* (Cambridge, UK: Cambridge University Press, 1981).

16. Herbert Simon, "A Behavioral Model of Rational Choice," in *Models of Man, Social and Rational: Mathematical Essays on Rational Human Behavior in a Social Setting* (New York: Wiley, 1957).

17. Henry P Sims and Dennis A. Gioia, *The Thinking Organization: Dynamics of Organizational Social Cognition* (San Francisco: Jossey-Bass, 1986): 49–74.

18. Sims and Gioia, *The Thinking Organization.*

19. Herbert A. Simon, "Rational Choice and the Structure of the Environment," *Psychological Review* 63, no. 2 (1956): 129–38.

20. Ian Hurd, "Constructivism" (Chapter 17), Northwestern University, January 18, 2008, accessed August 31, 2019, http://faculty.wcas.northwestern.edu/~ihu355/Home_files/17-Smit-Snidal-c17.pdf

21. Karl E. Weick, "Cognitive Processes in Organizations," *Research in Organizational Behavior*, 1 (1979): 41–74.

22. "Four Causal Hurdles" in Paul M. Kellstedt and Guy D. Whitten. *The Fundamentals of Political Science Research Second Edition* (Cambridge: Cambridge University Press, 2013), 55.

23. Weick, "Cognitive Processes in Organizations," 55.

24. Weick, "Cognitive Processes in Organizations," 55.

25. Weick, "Cognitive Processes in Organizations," 55.

26. John W. Creswell and Cheryl N. Poth, *Qualitative Inquiry and Research Design: Choosing Among Five Approaches* (Los Angeles; Sage publications, 2017), 51.

27. Colin Wright, *Agents, Structures and International Relations: Politics as Ontology* (Cambridge: Cambridge University Press, 2006).

28. Mark Juergensmeyer, *Gandhi's Way A Handbook of Conflict Resolution* (Berkeley: University of California Press, 2005).

29. Statement of Dr. Namrata Goswami Independent Senior Analyst and Author 2016-2017 Minerva Grantee before the U.S.-China Economic and Security Review Commission Hearing on "China in Space: A Strategic Competition?," April 25, 2019, accessed October 14, 2019, https://www.uscc.gov/sites/default/files/Namrata%20Goswami%20USCC%2025%20April.pdf

30. Peter Garretson, "The First Duty of a Space Force is to Protect Space Commerce," *Politico*, June 21, 2019, accessed October 14, 2019, https://www.politico.com/story/2019/06/21/opinion-space-force-commerce-1374229; Namrata Goswami, "The New Space Race Pits the U.S. Against China. The U.S. is Losing Badly," *The*

Washington Post, January 10, 2019, accessed October 14, 2019, https://www.washingtonpost.com/opinions/the-new-space-race-pits-the-us-against-china-the-us-is-losing-badly/2019/01/10/bcdcad10-14f9-11e9-b6ad-9cfd62dbb0a8_story.html

31. J. David Singer, "The Level of Analysis Problem in International Relations," *World Politics* 14, no. 1 (October 1961): 77–92. Also Owen Temby, "What are Levels of Analysis and What do They Contribute to International Relations Theory," *Cambridge Review of International Affairs* 28, no. 4 (2015): 721–42.

32. James D. Fearson, "Domestic Politics, Foreign Policy, and Theories of International Relations," *Annual Review of Political Science*, 1 (1998): 289–313, accessed September 12, 2019, http://www.rochelleterman.com/ir/sites/default/files/Fearon%201998.pdf

33. David Collier, "Understanding Process Tracing," *Political Science & Politics* 44, no. 4 (October 2011): 823–830.

34. Collier, "Understanding Process Tracing."

35. Ann Blandford, "Semi Structured Qualitative Studies," in Mads Soegaard and Rikke Friis Dam ed., *The Encyclopedia of Human-Computer Interaction* (Aarhus, Denmark: The Interaction Design Foundation, 2013), accessed September 12, 2019, http://www.interactiondesign.org/encyclopedia/semi-structured_qualitative_studies.html; Oisín Tansey, " Process Tracing and Elite Interviewing: A Case for Non Probability Sampling," *PS: Political Science and Politics* 40, no. 4 (October 2007): 765–72.

36. Collier, "Understanding Process Tracing."

37. PricewaterhouseCoopers, "The World in 2050 The Long View: How Will the Global Economic Order Change by 2050," accessed October 8, 2019, https://www.pwc.com/gx/en/issues/economy/the-world-in-2050.html

38. "Two Views of Culture: Etic and Emic," Lumen Learning, accessed August 31, 2019, https://courses.lumenlearning.com/culturalanthropology/chapter/two-views-of-culture-etic-emic/

39. B. G Glaser and A. Strauss, *The Discovery of Grounded Theory. Strategies for Qualitative Research* (Chicago: Aldine, 1967).

40. Norma M. Riccucci, *Public administration: Traditions of inquiry and philosophies of knowledge* (Washington. D.C: Georgetown University Press, 2010), chapter 5.

41. Jean Michel Montsion, "Ethnography and International Relations: Situating Recent Trends, Debates, and Limitations from an Interdisciplinary Perspective", *The Journal of Chinese Sociology* 5, no. 9 (2018), accessed September 12, 2019, https://doi.org/10.1186/s40711-018-0079-4

42. David N. Gibbs and Robert S. Snyder, "Confronting Bias in International Relations," *International Studies Perspective* 2, no. 4 (November 2001): 416–37.

43. John W. Creswell, *Qualitative Inquiry & Research Design* Choosing Among Five Approaches (Los Angeles: Sage, 2013): 306.

44. John W. Creswell, *Qualitative Inquiry & Research*, 214.

45. "Human Research Protection Board," *University of Pittsburg*, accessed September 12, 2019, https://www.irb.pitt.edu/content/department-defense

46. Robert K. Merton, "The Thomas Theorem and The Matthew Effect," *Social Forces* 74, no. 2 (1995): 379–424, available at http://garfield.library.upenn.edu/merton/thomastheorem.pdf

47. Henry P. Sims and Dennis A. Gioia, *The Thinking Organization: Dynamics of Organizational Social Cognition* (San Francisco: Jossey-Bass, 1986): 8.

48. Sims and Gioia, *The Thinking Organization: Dynamics.*

49. Thomas S. Kuhn, *The Structure of Scientific Revolutions* (Chicago, IL: University of Chicago Press, 1996).

50. Donald A. Schon and Martin Rein, *Frame Reflection: Toward the Resolution of Intractable Policy Controversies* (New York: Basic Books, 1994).

51. Halford Halford Mackinder, "The Geographical Pivot of History," *The Geographical Journal* (Royal Geographical Association), April 1904, available at: https://rgs-ibg.onlinelibrary.wiley.com/doi/abs/10.1111/j.0016-7398.2004.00132.x

52. WorldAtlas, "What is Heartland Theory?," accessed August 31, 2019, https://www.worldatlas.com/articles/what-is-the-heartland-theory.html

53. Timothy Dunne, "The Social Construction of International Society," *European Journal of International Relations* 1, no.3 (September 1995): 367–89.

54. Security dilemma implies that an action by one state to improve upon its own security results in another state perceiving that as an existential threat, and consequentially arming itself. Please see John H. Hertz, "Idealist Internationalism and the Security Dilemma," *World Politics* 2, no. 2 (January 1950): 157-180; Robert Jervis, "Cooperation Under the Security Dilemma," *World Politics* 30, no. 2 (January 1978): 167-214.

55. Clara Eroukhmanoff, "Securitisation Theory: An Introduction," *E-International Relations*, January 14, 2018, accessed August 31, 2019, https://www.e-ir.info/2018/01/14/securitisation-theory-an-introduction/

56. Robert Zubrin, *The Case for Space How the Revolution in Spaceflight Opens up a Future of Limitless Possibilities* (New York: Prometheus Books, 2019), 304.

57. Ray Bradbury, *Yestermorrow: Obvious Answers to Impossible Futures* (Central Valley, CA: Joshua Odell Editions/Capra Press, 1991), 155.

58. Joan Johnson-Freese, *Heavenly Ambitions: America's Quest to Dominate Space* (Philadelphia: University of Pennsylvania Press, 2009); James Clay Moltz, *The Politics of Space Security: Strategic Restraint and the Pursuit of National Interests* (Paolo Alto: Stanford University Press, 2011); Thomas Graham Jr., "Space Weapons and the Risks of Accidental Nuclear War," *Arms Control Today* 35, no. 10 (December 2005); Michael Krepon, "Weapons in the Heavens: A Radical and Reckless Option," *Arms Control Today* 34, no. 9 (November 2004); Mathew Mowthorpe, *The Militarization and Weaponization of Space* (Lanham: Lexington Books, 2004): 11, 13–18.

59. Natalie Bormann and Michael Sheehan, *Securing Outer Space: International Relations Theory and the Politics of Space* (London and New York: Routledge, 2012).

60. Astropolitik is defined as "the study of the relationship between outer space terrain and technology and the development of political and military policy and

strategy," Everett C. Dolman, *Astropolitik Classical Geopolitics in the Space Age* (London: Frank Cass, 2002), 15.

61. For relevance of case studies, please see Alexander L. George and Andrew Bennett, *Case Studies and Theory Development in the Social Sciences* (Cambridge, MA: MIT Press, 2005).

62. Nazli Choucri and Thomas W. Robinson, *Forecasting in International Relations: Theory, Methods, Problems, Prospects* (San Francisco: W.H. Freeman, 1978); Sean P. O'Brien, "Crisis Early Warning and Decision Support: Contemporary Approaches and Thoughts on Future Research," *International Studies Review* 1, no. 1 (2010): 87–104; John R. Freeman and Brain L. Job, "Scientific Forecasts in International Relations: Problems of Definition and Epistemology," *International Studies Quarterly* 23, no. 1 (March 1979): 113–49.

63. Michael T. Klare, *The Race for What's Left: The Global Scramble for the World's Last Resources* (New York: Macmillan, 2012).

64. Collier, "Understanding Process Tracing"; Alexander L. George, "Case Studies and Theory Development: The Method of Structured, Focused Comparison," in Paul Gordon Lauren, ed., *Diplomacy: New Approaches in History, Theory, and Policy* (New York: Free Press, 1979).

65. "H.R.2262 - U.S. Commercial Space Launch Competitiveness Act 114th Congress (2015-2016)," Congress.gov, November 25, 2015, accessed October 16, 2019, https://www.congress.gov/bill/114th-congress/house-bill/2262/text

66. Loren Grush, "Why you shouldn't compare Blue Origin's rocket landing to SpaceX," *The Verge*, November 24, 2015, accessed February 5, 2015 at https://www.theverge.com/2015/11/24/9793220/blue-origin-vs-spacex-rocket-landing-jeff-bezos-elon-musk

67. Jeff Foust (21 December 2015). "Falcon 9 Launches Orbcomm Satellites, Lands First Stage," *SpaceNews*. Retrieved 22 December 2015. The first time SpaceX had successfully landed the rocket's first stage.

68. Hanneke Weitering, "France Is Launching a 'Space Force' with Weaponized Satellites," *Space.com*, August 02, 2019, accessed February 5, 2020, https://www.space.com/france-military-space-force.html

69. Ajey Lele, "India needs its own space force," *Space News*, May 28, 2019, accessed February 5, 2020, https://spacenews.com/op-ed-india-needs-its-own-space-force/

70. Michael Peck, "Japan is Getting Its Own Space Force (With Better Uniforms?)," *National Interest*, January 21, 2020, accessed February 5, 2020, https://nationalinterest.org/blog/buzz/japan-getting-its-own-space-force-better-uniforms-115816

71. Max Blenkin, "ADF urged to establish Aussie space command," January 30, 2020, accessed February 5, 2020, https://www.spaceconnectonline.com.au/operations/4107-calls-for-adf-to-establish-space-command-to-manage-aussie-space-operations

72. John Christian, "Canada should start its own space force," *Futurism*, Janary 6, 2019, accessed February 5, 2020, https://futurism.com/the-byte/former-general-canada-should-start-space-force

73. Laura Winter, "NATO Declares Space an 'Operational Domain'," Al Jazeera, December 4, 2019, accessed February 5, 2020, https://www.aljazeera.com/ajimpact/nato-declares-space-operational-domain-191204155031893.html

74. Paul A. Sabatier and Christopher M. Weible, eds. *Theories of the Policy Process* (Boulder: Westview Press, 2019).

75. James L. True, Bryan D. Jones, and Frank R. Baumgartner, "Punctuated equilibrium theory," in, *Theories of the Policy Process* Paul A. Sabatier and Christopher M. Weible, ed. (Boulder: Westview Press, 2019), 55–101.

76. Frances Stokes Berry and William D. Berry, "Innovation and Diffusion Models in Policy Research," in *Theories of the Policy Process* Paul A. Sabatier and Christopher M. Weible, ed. (Boulder: Westview Press, 2019), 253–97.

77. Paul A. Sabatier and Christopher M. Weible, "The Advocacy Coalition Framework," in *Theories of the Policy Process* Paul A. Sabatier and Christopher M. Weible, ed. (Boulder: Westview Press, 2019), 135–71; Mark K. McBeth, Michael D. Jones, and Elizabeth A. Shanahan, "The Narrative Policy Framework," in *Theories of the Policy Process* Paul A. Sabatier and Christopher M. Weible, ed. (Boulder: Westview Press, 2019), 173–213.

78. McBeth, Jones, and Shanahan, "The Narrative Policy Framework."

79. McBeth, Jones, and Shanahan, "The narrative policy framework".

80. Linda Billings, "Humans to Mars: A Deeply Disturbing Idea," *blog*, September 28, 2016, accessed February 10, 2020, https://doctorlinda.wordpress.com/2016/09/28/humans-to-mars-a-deeply-disturbing-idea/; Linda Billings, "Should Humans Colonize Other Planets? No.," *Theology and Science* 15, no. 3 (2017): 321–32. https://www.tandfonline.com/doi/full/10.1080/14746700.2017.1335065; Marcie Bianco, "The Patriarchal Race to Colonize Mars is Just Another Example of Male Entitlement" *NBC News*, February 21, 2018, accessed February 10, 2019, https://www.nbcnews.com/think/opinion/patriarchal-race-colonize-mars-just-another-example-male-entitlement-ncna849681; Ethan Baron, "Elon Musk Slammed by Stanford Gender Researcher for Space Patriarchy," *The Mercury News*, February 21, 2018, accessed February 10, 2020, https://www.mercurynews.com/2018/02/21/elon-musk-slammed-by-stanford-gender-researcher-for-space-patriarchy/; Ryan F. Mandelbaum, "Decolonizing Mars: Are We Thinking About Space Exploration All Wrong?," *Gizmodo*, November 20, 2018, accessed February 10, 2020, https://gizmodo.com/decolonizing-mars-are-we-thinking-about-space-explorat-1830348568; Martin Elvis and Tony Milligan, "How much of the Solar System Should We Leave as Wilderness?," *Acta Astronautica* 162 (2019): 574–80.; Paris Marx, "We Don't Need Space Colonies, and We Definitely Don't Need Jeff Bezos," *Jacobin*, July 21, 2019, accessed January 15, 2020. https://jacobinmag.com/2019/07/space-colonies-jeff-bezos-blue-origin; Daniel Deudney, *Dark Skies: Space Expansionism, Planetary Geopolitics, and the Ends of Humanity* (New York: Oxford University Press, 2020).

81. Johan P. Olsen and James G. March, "The Logic of Appropriateness," in Robert E. Goodin, Michael Moran, and Martin Rein, eds., *The Oxford Handbook of Political Science* (Oxford: Oxford University Press, 2004); Mark J. Weber, Shirli Kopelman, and David M. Messick, "A Conceptual Review of Decision Making in Social Dilemmas: Applying a Logic of Appropriateness," *Personality and Social Psychology Review* 8, no. 3 (2004): 281–307; Harald Müller, "Arguing, Bargaining and All That: Communicative Action, Rationalist Theory and the Logic of Appropriateness in International Relations," *European Journal of International Relations* 10, no. 3 (2004): 395–435.

82. Carlos A. Martínez-Vela, "World Systems Theory," *Engineering System Division* 83 (2001): 1–5.

83. Philipp Mayring, "Qualitative content analysis," *A Companion to Qualitative Research* 1 (2004): 159–176.

84. Anthony Downs, "Up and Down with Ecology-the Issue-Attention Cycle," *Public Interest* 28 (Summer, 1972): 48.

85. For example, uncertainty avoidance, power distance, and masculinity were identified in Geert Hofstede, *Culture's Consequences: International Differences in Work-Related Values*. Vol. 5. (London and Beverly Hills: Sage, 1984); Alternately: openness, conscientiousness, extraversion, agreeableness, neuroticism is explored by: Robert R. McCrae and Oliver P. John, "An Introduction to The Five-Factor Model and its Applications," *Journal of Personality* 60, no. 2 (1992): 175–215.

86. For a comparison, see "Qualitative Data Analysis: Choosing QDA Software," *NYU*, accessed February 8, 2020, https://guides.nyu.edu/QDA/comparison

87. Stanley Wasserman and Katherine Faust, *Social Network Analysis: Methods and Applications*. Vol. 8 (Cambridge: Cambridge University Press, 1994); Peter J. Carrington, John Scott, and Stanley Wasserman, eds. *Models and Methods in Social Network Analysis*. Vol. 28 (Cambridge: Cambridge University Press, 2005).

88. Ina Möller, *The Emergent Politics of Geoengineering* [Doctoral Thesis] (Lund: Lund University, 2019). https://portal.research.lu.se/portal/files/63030851/Thesis_InaM_ller.pdf

89. E. Schaler, "An Assessment of The Institutional Analysis and Development Framework and Introduction of The Social-Ecological Systems Framework, in *Theories of the Policy Process*: 215-252

90. Athina Balta, "Space Property Rights: The Winds of Change," presentation to UN Office of Outer Space Affairs High Level Forum, Dubai, UAE 2016, accessed February 8, 2020, https://www.unoosa.org/documents/pdf/hlf/1st_hlf_Dubai/Presentations/101.pdf

91. Peter Garretson, Alfred B. Anzaldúa, and Hoyt Davidson, "Catalyzing Space Debris Removal, Salvage, and Use," *The Space Review*, December 9, 2019, accessed February 8, 2020, https://www.thespacereview.com/article/3847/1

92. For example, Kaveh Madani, "Game Theory and Water Resource," *Journal of Hydrology* 381, no. 3–4 (2010): 225–38. https://ethz.ch/content/dam/ethz/special-interest/mtec/necom-dam/documents/Applied%20Negotiation%20Seminar/10_Madani_Game_theory_and_water_resources.pdf

93. Amos Tversky and Daniel Kahneman, "Advances in Prospect Theory: Cumulative Representation of Uncertainty," *Journal of Risk and Uncertainty* 5, no. 4 (1992): 297–323.

94. Peter P. Perla, *The Art of Wargaming: A Guide for Professionals and Hobbyists* (Annapolis: Naval Institute Press, 1990); James F. Dunnigan, *The Complete Wargames Handbook: How to Play, Design, and Find Them* (New York: Morrow, 1992); Philip Sabin, *Simulating War: Studying Conflict Through Simulation Games* (New York and London: A&C Black, 2012); Matthew B. Caffrey Jr., "On Wargaming," *The Newport Papers* 43, Newport, Rhode Island, 2019, https://digital-commons.usnwc.edu/newport-papers/43

Appendix B

The Expected Path and Timeline for Space Missions

2020:

- China: First Mars probe (China White Paper 2016).
- China: First un-crewed test of lunar crew and service module
- China: Finish the industrial-level testing of in-orbit construction and wireless transmission precursors for space solar power (CAST).
- United States: First orbital flight of SpaceX Starship (~100MT payload).
- United States: U.S. Space Force established.
- Chang'e 5 sample return mission.
- India: Aditya-L1

2021:

- United States: Two commercial lunar payload system (CLPS) landers land on the Moon at Oceanus Procellarium (intuitive machines) and Lacus Mortis (Astrobotic).
- United States: First commercial flight of SpaceX Starship (fully reusable heavy launcher).
- United States: First launch of Blue Origin's New Glenn heavy lifter.
- China's ispace launches first private reusable spacecraft.

2022:

- India: Chandrayaan-3
- India: Gaganyaan (manned spaceflight)
- India: Mangalyaan-2 (MOM-2)
- China: Launches probe to asteroid(s) (CNSA, CAST, CALT).
- China: Permanent space station, *Tianhe-1* (China space policy papers).
- India: First manned orbital mission (Gaganyaan).
- India: Launches Venus probe.

- United States: First planned SpaceX Mars Cargo Mission to confirm water, resources and emplace infrastructure for power, mining, and life support.

2023:

- United States: First SpaceX private mission around the Moon (Yukazu Maezawa with 6–8 artists).

2024:

- China: Chang'e 6 to bring back samples from the South Pole (CNSA).
- India: Chandrayaan 3 explores the lunar South Pole.
- United States: NASA puts "first woman and next man" on the Moon (NASA).
- United States: SpaceX second planned mission to Mars, Cargo, and Crew to set up propellant production plant to create methane and liquid oxygen; humans would be tasked to collect 1 tonne of ice for every day to fuel the plant.

2025:

- China: Reusable suborbital carrier/suborbital tourism (CASC).
- China: Complete the first 100kW SBSP demonstration at LEO (CAST).
- India: Begins constructing its space station (estimate).
- US: SpaceX elements for a sustained Martian Colony in place.
- Global:
  - Global Space Product reaches $500B/year;
  - Population in orbit 20;
  - New activities include:
    - first commercial space habitats,
    - commercial research in orbit,
    - in-space manufacturing in low Earth orbit,
    - upper-stage reuse, and
    - commercial prospecting missions to the Moon and asteroids.

2026:

- United States: SpaceX sends more ships to Mars.

2027:

- India: Completes manned space station.

2028:

- China: Mars sample return (CNSA).
- United States: NASA goal of a permanent presence on the Moon.

2029:

- China: Jupiter probe (CNSA).
- China: Mission to capture a near-Earth asteroid (NEA; CAS).

2030:

- China: Probe to lunar South and North Pole (CNSA).
- China: Launch of *Long March 9* super heavy lifter (CASC).

- China: Megawatt-class space solar power demonstration in GEO (CAST).
- China: Approximate year Chinese economy expected to pass the United States in real GDP.

2031:

- Global:
  - Global space product reaches $900Bn,
  - Population in orbit 300,
  - New activities include:
    - low earth orbit tourism,
    - on-orbit propellant storage creates a reusable transportation system,
    - mining and manufacturing equipment landed on the Moon,
    - lunar propellant mining (lowers the cost of space transportation by a factor of 1,000).

2034:

- China: Return asteroid to Earth (CAS).

2035:

- China: Test key technologies such as 3D printing to lay groundwork for the construction of a lunar base (CAS).
- China: Completes 100 Megawatt space solar power prototype plant with electric-power-generating capacity (CAST).

2036:

- China: First Chinese settler lands on Moon and establishment of lunar research base (CNSA).

2037

- First expected commercial asteroid mining.

2040:

- China: Nuclear-powered space fleet to be ready for large-scale space mining and colonization (CASC).
- China: Fully reusable carrier rocket fleet.
- China: SBSP to orbit (36,000 kms) above Earth and start generating power (CAST).
- Global space economy reaches $1.1Trillion.

2045:

- China: Most advanced Space Nation (President Xi).
- Global:
  - Gross Space Product $2.7 Trillion;
  - On-orbit population reaches 1000;
  - New activities:
    - lunar regolith mining for structural materials,
    - large scale in-space manufacturing at EML1,
    - commercial asteroid mining,

- finished products transported to Earth orbits including the construction of solar power satellites.

2047:

- Global space economy reaches 2.7 Trillion.

2049:

- China: 100th-year celebration of the establishment of the PRC and the leading space technology power (President Xi).

2050:

- China: First commercial-level SBSP in operation in GEO (CAST).
- United States: SpaceX goal for a city on Mars is self-sustaining

2070

- United States: Population of SpaceX Martian colony reaches 1 million.

2117:

- UAE: Mars City established.

Appendix C

Timeline of Important Space Development Dates

1925 Tsiolkovsky conceptualizes solar power satellites
1941 Asimov uses solar power satellites in a story
1957 Sputnik 1
1958 Cox & Stoiko publish Spacepower: What it means to you
1958 Eisenhower signs National Aeronautics and Space Act of 1958
1959 General Schriever testifies all three services should consider a Moon Base
1959 U.S. Army completed *Project Horizon* Moon Base study
1960 Ralph Cordiner lectures on space development
1961 Kennedy proposes Moonshot goal to congress
1962 Kennedy announces NASA to go to Moon at Rice University
1963 Limited Nuclear Test Ban Treaty (effectively shuts down Projects Orion and NERVA)
1967 Outer Space Treaty
1968 Glaser invents solar power satellites
1968 U.S. National Committee on Space "human settlement … from the highlands of the moon to plains of Mars"
1969 Apollo 11 lands on the Moon, United States wins Moon race
1969 Space Task Group chaired by VP Agnew
1973 Glaser patents solar power satellites
1974 O'Neill's colonization of space article in Physics Today
1975 L5 Society founded
1975 NASA Ames Space Settlements Design Study
1977 O'Neill's *High Frontier* published
1977 Herman Kahn's Long-Term Prospects for Developments in Space
1977 Space Studies Institute founded
1977 NASA and DOE begin solar power studies (until 1981)

1977 H.R. 451 Pro-High Frontier Resolution
1977 Senator Proxmire "not a penny for this nutty fantasy"
1979 Moon Treaty (failed treaty)
1980 Planetary Society founded
1980 SEDS founded
1981 NASA completes and NRC reviews solar power satellite research
1981 Rep Gingrich sponsors H.R. 4286 National Space and Aeronautics Policy Act & "Northwest Ordinance for Space"
1982 Graham publishes *High Frontier: A New National Strategy*
1982 L5 defeats Moon Treaty in the United States
1987 NSS founded by merger of L5 & NSI
1987 ISU founded
1988 Rep Brown introduces H.R. 4218 "Space Settlement Act of 1988," became law until 1995
1995 Nansen publishes *Sun Power*
1996 Ansari X-prize begins (ends 2004)
1996 Lewis publishes *Mining the Sky*
1998 Mars Society Founded
2000 Moon Society founded
2002 Dolman publishes *Astropolitik*
2002 Worden and Shaw publish *Whither Space Power?*
2002 Ouyang Ziyuan "Whoever first conquers the Moon will benefit first"
2002 AIAA Space Colonization Technical Committee Founded
2004 Ansari X-Prize won
2004 President G.W. Bush announces Vision for Space Exploration (VSE)
2004 Marburger "incorporate the solar system into our economic sphere" speech
2004 Wingo publishes *MoonRush*
2005 NASA Admin Griffin "goal … also … extending the rand of human habitat"
2006 Marburger "Exploration … is not … Occupation by many is grander."
2006 India's Dr. Kakodkar (AEC) on Helium-3
2007 India's President Dr. APJ Kalam Space Vision talk in Boston
2007 Google Lunar X-Prize begins
2007 Pentagon publishes *Space Solar Power: An Opportunity for Strategic Security*
2008 Obama Space Solar Transition Team white paper
2008 India ISRO Chairman Nair on He-3, water for lunar colony and hedging its claim
2008 India's Chandrayaan-1 reaches lunar orbit
2009 Space Development Steering Committee founded

2009 Ralph Nansen publishes *Energy Crisis Solution from Space*
2010 Space Solar is #1 Open Gov idea at OSTP, DOE, NASA
2010 CAST releases Space Solar Power Roadmap
2010 TeamIndus formed for Google Lunar X-prize, beginning of India's NewSpace sector
2011 Zubrin updates *The Case for Mars*
2012 Earthlight Foundation founded
2012 China proposes to India's Dr. APJ Kalam cooperation on Space Solar Power
2012 Keck Institute Asteroid Mining Study
2012 Planetary Resources (PRI) founded
2012 SpaceX Dragon Spacecraft is the first commercial spacecraft to dock with the ISS
2013 Deep Space Industries founded
2013 Dr. APJ Kalam "Space Solar Power—Key to a Livable Planet" & "World Space Vision"
2013 Chang'e 3 lands on the Moon
2013 India's N.R. Sonkavday publishes *Space Based Solar Power*
2014 Mankins publishes *The Case for Space Solar Power*
2014 White House "Bootstrapping a Solar system civilization" blog post
2014 H.R. 5063 ASTEROIDS Act introduced (did not pass)
2015 Obama SOU "Not just to visit but to stay"
2015 Pioneering Space National Summit "long term goal … enable human settlement"
2015 Alliance for Space Development (ASD) founded
2015 Lewis publishes *Asteroid Mining 101*
2015 Obama signs U.S. Commercial Space Launch Competitiveness Act (CSLCA)
2015 International Institute of Space Law pro-space mining America
2015 Ziarnick publishes *Developing National Power in Space*
2015 TransAstra founded
2015 First private suborbital vertical launch and landing of Blue Origin's New Shepard
2015 First successful Falcon 9 landing after orbital launch
2015 H.R.4752 The Space Exploration, Development, and Settlement Act of 2016 (did not pass)
2015 China establishes its Strategic Support Force (PLASSF)
2016 Gen Zhang Yulin "Industrialize the moon to build solar power satellites"
2016 Elon Musk presents "Making Humans a Multiplanetary Species"
2016 Paul Spudis publishes *The Value of the Moon*
2016 ULA releases *cislunar 1000 vision*

2016 Space Solar Power wins D3 competition
2016 Rep Bridenstine introduces H.R.4945—American Space Renaissance Act (did not pass)
2016 Rep Bridenstine speech "alter the economic and geopolitical balance of power on Earth."
2016 CNAS publishes *Blue To Black*
2016 CNAS publishes *A Space Policy for the Trump Administration*
2017 Elon Must presents "Making Life Multiplanetary"
2017 India's Dr. Kasturirangan "humans definitely need another habitation beyond Earth."
2017 CASC Roadmap (full RLV fleet, Nuclear Shuttles for large-scale mining)
2017 India's Dr. Pillai on ISRO's plans to mine the Moon for He-3
2017 Ye Peijian "the moon is the Diaoyu Islands, Mars is Huangyan Island."
2017 Luxembourg passes space mining law
2017 Air University publishes *FAST Space Study*
2017 Sen Cruz hearings on the OST
2017 NASA Ames *Stepping Stones* (public-private partnership asteroid study) released
2017 National Space Council revived
2017 H.R. 2807 American Space Commerce Free Enterprise Act introduced
2017 SPD-1 "lead … exploration with commercia … to enable human expansion across solar system"
2017 Scott Pace "private sector must have confidence … develop and utilize in-situ resources"
2017 National Security Strategy "consider extending national security protections … to private"
2018 Aerospace publishes *Cislunar Development*
2018 Wilbur Ross "turn moon into a gas station"
2018 new National Space Strategy, first mention of Space Force idea
2018 Ross "for colonization or mining or anything else"
2018 SPD-2 (streamline regulations)
2018 SPD-3 Space Traffic Management, Trump orders immediately begin Space Force
2018 Google Lunar X-Prize ends with no winner
2018 Davenport publishes *The Space Barons*
2018 Gingrich circulates "Space Force Planning"
2018 Space Horizons Papers (Movement and Maneuver, Space Force draft legislation)
2018 Gen Kwast Politico piece links grand vision with military power
2018 \$3.9Bn in private space investment in one year; \$19Bn in private funding last ten years

2019 Chang'e 4 lands on the far side of the moon
2019 India establishes its Defense Space Agency
2019 Zubrin Publishes *The Case for Space*
2019 NASA Ames *Evolvable Lunar Architecture* (public–private partnership study)
2019 SPD-4 (Space Force)
2019 Space Development Agency "As commercial activities (such as resource extraction) expand outward from Earth to the Moon ... a need for the equivalent to a Navy or Air Force"
2019 Ross at Davos ($1.1–3T space economy)
2019 National Space Council, VP Pence directs Moon by 2024 South Pole
2019 NASA selects commercial lunar lander providers for CLPS program
2019 NASA Bridenstine frames lunar program as industrialization
2019 Air Force Space Command releases *The Future of Space 2060 and Implications for U.S. Strategy*
2019 AFRL-DIU report on PRC predation on U.S. space industrial base
2019 Jeff Bezos presents "Going to Space to Benefit Earth"
2019 USCC report calls for strategy on space resources, cislunar
2019 Pyle publishes *Space 2.0*
2019 Bluewater school pushes for a Space Force to protect commerce
2019 Sadat "the great game for cislunar leadership"
2019 SpaceX orbits 300 Starlink Satellites, requests 30,000 more, shows Starship
2019 AFRL SSPDR power beaming
2019 Wayne White circulates "Space Pioneer Act"
2019 DIU solicitation for "outpost"
2019 Trump signs U.S. Space Force into Law
2019 NATO declares space an "Operational Domain"
2020 Air Force Small Business Research call for cislunar solutions
2020 Trump adds to space council: DOE, Economic Policy, Domestic Policy
2020 DIU call for multi-orbit l
2020 White House Releases "A New Era in Deep Space Exploration and Development
2020 White House Executive order on Encouraging Space Resources
2020 NASA releases Artemis Accord
2020 China launches to Mars
2020 UAE Launches to Mars
2020 USSF, DIU, AFRL releases 'State of the Space Industrial Base 2020'

Bibliography

Abrahamian, Atossa Araxia. "How a Tax Haven is Leading the Race to Privatise Space," *The Guardian*, September 15, 2017. https://www.theguardian.com/news/2017/sep/15/luxembourg-tax-haven-privatise-space

Acuthan, Jayan Panthamakkada. "China's Outer Space Program: Diplomacy of Competition or Cooperation?," *China's Perspectives*, 2006. Adopted by consensus by the Board of Director. December 20, 2015. http://iislwebo.wwwnlss1.a2hosted.com/wp-content/uploads/2015/12/SpaceResourceMining.pdf

Agarwal, Prateek. "Production Possibilities Frontier," *Intelligent Economist*, December 25, 2017. https://www.intelligenteconomist.com/production-possibilities-frontier/

Agnikul Cosmos Website. http://www.agnikul.in/#

Agreement Governing the Activities of States on the Moon and Other Celestial Bodies, UNOOSA, December 5, 1979. https://www.unoosa.org/pdf/gares/ARES_34_68E.pdf

Air Force Space Command. "The Future of Space 2060 and Implications for U.S. Strategy: Report on the Space Futures Workshop," September 5, 2019. https://www.afspc.af.mil/Portals/3/documents/Future%20of%20Space%202060%20v2%20(5%20Sep).pdf?ver=2019-09-06-184933-230 with an graphically updated version released on 3 October 2019 at: https://www.afspc.af.mil/Portals/3/The%20Future%20of%20Space%202060%20-%203Oct19.pdf

Air Force, Department of Defense. "Fiscal Year (FY) 2019," Budget Estimates for Research, Development, Test & Evaluation, *Air Force Vol-II*, February 2018. https://www.saffm.hq.af.mil/Portals/84/documents/FY19/RDT_E/Air%20Force%20Research%20Development%20Test%20and%20Evaluation%20Vol%20II%20FY19%20v2.pdf?ver=2018-02-20-125551-480

Air Force, Department of Defense. "Fiscal Year (FY) 2020," Budget Estimates for Research, Development, Test & Evaluation, *Air Force Vol-II*, March 2019. https://www.saffm.hq.af.mil/Portals/84/documents/FY20/RDTE/FY20_PB_RDTE_Vol-II.PDF?ver=2019-03-18-153506-683

Alexander VI, Pope. "Inter Caetera Division of the Undiscovered World Between Spain and Portugal," 1493, *Papal Encyclicals Online*. https://www.papalencyclicals.net/alex06/alex06inter.htm

Aliberti, Marco. *When China Goes to the Moon*. Switzerland: Springer, 2015.

Alliance for Space Development, "Alliance Objectives for 2019," n.d. http://allianceforspacedevelopment.org/about-us/

Alliance for Space Development. "March Storm 2020 Congressional Visits," n.d. http://allianceforspacedevelopment.org/blitzes/

Anderson, Scot W., Korey Christensen and Julia LaManna. "The Development of Natural Resources in Outer Space," *Journal of Energy & Natural Resources Law*, 2018. https://www.hoganlovells.com/~/media/hogan-lovells/pdf/2018/the_development_of_natural_resouces_in_outer_space_august_2018.pdf

Arc Parabolic. "OrbitBeyond Teams TeamIndus Honeybee Robotics NASA Lunar Program," November 29, 2018. http://www.parabolicarc.com/2018/11/29/orbitbeyond-teams-team-indus-honeybee-robotics-nasa-lunar-program/

ARDL Website. no date (n.d). https://www.ardl.co.in/

Ashok, G. V. "The Space Activities Bill- Does it deliver?," *ORF Online*, April 19, 2019. https://www.orfonline.org/expert-speak/the-space-activities-bill-does-it-deliver-50019/

Asterank. n.d. http://www.asterank.com/

Astrobotic Website. n.d. https://www.astrobotic.com/

Astrome Website. n.d. http://www.astrome.co/

Astropreneurs. "New Space in India," August 19, 2018. http://astropreneurs.space/2018/07/19/new-space-in-india/

Atkinson, Nancy. "How Many People Have Walked on the Moon?," *Universe Today*, August 1, 2013. https://www.universetoday.com/55512/how-many-people-have-walked-on-the-moon/

Atlantic Council. "One Hundred Years of American Grand Strategy," n.d. https://www.atlanticcouncil.org/blogs/new-atlanticist/one-hundred-years-of-american-grand-strategy

Autry, Greg. "Commercial Orbital Transportation Services: A Case Study for National Industrial Policy," *NewSpace*, September 2018. https://www.liebertpub.com/doi/10.1089/space.2017.0043

Autry, Greg. "Robert Zubrin Makes a Strong Case for Space Development," *Forbes*, May 16, 2019. https://www.forbes.com/sites/gregautry/2019/05/16/robert-zubrin-makes-a-strong-case-for-space-development/

Avey, Paul C., Jonathan N. Markowitz and Robert J. Reardon. "Disentangling Grand Strategy: International Relations Theory and U.S. Grand Strategy," *Texas National Security Review*, November 21, 2018. https://tnsr.org/2018/11/disentangling-grand-strategy-international-relations-theory-and-u-s-grand-strategy/

Aviation Quotations. "Great Aviation Quotes: Predictions." http://www.aviationquotations.com/predictions.html

Axe, David. "Wilbur Ross: We've Gotta Get Back to the Moon—to Exploit the Hell Out of It," *Daily Beast*, April 9, 2019. https://www.thedailybeast.com/wilbur-ross-weve-gotta-get-back-to-the-moonto-exploit-the-hell-out-of-it

Badescu, Viorel, ed. *Asteroids: Prospective Energy and Material Resources*. New York: Springer Science & Business Media, 2013.

Baele, Stéphane J. and Catarina P. Thomson. "An Experimental Agenda for Securitization Theory," *International Studies Review* 19, no.4 (December 2017): 646–66.

Bagla, Pallava. "India Plans Tricky and Unprecedented Landing Near Moon's South Pole," *Science Magazine*, January 31, 2018. https://www.sciencemag.org/news/2018/01/india-plans-tricky-and-unprecedented-landing-near-moon-s-south-pole

Bagla, Pallava. "India Planning to Launch Own Space Station By 2030, Says ISRO Chief K Sivan," *NDTV*, June 13, 2019. https://www.ndtv.com/india-news/indian-space-station-india-planning-to-launch-own-space-station-says-isro-chief-2052686

Bagla, Pallava. "India's Eyes a Return to Mars and a First Run at Venus," *Science*, February 17, 2017. http://www.sciencemag.org/news/2017/02/india-eyes-return-mars-and-first-run-venus

Bajoria, Jayshree. "RAW: India's External Intelligence Agency," *CFR*, November 7, 2008. https://www.cfr.org/backgrounder/raw-indias-external-intelligence-agency

Bajpai, Kanti. "Indian Conceptions of Order and Justice: Nehruvian, Gandhian, Hindutva, Neo-Liberal," in *Order and Justice in International Relations*. Edited by Rosemary Foot, John Gaddis and Andrew Hurrell, 236–61. Oxford: Oxford University Press, 2003.

Balta, Athina. "Space Property Rights: The Winds of Change," presentation to UN Office of Outer Space Affairs High Level Forum, Dubai, UAE, 2016. https://www.unoosa.org/documents/pdf/hlf/1st_hlf_Dubai/Presentations/101.pdf

Barbosa, Rui C. "China Returns Long March 4 to Service with Ziyuan Launch," *NASA Spaceflight.com*, September 12, 2019. https://www.nasaspaceflight.com/2019/09/china-returns-long-march-4-ziyuan-2d/

Baron, Ethan. "Elon Musk slammed by Stanford Gender Researcher for Space Patriarchy," *The Mercury News*, February 21, 2018. https://www.mercurynews.com/2018/02/21/elon-musk-slammed-by-stanford-gender-researcher-for-space-patriarchy/

Barrabi, Thomas. "'New Space Race' Could Yield $1 Trillion Industry: Wilbur Ross," *Fox Business News*, November 29, 2018. https://www.foxbusiness.com/business-leaders/new-space-race-could-yield-1-trillion-industry-wilbur-ross

Barron Hilton. "Rocket Societies," *Pioneers of Flight*, July 2, 2014. https://pioneersofflight.si.edu/culture/rocket_societies

Barry, Ellen. "India Launches 104 Satellites from a Single Rocket, Ramping Up a Space Race," *The New York Times*, February 15, 2017. https://www.nytimes.com/2017/02/15/world/asia/india-satellites-rocket.html

Bazzi, Samuel, Martin Fiszbein, Mesay Gebresilasse. "Frontier Culture: The Roots and Persistence of 'Rugged Individualism' in the United States," *National Bureau of Economic Research* No. w23997, 2017. https://www.bu.edu/econ/files/2017/03/Frontier-Culture-The-Roots-and-Persistence-of-%E2%80%9CRugged-Individualism%E2%80%9D-in-the-United-States-2.pdf

Beck, Robert. "China's Mystery Satellite Could be a Dangerous New Weapon," *Medium*, August 22, 2013. https://medium.com/war-is-boring/chinas-mystery-satellite-could-be-a-dangerous-new-weapon-630a858923ec; Defense Intelligence Agency

Becker, Jeffrey. "Militarizing Musk," War on the Rocks, May 17, 2016. https://warontherocks.com/2016/05/militarizing-musk/

Beckley, Michael. "Economic Development and Military Effectiveness," *The Journal of Strategic Studies* 33, no. 1 (2010): 43–79.

Bellatrix Aerospace Website. n.d. http://www.bellatrixaerospace.com/

Bender, Bryan. "Newt Gingrich trying to sell Trump on a cheap moon plan: A general, Gingrich and Michael Jackson's publicist are proposing a $2 billion contest to return Americans to the moon," *Politico*, August 19, 2019. https://www.politico.com/story/2019/08/19/newt-gingrich-michael-jackson-moon-1466853

Bennett, Jay. "One Chart Shows How Much SpaceX Has Come to Dominate Rocket Launches," *Popular Mechanics*, July 13, 2017. https://www.popularmechanics.com/space/rockets/a27290/one-chart-spacex-dominate-rocket-launches/

Berger, Eric. "NASA is Working to Bring Down the Cost of a Single SLS Launch," *ArsTechnica*, November 8, 2019. https://arstechnica.com/science/2019/11/nasa-does-not-deny-the-over-2-billion-cost-of-a-single-sls-launch/

Berger, Eric. "How Can NASA Return to the Moon? By Making Everything Reusable, Chief Says," *ArsTechnica*, August 2, 2018. https://arstechnica.com/science/2018/08/nasa-chief-seeks-a-sustainable-reusable-pathway-for-lunar-return/

Bernhardt, George, Sandra M. Gresko and Thomas R. Merry. "Star Wars versus Star Laws: Does SDI Conform to Outer Space Law; The Reagan Legacy and the Strategic Defense Initiative," *Journal of Legislation* 15 no. 2 Article 14, May 1, 1989. https://scholarship.law.nd.edu/cgi/viewcontent.cgi?article=1319&context=jleg

Berry, Frances Stokes and William D. Berry. "Innovation and Diffusion Models in Policy Research," in *Theories of the Policy Process*. Edited by Paul A. Sabatier and Christopher M. Weible, 253–97. Boulder: Westview Press, 2019.

Bezos, Jeff. "Going to Space to Benefit Earth," *YouTube*, May 9, 2019. https://www.youtube.com/watch?v=GQ98hGUe6FM

Bezos, Jeff. "Jeff Bezos discusses space flight and his vision for Blue Origin." (at 12 min 01 seconds into talk), filmed October 23, 2016 at the 2016 Pathfinder Awards at Seattle's Museum of Flight. *YouTube video*. posted October 23, 2016, 13:14. https://www.youtube.com/watch?v=VNwE3sRWxHw#action=share

Bezos, Jeff. "Jeff Bezos discusses space flight and his vision for Blue Origin" (at 12 min 01 seconds into talk), filmed October 23, 2016 at the 2016 Pathfinder Awards at Seattle's Museum of Flight," YouTube video, posted October 23, 2016, 13:14, https://www.youtube.com/watch?v=VNwE3sRWxHw#action=share Power.

Bhaskar, Utpal. "ISRO Plans to Mine Energy from Moon by 2030 to Help Meet India needs," *LiveMint*, April 21, 2017. https://www.livemint.com/Science/W5WjJCdqqxXYpHvrB2TTHP/Isro-plans-to-mine-energy-from-Moon-by-2030-to-help-meet-Ind.html

Bianco, Marcie. "The Patriarchal Race to Colonize Mars is Just Another Example of Male Entitlement" *NBC News*, February 21, 2018. https://www.nbcnews.com/think/opinion/patriarchal-race-colonize-mars-just-another-example-male-entitlement-ncna849681

Bidkar, Rugved. "Space Based Solar Power (SBSP): An Emerging Technology," December 8, 2012. https://ieeexplore.ieee.org/abstract/document/6450440/authors#authors

Biesecker, Cal. "Commercial Space Bill Reintroduced in House," *Defense Daily*, July 3, 2019. https://www.defensedaily.com/commercial-space-bill-reintroduced-house/congress/

Billings, Linda. "Humans to Mars: A Deeply Disturbing Idea," *blog*, September 28, 2016. https://doctorlinda.wordpress.com/2016/09/28/humans-to-mars-a-deeply-disturbing-idea/

Billings, Linda. "Should Humans Colonize Other Planets? No," *Theology and Science* 15, no. 3 (2017): 321–32. https://www.tandfonline.com/doi/full/10.1080/14746700.2017.1335065

BillionYearPlan Blog. "Citizens for Space Based Solar Power Open Letter to OSTP," April 10, 2010. https://billionyearplan.blogspot.com/search?q=open energy

Bisht, Medha. *Kautilya's Arthashastra: Philosophy of Strategy*. New York and London: Routledge, 2020.

Blandford, Ann. "Semi Structured Qualitative Studies," in *The Encyclopedia of Human-Computer Interaction*. Edited by Mads Soegaard and Rikke Friis Dam. Aarhus, Denmark: The Interaction Design Foundation, 2013. http://www.interactiondesign.org/encyclopedia/semi-structured_qualitative_studies.html

Blank, Jonah. "Pakistan and China's Almost Alliance," *Foreign Affairs*, October 16, 2015. https://www.rand.org/blog/2015/10/pakistan-and-chinas-almost-alliance.html

Blenkin, Max. "ADF Urged to Establish Aussie Space Command," January 30, 2020. https://www.spaceconnectonline.com.au/operations/4107-calls-for-adf-to-establish-space-command-to-manage-aussie-space-operations

Bloom, Howard. "China's New Silk Road vs. America's Platinum Highway in the Sky," *Over the Horizon Journal*, September 24, 2018. https://othjournal.com/2018/09/24/chinas-new-silk-road-vs-americas-platinum-highway-in-the-sky/

Blue Origin, Blue Moon. https://www.blueorigin.com/blue-moon

Blue Origin. "Our Vision," https://www.blueorigin.com/

Blue Sky Analytics Website. https://blueskyhq.in/about

Boardman, Mark. "Go West, Young Man? Did Anybody Actually Say that Famous Phrase?," *True West Magazine*, June 5, 2015. https://truewestmagazine.com/go-west-young-man/

Boot, Max. The *Savage Wars of Peace: Small Wars and the Rise of American Power*. New York: Basic Books, 2002.

Bormann, Natalie and Michael Sheehan. *Securing Outer Space: International Relations Theory and the Politics of Space*. London and New York: Routledge, 2012.

Bowe, Alexander. "China's Pursuit of Space Power Status and Implications for the United States," U.S-China Economic and Security Review Commission, April 11, 2019. https://www.uscc.gov/sites/default/files/Research/USCC_China%27s%20Space%20Power%20Goals.pdf

Boyd, John R. *A Discourse on Winning and Losing*. Maxwell AFB, Alabama: Air University Press, 2018.

Boyle, Alan. "Jeff Bezos: 'We Will Have to Leave this Planet…and it's Going to Make this Planet Better'," *GeekWire*, May 29, 2018. https://www.geekwire.com/2018/jeff-bezos-isdc-space-vision/

Bradford Kristi J. "A Model for Space Sector Growth: A Luxembourg Case Study," *The Aerospace Corporation*, October, 2018. https://aerospace.org/sites/default/files/2018-10/Bradford_Luxembourg_10292018.pdf

Brady, Anne Marie. *China as a Polar Great Power*. Cambridge: Cambridge University Press, 2017.

Brands, Hal. *American Grand Strategy and the Liberal Order Continuity, Change, and Options for the Future*. Santa Monica: RAND, 2016. https://www.rand.org/content/dam/rand/pubs/perspectives/PE200/PE209/RAND_PE209.pdf;

Brennan, Louis. "How Luxembourg is Positioning Itself to be the Centre of Space Business," *The Conversation*, July 16, 2019. https://theconversation.com/how-luxembourg-is-positioning-itself-to-be-the-centre-of-space-business-120436

Brid-Aine, Parnell. "NASA will Reach Unique Metal Asteroid Worth $10,000 Quadrillion Four Years Early," *Forbes*, May 26, 2017. https://www.forbes.com/sites/bridaineparnell/2017/05/26/nasa-psyche-mission-fast-tracked/#506d52db4ae8

Bridenstine, Jim. "This is Our Sputnik Moment" (speech to the Lunar Exploration Group), Bridenstine.house.gov, November 6, 2016. https://bridenstine.house.gov/blog/?postid=758 archived here: https://web.archive.org/web/20171126063553/https://bridenstine.house.gov/blog/?postid=758 also available at OKG News, November 6, 2016. https://okgrassroots.com/?p=642815

Brinker, Scott. "One Thing Everybody Forgets About Gartner's Hype Cycle: The Gaps Between Hype and Reality Are Opportunities," *Think Growth*, April 2, 2018. https://thinkgrowth.org/one-thing-everybody-forgets-about-gartners-hype-cycle-ecfe7e9de8ff

Britton-Purdy, Jedediah, "Infinite Frontier: The Eternal Return of American Expansionism," *The Nation*, April 1, 2019. https://www.thenation.com/article/archive/greg-grandin-end-of-the-myth-frontier-border-wall-book-review/

Brock, Bernard L., et al., *Making Sense of Political Ideology the Power of Language in Democracy*. Lanham: Rowman & Littlefield, 2005.

Brown, Trevor. "A Space Joint Stock Company," *The Space Review*, April 30, 2012. https://www.thespacereview.com/article/2073/1

Brown, William M and Herman Kahn. "Long-term Prospects for Developments in Space (A Scenario Approach)," *The Hudson Institute*, Washington, DC (prepared for National Aeronautics and Space Administration under contract NASW-2924), 1977. https://apps.dtic.mil/dtic/tr/fulltext/u2/b310563.pdf

Buckley, Christopher. "Is Trump's 'Space Force' Really Such an Insane Idea?," Spectator USA, June 27, 2018. https://spectator.us/is-trumps-space-force-really-such-an-insane-idea/

Bukharin, Nikolaĭ and Vladimir Il'ič Lenin. *Imperialism and World Economy*. Vol. 4. New York: International Publishers, 1929.

Bull, Hedley. *The Anarchical Society A Study of Order in World Politics*. New York: Columbia University Press, 2012.

Bullock Museum. "Cowboys In Space And Fantastic Worlds: A Journey Through the History of Westerns in Science Fiction," *The Story of Texas*, December 1, 2019. https://www.thestoryoftexas.com/press/media-kits/cowboys-in-space-2019 0715

Bureau of Arms Control, Verification, and Compliance. "Treaty on Principles Governing the Activities of States in the Exploration and Use of Outer Space, Including the Moon and Other Celestial Bodies," *U.S. State Department*, January 21, 2017. https://2009-2017.state.gov/t/isn/5181.htm

Burgess, Matt. "China's Latest Rocket Launch Fails to Get to Orbit after "Anomaly," *Wired*, July 3, 2017, https://www.wired.co.uk/article/china-rocket-long-march-5 -fail

Burton, Rachael and Mark Stokes. "The People's Liberation Army's Strategic Support Force Leadership and Structure," *Project 2049 Institute*, September 25, 2018. https://project2049.net/wp-content/uploads/2018/09/180925_PLA_SSF _Leadership-and-Structure_Stokes_Burton.pdf

Buzan, Barry and Richard Little. *International Systems in World History Remaking the Study of International Relations*. Oxford: Oxford University Press, 2000.

Caffrey, Matthew B. Jr. "On Wargaming," *The Newport Papers* 43, Newport, Rhode Island, 2019. https://digital-commons.usnwc.edu/newport-papers/43

Cahan, Bruce B, Bruce Pittman, Sarah Cooper and John Cumbers. "Space Commodities Futures Trading Exchange: Adapting Terrestrial Market Mechanisms to Grow a Sustainable Space Economy," *New Space*, September 1, 2018. https://www.lie bertpub.com/doi/10.1089/space.2017.0047

Cahan, Bruce. "A Space Commodities Futures Trading Exchange to Grow the Lunar Economy," 2017 Annual Meeting of the Lunar Exploration Analysis Group, held 10-12 October, 2017 in Columbia, Maryland. LPI Contribution No. 2041, id.5008

Cahan, Bruce. "Space Commodities Exchange – A New Frontier," *Urban Logic*, October 8, 2016. https://urbanlogic.org/space-commodities-exchange/

Callahan, William A. "National Insecurities: Humiliation, Salvation, and Chinese Nationalism," *Alternatives* 29, no. 2 (March-May 2004): 199–218.

Callahan, William A. *The Cartography of National Humiliation and the Emergence of China's Geobody, Public Culture* 21, no. 1 (2009): 141–73.

Caltech. "Space Solar Power Project." n.d. https://www.spacesolar.caltech.edu/

Caltech. "Space-Based Solar Power Project Funded," April 28, 2015. https://www.cal tech.edu/about/news/space-based-solar-power-project-funded-46644

Calum, Matheson. "The Sublime Rhetoric of Pascal's Wager," *Argumentation and Advocacy* 53, no. 4 (2017): 271–86. doi: 10.1080/00028533.2017.1375737

Campbell, Charlie. "From Satellites to the Moon and Mars, China is Quickly Becoming a Space Superpower," *Time*, July 17, 2019. https://time.com/5623537/china-space/

Campbell, James T. "'Print the Legend': John Wayne and Postwar American Culture," *Reviews in American History* 28, no. 3 (2000): 465–77.; 05

Can, Yuan. "China's Solar Power Station in Space is Under Research," *People's Daily Online*, March 08, 2016. http://en.people.cn/n3/2016/0308/c90000-9027131.html

Cantir, Christian and Juliet Karrbo. "Contested Roles and Domestic Politics: Reflections on Role Theory in Foreign Policy Analysis and IR Theory," *Foreign Policy Analysis* 8, no. 1 (January 2012): 5–24.

Cao, Siqi. "Heroes to Receive Highest Honor in China," *The Global Times*, August 08, 2019, http://www.globaltimes.cn/content/1162994.shtml

Carrington, Peter J., John Scott, Stanley Wasserman, eds. *Models and Methods in Social Network Analysis*. Vol. 28. Cambridge: Cambridge University Press, 2005.

Carter, Jamie. "China Hints at Surprise Space Station Launch, Says its New Rocket Can Challenge SpaceX," *Forbes*, March 5, 2019. https://www.forbes.com/sites/jamiecartereurope/2019/03/05/china-hints-at-surprise-space-station-launch-says-its-new-rocket-can-challenge-spacex/#32a313f43e80

Cartright, Mark. "Surya," June 23, 2016. https://www.ancient.eu/Surya/

Chaikin, Andrew. *A Man on the Moon: The Voyages of the Apollo Astronauts*. New York: Penguin, 2007.

Chan, Mitchell. "Communism with Chinese Characteristics," Confucius and China's Future," *Penn Political Review*, February 16, 2018. https://pennpoliticalreview.org/2018/02/communism-chinese-characteristics-confucius-and-chinas-future/

Chandler, David. "Shining Brightly," *MIT News*, October 26, 2011. http://news.mit.edu/2011/energy-scale-part3-1026

Chang, Deng. "Look Out, America: China's New Military Force are Awakening," *The National Interest*, January 11, 2016. http://nationalinterest.org/blog/the-buzz/look-out-america-chinas-new-military-forces-are-awakening-14872

Chang, Kenneth. "Elon Musk Sets Out SpaceX Starship's Ambitious Launch Plan," *The New York Times*, September 28, 2019. https://www.nytimes.com/2019/09/28/science/elon-musk-spacex-starship.html

Chang, Kenneth. "Trump's NASA Nominee, Jim Bridenstine, Confirmed by Senate on Party-Line Vote," *The New York Times*, April 19, 2018. https://www.nytimes.com/2018/04/19/science/jim-bridenstine-nasa.html

Chau, Donovan C. and Thomas M. Kane, eds. *China and International Security History, Strategy and 21st Century Policy*. Santa Barbara: Praeger, 2014.

Chen, Stephen. "China's Nuclear Spaceships Will be 'Mining Asteroids and Flying Tourists' As It Aims to Overtake US in Space Race," *South China Morning Post*, November 17, 2017. https://www.scmp.com/news/china/policies-politics/article/2120425/chinas-nuclear-spaceships-will-be-mining-asteroids

Chen, Stephen. "Months After Rocket failure, China Sends Spy Satellites into Space," *The South China Morning Post*, September 29, 2017. https://www.scmp

.com/news/china/society/article/2113427/months-after-rocket-failure-china-sends-spy-satellites-space

Cheng, Dean. "Space and Information Warfare: A Key Battleground for Information Dominance," in *Outer Space; Earthly Escalation? Chinese Perspectives on Space Operations an Escalation*. Washington, DC: Joint Staff Strategic Multilayer Assessment, 2018. https://nsiteam.com/social/wp-content/uploads/2018/08/SMA-White-Paper_Chinese-Persepectives-on-Space_-Aug-2018.pdf

Chi, Ma. "China Aims to be World-Leading Space Power by 2045," *China Daily*, November 17, 2017. http://www.chinadaily.com.cn/china/2017-11/17/content_34653486.htm

Choi Chi-Yuk and Laura Zhou. "China's Leaders Reaffirm Loyalty to party Under Xi Jinping at top Level Gathering," *South China Morning Post*, December 28, 2018. https://www.scmp.com/news/china/politics/article/2179722/chinas-leaders-reaffirm-loyalty-party-under-xi-jinping-top-level

Choucri, Nazli and Thomas W. Robinson. *Forecasting in International Relations: Theory, Methods, Problems, Prospects*. San Francisco: W.H. Freeman, 1978.

Choudhury, Dipanjan Roy. "India Needs Dedicated Military Space Program: Lt. Gen. PM Bali," *The Economic Times*, February 18, 2017. http://economictimes.indiatimes.com/news/defence/india-needs-dedicated-military-space-programme-lt-gen-pm-bali/articleshow/57225348.cms

Christian, John. "Canada Should Start Its Own Space Force," *Futurism*, January 6, 2019. https://futurism.com/the-byte/former-general-canada-should-start-space-force

Chung, Winchell. "Mission Delta-V and Flight Times," *ProjectRho*, June 7, 2012. http://www.projectrho.com/public_html/rocket/appmissiontable.php

Ciaccia, Chris. "Elon Musk Lauds Newt Gingrich's $2B Moon Competition Prize," *Fox News*, August 20, 2019. https://www.foxnews.com/science/elon-musk-newt-gingrich-moon-prize

Clark, Colin. "'This Is Our Sputnik Moment:' Rep. Bridenstine Offers Sweeping Space Bill," *Breaking Defense*, April 12, 2016. https://breakingdefense.com/2016/04/this-is-our-sputnik-moment-rep-bridenstine-offers-sweeping-space-bill/

Clark, Stephen "NASA Picks Three Companies to Send Commercial Landers to the Moon," *Spaceflight Now*, June 4, 2019. https://spaceflightnow.com/2019/06/04/nasa-picks-three-companies-to-send-commercial-landers-to-the-moon/

Clark, Stephen, "NASA Terminates Lunar Lander Contract with OrbitBeyond," *Spaceflightnow*, July 30, 2019. https://spaceflightnow.com/2019/07/30/nasa-terminates-lunar-lander-contract-with-orbitbeyond/

Clifford, Catherine. "Jeff Bezos Dreams of a World With a Trillion People Living in Space," *CNBC*, May 1, 2018. https://www.cnbc.com/2018/05/01/jeff-bezos-dreams-of-a-world-with-a-trillion-people-living-in-space.html

CNN. "State of the Union 2015: Full Transcript," *CNN*, January 20, 2015. https://www.cnn.com/2015/01/20/politics/state-of-the-union-2015-transcript-full-text/index.html

Cohen, Rachel S. "AFSPC Mulls Intel, Personnel Questions of the New Space Age," *Air Force Magazine*, September 5, 2019. https://www.airforcemag.com/afspc-mulls-intel-personnel-questions-of-the-new-space-age/

Cohen, Warren I. "China's Strategic Culture," *The Atlantic*. March 1997. http://www.theatlantic.com/magazine/archive/1997/03/chinas-strategic-culture/376815/

Cohen., Marc M. et al. "Robotic Asteroid Prospector (RAP)," *Final Report*, July 09, 2013. https://www.nasa.gov/sites/default/files/files/Cohen_2012_PhI_RAP.pdf

Collier, David. "Understanding Process Tracing," *Political Science & Politics* 44, no. 4 (October 2011): 823–30.

Colucci, Lamont. "A Space Service in Support of American Grand Strategy," *The Space Review*, February 25, 2019. http://www.thespacereview.com/article/3664/1

Columbus, Christopher. *The Four Voyages of Christopher Columbus,* translated by J. M. Cohen. New York: Penguin, 2004.

Combs, Mike. "The Case for Space," *The Space Front*, Fall 1999. https://space.nss.org/settlement/MikeCombs/case_spc.htm

Congress.gov. "S.1790 - National Defense Authorization Act for Fiscal Year 2020" (Subtitle D—United States Space Force Chapter 908, section 9083 Became Public Law No: 116-92). https://www.congress.gov/bill/116th-congress/senate-bill/1790/text?format=txt

Cooley, Thomas, Eric Felt and Steven J. Butow. "State of the Space Industrial Base: Threats, Challenges and Actions," *AFRL & DIU*, May 30, 2019. https://aerospace.csis.org/wp-content/uploads/2019/08/AFRL_DIU_Report_State_of_Space_Ind_Base_30May2019_Final.pdf

Cordiner, Ralph J. "Competitive Private Enterprise in Space," in *Peacetime Uses of Outer Space*. New York: McGraw-Hill, 1961. https://denniswingo.files.wordpress.com/2014/04/cordiner-article-1961.pdf

Cowling, Keith. "Who is DIU and Do They Want Their Own Space Station?," *NASA Watch*, July 5, 2019. http://nasawatch.com/archives/2019/07/who-is-diu-and.html

Cox, Donald William and Michael Stoiko. *Spacepower: What It Means to You*. Philadelphia and Toronto: Winston, 1958.

Cox, Tim. "In Space Force Debate, The Military's Space Experts Are Missing in Action," *The Hill*, February 12, 2019. https://thehill.com/opinion/national-security/429665-in-the-space-force-debate-the-militarys-space-experts-are-missing

Cramer, John G. "2001 - Then and Now," *Analog Science Fiction & Fact Magazine*, July, 2001. https://www.npl.washington.edu/AV/altvw107.html

Crane, Leah. "Distant Space Rock Known as Ultima Thule Renamed to Avoid Nazi Links," *New Scientist*, November 12, 2019. https://www.newscientist.com/article/2223176-distant-space-rock-known-as-ultima-thule-renamed-to-avoid-nazi-links/

Creswell John W and Cheryl N. Poth. *Qualitative Inquiry and Research Design: Choosing Among Five Approaches*. Los Angeles: Sage publications, 2017.

Creswell, John W., *Qualitative Inquiry & Research Design: Chosing Among Five Approaches*. Los Angeles: Sage, 2013.

Criswell, David R. "Return on Investment (ROI) from 'New Worlds & Lunar Solar; Jeff Bezos, Power,' (7pp.) International Conference: Moon Base—A Challenge for Humanity," Washington Workshop: Jamestown on the Moon, Washington Academy of Sciences, Washington, DC, October 12, 2005.

Criswell, David R., ''Lunar Solar Power System and Lunar Exploration,'' Workshop on Moon Beyond 2002. http://www.1pi.usro.edu/meetings/moon2002/pdf/3004.pdf

Cronin, Patrick M., Masashi Murano and H.R. McMaster. "Transcript: U.S. Space Strategy and Indo-Pacific Cooperation," *The Hudson Institute*, November 13, 2019. https://www.hudson.org/research/15481-transcript-u-s-space-strategy-and-indo-pacific-cooperation

CSDC. "Welcome to Cislunar Space Development Company," n.d. https://csdc.space/about-csdc/

Current World Population. Worldometers. n.d. http://www.worldometers.info/world-population/

Curtis, Lisa. "U.S.-India Relations: The China Factor," *The Heritage Foundation*, November 25, 2008. https://www.heritage.org/asia/report/us-india-relations-the-china-factor

Czysz, Paul Berndt Chudoba and Christopher Rahaim. "Low Earth Orbit Infrastructure Configuration and Requirements Translated into Support Vehicles Requirements," in 15th AIAA International Space Planes and Hypersonic Systems and Technologies Conference, 2008.

D3SSP. D3 Space Solar Power Proposal, n.d. http://d3ssp.org/ see also YouTube link: https://www.youtube.com/watch?v=M9dQsRv1XDg

Dadush, Uri and Bennett Stancil. "The World Order in 2050," Carnegie Endowment for International Peace. April 2010. https://carnegieendowment.org/files/World_Order_in_2050.pdf

Daniels, Max. "The Need to Consider Chinese Tourism in Antarctica," The Polar Connection, April 18, 2017. http://polarconnection.org/chinese-tourism-antarctica/

Daudi Barne's LinkedIn site. n.d., https://www.linkedin.com/in/daudi-barnes-0b2737132/ .

Davenport, Christian. *The Space Barons: Elon Musk, Jeff Bezos, and the Quest to Colonize the Cosmos*. U.S: Public Affairs, 2018.

Davenport, Kelsey. "India's ASAT Test Raises Space Risks," *Arms Control Today*. May 2019. https://www.armscontrol.org/act/2019-05/news/indian-asat-test-raises-space-risks

David, Leonard. "The US Geological Survey is Getting Serious about Space Resources and Mining," *Space.com*, September 04, 2018. https://www.space.com/41707-space-mining-usgs-resource-survey.html

David, Leonard. "CNSA Boss Outlines China's Space Exploration Agenda," *Spacenews.com*, April 5, 2017, http://spacenews.com/cnsa-boss-outlines-chinas-space-exploration-agenda/

David, Leonard. "Inside ULA's Plan to Have 1,000 People Working in Space by 2045," *Space.com*, June 29, 2016. https://www.space.com/33297-satellite-refueling-business-proposal-ula.html

David, Leonard. "Peter Glaser, Father of Solar-Power Satellite Idea, Dies at 90," *Space.com*, June 9, 2014. https://www.space.com/26175-peter-glaser-solar-power-satellite-obituary.html

David, Leonard. "US Military Eyes Strategic Value of Earth-Moon Space," *Space.com*, August 29, 2019. https://www.space.com/us-military-strategic-value-earth-moon-space.html

Davis, Jason. "Chandrayaan-2 Enters Lunar Orbit," *The Planetary Society*, August 20, 2019. http://www.planetary.org/blogs/jason-davis/chandrayaan-2-successful-loi.html

de Leon, Janice Ponce. "Mars Science City to be Fully Operational in 4 Years," *Gulf News*, January 22, 2019. https://gulfnews.com/uae/science/mars-science-city-to-be-fully-operational-in-4-years-1.61596317

Dean, Jason and Stephen Miller. "Trained in the U.S., Scientist Became China's Rocket King," *The Wall Street Journal*, November 4, 2009. https://www.wsj.com/articles/SB125721495250424443

Declaration on the Conduct of Parties in the South China Sea. "Association of Southeast Asian Nations," October 17, 2012. https://asean.org/?static_post=declaration-on-the-conduct-of-parties-in-the-south-china-sea-2

Deep Space Industries, "Mission," January 24, 2013. https://web.archive.org/web/20130124211904/http://deepspaceindustries.com/mission/

Deep Space Industries. Youtube. January 26, 2013. https://www.youtube.com/watch?v=zXXJtSZffVg

Defense Innovation Unit (DIU). "Multi-Orbit Logistics Services," *DIU*, February 12, 2020. https://www.diu.mil/work-with-us/open-solicitations

Defense Intelligence Agency. "China Military Power Modernizing a Force to Fight and Win." 2019. http://www.dia.mil/Portals/27/Documents/News/Military%20Power%20Publications/China_Military_Power_FINAL_5MB_20190103.pdf?utm_source=Daily%20on%20Defense%20(Template)_01/16/2019&utm_medium=email&utm_campaign=WEX_Daily%20on%20Defense

Dembling, Paul G. and Daniel M. Arons. "The Evolution of the Outer Space Treaty," *Journal of Air Law and Commerce* 33 (1967): 419–56

Department of Space, Indian Space Research Organisation. "Chandrayaan-2." n.d. http://www.isro.gov.in/chandrayaan-2.

Deringer, Freshfields Bruckhaus. "*China Establishes International Commercial Courts to handle Belt and Road Initiative Disputes*." University of Oxford, Faculty of Law, August 17, 2018. https://www.law.ox.ac.uk/business-law-blog/blog/2018/08/china-establishes-international-commercial-courts-handle-belt-and

Desjardins, Jeff. "There's Big Money to be Made In Asteroid Mining," *Business Insider*. November 3, 2016. http://www.businessinsider.com/the-value-of-asteroid-mining-2016-11.

Deudney, Daniel H. *Bounding Power: Republican Security Theory from the Polis to the Global Village*. Princeton: Princeton University Press, 2007.

Deudney, Daniel H. *Dark Skies: Space Expansionism, Planetary Geopolitics, and the Ends of Humanity*. New York: Oxford University Press, 2020.

Devlin, Hannah. "Giant Leaf for Mankind? China Germinates First Seed on Moon," *The Guardian*, January 15, 2019. https://www.theguardian.com/science/2019/jan/15/china-germinates-first-seed-on-moon-cotton-shoot-change4?CMP=share_btn_fb&fbclid=IwAR2SUPboveO7T06r7SakHqJ3eFUeDSbwouDJDMAM0-fprXiLAMuMdE458nk

Dewar, James A. *To the End of the Solar System: The Story of the Nuclear Rocket*. Ontario, Canada: Apogee Books, 2008.

Dhruva Space Website. n.d. https://www.dhruvaspace.com/

Disha, Missal. "9 Indian Startups That Are Taking India's Space Industry to The Next Level," *Curious Dose*, March 11, 2019. https://curiousdose.com/2019/03/9-indian-startups-that-are-taking-indias-space-industry-to-the-next-level/

Dolman, Everett C. *Astropolitik: Classical Geopolitics in the Space Age*. Portland, OR: Frank Cass, 2002.

Dorminey, Bruce. "Golden Spike Still Aims for Human Lunar Surface Expeditions by Decade's End," *Forbes*, January 30, 2014. https://www.forbes.com/sites/brucedorminey/2014/01/30/golden-spike-still-aims-for-human-lunar-surface-expeditions-by-decades-end/#54695f4d7afe

Dorminey, Bruce. "Trump Space Policy to Aim for Mars and Beyond," *Forbes*, November 14, 2016. https://www.forbes.com/sites/brucedorminey/2016/11/14/trump-space-policy-to-aim-for-mars-and-beyond/#2756651a250d

Downs, Anthony. "Up and Down with Ecology-the Issue-Attention Cycle," *Public Interest* 28 (Summer, 1972): 38–50.

DRDO. Official video on Mission Shakti (Project XSV1). Youtube. April 6, 2019. https://www.youtube.com/watch?v=1-J4jVlx5Do

Dreier, Casey. "How NASA's Planetary Defense Budget Grew by More Than 4000% in 10 years," *The Planetary Society*, September 26, 2019. https://www.planetary.org/blogs/casey-dreier/2019/nasas-planetary-defense-budget-growth.html

Dreyer, Edward L. *Zheng He: China and the Oceans in the Early Ming Dynasty, 1403-1433*. New York: Pearson Longman, 2007.

Dudley, Dominic. "Dates in Space: UAE Floats Idea of Planting Palm Trees on Mars," *Forbes*, August 5, 2019, https://www.forbes.com/sites/dominicdudley/2019/08/05/dates-in-space/#52c83a734ef6

Dunlap, Charles. "Law and Military Interventions: Preserving Humanitarian Values in 21st Conflicts," prepared for the Humanitarian Challenges in Military Intervention Conference Carr Center for Human Rights Policy, Kennedy School of Government, Harvard University Washington, DC, November 29, 2001. https://people.duke.edu/~pfeaver/dunlap.pdf

Dunne, Timothy. "The Social Construction of International Society," *European Journal of International Relations* 1, no. 3 (September 1995): 367–89.

Dunnigan, James F. *The Complete Wargames Handbook: How to Play, Design, and Find Them*. New York: Morrow, 1992.

Dutton, Jack. "UAE Space Agency to Launch Satellite Developed by Students," *The National*, May 12, 2019. https://www.thenational.ae/uae/science/uae-space-agency-to-launch-satellite-developed-by-students-1.860640

Dyson, George. *Project Orion: The True Story of the Atomic Spaceship*. New York: Henry Holt and Co., 2002.

Earth2Orbit Website. n.d. http://www.earth2orbit.com/

Editorial Board. "The UN Can't Ignore Kashmir Anymore," *The New York Times*, October 2, 2019. https://www.nytimes.com/2019/10/02/opinion/editorials/kashmir-india-pakistan-un.html

Editors. "NASA's Griffin: 'Humans Will Colonize the Solar System'," *The Washington Post*, September 25, 2005. http://www.washingtonpost.com/wp-dyn/content/article/2005/09/23/AR2005092301691.html?noredirect=on

Edwards, Jim. "500 Years Ago, China Destroyed its World-Dominating Navy Because its Political Elite was Afraid of Free Trade," *Business Insider*, February 26, 2017. https://www.businessinsider.com/china-zhenge-he-treasure-fleet-elite-free-tr

Egan, Brian J. "The Next Fifty Years of the Outer Space Treaty" [to Galloway Symposium on Critical Issues in Space Law in Washington, DC on December 7, 2016], State Department, December 7, 2016. http://carbondate.cs.odu.edu/#https://2009-2017.state.gov/s/l/releases/remarks/264963.htm

Egan, Matt. "America is Now the World's Largest Oil Producer," *CNN Money*, September 12, 2018. https://money.cnn.com/2018/09/12/investing/us-oil-production-russia-saudi-arabia/index.html

Elmana, Colin and Miriam Fendius Elman. *Bridges and Boundaries Historians, Political Scientists, and the Study of International Relations*. Cambridge, MA: The MIT Press, 2001.

Elsa, Evangeline. "UAE in Space-as it Happened: First Emirati Astronaut Hazzaa AlMansoori and Team Enter ISS: Makes History for UAE," *Gulf News*, September 25, 2019. https://gulfnews.com/uae/uae-in-space---as-it-happened-first-emirati-astronaut-hazzaa-almansoori-and-team-enter-iss-make-history-for-uae-1.1569396150288

Elvis, Martin and Tony Milligan. "How Much of the Solar System Should We Leave As Wilderness?," *Acta Astronautica* 162 (2019): 574–80.

Elvis, Martin, Tony Milligan, Alanna Krolikowski. "The Peaks of Eternal Light: A Near-term Property Issue on the Moon," (Postdoctoral NSF Grant#1066293), 2016. https://arxiv.org/ftp/arxiv/papers/1608/1608.01989.pdf

Erickson, Amanda. "Here They Come Again': Chinese Police Arrest Dissident Professor During On-Air Interview," *The Washington Post*, August 3, 2018. https://www.washingtonpost.com/news/worldviews/wp/2018/08/03/here-they-come-again-chinese-police-arrest-dissident-professor-during-on-air-interview/

Eroukhmanoff, Clara. "Securitisation Theory: An Introduction," *E-International Relations*. January 14, 2018. https://www.e-ir.info/2018/01/14/securitisation-theory-an-introduction/

Erwin, Sandra. "Air Force, DoD Research Organizations Call for Space Industrial Policies to Counter China," *SpaceNews*, June 17, 2019. https://spacenews.com/air-force-dod-research-organizations-call-for-space-industrial-policies-to-counter-china/

Erwin, Sandra. "Air Force Seeking Commercial Technologies for Cislunar Space Operations," *Space News*, December 12, 2019. https://spacenews.com/air-force-seeking-commercial-technologies-for-cislunar-space-operations/

Etherington, Darrel. "China launches a Rocket to Space from a Ship at Sea for the First time," *TechCrunch*, June 5, 2019. https://techcrunch.com/2019/06/05/china-launches-a-rocket-to-space-from-a-ship-at-sea-for-the-first-time/

Ewart, Roberta. "XST: On the Shoulders of Giants," November 13, 2018, 2019. http://afcea-la.org/afceala-wp/wp-content/uploads/XST-Dr-Ewart-AFCEA-15-Nov-2018.pdf

Express News Service. "India Needs to Create Solar Power Satellite: Indian Space Research Organization," *The New Indian Express*, July 8, 2018. http://www.newi

ndianexpress.com/states/karnataka/2018/jul/08/india-needs-to-create-solar-power-satellite-indian-space-research-organisation-1840117.html

Exseed Space. Website. n.d. https://www.exseedspace.com/

Facebook. March Storm 2019. https://www.facebook.com/events/351030189012751/

Farrell, Theo. "Strategic Culture and American Empire," SAIS Review XXV, no. 2 (Summer-Fall 2005): 3–18.

FAST SPACE: Leveraging Ultra low-cost Space Access for 21st Century Challenges. Air University. January 13, 2017. https://www.airuniversity.af.edu/Portals/10/Research/Space-Horizons/documents/Fast%20Space_Public_2017.pdf

Fearson, James D. "Domestic Politics, Foreign Policy, and Theories of International Relations," *Annual Review of Political Science* 1 (1998): 289–313. http://www.rochelleterman.com/ir/sites/default/files/Fearon%201998.pdf

Fecht, Sarah. "Now That We've Explored All The Planets, What's The Next Frontier?," *Popular Science*, July 17, 2015. https://www.popsci.com/now-weve-explored-all-planets-whats-next-frontier

Fecht, Sarah. "U.S. Lawmakers Want To Kill NASA's Mission To Send Astronauts To An Asteroid: Instead They Favor Going Back to the Moon," *Popular Science*, May 24, 2016. https://www.popsci.com/house-bill-would-provide-zero-funding-for-nasas-asteroid-redirect-mission/

Fei, Yu. "China's Satellite Sends Unbreakable Cipher from Space," *Xinhua,* August 10, 2017. http://www.xinhuanet.com//english/2017-08/10/c_136514705.htm

Ferholz, Tim. "The Fastest Object Made by Humans is Just Trying to Slow Down," *Quartz*, August 10, 2018. https://qz.com/1353429/the-parker-solar-probe-is-the-fastest-man-made-object-ever/

Fernholz, Tim. "America's First Private Moon Lander Will be Engineered in India," *Quartz*, June 3, 2019. https://qz.com/1633918/americas-first-private-moon-lander-will-be-made-in-india/

Finnemore, Martha and Kathryn Sikkink, "International Norm Dynamics and Political Change," *International Organization* 52, no. 4 (Autumn 1998): 887–918.

Firebaugh, Glenn, *Seven Rules for Social Research.* Princeton and Oxford: Princeton University Press, 2008.

Foreman, Amanda. "The British View the War of 1812 Quite Differently than Americans Do," *Smithsonian Magazine*, July 2104. https://www.smithsonianmag.com/history/british-view-war-1812-quite-differently-americans-do-180951852/

Foster, L. M. and Namrata Goswami. "What China's Antarctic Behavior Tells us About the Future of Space," *The Diplomat*, January 11, 2019. https://thediplomat.com/2019/01/what-chinas-antarctic-behavior-tells-us-about-the-future-of-space/

Foust, Jeff. "Asteroid Mining Company Planetary Resources Acquired by Blockchain Firm," *SpaceNews*, October 31, 2018. https://spacenews.com/asteroid-mining-company-planetary-resources-acquired-by-blockchain-firm/

Foust, Jeff. "Deep Space Industries Acquired by Bradford Space," *SpaceNews*, January 2, 2019. https://spacenews.com/deep-space-industries-acquired-by-bradford-space/

Foust, Jeff. "Falcon 9 Launches Orbcomm Satellites, Lands First Stage," *SpaceNews*, December 21, 2015. https://spacenews.com/falcon-9-launches-orbcomm-satellites-first-stage-lands/

Foust, Jeff. "Former Google Lunar X-Prize Teams Focused on New Commercial and Government Opportunities," *Space.com*, May 25, 2018. https://spacenews.com/former-google-lunar-x-prize-teams-focused-on-new-commercial-and-government-opportunities/

Foust, Jeff. "Lunar Exploration Providing New Impetus for Space Resources Legal Debate," *SpaceNews*, September 7, 2019. https://spacenews.com/lunar-exploration-providing-new-impetus-for-space-resources-legal-debate/

Foust, Jeff. "Luxembourg Establishes Space Agency and New Fund," *SpaceNews*, September 13, 2018. https://spacenews.com/luxembourg-establishes-space-agency-and-new-fund/

Foust, Jeff. "NASA closing out Asteroid Redirect Mission," *SpaceNews,* June 14, 2017. https://spacenews.com/nasa-closing-out-asteroid-redirect-mission/

Foust, Jeff. "NASA Seeks 'Industrialization' of Low Earth Orbit with ISS Commercialization Strategy," *SpaceNews*, August 1, 2019. https://spacenews.com/nasa-seeks-industrialization-of-low-earth-orbit-with-iss-commercialization-strategy/

Foust, Jeff. "NASA Selects Nine Companies for Commercial Lunar Lander Program," *SpaceNews*, November 29, 2018. https://spacenews.com/nasa-selects-nine-companies-for-commercial-lunar-lander-program/

Foust, Jeff. "Proposed Legislation Would Make Space Settlement a National Goal," *SpaceNews*, May 1, 2015. https://spacenews.com/proposed-legislation-would-make-space-settlement-a-national-goal/

Foust, Jeff. "White House Releases Near Earth Object Action Plan," *SpaceNews*, June 21, 2018. https://spacenews.com/white-house-releases-near-earth-object-action-plan/

Foust, Jeff. "NASA Awards Contracts to Three Companies to Land Payloads on the Moon," *SpaceNews*, May 31, 2019. https://spacenews.com/nasa-awards-contracts-to-three-companies-to-land-payloads-on-the-moon/

Freeman, John R. and Brain L. Job. "Scientific Forecasts in International Relations: Problems of Definition and Epistemology," *International Studies Quarterly* 23, no. 1 (March 1979): 113–49.

Froehlich, Annette, ed. *Space Resource Utilization: A View from an Emerging Space Faring Nation*. Vol. 12. New York: Springer, 2018.

Gadbury, Deborah. *The Space Race: The Epic Battle between America and the Soviet Union for Dominion of Space*. New York: Harper Collins, 2006.

Gallahar, Sean. "India ASAT Test Debris Poses Danger to International Space Station, NASA Says," *Ars Technica*, April 2, 2019. https://arstechnica.com/tech-policy/2019/04/india-asat-test-debris-poses-danger-to-international-space-station-nasa-says/

Galli, Anthony. "Elon Musk Life Purpose Explained," *Mission.org*, October 10, 2017. https://medium.com/the-mission/elon-musk-life-purpose-explained-d08e7f37ce28

Gamble, Jessa. "How do you build a city in space?," *The Guardian*, May 16, 2014. https://www.theguardian.com/cities/2014/may/16/how-build-city-in-space-nasa-elon-musk-spacex

Garretson, Peter A. "OPINION: The first duty of a Space Force is to protect space commerce," *Politico*, July 21, 2019. https://www.politico.com/story/2019/06/21/opinion-space-force-commerce-1374229

Garretson, Peter A. "Sky's No Limit: Space-Based Solar Power, The Next Major Step in the Indo-US Strategic Partnership?," IDSA Occasional Paper No.9, Institute for Defence Studies and Analyses (IDSA), New Delhi, August 2010. https://idsa.in/sites/default/files/OP_SkysNoLimit.pdf also available at https://spacejournal.ohio.edu/issue16/papers/OP_SkysNoLimit.pdf and at https://space.nss.org/media/Skys-No-Limit-space-based-solar-power-IDSA-Garretson-2010.pdf

Garretson, Peter A. "The Future of the Asian Order Will Be Decided in Space," *The Diplomat*, July 3, 2019. https://thediplomat.com/2019/07/the-future-of-the-asian-order-will-be-decided-in-space/

Garretson, Peter A. "US Space Command: A Vision for the Final Frontier," *The Hill*, August 28, 2019. https://thehill.com/opinion/technology/459066-us-space-command-a-vision-for-the-final-frontier

Garretson, Peter and Namrata Goswami. "Is India Looking Towards Space-Based Resources?," *The Space Review*, October 2, 2017. ttp://www.thespacereview.com/article/3338/1

Garretson, Peter authored on behalf of the NSS. "The National Space Society Pays Tribute to Dr. Kalam — One of Our Leading Lights Has Joined the Stars," *NSS*, July 31, 2015. https://space.nss.org/the-national-space-society-pays-tribute-to-dr-kalam-one-of-our-leading-lights-has-joined-the-stars/

Garretson, Peter, Alfred B. Anzaldúa, Hoyt Davidson. "Catalyzing Space Debris Removal, Salvage, and Use," *The Space Review*, December 9, 2019. https://www.thespacereview.com/article/3847/1

Garretson, Peter, Brent Ziarnick, M. V. "Coyote" Smith, and Everett Dolman. "A National Space Policy for this Century," *Wired*, November 7, 2016. http://www.thespacereview.com/article/3099/1

Garretson, Peter. "A Historic National Vision for Spacepower," *War on the Rocks*, September 9, 2019. https://warontherocks.com/2019/09/a-historic-national-vision-for-spacepower/

Garretson, Peter. "Air Force Suppressed Space Force Debate; Lt. Gen. Kwast Spoke Truth To Power," *Breaking Defense*, August 08, 2019. https://breakingdefense.com/2019/08/air-force-suppressed-space-force-debate-lt-gen-kwast-spoke-truth-to-power/

Garretson, Peter. "Scenario Building: A Suggested Methodology," *Institute for Defence Studies and Analyses (IDSA)*, November 25, 2009. https://web.archive.org/web/20121122211758/http://www.idsa.in/asc/opinion11.html

Garretson, Peter. "Space Force's Jupiter-Sized Culture Problem," *War on the Rocks*, July 11, 2019. https://warontherocks.com/2019/07/space-forces-jupiter-sized-culture-problem/

Garretson, Peter. "The Purpose of a Space Force is a Space Faring Economy," *The Hill*, June 26, 2019. https://thehill.com/opinion/technology/450519-the-purpose-of-a-space-force-is-a-spacefaring-economy.

Garretson, Peter. "A Speech for The Next SECAF To Launch A New Era of Spacepower," *War on the Rocks*, September 20, 2019. https://warontherocks.com/2019/09/a-speech-for-the-next-secaf-to-launch-a-new-era-of-spacepower/

Garretson, Peter. "Guess What Could Be Totally Missing from the New U.S. President's Intel Briefing: Oh, Just the Entire Outer-Space Thing," *War is Boring*, September 29, 2016. https://warisboring.com/guess-what-could-be-totally-missing-from-the-new-u-s-presidents-intel-briefing/

Garretson, Peter. "Re-Opening the American Frontier: Recent Congressional Hearings on Space," *The Space Review*, July 3, 2017. http://www.thespacereview.com/article/3275/1

Garretson, Peter. "Solar Power in Space?," *Strategic Studies Quarterly*, Spring 2012. http://www.au.af.mil/au/ssq/2012/spring/garretson.pdf

Garretson, Peter. "Space Force Defenses Must Stretch to the Moon," *The Hill*, August 5, 2019. https://thehill.com/opinion/national-security/456214-space-force-defenses-must-stretch-to-the-moon

Garretson, Peter. "Tanham in Retrospect: 18 Years of Evolution in Indian Strategic Culture." *South Asia Journal*, January 22, 2013. http://southasiajournal.net/tanham-in-retrospect-18-years-of-evolution-in-indian-strategic-culture/

Garretson, Peter. "The First Duty of a Space Force is to Protect Space Commerce," *Politico*, June 21, 2019. https://www.politico.com/story/2019/06/21/opinion-space-force-commerce-1374229

Garretson, Peter. "The Next Great White Fleet: Extending the Benefits of the International System into Space," *Astropolitics* 6, no. 1 (2008): 50–70.

Gautam, P.K., *One Hundred Years of Kautilya's Arthasashtra*, IDSA Monograph, Series No. 20, 2013. https://idsa.in/monograph/OneHundredYearsofKautilyasArthasastra

Gautam, Pradeep Kumar, "A Longue Durée Perspective on Military Science in India," *Journal of* Defence Studies 12, no.1 (January–March 2018): 15–38. https://idsa.in/system/files/jds/jds-12-1-2018-military-science.pdf

George, Alexander L. and Andrew Bennett. *Case Studies and Theory Development in the Social Sciences*. Cambridge, MA: MIT Press, 2005.

George, Alexander L. "Case Studies and Theory Development: The Method of Structured, Focused Comparison," in *Diplomacy: New Approaches in History, Theory, and Policy*. Edited by Paul Gordon Lauren, 43–68. New York: Free Press, 1979.

Gesl, Paul M. "Preparing for The Next Space Race: Legislation and Policy Recommendations for Space Colonies," *Air Command and Staff College Maxwell AFB United States*, April 1, 2018. https://apps.dtic.mil/dtic/tr/fulltext/u2/1053024.pdf

Getlen, Larry. "Could Private Armies Fight the Next Major Conflict?," *New York Post*, December 28, 2014. https://nypost.com/2014/12/28/could-private-armies-fight-the-next-major-conflict/

Ghosh, Palash. "China Proposes Space Solar Power Collaboration with India," *International Business Times*, November 2, 2012. https://www.ibtimes.com/china-proposes-space-solar-power-collaboration-india-858330

Gibbs, David N. and Robert S. Snyder. "Confronting Bias in International Relations," *International Studies Perspective* 2, no. 4 (November 2001): 416–37.

Gibney, Elizabeth. "Chinese Satellite is One Giant Step for The Quantum Internet," *Nature*, July 27, 2016. https://www.nature.com/news/chinese-satellite-is-one-giant-step-for-the-quantum-internet-1.20329

Gilbert, Alex and Morgan D. Bazilian. "We Need a Space Resources Institute," *Scientific American*, April 19, 2019. https://blogs.scientificamerican.com/observations/we-need-a-space-resources-institute/

Gilpin, Robert. *War and Change in World Politics*. Cambridge: Cambridge University Press, 1983.

Gingrich, Newt. "Gingrich, Kwast, Walker and Autry: A $2 billion prize will bring US permanent moon and Mars settlements," *Fox News*, August 22, 2019. https://www.foxnews.com/opinion/newt-gingrich-steve-kwast-bob-walker-greg-autry-we-have-an-innovative-plan-for-getting-back-to-space

GLAE. n.d., https://glae.lu/

Glaser, B. G. and A. Strauss. *The Discovery of Grounded Theory. Strategies for Qualitative Research*. Chicago: Aldine, 1967.

Gofus, Andreas et al. eds *Sage Handbook of History, Philosophy and Sociology of International Relations*. Los Angeles: Sage, 2018.

Gong, Xue. "Welcome to Conflict Tourism: How Chinese State Firms are Using the South China Sea," *The South China Morning Post*, September 19, 2016. https://www.scmp.com/week-asia/geopolitics/article/2020096/welcome-conflict-tourism-how-chinese-state-firms-are-using

Gopalaswami, Raghavan. "Sustaining India's Economic Growth," *Online Journal of Space Communication*, Winter 2010. https://spacejournal.ohio.edu/issue16/gopal.html

Gopalaswami, Raghavan. "The Spaceplane Equation," *Online Journal of Space Communication*, Winter 2010. https://spacejournal.ohio.edu/issue16/gopal2.html

Gorka, Alex. "It Starts: EU Founders to Form Federal Union of European States," *Strategic Culture Foundation*, March 3, 2017. https://www.strategic-culture.org/news/2017/03/03/it-starts-eu-founders-form-federal-union-european-states/

Goswami, Namrata and Peter Garretson. "Critical Shifts in India's Outer Space Policy," *The Diplomat*, April 16, 2019. https://thediplomat.com/2019/04/critical-shifts-in-indias-outer-space-policy/

Goswami, Namrata. "Can China be Taken Seriously on its 'Word' to Negotiate Disputed Territory?," *The Diplomat*, August 18, 2017. https://thediplomat.com/2017/08/can-china-be-taken-seriously-on-its-word-to-negotiate-disputed-territory/.

Goswami, Namrata. "Chang'e 4 and Chandrayaan 2: To the Moon and Beyond," *The Diplomat*, September 9, 2019, https://thediplomat.com/2019/09/change-4-and-chandrayaan-2-to-the-moon-and-beyond/

Goswami, Namrata. "China in Space: Ambitions and Possible Conflicts," *Strategic Studies Quarterly*, 12, no. 1 (Spring 2018): 74–97, https://www.jstor.org/stable/26333878?seq=1#metadata_info_tab_contents

Goswami, Namrata. "China's Grand Strategy in Outer Space: To Establish Compelling Standards of Behavior," *The Space Review*, August 5, 2019. https://www.thespacereview.com/article/3773/1

Goswami, Namrata. "China's Unique Space Ambitions," *The Diplomat*, August 3, 2016. http://thediplomat.com/2016/08/chinas-unique-space-ambitions/

Goswami, Namrata. "India's Strategic Culture is Plain to See," *Asia Times*, April 5, 2013. http://www.atimes.com/atimes/South_Asia/SOU-01-050413.html

Goswami, Namrata. "The Competition for Outer Space Resources and the Role of Middle Powers," *Live Encounters Magazine*, June, 2019. https://liveencounters.net/2019-le-mag/06-june-2019/dr-namrata-goswami-the-competition-for-outer-space-resources/?fbclid=IwAR0DUohKoqUMf2fZZMPl-dyxXDP_4P1YVfnGpmfL7EUPRO6zIG8HdGm3h_w

Goswami, Namrata. "The New Space Race Pits the U.S. Against China. The U.S. is Losing Badly," *The Washington Post*, January 10, 2019. https://www.washingtonpost.com/opinions/the-new-space-race-pits-the-us-against-china-the-us-is-losing-badly/2019/01/10/bcdcad10-14f9-11e9-b6ad-9cfd62dbb0a8_story.html

Goswami, Namrata. "Waking up to China's Space Dream," *The Diplomat*, October 15, 2018. https://thediplomat.com/2018/10/waking-up-to-chinas-space-dream/

Goswami, Namrata. "Why China's Long March 11 Launch Matters," *The Diplomat*, June 10, 2019. https://thediplomat.com/2019/06/why-chinas-long-march-11-launch-matters/

Gouyon, Louis de., "Starfish Prime or The Legality Of High-Altitude Nuclear Explosions," *Space Legal Issues*, February 14, 2019. https://www.spacelegalissues.com/space-law-starfish-prime-or-the-legality-of-high-altitude-nuclear-explosions/

Governance Working Group Annex I – Participants of The Hague Space Resources Governance Working Group, December 18, 2017. https://www.universiteitleiden.nl/binaries/content/assets/rechtsgeleerdheid/instituut-voor-publiekrecht/lucht--en-ruimterecht/space-resources/final-report_the-hague-space-resources-governance-working-group.pdf

Government of Dubai, Media Office. "UAE Cabinet Approves National Space Strategy 2030," March 11, 2019. http://www.mediaoffice.ae/en/media-center/news/11/3/2019/smnationalstrategy.aspx.

Graham, Daniel O. *High Frontier: A New National Strategy*. Washington, DC: High Frontier, 1982.

Graham, Daniel O. *High Frontier: A strategy for National Survival*. New York: Tom Doherty Associates, 1984.

Graham, Melanie. *Redefining Strategic Culture*. Canada: The University of Northern British Columbia, June 2014.

Graham, Thomas, Jr. "Space Weapons and the Risks of Accidental Nuclear War," *Arms Control Today* 35, no. 10 (December 2005).

Grant, Dustin L. and Matthew J. Neil. "The Case for Space: A Legislative Framework For An Independent United States Space Force," *United States: Air Command and Staff College Maxwell AFB*, April 1, 2018. https://apps.dtic.mil/dtic/tr/fulltext/u2/1053020.pdf

Gray, Colin S. *The Geopolitics of Super Power*. Lexington, KY: The University Press of Kentucky, 1988.

Greene, Kate. "Why the United Arab Emirates is Building a Space Program from Scratch," *Slate*, March 30, 2017. https://slate.com/technology/2017/03/why-the-united-arab-emirates-is-building-a-space-program.html.

Gregor, James A. and Maria Hsia Chang. "Anti-Confucianism: Mao's Last Campaign," *Asian Survey* 19, no. 11 (November 1979):1073–92.

Grush, Loren. "Why You Shouldn't Compare Blue Origin's Rocket Landing to SpaceX," *The Verge*, November 24, 2015. https://www.theverge.com/2015/11/24/9793220/blue-origin-vs-spacex-rocket-landing-jeff-bezos-elon-musk

Gruss, Mike. "Pentagon Says 2013 Chinese Launch May Have Tested Antisatellite Technology," *Space News*, May 14, 2015. https://spacenews.com/pentagon-says-2013-chinese-launch-may-have-tested-antisatellite-technology/

Gruss, Mike. "U.S. State Department: China Tested Anti-Satellite Weapon," *Space News*, July 28, 2014. https://spacenews.com/41413us-state-department-china-tested-anti-satellite-weapon/

Gump, David. *Space Enterprise: Beyond NASA*. New York: Praeger Publishers, 1990.

H.R. 4218. NSS. https://space.nss.org/media/Space-Settlement-Act-Of-1988.pdf

H.R. 4286 97th Congress, *SCRIBD*, July 28, 1981. https://www.scribd.com/doc/79483990/H-R-4286-97th-Congress

H.R.2262 - U.S. Commercial Space Launch Competitiveness Act 114th Congress (2015-2016. Congress.gov. November 25, 2015. https://www.congress.gov/bill/114th-congress/house-bill/2262/text

H.R.2809 - American Space Commerce Free Enterprise Act, 115th Congress., introduced June 7, 2017. https://www.congress.gov/bill/115th-congress/house-bill/2809

H.R.4945 - American Space Renaissance Act in the 114th Congress (2015-2016), Congress.gov, January 6, 2017, https://www.congress.gov/bill/114th-congress/house-bill/4945/text

H.R.5063 – ASTEROIDS Act. 113th Congress, Congress.gov. introduced July 10, 2014. https://www.congress.gov/bill/113th-congress/house-bill/5063

Han Fei Tzu: Basic Writings, translated by Burton Watson. New York: Columbia University Press, 1964.

"Han Feizi," n.d. http://www2.hawaii.edu/~freeman/courses/phil301/13.%20Han%20Feizi.pdf

Handmer, Casey. "Starlink is a Very Big Deal," *Casey Handmer's Blog*, November 2, 2019. https://caseyhandmer.wordpress.com/2019/11/02/starlink-is-a-very-big-deal/

Handmer, Casey. "The SpaceX Starship is a very big deal," Casey Handmer's Blog, October 29, 2019. https://caseyhandmer.wordpress.com/2019/10/29/the-spacex-starship-is-a-very-big-deal/

Hans, Brian, Christopher Jefferson, Joshua Wehrle. *Movement and Maneuver in Deep Space: A Framework to Leverage Advanced Propulsion*. Maxwell Air Force Base: Air University Press, 2018. https://apps.dtic.mil/dtic/tr/fulltext/u2/1042207.pdf, updated here: https://media.defense.gov/2019/May/23/2002135778/-1/-1/0/WF_0067_HANS_JEFFERSON_WEHRLE_MOVEMENT_AND_%20MANEUVER_IN_DEEP%20SPACE_A_FRAMEWORK_TO_LEVERAGE_ADVANCED_PROPULSION.PDF

Harris, Francis. "Beijing Secretly Fires Lasers to Disable U.S. Satellites," *The Telegraph*, September 26, 2006. https://www.telegraph.co.uk/news/worldnews/1529864/Beijing-secretly-fires-lasers-to-disable-US-satellites.html

Harry, G. Stine. *The Third Industrial Revolution*. New York: Ace Books, 1979.

Hart, Jonathan. *Contesting Empires: Opposition, Promotion and Slavery*. New York: Palgrave, MacMillan, 2005.

Headrick, Daniel R. "The Tools of Imperialism: Technology and the Expansion of European Colonial Empires in the Nineteenth Century," *The Journal of Modern History* 51, no. 2 (June 1979): 231–63.

Heater, Brian. "A Blockchain Firm Bought Asteroid Mining Company Planetary Resources," *Techcrunch*, November 3, 2018. https://techcrunch.com/2018/11/03/a-blockchain-firm-bought-asteroid-mining-company-planetary-resources/

Heaver, Stuart. "Chairman Mao's Historic Swim-Glorified in China but Ridiculed by the Rest of the World," *South China Morning Post*, August 4, 2016. https://www.scmp.com/magazines/post-magazine/long-reads/article/1999098/chairman-maos-historic-swim-glorified-china

Heilman, Wayne. "Space Symposium 2019: Commerce Secretary Warns of Moon Becoming 'Wild West'," *Gazette*, April 9, 2019. https://gazette.com/business/space-symposium-commerce-secretary-warns-of-moon-becoming-wild-west/article_f24f7b62-5ad6-11e9-abd9-97c2e5f7da0f.html

Hellgren, Vide. *Asteroid Mining: A Review of Methods and Aspects*. Lund, Sweden: Lund University, 2016. https://lup.lub.lu.se/luur/download?func=downloadFile&recordOId=8882371&fileOId=8884121

Hendrix, Jerry and Adam Routh. "A Space Policy for the Trump Administration," *Center for New American Security (CNAS)*, November 2016. https://s3.amazonaws.com/files.cnas.org/documents/CNASReport-SpacePolicy-Final5.pdf?mtime=20171023151801

Hendrix, Jerry and Michelle Shevin-Coetzee. "From Blue to Black: Applying the Concepts of Sea Power to the Ocean of Space," *Center for New American Security (CNAS)*, November 2016. https://s3.amazonaws.com/files.cnas.org/documents/CNAS-Report-FromBluetoBlack-2016.pdf?mtime=20161118095339

Henry, Caleb. "Indian Startup Bellatrix Aerospace Raises $3 Million," *SpaceNews*, June 21, 2019. https://spacenews.com/indian-startup-bellatrix-aerospace-raises-3-million/

Henry, Caleb. "SpaceX Submits Paperwork for 30,000 more Starlink Satellites," *Space News*, October 15, 2019. https://spacenews.com/spacex-submits-paperwork-for-30000-more-starlink-satellites/

Henson, H. and Arel Lucas. "Star Laws," *Reason*, August 1982. https://reason.com/1982/08/01/star-laws

Hertz, John H. "Idealist Internationalism and the Security Dilemma," *World Politics* 2, no. 2 (January 1950): 157–80.

Heydarian, Richard Javed. "Hide your Strength, Bid your Time," *al Jazeera*, November 21, 2014. http://www.aljazeera.com/indepth/opinion/2014/11/china-hide-your-strength-bide-y-201411198028498329.html

Higgenbotham, Brian. "The Space Economy: An Industry Takes Off," *U.S. Chamber of Commerce*, November 11, 2018. https://www.uschamber.com/series/above-the-fold/the-space-economy-industry-takes

Hitchens, Theresa. "Commercial Satellites: Will They Be Military Targets?" *Breaking Defense*, July 16, 2019. https://breakingdefense.com/2019/07/commercial-satellites-will-they-be-military-targets/

Hitchens, Theresa. "Pentagon Eyes Military Space Station," *Breaking Defense*, July 02, 2019. https://breakingdefense.com/2019/07/pentagon-eyes-military-space-station/

Hitchens, Theresa. "SDA's Kennedy: Cislunar Space The Next Military Frontier," *Breaking Defense*, April 17, 2019, https://breakingdefense.com/2019/04/sdas-kennedy-cislunar-space-the-next-military-frontier/

Hitopadesha. translated by Sir Edwin Arnold, London, 1861. http://www.columbia.edu/itc/mealac/pritchett/00litlinks/hitopadesha_arnold/index.html

Hoey, Matt. "Space Alert 21: India Developing Space Weapons," *Military Space Transparency Project*, June 2010. http://www.space4peace.org/articles/india_developing_space_weapons.htm

Hofstede, Geert. *Culture's Consequences: International differences in Work-Related Values*. Vol. 5. London and Beverly Hills: Sage, 1984.

Holland, Kris. "Space Based Solar Power: A Solution to the Carbon Crisis," *You Tube*, April 2016. https://www.youtube.com/watch?v=zrcoD_vHzxU

Hong, Brendon. "China's Looming Land Grab in Outer Space," *Newsweek*, June 22, 2016. https://www.thedailybeast.com/chinas-looming-land-grab-in-outer-space

Hongpei, Zhang and Yang Kunyi. "Luxembourg Joins the Ranks of BRI Participants After Italy to be a Part of Inclusive Initiative," *Global Times*, March 28, 2019. http://www.globaltimes.cn/content/1143864.shtml

Hooker, Jr, R. D., *The Grand Strategy of the United States*, INSS Strategic Monograph, Institute for National Strategic Studies, National Defense University, Washington, DC, October 2014. https://inss.ndu.edu/Portals/68/Documents/Books/grand-strategy-us.pdf

Howell, Elizabeth. "China National Space Administration: Facts & Information," *Space.com*, May 25, 2016. http://www.space.com/22743-china-national-space-administration.html

Howell, Elizabeth. "Tiangong-1: China's First Space Station," *Space.com*, March 26, 2018. https://www.space.com/27320-tiangong-1.html

Howell, Elizabeth. "Voyager 1: Earth's Farthest Spacecraft," *Space.com*, March 1, 2018. https://www.space.com/17688-voyager-1.html

Howlett, Darrl. *The Future of Strategic Culture. SAIC* (Prepared for Defense Threat Reduction Agency), 2006. http://www.socsci.uci.edu/~bskyrms/bio/papers/StagHunt.pdf

HQ Integrated Defense Staff PP&FD/WSOI Division Maritime DTE. *Technology Perspective and Capability Roadmap* (TPCR 01/10). May 20, 2010.

Huang, Echo. "China lays Out its Ambitions to Colonize the Moon and Build a Lunar Palace," *Quartz*, April 26, 2018. https://qz.com/1262581/china-lays-out-its-ambitions-to-colonize-the-moon-and-build-a-lunar-palace/

Huang, Echo. "Chinese Researchers are Figuring Out How to make Dumplings, Noodles and Stir-Fries in Space." *Quartz*, July 20, 2017. https://qz.com/1034097

/a-chinese-university-is-putting-students-on-a-grueling-space-diet-of-food-grown-with-their-own-poop/

Hui, Jiang. "The Spatial Information Corridor Contributes to UNISPACE + 50," *China National Space Administration.*, n.d. http://www.unoosa.org/documents/pdf/copuos/stsc/2018/tech-08E.pdf

Hui, Mary. "These are all the Crimes Hong Kong is Considering Extraditing People for Under a New Law," *Quartz*, June 7, 2019. https://qz.com/1636663/the-37-crimes-included-in-hong-kongs-proposed-extradition-law/

Huixian, Sun, et al. "Scientific Objectives and Payloads of Chang'e-1 Lunar Satellite," *Journal of Earth System Science* 114, no. 6 (December 2005): 789–94. https://link.springer.com/article/10.1007%2FBF02715964.

Human Research Protection Board. University of Pittsburg. n.d. https://www.irb.pitt.edu/content/department-defense

Hurd, Ian. "Construtivism" (Chapter 17), Northwestern University, January 18, 2008. http://faculty.wcas.northwestern.edu/~ihu355/Home_files/17-Smit-Snidal-c17.pdf

IISS. "The Military Balance 2018," February 1, 2018. https://www.iiss.org/publications/the-military-balance/the-military-balance-2018

Ikenberry, G. John. *After Victory: Institutions, Strategic Restraint, and the Rebuilding of Order after Major Powers*. New Jersey: Princeton University Press, 2001.

Ikenberry, G. John and Charles Kupchan. "Socialization and Hegemonic Power," *International Organization* 44, no. 3 (Summer 1990): 283–316.

Ikenberry, G. John, "Institutions, Strategic Restraint, and the Persistence of American Postwar Order," *International Security* 23, no. 3 (Winter 1998/1999): 43–78.

Ikenberry, John G. *Liberal Leviathan: The Origins, Crisis, and Transformation of the American World Order*. Princeton, NJ: Princeton University Press, 2012.

Impey, Chris. "How China Entered the Space Race," *The Wired*, April 13, 2015. https://www.wired.com/2015/04/how-china-entered-the-space-race/

Imster, Eleanor. "Parker Solar Probe Becomes Closest-Yet Spacecraft to Sun," *EarthSky*, October 20, 2018. https://earthsky.org/space/parker-solar-record-closest-yet-to-sun-oct2018

Indian Express. "Explained: What Makes PSLV-C45 Special," *The Indian Express*, April 2, 2019. https://indianexpress.com/article/explained/pslv-c45-rocket-launch-indian-space-research-organisation-isroc-5653610/

Indian Space Research Organisation (ISRO). *From Fishing Hamlet to Red Planet*. Bengaluru: ISRO, 2015.

Ingber, Sasha. "Cotton Seeds Sprout in China's Lunar Lander," *NPR*, January 15, 2019. https://www.npr.org/2019/01/15/685449027/cotton-seeds-have-sprouted-in-chinas-lunar-lander

InsightsIAS. "Space Activities Bill, 2017," January 1, 2019. https://www.insightsonindia.com/2019/01/01/space-activities-bill-2017/

Institute of Air and Space Law of Leiden University. "The Hague International Space Resources Governance Working Group: The Consortium," (The Netherlands), n.d. https://www.universiteitleiden.nl/en/law/institute-of-public-law/institute-of-air-space-law/the-hague-space-resources-governance-working-group

Institute for Defence Studies & Analyses. *Strategic Trends 2050 Project*, 2013. https://www.facebook.com/indiaon15august2047/posts/idsa-drdo-strategic-trends-2050projectthe-strategic-trends-2050-projectst2050-jo/1590640924502650/

International Academy of Astronautics (IAA). *Space Mineral Resources: A Global Assessment of Challenges and Opportunities.* 2015.

International Institute of Space Law. *Does International Space Law Either Permit or Prohibit the Taking of Resources in Outer Space and on Celestial Bodies, and how is this Relevant for National Actors? What is the Context, and what are the Contours and Limits of this Permission or Prohibition? (Background Paper).* International Institute of Space Law, 2016. http://iislweb.org/docs/IISL_Space_Mining_Study.pdf

International Institute of Space Law. *Position Paper on Space Resource Mining.* December 20, 2018. https://iislweb.org/iisl-position-paper-on-space-resource-mining/

Jaffe, Paul, Kaylin Borders, Colin Browne, Chris DePuma, Lewis Longbottom, Hassan Nisar, Vinay Simlot, Avram Bar-Cohen, James McSpadden, Ryan Brandt, Eric Conrad, Katherine Duncan, William Rowley, Tony Thampan, Peter Garretson, John C. Mankins, Seth Potter, Eric Sundberg, Alexander Walts. *Opportunities and Challenges for Space Solar for Remote Installations.* Washington, DC: Naval Research Lab, October 21, 2019.

Jain, Naveen. 'If We Can Learn to Live on the Moon We Can Live Anywhere in Space,' *LSE Business Review*, May 11, 2017. https://blogs.lse.ac.uk/businessreview/2017/05/11/naveen-jain-if-we-can-learn-to-live-on-the-moon-we-can-live-anywhere-in-space/

Jain, Rupam and Tom Lasseter. "By Rewriting History, Hindu Nationalists Aim to Assert their Dominance over India," *Reuters*, March 6, 2018. https://www.reuters.com/investigates/special-report/india-modi-culture/

Jakhu, Ram S., Joseph N. Pelton, Yaw Otu Mankata Nyampong. *Space Mining and Its Regulation.* New York: Springer International Publishing, 2017.

James, Tom. *Deep Space Commodities: Exploration, Production and Trading.* Switzerland: Palgrave Macmillan, 2018.

Jayaraman, K. S. "Indian Government Set To Create Aerospace Command," *SpaceNews*, October 31, 2005. https://spacenews.com/indian-government-set-create-aerospace-command/

Jervis, Robert. "Cooperation Under the Security Dilemma," *World Politics*, 30, no. 2 (January 1978): 167–214.

Ji, Gao et al. "Solar Power Satellites Research in China," *Online Journal of Space Communication*, no. 16, (Winter 2010). https://spacejournal.ohio.edu/issue16/ji.html

Jinks, Beth. "Google Chiefs Back Startup Mining Asteroids for Metals," *Bloomberg*, April 6, 2012. https://www.bloomberg.com/news/articles/2012-04-24/google-chiefs-back-startup-mining-asteroids-for-metals

Johnson, Jeannie L. *Strategic Culture: Refining the Theoretical Construct.* SAIC (for Defense Threat Reduction Agency), 2006.

Johnson, Richard D. and Charles H. Holbrow, eds. "Space Settlements: A Design Study," 413, Scientific and Technical Information Office, National Aeronautics and Space Administration, 1977. https://settlement.arc.nasa.gov/75SummerStudy/Chapt7.html

Johnson-Freese, Joan. *Heavenly Ambitions: America's Quest to Dominate Space.* Philadelphia, PA: University of Pennsylvania Press, 2009.

Johnson-Freese, Joan. *Space as a Strategic Asset.* New York: Columbia University Press, 2007.

Johnston, Alastair Iain. "Thinking about Strategic Culture," *International Security* 19, no. 4, Spring (1995): 32–64. http://www.fb03.uni-frankfurt.de/45431264/Johnston-1995-Thinking-about-Strategic-Culture.pdf

Johnston, Alastair Iain. *Cultural Realism: Strategic Culture and Grand Strategy in Chinese History.* Princeton: Princeton University Press, 1998.

Jordaan, Eduard. "The Concept of a Middle Power in International Relations: Distinguishing Between Emerging and Traditional Middle Powers," *Politikon: South African Journal of Political Studies* 30, no. 1 (2003):165–81. https://ink.library.smu.edu.sg/cgi/viewcontent.cgi?article=1393&context=soss_research

Joshi, Manoj. "India- A Counterweight to the Rise of China," *Observer Research Foundation*, September 17, 2014. https://www.orfonline.org/research/india-a-counterweight-to-the-rise-of-china/

Juergensmeyer, Mark. *Gandhi's Way A Handbook of Conflict Resolution.* Berkeley: University of California Press, 2005.

Kader, Binsal Abdul. "UAE Space Agency Unveils Ambitious Plans," *Gulfnews*, May 25, 2017, http://gulfnews.com/news/uae/government/uae-space-agency-unveils-ambitious-plans-

Kaiman, Jonathan. "Chinese Passport Maps Causes Diplomatic Row," *The Guardian*, November 27, 2012. https://www.theguardian.com/world/2012/nov/27/chinese-passport-row-diplomatic-dispute

Kalam, A. P. J. "Space Solar Power—Key to a Livable Planet," Address at the 32nd International Space Development Conference (ISDC) and Acceptance speech on receiving Wernher Von Braun Memorial Award. San Diego, California, May 24, 2013. https://space.nss.org/media/Kalam-Address-ISDC-2013.pdf video here: https://space.nss.org/2013-isdc-video-dr-abdul-kalam-space-solar-power-key-to-a-livable-planet-earth/

Kalam, A. P. J. "The Future of Space Exploration and Human Development," *The Pardee Papers*, No. 1, Boston University, August 2008. https://www.bu.edu/csp/Conferences/Space_Exploration/Pardee_ThinkBooklet_andCover.pdf http://www.bu.edu/csp/Conferences/Space_Exploration/Day1/Presentations/Kalam_Space%20Exploration%20and%20Human%20Life.pdf see video of talk here: https://www.bu.edu/pardee/space-abdul-kalam/

Kalil, Tom. "Bootstrapping a Solar System Civilization," *The White House Blog*, October 14, 2014. https://obamawhitehouse.archives.gov/blog/2014/10/14/bootstrapping-solar-system-civilization

Kangle, R. P. *The Kautilya Arthasashtra An English Translation with Critical and Explanatory Notes.* New Delhi: Motilal Banarsidass Publ., 1986.

Kaplan, Morton A. *System and Process in International Politics*. New York: John Wiley, 1957.

Kaplan, Sarah. "NASA's New Horizons Spacecraft Just Visited the Farthest Object Ever Explored," *The Washington Post*, January 1, 2019. https://www.washingtonpost.com/science/2018/12/31/most-distant-space-encounter-history-is-happening-now/?noredirect=on&utm_term=.54345095fc60

Karthik, Vijay. "ISRO Working on Launching Spacecraft to Venus, Asteroids: A.S. Kiran Kumar," *LiveMint*, September 8, 2016. http://www.livemint.com/Science/1Qn4uyqO0pe8iwIbqKX9AN/Isro-working-on-launching-spacecraft-to-Venus-asteroids-A.html

Kassab, Hanna Samir. *Grand Strategies of Weak States and Great Powers*. Basingstoke: Palgrave Macmillan, 2018.

Katzenstein, Peter. "International Relations and Domestic Structures: Foreign Economic Policies of Advanced Industrial States," *International Organization* 30, no. 1 (Winter 1976): 1–45.

Kautilya, *Arthasashtra*. New Delhi: Penguin, 1992.

Kawa Space Website. n.d., https://kawa.space/about.html

Kaza, Lalitendra. *Militarisation of Space*. New Delhi: Center for Air Power Studies, 2010.

Keck Institute for Space Studies (KISS). *Asteroid Retrieval Study*. Keck Institute, 2012. http://kiss.caltech.edu/final_reports/Asteroid_final_report.pdf

Kellstedt, Paul M. and Guy D. Whitten. *The Fundamentals of Political Science Research* 2nd edition. Cambridge, UK: Cambridge University Press, 2013.

Kennan, George F. *American Diplomacy: Expanded Edition*. Chicago: University of Chicago Press, 1984.

Keohane, Robert O. *After Hegemony: Cooperation and Discord in the World Political Economy*. Princeton, NJ: Princeton University Press, 1984.

Khan, Taimur. "Mars Mission 'Arab World's Kennedy Moon Shot', Says UAE," *The National*, December 3, 2015. https://www.thenational.ae/world/mars-mission-arab-world-s-kennedy-moon-shot-says-uae-ambassador-to-us-1.103660

King Jr., Martin Luther. "Letter from the Birmingham Jail" (April 16, 1963), Charter for Compassion, n.d. https://charterforcompassion.org/images/SocialJustice/birmingham.pdf

King, Gary, Robert O. Keohane and Sidney Verba. *Designing Social Inquiry: Scientific Inference In Qualitative Research*. Princeton, NJ: Princeton University Press, 1994.

King, Ledyard. "President Obama Proposes $19 Billion for NASA in Fiscal 2017 Budget," *USA TODAY*, February 9, 2016. https://www.usatoday.com/story/news/politics/2016/02/09/president-obama-proposes-19-billion-nasa-fiscal-2017/80053964/

Kingdon, John W. *Agenda's, Alternatives, and Public Policies*. New York: Longman, 2003.

Kirk, G. S. "Greek Mythology: Some New Perspectives," *The Journal of Hellenic Studies* 92 (1972): 74–85. https://www.jstor.org/stable/629974?seq=1#page_scan_tab_contents

Kirtland Public Affairs. "U.S. Air Force Research Laboratory Developing Space Solar Power Beaming," *Air Force Research Laboratory*, October 24, 2019. https:/

/www.robins.af.mil/News/Article-Display/Article/1998062/us-air-force-research-laboratory-developing-space-solar-power-beaming/

Kissinger, Henry. *On China*. New York/London: Penguin/Random House, 2012.

Kittrie, Orde F. “The Chinese Government Adopts and Implements a Lawfare Strategy,” in Kittrie, Orde, F., *Lawfare: Law as a Weapon of War*. Oxford: Oxford University Press, 2016.

Kivimaki, Timo. “The Long Peace of ASEAN,” *Journal of Peace Research* 38, no. 1 (January 2001): 5–25. https://www.jstor.org/stable/425780?seq=1#page_scan_tab_contents

Klare, Michael T. *The Race for What’s Left: The Global Scramble for the World’s Last Resources*. London, UK: Picador, 2012.

Klein, John. *Space Warfare: Strategy, Principles, and Policy*. London & New York: Routledge, 2006.

Klein, John. *Understanding Space Strategy: The Art of War in Space (Space Power and Politics)*. London & New York: Routledge, 2019.

Klerkx, Greg. *Lost in Space: The Fall of NASA and the Dream of a New Space Age*. New York: Vintage Books, 2004 & 2005.

Klotz, Irene. “Tech Billionaires Bankroll Gold Rush to Mine Asteroids,” *Reuters*, April 24, 2012. https://www.reuters.com/article/space-asteroid-mining/tech-billionaires-bankroll-gold-rush-to-mine-asteroids-idUSB27208920120424

Kluger, Jeffrey. “Elon Musk Told Us Why He Thinks We Can Land on the Moon in ‘Less Than 2 Years’,” *Time*, July 18, 2019. https://time.com/5628572/elon-musk-moon-landing/

Knapton. Sarah. “Elon Musk: We’ll create a city on Mars with a million inhabitants,” *The Telegraph UK*, June 21, 2018. https://www.telegraph.co.uk/science/2017/06/21/elon-musk-create-city-mars-million-inhabitants/

Kocs, Stephen A. “Explaining the Strategic Behaviour of States: International Law as System Structure,” *International Studies Quarterly* 38, no. 4 (December 1994): 535–56.

Koren, Marina. “What Does Trump Mean By 'Space Force'?,” *The Atlantic*, March 13, 2018. https://www.theatlantic.com/science/archive/2018/03/trump-space-force-nasa/555560/

Kotoky, Anurag. “ISRO in Quest to Find a Trillion-Dollar Nuclear Fuel on the Moon,” *Hindustan Times*, June 27, 2018. https://www.hindustantimes.com/science/isro-s-quest-to-find-a-trillion-dollar-nuclear-fuel-on-the-moon/story-2JXQrUGaUbhfUDe6Wu8NKO.html

Kotoky, Anurag. “The Quest to Find a Trillion-Dollar Nuclear Fuel on the Moon,” *Bloomberg*, June 26, 2018. https://www.bloomberg.com/news/articles/2018-06-26/the-quest-to-find-a-trillion-dollar-nuclear-fuel-on-the-moon

Kramer, Mark. “Ideology and the Cold War,” *Review of International Studies* 25, no.4 (October 1999): 539–76

Krepon, Michael. “Weapons in the Heavens: A Radical and Reckless Option,” *Arms Control Today* 34, no. 9 (November 2004).

Kuhn, Thomas S. *The Structure of Scientific Revolutions*. Chicago, IL: University of Chicago Press, 1996.

Kumar, Rajesh. "Negotiating with The Complex, Imaginative Indian," *Ivey Business Journal*, March 2005. https://iveybusinessjournal.com/publication/negotiating-with-the-complex-imaginative-indian/

Kumar, Sanjay. "India Joins Elite Mars Club," *NATURE*, September 24, 2014. https://www.nature.com/news/india-joins-elite-mars-club-1.15997

Kuo, Lily. "In China, They are Closing Churches, Jailing Pastors-and Even Rewriting Scriptures," *The Guardian*, January 13, 2019. https://www.theguardian.com/world/2019/jan/13/china-christians-religious-persecution-translation-bible

Kutter, Barnard F. and George F. Sowers. "Cislunar-1000: Transportation Supporting a Self-Sustaining Space Economy," ULA, AIAA Space 2016. available at AIAA. https://arc.aiaa.org/doi/10.2514/6.2016-5491

Kwast, Steve. "The Real Stakes in the New Space Race," *War on the Rocks*, August 19, 2019. https://warontherocks.com/2019/08/the-real-stakes-in-the-new-space-race/

Kwast, Steve. "There Won't Be Many Prizes for Second Place," *Politico*, August 10, 2018. https://www.politico.com/story/2018/08/10/space-race-kwast-768751

Kwast, Steven L. "The Urgent Need for a United States Space Force," *Imprimis* 49, no. 1 (January 2020). https://imprimis.hillsdale.edu/urgent-need-united-states-space-force/

Kyle, Ed. "Space Launch Report," December 29, 2018. http://www.spacelaunchreport.com/log2018.html

L5 Society. "UN Moon Treaty Falling to US Opposition Groups," *L5 News*, March 1982. https://space.nss.org/l5-news-un-moon-treaty-falling-to-us-opposition-groups/

Lamber, Andrew. *Seapower States*. New Haven, CT: Yale University Press, 2019.

Lee, Ricky. *Law and Regulation of Commercial Mining of Minerals in Outer Space*. Vol. 7. New York: Springer Science & Business Media, 2012.

Lee, Valentine. "A Space Roadmap: Mine the Sky, Defend the Earth, Settle the Universe." *Space Studies Institute (SSI)*. May 2002. http://ssi.org/reading/papers/space-studies-institute-roadmap/

Legro, Jeffrey W. "Which Norms Matter? Revisiting the 'Failure' of Internationalism," *International Organization* 51, no. 1, (1997): 31–63.

Lei, Zhao. "Asteroid Mission Planned by 2025," *China Daily*, March 4, 2017. http://www.chinadaily.com.cn/china/2017-03/04/content_28430297.htm

Lei, Zhao. "Civil-Military Integration will Deepen," *China Daily*, March 3, 2018. http://www.chinadaily.com.cn/a/201803/03/WS5a99d67ca3106e7dcc13f437.html

Lei, Zhao. "Scientists Envision Solar Power Station in Space," *China Daily*, February 27, 2019. http://www.chinadaily.com.cn/a/201902/27/WS5c75c8b3a3106c65c34eb8e3.html

Lei, Zhao. "Super-Powerful Long March 9 said to begin missions around 2030," *China Daily*, March 11, 2019. http://www.chinadaily.com.cn/a/201903/11/WS5c859b62a3106c65c34edcc0.html

Lele, Ajay. "Space Activities Bill: India's Great Galactic Leap," *The Financial Express*, July 2, 2019. https://www.financialexpress.com/lifestyle/science/space-activities-bill-indias-great-galactic-leap/1625980/

Lele, Ajey. "Kalam-NSS Indian-American Energy Initiative," *IDSA*, November 9, 2010. https://idsa.in/idsacomments/KalamNSSIndianAmericanEnergyInitiative_alele_091110

Lele, Ajey. "Op-ed | India Needs Its Own Space Force," *Space News*, May 28, 2019. https://spacenews.com/op-ed-india-needs-its-own-space-force/

Lele, Ajey. *Mission Mars India's Quest for the Red Planet*. New Delhi: Springer Briefs in Applied Sciences and Technology, 2014.

Lenin, Vladimir Il'ič. "Imperialism: The Highest Stage of Capitalism," *The New Imperialism: Analysis of late Nineteenth-century Expansion* (1939), available from Resistance Books, 1999. https://www.readingfromtheleft.com/Books/Classics/LeninImperialism.pdf

Levathes, Louise. *When China Ruled the Seas: The Treasure Fleet of the Dragon Throne, 1405-1433*. New York: Oxford University Press, 1997.

Levy, Jack S. "Too Important to Leave to the Other: History and Political Science in the Study of International Relations," *International Security* 22, no. 1 (Summer 1997): 22–33.

Lewis, Jeffrey. "An Aerospace Command for India?," *Arms Control Wonk*, May 1, 2005. https://www.armscontrolwonk.com/archive/200570/an-aerospace-command-for-india/

Lewis, John S. "'Demandite' and Resources in Space," Dr. John S. Lewis website (author of *Mining the Sky* and *Asteroid Mining 101* and Chief Technical Advisor Deep Space Industries (DSI)), January 13, 2011. http://www.johnslewis.com/2011/01/demandite-and-resources-in-space.html

Lewis, John S. *Asteroid Mining 101: Wealth for the New Space Economy*. Mountain View, CA: Deep Space Industries, 2014.

Lewis, John S. *Mining the Sky: Untold Riches from the Asteroids, Comets, and Planets*. Reading, MA: Addison-Wesley Pub. Co., 1996.

Li, Chenyang. "The Confucian Ideal of Harmony," *Philosophy East and West* 56, no. 4 (October 2006): 583–603. http://doi.org/ftvnxn

Liang, Qiao and Wang Xiangsui. *Unrestricted Warfare: China's Master Plan to Destroy America*. Panama City, Panama: Pan American Publishing Company, 2002.

Lin, Jeffrey and Peter W. Singer. "China Shot Down Another Missile in Space," *Popular Science*, February 13, 2018. https://www.popsci.com/china-space-missile-test

Lipp, Charles D. et al. *Towards a Theory of Spacepower Selected Essays*. Washington, DC: National Defense University, 2001. https://ndupress.ndu.edu/Portals/68/Documents/Books/spacepower.pdf

Lippmann, Walter. *Public Opinion and Foreign Policy in the United States*. London, UK: Allen and Unwin, 1952.

Listner, Michael. "The Art of Lawfare And The Real War In Outer Space," *The Space Review*. September 17, 2018. http://www.thespacereview.com/article/3571/1

Litner, Bertil. *China's India War: Collison Course on the Roof of the World*. New Delhi: Oxford University Press, 2018.

Lo, Pin-Cheung. "The 'Art of War' Corpus and Chinese Just War Ethics Past and Present," *The Journal of Religious Ethics*, 40, no. 3 (September 2012): 404–46.

Long, Wei. "New Details of Shenzhou and its Launcher Revealed," *Space Daily*, April 1, 2002. http://www.spacedaily.com/news/china-02zf.html

Lord, Carnes. "American Strategic Culture," *Comparative Strategy* 5, no. 3, (1985): 269–93. doi: 10.1080/01495938508402693

Lovelace, Berkeley. "Wilbur Ross on Space Race: We Want to Turn the Moon into a Gas Station," *CNBC*, February 22, 2018. https://www.cnbc.com/2018/02/22/wilbur-ross-on-space-race-we-want-to-turn-the-moon-into-a-gas-station.html

Lunarpedia. "Space Colonization Technical Committee," December 7, 2019. http://carbondate.cs.odu.edu/#http://lunarpedia.org/w/American_Institute_of_Aeronautics_and_Astronautics

Lutes, Charles D. et al. *Towards a Theory of Spacepower Selected Essays*, Washington, DC: National Defense University, 2001. https://ndupress.ndu.edu/Portals/68/Documents/Books/spacepower.pdf

Luxembourg Space Agency. "UNOOSA and Luxembourg Launch New 'Space Law for New Space Actors" *Project*," November 13, 2019. https://space-agency.public.lu/en/news-media/news/2019/UN_and_Luxembourg_sign_project.html

Luxembourg Space Agency. "Space Resources," n.d. https://space-agency.public.lu/en/space-resources.html

Luxembourg Space Agency. "UNOOSA and Luxembourg Launch New 'Space Law for New Space Actors' Project," November 13, 2019. https://space-agency.public.lu/en/news-media/news/2019/UN_and_Luxembourg_sign_project.html

Luxembourg Trade & Invest. "Luxembourg and China Agreed to Cooperate on Space Exploration," January 17, 2018. https://www.tradeandinvest.lu/news/luxembourg-china-sign-agreement/

Luxemburg, Rosa. *Accumulation of Capital*, originally published 1913, London: Routledge and Kegan Paul, 1951; translated by Agnes Schwarzschild and with an introduction by Joan Robinson.

Ma, Alexandra. "Barging into Your Home, Threatening Your Family, or Making You Disappear: Here's What China does to People Who Speak Out Against Them," *Business Insider*, August 19, 2018. https://www.businessinsider.com/how-china-deals-with-dissent-threats-family-arrests-2018-8

Machi, Vivienne. "U.S. Space Sectors Must Work Together to Combat Emerging Threats, Official Says," *Defense Daily*, October 12, 2018. https://www.defensedaily.com/u-s-space-sectors-must-work-together-combat-emerging-threats/space/

Mackinder, Halford. "The Geographical Pivot of History," *The Geographical Journal* (Royal Geographical Association), April 1904. https://rgs-ibg.onlinelibrary.wiley.com/doi/abs/10.1111/j.0016-7398.2004.00132.x

Madani, Kaveh. "Game Theory and Water Resource," *Journal of Hydrology* 381, no. 3–4 (2010): 225–38. https://ethz.ch/content/dam/ethz/special-interest/mtec/necom-dam/documents/Applied%20Negotiation%20Seminar/10_Madani_Game_theory_and_water_resources.pdf

Madhumathi, D. S. "India Needs at Least Three More Spy Satellites," *The Hindu*, April 1, 2019. https://www.thehindu.com/news/national/india-needs-at-least-three-more-spy-satellites/article26705497.ece

Madhumathi, D. S. "Starting May, ISRO to Launch a String of 'Defence' Satellites," *The Hindu*, April 2, 2019. https://www.thehindu.com/news/national/starting-may-isro-to-launch-a-string-of-defence-satellites/article26714439.ece

Magoulick, Mary. "What is a Myth?," September 15, 2014. https://faculty.gcsu.edu/custom-website/mary-magoulick/defmyth.htm

Mahaskey, M. Scott. "We Want Access to Every Part of the Moon, At Any Time," *Politico*, June 13, 2019. https://www.politico.com/agenda/story/2019/06/13/jim-bridenstine-nasa-interview-000900

Mahnken, Thomas G. "United States Strategic Culture," *SAIC* (Prepared for Defense Threat Reduction Agency), 2006. http://www.au.af.mil/au/awc/awcgate/dtra/mahnken_strat_culture.pdf

Malik, Tariq. "How Trump's Space Force Would Help Protect Earth from Future Asteroid Threats," *Space.com*, June 20, 2018. https://www.space.com/40949-trump-space-force-asteroid-defense.html (Plan available here: https://www.whitehouse.gov/wp-content/uploads/2018/06/National-Near-Earth-Object-Preparedness-Strategy-and-Action-Plan-23-pages-1MB.pdf)

Malik, Tariq. "India Delays Chandrayaan-2 Moon Lander Launch Over 'Technical Snag'," *Space.com*, July 14, 2019. https://www.space.com/india-delays-moon-lander-chandrayaan-2-launch.html

Malik, Tariq. "Obama Budget Scraps NASA Moon Plan for '21st Century Space Program'," *Space.com*, February 01, 2010. https://www.space.com/7849-obama-budget-scraps-nasa-moon-plan-21st-century-space-program.html

Mallick, Senjuti and Rajeswari Pillai Rajagopalan. "If Space is 'the Province of Mankind', Who Owns its Resources? The Potential of Space Mining and its Legal Implications," *ORF Occasional Paper No. 182*, January 2019. https://www.orfonline.org/research/if-space-is-the-province-of-mankind-who-owns-its-resources-47561/

Mandelbaum, Ryan F. "Decolonizing Mars: Are We Thinking About Space Exploration All Wrong?," *Gizmodo*, November 20, 2018. https://gizmodo.com/decolonizing-mars-are-we-thinking-about-space-explorat-1830348568 ;

Mankins, John and Nobuyuki Kaya, eds. *Space Solar Power: The First International Assessment of Space Solar Power: Opportunities, Issues and Potential Pathways Forward*. Paris: International Academy of Astronautics, 2011.

Mankins, John. *The Case for Space Solar Power*. Houston, TX: Virginia Edition Publishing, 2014.

Mann, Adam. "Golden Spike Company Unveils Plans to Fly Commercial Crews to the Moon," *Wired*, December 6, 2012. https://www.wired.com/2012/12/golden-spike-announcement/

Mann, Adam. "Tech Billionaires Plan Audacious Mission to Mine Asteroids," *Wired*, April 23, 2012. https://www.wired.com/2012/04/planetary-resources-asteroid-mining/

Mansharamani, Vikram. "China Wants to Mine the Moon for "Space Gold," *PBS*, March 31, 2016. https://www.pbs.org/newshour/economy/china-wants-to-mine-the-moon-for-space-gold

Marburger, John. "Keynote Address," 44th Robert H. Goddard Memorial Symposium Greenbelt, Maryland, March 15, 2006. https://www.lpi.usra.edu/lunar/strategies/WhiteHouseDocuments/marburger_goddard06.pdf

Martínez-Vela, Carlos A. "World Systems Theory," *Engineering System division* 83 (2001): 1–5.

Marx, Karl. *Capital: A Critique of Political Economy, Volume 1*. Originally published 1867. Arkose Press, 2015.

Masten Space Systems n.d. https://www.masten.aero/Matignon,

Maxwell, Neville. *India's China War*. New York: Random House, 2000.

Mayring, Philipp. "Qualitative Content Analysis," *A Companion to Qualitative Research* 1 (2004): 159–76.

McBeth, Mark K., Michael D. Jones, Elizabeth A. Shanahan. "The Narrative Policy Framework," in *Theories of the Policy Process*. Edited by Paul A. Sabatier and Christopher M. Weible, 173–213 (Boulder: Westview Press, 2019).

McClain, Sean (Major, USAF). *Celestial Sentinels: A Framework for Cis-Lunar Space Domain Awareness In 2035* (Master's Thesis), Air Command and Staff College, Air University, March 2020.

McCrae, Robert R. and Oliver P. John. "An Introduction to The Five-Factor Model and its Applications," *Journal of Personality* 60, no. 2 (1992): 175–215.

McDougall, Walter. *The Heavens and the Earth: A Political History of the Space Age*. Washington, DC: Johns Hopkins University Press, 1997.

McGill, Kevin. "NASA Administrator Supports Trump's 'Space Force' Proposal," *Bloomberg*, August 13, 2018. https://www.bloomberg.com/news/articles/2018-08-13/nasa-administrator-supports-trump-space-force-proposal

McLellan, David, ed. *Karl Marx: Selected Writings*. Oxford: Oxford University Press, 2000.

Meadows, Donella H. et al. Limits of Growth, 1972. https://www.clubofrome.org/report/the-limits-to-growth/

Mearsheimer, John J. *The Tragedy of Great Power Politics*. New York: WW Norton & Company, 2001.

Mearsheimer, John. "Structural Realism," in *International Relations Theories* 3rd ed. Edited by Tim Dunne et al., 77–93. Oxford: Oxford University Press, 2013.

Melton, J. Gordon, ed. *Religious Celebrations: An Encyclopedia of Holidays, Festivals, Solemn Observances, and Spiritual Commemorations*. Santa Barbara: ABC-CLIO, 2011.

Merriam-Webster. "Territoriality," n.d. https://www.merriam-webster.com/dictionary/territoriality

Merton, Robert K. "The Thomas Theorem and The Matthew Effect," *Social Forces* 74, no. 2 (December 1995): 379–424. http://garfield.library.upenn.edu/merton/thomastheorem.pdf

Messier, Doug. "ISRO Budget to Increase 23 Percent, Venus & Mars Missions Planned," *Parabolic*, February 26, 2017. http://www.parabolicarc.com/2017/02/26/isro-budget-increase-23-percent-venus-mars-missions-planned/

Messier, Doug. "Northrop Grumman, AFRL Partner to Provide Critical Advanced Technology in Space Solar Power," *Parabolic Arc*, December 11, 2019, http:

//www.parabolicarc.com/2019/12/11/northrop-grumman-afrl-partner-to-provide-critical-advanced-technology-in-space-solar-power/#more-71617

Metzger, Philip T, Anthony Muscatello, Robert P. Mueller, James Mantovani. "Affordable, Rapid Bootstrapping of Space Industry and Solar System Civilization," *ARVIX* (preprint), December 10, 2016. https://arxiv.org/pdf/1612.03238v1.pdf (paper is also available at https://ascelibrary.org/doi/10.1061/%28ASCE%29AS.1943-5525.0000236)

Meyer, David. "India's Mission to Find and Mine Futuristic Nuclear Fuel on the Moon Is About to Launch," *Fortune*, June 27, 2018. http://fortune.com/2018/06/27/india-moon-helium-3/

Doyle, Michael W., *Liberal Peace: Selected Essays*. New York: Routledge, 2011.

Michaud, Michael A. G. *Reaching for the High Frontier: The American Pro-Space Movement*, 1972-84. Westport, CT: Praeger, 1986. https://space.nss.org/reaching-for-the-high-frontier-the-american-pro-space-movement-1972-84/

Miller, Charles. "Ex-NASA Exec: Gingrich Moon Colony Lost in the Laughter," *CNN*, February 7, 2012. https://www.cnn.com/2012/02/07/opinion/miller-gingrich-space-policy/index.html

Ming, Li. "Using Resources on Asteroid for Manufacturing of SSPS-A New Attempt and its Potential," 2016 International Astronomical Congress, Guagalajara, Mexico, September 26, 2016. https://www.youtube.com/watch?v=6kum9VbVmN8

Ming, Wan. "Reflections on the Study of Zheng He's Expeditions." *Ming Studies* 1 (2004) https://www.tandfonline.com/doi/abs/10.1179/014703704788762899

Mingfu, Liu. *The China Dream: Great Power Thinking and Strategic Posture in the Post-American Era.* Beijing: Beijing Mediatime Books. Co. Ltd, 2015.

Ministry of Defense Press Release (Sitanshu Kar/RAJ (Release ID :39503). "Special cell set up to counter growing threat to space assets." Press Information Bureau, Government of India. June 10, 2008. http://pib.nic.in/newsite/erelease.aspx?relid=39503

Ministry of External Affairs, Government of India, "History and Evolution of Non-Aligned Movement," August 22, 2012. https://mea.gov.in/in-focus-article.htm?20349/History+and+Evolution+of+NonAligned+Movement

Ministry of Foreign Affairs of the People's Republic of China. "Foreign Ministry Spokesperson Lu Kang's Regular Press Conference on June 29, 2017," http://www.fmprc.gov.cn/mfa_eng/xwfw_665399/s2510_665401/t1474166.shtml

Ministry of Foreign Affairs. Government of Bhutan. "Press Release." June 29, 2017. http://www.mfa.gov.bt/?p=4799.

Ministry of National Defense The People's Republic of China, "China Establishes Rocket Force and Strategic Support Force," n.d. http://eng.mod.gov.cn/ArmedForces/ssf.htm

Ministry of Science and Technology of the People's Republic of China. "National High-tech R&D Program (863 Program)," n.d. http://www.most.gov.cn/eng/programmes1/

Mirza, Anzish. "What Are the National Debt, Debt Ceiling and Budget Deficit?," *US News and World Report*, April 13, 2017. https://www.usnews.com/news/economy/articles/2017-04-13/what-are-the-national-debt-debt-ceiling-and-budget-deficit

Möller, Ina. *The Emergent Politics of Geoengineering* [Doctoral Thesis] Lund: Lund University, 2019. https://portal.research.lu.se/portal/files/63030851/Thesis_InaM_ller.pdf

Moltz, James Clay. *Asia's Space Race: National Motivations, Regional Rivalries and International Risk*. New York: Columbia University Press, 2011.

Moltz, James Clay. *The Politics of Space Security: Strategic Restraint and the Pursuit of National Interests*. Paolo Alto: Stanford University Press, 2011.

Momentus Space. "Powered by Water Plasma Propulsion," n.d. https://momentus.space/

Montsion, Jean Michel. "Ethnography and International Relations: Situating Recent Trends, Debates, and Limitations from an Interdisciplinary Perspective," *The Journal of Chinese Sociology* 5, no. 9 (2018): 1–21. https://doi.org/10.1186/s40711-018-0079-4

Moon Express. "Moon Express," n.d. http://www.moonexpress.com/

Moon, Mariella. "Luxembourg's Asteroid Mining Law Takes Effect August 1st It's the first of Its Kind in Europe," *Engadget*, July 30, 2017. https://www.engadget.com/2017/07/30/luxembourg-asteroid-mining-law-august-1/

Moore, Russell A. "Strategic Culture–How It Affects Strategic "Outputs," *CSC*, 1998. https://apps.dtic.mil/dtic/tr/fulltext/u2/a525913.pdf

Morgenthau, Hans. *Politics Among Nations The Struggle for Power and Peace*. New York: Knopf, 1988.

Morris, David Z. "Luxembourg to Invest $227 Million in Asteroid Mining," *Fortune*. June 5, 2016. https://fortune.com/2016/06/05/luxembourg-asteroid-mining/

Mourdoukoutas, Panos. "China Sends America Another Warning on Rare Earth – Seriously!," *Forbes*, July 18, 2019. https://www.forbes.com/sites/panosmourdoukoutas/2019/07/18/china-sends-america-another-warning-on-rare-earths-seriously/#104dac07748c;

Mowthorpe, Mathew. *The Militarization and Weaponization of Space*. Lanham: Lexington Books, 2004.

Mrityunjay, Nigam. "Many Population Analysts Predict India will Overtake China in 4 to 3 Decades Your Thoughts?," *Quora*, March 5, 2017. https://www.quora.com/Many-population-analysts-predict-India-will-overtake-China-in-4-to-3-decades-your-thoughts

Muhammad Bin Rashid Space Center. n.d. https://mbrsc.ae/en/page/aboutus

Mukunth, Vasudevan. "Don't Let the Excitement Over Chandrayaan 2 Distract from the Bigger Picture," *The Wire*, July 4, 2019. https://thewire.in/space/isro-chandrayaan-2-gslv-mk-iii-gaganyaan

Müller, Harald. "Arguing, Bargaining and All That: Communicative Action, Rationalist Theory and the Logic of Appropriateness in International Relations," *European journal of international relations* 10, no. 3 (2004): 395–435.

Murphy, Tom. "Galactic-Scale Energy," *Do the Math (UCSD)*, July 7, 2012. https://dothemath.ucsd.edu/2011/07/galactic-scale-energy/

Murray, Charles and Catherine Bly Cox. *Apollo: The Race to the Moon*. New York: Simon & Schuster, 1989.

Musk, Elon. "The Future We're Building – And Boring." *TED*, 2017. https://www.ted.com/talks/elon_musk_the_future_we_re_building_and_boring/transcript?language=en

Nankievell, Kirk. “Why the Future of Solar Power Is from Space,” *Singularity Hub*, December 31, 2018. https://singularityhub.com/2018/12/31/why-the-future-of-solar-power-is-from-space/

Nansen, Ralph. *Energy Crisis: Solution from Space*. Ontario, Canada: Apogee Books, 2009.

NASA STMD, STMD: Centennial Challenges, October 3, 2018. https://www.nasa.gov/directorates/spacetech/centennial_challenges/history/index.html

NASA. “Sputnik and The Dawn of the Space Age,” April 24, 2001, https://history.nasa.gov/sputnik/

NASA. “John F. Kennedy Moon Speech - Rice Stadium,” (September 12, 1962), *NASA*, January 7, 2004. https://er.jsc.nasa.gov/seh/ricetalk.htm

NASA. “About CLPS,” n.d. https://www.nasa.gov/content/commercial-lunar-payload-services

NASA. “Ansari X-Prize: A Brief History and Background,” n.d. https://history.nasa.gov/x-prize.htm

NASA. “Commercial Lunar Payload Service CLPS,” n.d. https://www.nasa.gov/subject/16440/commercial-lunar-payload-services-clps/

NASA. “Commercial Lunar Payload Services Update,” July 29, 2019. https://www.nasa.gov/feature/commercial-lunar-payload-services-update

NASA. “COTS: Final Report.” May 2014. https://www.nasa.gov/content/cots-final-report

NASA. “Looking at Earth from Space,” *NASA*, August 1994. https://er.jsc.nasa.gov/seh/earth_glossary.pdf

NASA. “NASA Announces US Industry Partnerships to Advance Moon, Mars Technology,” July 30, 2019. https://www.nasa.gov/press-release/nasa-announces-us-industry-partnerships-to-advance-moon-mars-technology

NASA. “NASA Office to Coordinate Asteroid Detection, Hazard Mitigation,” January 7, 2016. https://www.nasa.gov/feature/nasa-office-to-coordinate-asteroid-detection-hazard-mitigation

NASA. “NASA-UAE Space Agency Sign Historic Implementing Arrangement for Cooperation in Human Spaceflight,” October 8, 2018. https://www.nasa.gov/press-release/nasa-uae-space-agency-sign-historic-implementing-arrangement-for-cooperation-in-human

NASA. “National Aeronautics and Space Act of 1958 (Unamended),” April 24, 2001. https://history.nasa.gov/spaceact.html

NASA. “National Space Council Users' Advisory Group,” October 5, 2017. https://www.nasa.gov/content/national-space-council-users-advisory-group

NASA. “NEAs as Resources,” May 2018, https://cneos.jpl.nasa.gov/about/nea_resource.html; https://web.archive.org/web/20180518005833/https://cneos.jpl.nasa.gov/about/nea_resource.html

NASA. “New Horizons Spacecraft Return its Sharpest Views of Ultima Thule,” February 22, 2019. https://www.nasa.gov/feature/new-horizons-spacecraft-returns-its-sharpest-views-of-ultima-thule

NASA. “Orion Overview,” n.d. https://www.nasa.gov/exploration/systems/orion/about/index.html

NASA. "President Obama's Speech at Kennedy," April 15, 2010. https://www.nasa.gov/about/obama_ksc_pod.html

NASA. "Report of the Task Force on Space," *The Evolution of US Space Policy and Plans* 1969. https://history.nasa.gov/SP-4407/vol1/chapter3-3.pdf

NASA. "Space Launch System," n.d. https://www.nasa.gov/exploration/systems/sls/index.html and https://www.nasa.gov/sites/default/files/atoms/files/0080_sls_fact_sheet_10092018.pdf

NASA. "State of the Union and NASA," *Youtube*, January 20, 2015. https://www.youtube.com/watch?v=xhyVEbZls54

NASA. "The Vision for Space Exploration," February 2004. https://www.nasa.gov/pdf/55583main_vision_space_exploration2.pdf

NASA. "Vanguard-A History: Background of Space Exploration," *NASA*, April 24, 2001. https://history.nasa.gov/SP-4202/chapter1.html

NASA. "Vikram Lander Found," December 2, 2019. https://www.nasa.gov/image-feature/goddard/2019/vikram-lander-found

NASA. "What is Artemis?," June 28, 2019, n.d. https://www.nasa.gov/feature/what-is-artemis/

NASA. "What Is NASA's Asteroid Redirect Mission?," December 11, 2017. https://www.nasa.gov/content/what-is-nasa-s-asteroid-redirect-mission

NASA. Lunar Oxygen Production or MoonROx Challenge, August 3, 2017. https://www.nasa.gov/directorates/spacetech/centennial_challenges/moonrox/index.html

NASA. Q&A NASA's New Spaceship, November 13, 2018. https://www.nasa.gov/feature/questions-nasas-new-spaceship

NASA. What is Artemis?, July 25, 2019. https://www.nasa.gov/what-is-artemis

National Aeronautics and Space Act of 1958, Pub. L. No. 85-568, Stat. 426 (1958).

National Air & Space Intelligence Center (NASIC), "Competing in Space," December 2018. https://media.defense.gov/2019/Jan/16/2002080386/-1/-1/1/190115-F-NV711-0002.PDF: 20.

National Defense Authorization Act for Fiscal Year 2020, Congress.Gov. https://www.congress.gov/bill/116th-congress/senate-bill/1790/text?format=txt

National Defense Industries Association (NDIA). "AFSPC Space Futures Planning Meeting." March 5, 2019. https://ndiarmc.org/product/afspc-space-futures-planning-meeting-march-5-2019/

National Research Council, Electric Power from Orbit: A Critique of a Satellite System, National Academy Press, 1981. https://space.nss.org/media/1981-NRC-Electric-Power-From-Orbit-1-Report.pdf

National Security Space Office (NSSO). *Space-Based Solar Power as an Opportunity for Strategic Security*, October 10, 2007. https://space.nss.org/space-based-solar-power-as-an-opportunity-for-strategic-security/

National Space Intelligence Center (NASIC). "Competing in Space." December 2018. https://media.defense.gov/2019/Jan/16/2002080386/-1/-1/1/190115-F-NV711-0002.PDF.

National Space Society. "About the National Space Society." n.d. https://space.nss.org/about-national-space-society/

National Space Society. "National Space Society Position Papers," n.d. https://space.nss.org/national-space-society-position-papers/

National Space Society. "NSS Library." n.d. https://space.nss.org/national-space-society-library/

National Space Society. "NSS Roadmap to Space Settlement (3rd Edition 2018-2019)," n.d. https://space.nss.org/about-national-space-society/

National Space Society. "Space Settlement Summit," n.d. https://space.nss.org/space-settlement-summit/ see also https://spacesettlementsummit2019.nss.org/

National Space Society. NSS Statement of Philosophy. n.d. https://space.nss.org/nss-statement-of-philosophy/

National Space Society. "Space Settlement Contest," n.d. https://space.nss.org/settlement/nasa/Contest/

National Space Society's "Space Settlement Library," NSS, August 3, 2017. 2020. https://space.nss.org/space-settlement-library/

Nayef, R. F Al-Rodhan. "U.S. Space Policy and Strategic Culture," *Colombia Journal of International Affairs*, April 16, 2018. https://jia.sipa.columbia.edu/online-articles/us-space-policy-and-strategic-culture

NDIA. "AFSPC Space Futures Planning Meeting," March 5, 2019. https://ndiarmc.org/product/afspc-space-futures-planning-meeting-march-5-2019/

News18. "Sitharaman Launches New Space India Limited, ISRO's Commercial Arm to Market, Sell Products," July 5, 2019. https://www.news18.com/news/business/sitharaman-launches-new-space-india-limited-isros-commercial-arm-to-market-sell-products-2219263.html

NextGen LLC. "Economic Assessment and Systems Analysis of an Evolvable Lunar Architecture that Leverages Commercial Space Capabilities and Public-Private-Partnerships," July 13, 2015. https://space.nss.org/media/Evolvable-Lunar-Architecture.pdf

Niebuhr, Reinhold. *Moral Man and Immoral Society: A Study in Ethics and Politics*. Louisville: Westminster John Knox Press, 2013.

Nikitin, Mary Beth. "Proliferation Security Initiative," *Congressional Research Service*, August 9, 2018. https://fas.org/sgp/crs/nuke/RL34327.pdf

Nitisha. "5 Factors that Affect the Economic Growth of a Country," *Economic Discussion Net*, January 10, 2015. http://www.economicsdiscussion.net/economic-growth/5-factors-that-affect-the-economic-growth-of-a-country/4199

No author, "Chinese Scientists Mull Power Station in Space," *Xinhua*, March 30, 2015. http://news.xinhuanet.com/english/2015-03/30/c_134109115.htm

No author, "Dr. Vikram Ambalal Sarabhai (1963–1971)" *ISRO*, n.d. https://www.isro.gov.in/about-isro/dr-vikram-ambalal-sarabhai-1963-1971

No author. "Chandrayaan 2: Vikram Lander in Tilted Position after Hard Landing, Says ISRO," *The Hindu Business Line*, September 09, 2019. https://www.thehindubusinessline.com/news/chandrayaan-2-vikram-lander-tilted-after-hard-hit-says-isro/article29374671.ece

No author. "16 Psyche," *NASA*, May 16, 2019. https://solarsystem.nasa.gov/asteroids-comets-and-meteors/asteroids/16-psyche/in-depth/

No author. "2014 Two Sessions NPC & CPPCC: Annual Legislative and Political Advisory Sessions," *China Daily*, March 5, 2014. http://www.chinadaily.com.cn/china/2014npcandcppcc/2014-03/05/content_17324203.htm

No author. "2018 Military Strength Ranking," *Global Firepower*, November 22, 2018. https://www.globalfirepower.com/countries-listing.asp

No author. "2222 (XXI). Treaty on Principles Governing the Activities of States in the Exploration and Use of Outer Space, including the Moon and Other Celestial Bodies," *United Nations Office of Outer Space Affairs (UNOOSA)*, May 26, 2015. https://www.unoosa.org/oosa/en/ourwork/spacelaw/treaties/outerspacetreaty.html & https://www.unoosa.org/pdf/publications/STSPACE11E.pdf

No author. "7 Chinese Military Officers Promoted to Rank of General," *Xinhua*, December 12, 2019. http://www.xinhuanet.com/english/2019-12/12/c_138626138.htm

No author. "All You Need to Know about Mission Shakti," *Hindu Business Line*, March 27, 2019. https://www.thehindubusinessline.com/news/all-you-need-to-know-about-mission-shakti/article26652887.ece

No author. "Amount of Energy the Earth Gets from the Sun," *Ask Us: Sun, NASA GFC Cosmicopia*. https://cosmicopia.gsfc.nasa.gov/qa_sun.html

No author. "An Interview with Dan Deudney," *World Government Research Network (WGRN)*, January 4, 2016, http://wgresearch.org/an-interview-with-daniel-h-deudney/

No author. "Artemis Greek Goddess of the Hunt, Forests and Hills, the Moon, Archery," n.d., https://greekgodsandgoddesses.net/goddesses/artemis/

No author. "Artemis Greek Goddess," *Encyclopedia Britannica*, n.d. https://www.britannica.com/topic/Artemis-Greek-goddess

No author. "Aryabhata," n.d. https://www.britannica.com/biography/Aryabhata-I

No author. "Asteroids," *NASA*. https://solarsystem.nasa.gov/asteroids-comets-and-meteors/asteroids/in-depth/

No author. "Bahrain Boldly Going into Space, with the help of the UAE," *The National*, November 17, 2018. https://www.thenational.ae/uae/science/bahrain-boldly-going-into-space-with-the-help-of-the-uae-1.792920

No author. "Bank of China Lists 500-mln-dollars USD BRI Bond in Luxembourg," *Xinhua,* April 25, 2019. http://www.xinhuanet.com/english/2019-04/25/c_138006422.htm

No author. "Barrett, Air Force Leaders Applauds Space Force Formal Creation," *Secretary of the Air Force Public Affairs, U.S. Air Force*, December 20, 2019. https://www.af.mil/News/Article-Display/Article/2045939/barrett-air-force-leaders-applaud-space-forces-formal-creation/

No author. "Can India become a Great Power?," *The Economist*, March 30, 2013. http://www.economist.com/news/leaders/21574511-indias-lack-strategic-culture-hobbles-its-ambition-be-force-world-can-india

No author. "Chandrayaan-2: India Launches Second Moon Mission," *BBC*, July 22, 2019. https://www.bbc.com/news/world-asia-india-49032603

No author. "China Aerospace Science and Technology Corporation," n.d. http://english.spacechina.com/n16421/index.html

No author. "China at Technical Preparation Stage for Mars, Asteroid Mining," *Xinhua*, March 3, 2015. http://www.globaltimes.cn/content/909948.shtml

No author. "China Establishes Rocket Force and Strategic Support Force." *Ministry of National Defense, The People's Republic of China*, 2015. http://eng.mod.gov.cn/ArmedForces/ssf.htm

No author. "China Establishes Rocket Force and Strategic Support Force," *Ministry of National Defense, The People's Republic of China*, n.d. http://eng.mod.gov.cn/ArmedForces/ssf.htm

No author. "China Expects to Introduce Space Law by 2020," *China Daily*, November 17, 2014. http://www.chinadaily.com.cn/china/2014-11/17/content_18930721.htm

No author. "China Focus: Capture an Asteroid, Bring it Back to Earth?," *Xinhua*, July 23, 2018. http://www.xinhuanet.com/english/2018-07/23/c_137342866.htm

No author. "China Focus: Chang'e 4 Lands on Largest Crater in Solar System," *Xinhuanet*, January 3, 2019. http://www.xinhuanet.com/english/2019-01/03/c_137717410.htm

No author. "China Focus: China Sets Record for Longest Stay in Self-Contained Moon Lab," *Xinhuanet*, May 15, 2018. http://www.xinhuanet.com/english/2018-05/15/c_137180976.htm

No author. "China Focus: Moons Sees First Cotton Seed Sprout," *Xinhua*, January 15, 2019. http://www.xinhuanet.com/english/2019-01/15/c_137745505.htm,.

No author. "China International Commercial Court," n.d. http://cicc.court.gov.cn/html/1/219/index.html.

No author. "China is Expected to take the Lead in Building a Space Solar Power Station," *Science and Technology Daily.com*, February 14, 2019. http://www.stdaily.com/index/kejixinwen/2019-02/14/content_750019.shtml

No author. "China Launches First Cargo Spacecraft *Tianzhou* 1," *Xinhua*, April 20, 2017. http://www.xinhuanet.com//english/2017-04/20/c_136223932.htm

No author. "China Maritime Museum," n.d. http://www.timeoutshanghai.com/venue/Things_to_Do-Museums/7542/China-Maritime-Museum.html

No author. "China Military Power Modernizing a Force to Fight and Win," n.d., 39—45. http://www.dia.mil/Portals/27/Documents/News/Military%20Power%20Publications/China_Military_Power_FINAL_5MB_20190103.pdf?utm_source=Daily%20on%20Defense%20(Template)_01/16/2019&utm_medium=email&utm_campaign=WEX_Daily%20on%20Defense

No author. "China Plans for the Moon, Mars and Beyond," *BBC*, April 19, 2016. http://www.bbc.com/news/world-asia-36085659

No author. "China Promotes 5 Officers to General," *Xinhuanet*, July 28, 2017. http://www.xinhuanet.com/english/2017-07/28/c_136480993.htm

No author. "China Scientists Develop space 3D printer," *China Daily*, April 19, 2016. http://www.chinadaily.com.cn/china/2016-04/19/content_24676024.htm

No author. "China Space Activities in 2011." December 29, 2011. http://www.gov.cn/english/official/2011-12/29/content_2033200.htm

No author. "China Space Probe Flies by Asteroid Toutatis," *China Academy of Sciences*, December 18, 2012. http://english.cas.cn/news/ric/201212/t20121218_97190.shtml

No author. "China to Achieve "Major Breakthroughs" in Nuclear Powered Space Shuttle around 2040: Report." *Xinhuanet*, November 16, 2017. http://www.xinhuanet.com/english/2017-11/16/c_136757737.htm

No author. "China to Build Scientific Research Station on Moon's South Pole," *Xinhua*, April 24, 2019. http://www.xinhuanet.com/english/2019-04/24/c_138004666.htm

No author. "China to Launch Long March-9 Rocket in 2028," *Xinhuanet*, September 19, 2018, http://www.xinhuanet.com/english/2018-09/19/c_137477227.htm

No author. "China Unveils Follow-up Lunar exploration Missions," *Xinhua*, January 14, 2019. http://xinhuanet.com/english/2019-01/14/c_137743306.htm

No author. "China, ASEAN Agree on Framework for South China Sea Code of Conduct," *Reuters*, May 18, 2017. https://www.reuters.com/article/us-southchinasea-china-philippines-idUSKCN18E1FS .

No author. "China's Largest Carrier Rocket Long March-5 Makes New Flight," *Xinhua*, December 27, 2019. http://www.xinhuanet.com/english/2019-12/27/c_138662139.htm

No author. "China's Long March Rockets Complete 300 Launches," *Xinhua*, March 3, 2019, http://www.xinhuanet.com/english/2019-03/10/c_137883443.htm

No author. "China's Lunar Exploration Program," *People's Daily Online*. n.d. http://en.people.cn/90002/91752/index.html

No author. "China's Peaceful Rise: Speeches of Zheng Bijian 1997–2004," *The Brookings Institution*. June 15, 2005. https://www.brookings.edu/wp-content/uploads/2012/04/20050616bijianlunch.pdf

No author. "China's Planned Space Station to Support Hundreds of Experiments," *Xinhua*, April 24, 2019, http://www.xinhuanet.com/english/2019-04/24/c_138005075.htm

No author. "China's Space Dream Crystallized with Shenzhou-10 Launch," Economic and Commercial Counsellor's Office of the Embassy of the People's Republic of China in the Republic of Cameroon, June 13, 2013. http://cm2.mofcom.gov.cn/article/headnews/201306/20130600160022.shtml.

No author. "Chinese Nuclear Program, "*Atomic Heritage Foundation*, July 19, 2018. https://www.atomicheritage.org/history/chinese-nuclear-program

No author. "Chinese Red Hacker Army // Red Hacker Alliance // Honkers // Honker's Union // HUC // 中国红客队 // 红客, " Steemit, June 29, 2017. https://steemit.com/hacking/@odinthelibrarian/chinese-red-hacker-army-red-hacker-alliance-honkers-honker-s-union-huc

No author. "DF-1," *Astronautix*, December 28, 2016. http://www.astronautix.com/d/df-1.html

No author. "Discovery Statistics," NASA. n.d. https://cneos.jpl.nasa.gov/stats/totals.html

No author. "DoD Releases Fiscal Year 2019 Budget Proposal." Department of Defense, February 12, 2018. https://www.defense.gov/News/News-Releases/News-Release-View/Article/1438798/dod-releases-fiscal-year-2019-budget-proposal/

No author. "Electric Power Consumption (kWh per capita)," *World Bank*, March 22, 2010. https://data.worldbank.org/indicator/EG.USE.ELEC.KH.PC

No author. "Emirates Mars Mission," n.d. *UAE Space Agency*. https://www.mbrsc.ae/emirates-mars-mission

No author. "EO Portal Directory, "RISAT-2 (Radar Imaging Satellite-2)," https://earth.esa.int/web/eoportal/satellite-missions/r/risat-2

No author. "Executive Office of the President, "National Near-Earth Object Preparedness Strategy," NASA, December 2016. https://www.nasa.gov/sites/default/files/atoms/files/national_near-earth_object_preparedness_strategy_tagged.pdf

No author. "Exploiting Earth-Moon Space: China's Ambition After Space Station," *Xinhua*, March 7, 2016. http://news.xinhuanet.com/english/2016-03/07/c_135164574.htm

No author. "Exploring the Roots of India's Strategic Culture, Institute for Defence and Analyses Conference, October 5, 2017. https://idsa.in/event/exploring-the-roots-of-indias-strategic-culture

No author. "FCC Approves SpaceX's Plan to Launch more than 7,000 Internet-Beaming Satellites," *The Verge*, November 15, 2018. https://www.theverge.com/2018/11/15/18096943/spacex-fcc-starlink-satellites-approval-constellation-internet-from-space

No author. "First Industry Built PSLV by 2020, Says VSSC Director Sivan," *The Hindu*, November 17, 2017. https://www.thehindu.com/sci-tech/science/first-industry-built-pslv-by-2020/article20535383.ece

No author. "Flowers on the Moon? China's Chang'e 4 to Launch Lunar Spring," *Xinhuanet,* April 12, 2018. http://www.xinhuanet.com/english/2018-04/12/c_137106440.htm

No author. "Full Text of Proxmire's Letter to "60 Minutes," *L-5 News* 2 No. 12, December 1977. https://space.nss.org/media/L5-News-1977-12.pdf

No author. "Global Energy Statistical Yearbook 2018," *Enerdata*, July 9, 2019. https://yearbook.enerdata.net/total-energy/world-consumption-statistics.html

No author. "Government Spending in Space Programs Reaches $62 Billion in 2016," *Euroconsult*, May 30, 2017. http://www.euroconsult-ec.com/30_May_2017

No author. "India to Explore Moon's Surface for Helium-3," *Hindustan Times*, September 1, 2006. https://web.archive.org/web/20170528033036/https://www.hindustantimes.com/india/india-to-explore-moon-s-surface-for-helium-3/story-K4HAfpqIhRCJm6zeKVsGxM.html

No author. "India to Set Up Strategic Aerospace Command," *India Daily*, April 27, 2005. https://web.archive.org/web/20060902214721/http://www.indiadaily.com/breaking_news/33107.asp

No author. "India Working on Space Weapons: IAF Chief," *Rediff India Abroad*, October 07, 2003. https://www.rediff.com/news/2003/oct/06iaf1.htm

No author. "India: Subjects: Nehru Correspondence, November 1962: Nov 11–19," John F. Kennedy Presidential Library and Museum. https://www.jfklibrary.org/asset-viewer/archives/JFKNSF/111/JFKNSF-111-016

No author. "India's NAVIC GNSS Capability Declared an Allied System by U.S. Congress," *Spaceflight Asia Pacific*, n.d. https://spacewatch.global/2019/12/indias-navic-gnss-capability-declared-an-allied-system-by-u-s-congress/

No author. "Indian Judges are Ignoring the Government's Abuses in Kashmir," *The Economist*, October 5, 2019. https://www.economist.com/asia/2019/10/05/indias-judges-are-ignoring-the-governments-abuses-in-kashmir

No author. "InSight: Futron The Space Competitiveness Index—Update, by Futron," *SatMagazine*, December 9, 2009, http://www.satmagazine.com/story.php?number=1856883389

No author. "INTEL: Global Military Space by Futron," *MilSat Magazine* (September 2009 Edition), September 9, 2009. http://www.milsatmagazine.com/story.php?number=875113067

No author. "International Space Exploration Coordination Group," Annual Report 2015. https://www.dlr.de/Portaldata/28/Resources/dokumente/rm/ISECG_AnnualReport_2015.pdf

No author. "International Space Settlement Design Competition," n.d. https://spaceset.org/

No author. "ISRO to Increase Frequency of Launches to 12 per year," *The Times of India*, April 23, 2017. http://timesofindia.indiatimes.com/india/isro-to-increase-frequency-of-launches-to-12-per-year/articleshow/58326033.cms

No author. "IW2 Research Team, "Bangalore Startup Alpha Design Signs Pact with ISRO to Assemble Satellites," *Indian Web2*, June 18, 2018. https://www.indianweb2.com/2018/07/18/bangalore-startup-alpha-design-signs-pact-with-isro-to-assemble-satellite/

No author. "Legalism in Chinese Philosophy," *Stanford Encyclopedia of Philosophy*, December 10, 2014. https://plato.stanford.edu/entries/chinese-legalism/

No author. "Lewis And Clark Expedition," *History.net*, May 6, 2013. https://www.historynet.com/lewis-and-clark-expedition

No author. "List of Countries by Projected GDP," *Statistics Times*, May 6, 2018. http://statisticstimes.com/economy/countries-by-projected-gdp.php

No author. "Lunar Palace 365 Experiment Sets Record of 200 Days and Volunteers Change Shifts Again," *Beihang University*, January 30, 2018. https://ev.buaa.edu.cn/info/1013/1534.htm

No author. "Luxembourg and the United Arab Emirates Sign MoU on Space Resources." *Space Resources.Lu*, October 10, 2017. https://spaceresources.public.lu/en/actualites/2017/MoU-UAE.html

No author. "Luxembourg Cooperates with China in the Exploration and Use of Outer Space Resources for Peaceful Purpose." *Space Resources.LU*. n.d. https://spaceresources.public.lu/en/actualites/2018/Luxembourg-cooperates-with-China-in-the-exploration-and-use-of-outer-space-for-peaceful-purpose.html

No author. "Luxembourg Space Capabilities Turning Innovation into Business 2017." https://clustermembers.luxinnovation.lu/space/wp-content/uploads/sites/4/2016/10/07697_LUXINNOVATION_SpaceCapabilities_05-2017-Web.pdf

No author. "Luxembourg, A Member of the BRI," *OBOReurope*. https://www.oboreurope.com/en/luxembourg-member-bri/

No author. "Mangalyaan India's Mission to Mars," *National Geographic*, November 24, 2016. https://www.youtube.com/watch?v=Lv-4fiVadVk

No author. "Manifest Destiny," *History.com*, November 15, 2019. https://www.history.com/topics/westward-expansion/manifest-destiny

No author. "Minerva Research Initiative," n.d. https://minerva.defense.gov/

No author. "Missile Technology Control Regime," n.d. https://mtcr.info/

No author. "Multi-Orbit Logistics Services," DIU, February 12, 2020, https://www.diu.mil/work-with-us/open-solicitations

No author. "Multi-Rotary Joints SPS: 2015 SunSat Design Competition," *You Tube*, June 22, 2015. https://www.youtube.com/watch?v=XhgJwnpYRGc

No author. "PLA Strategic Support Force," *SinoDefence*, January 1, 2017. https://sinodefence.com/organisation/pla-strategic-support-force/

No author. "PLA Welcomes Launches of Private Rockets," *China Daily*, February 18, 2019. http://eng.chinamil.com.cn/view/2019-02/18/content_9429066.htm

No author. "Planting Tricolour may help India's claim on Moon: Nair," *One India*, October 20, 2008. https://www.oneindia.com/2008/10/20/planting-tricolour-may-help-indias-claim-on-moon-nair-1224511741.html

No author. "Power Rankings," *US News & World Report*, January 23, 2019. https://www.usnews.com/news/best-countries/power-full-list

No author. "President Trump Designates Secretary Brouillette a Member of the National Space Council," *Department of Energy*, February 14, 2020. https://www.energy.gov/articles/president-trump-designates-secretary-brouillette-member-national-space-council

No author. "Project 921-1," *Astronautix*, August 20, 2016, http://www.astronautix.com/p/project921.html

No author. "Protecting Antarctica from Ugly Tourists," *The Star Online*, February 4, 2018. https://www.thestar.com.my/news/regional/2018/02/04/protecting-antarctica-from-ugly-tourists-the-growing-popularity-of-the-icy-continent-among-chinese-v/

No author. "Qualitative Data Analysis: Choosing QDA Software," NYU. n.d. https://guides.nyu.edu/QDA/comparison

No author. "Query of Most Valuable Asteroids," *Asterank*, n.d. http://www.asterank.com/

No author. "Quotes about PREPARATION," Leadership Now, n.d. https://www.leadershipnow.com/preparationquotes.html

No author. "Recommendations Approved by the National Space Council to President Trump," *SpaceRef*, March 26, 2019. http://spaceref.com/news/viewsr.html?pid=52358

No author. "Richest People in the World," *CBS News*. n.d. https://www.cbsnews.com/pictures/richest-people-in-world-forbes/21/

No author. "Riding an Asteroid: China's Next Goal in Space," *Chinese Academy of Sciences*, March 1, 2017. http://english.cas.cn/newsroom/china_research/201703/t20170301_174455.shtml

No author. "Saudi Arabia, Qatar, and the UAE," SIPRI Fact Sheet, May 2019. https://www.sipri.org/publications/2019/sipri-fact-sheets/military-spending-and-arms-imports-iran-saudi-arabia-qatar-and-uae

No author. "Shaping the Space Age: The First International Geophysical Year," *NASA, Appel News Digest*, 1/12, December 2008. https://appel.nasa.gov/2010/02/23/aa_1-12_f_shaping-html/

No author. "Shell Scenarios," n.d. https://www.shell.com/energy-and-innovation/the-energy-future/scenarios.html

No author. "Singapore Topples United States as World's Most Competitive Economy," *IMD*, May 2019. https://www.imd.org/news/updates/singapore-topples-united-states-as-worlds-most-competitive-economy/

No author. "Space Debris from Anti Satellite Weapons," *Union of Concerned Scientist*, Fact Sheet. https://www.ucsusa.org/sites/default/files/legacy/assets/documents/nwgs/debris-in-brief-factsheet.pdf

No author. "Space Foundation Report Reveals Global Space Economy at $383.5 Billion in 2017," *The Space Foundation*, July 19, 2018. https://www.spacefoundation.org/news/space-foundation-report-reveals-global-space-economy-3835-billion-2017

No author. "Space Resources Program Approved for fall 2018 Launch," *Colorado School of Mines*. n.d. http://www.minesnewsroom.com/news/space-resources-program-approved-fall-2018-launch

No author. "Space Resources," Luxembourg Space Agency. n.d. https://space-agency.public.lu/en/space-resources.html

No author. "Space Resources.Lu," n.d. https://spaceresources.public.lu/en.html

No author. "Space Robotic Arm Inspires Scientists for Sorting Garbage," *Xinhua*, July 26, 2019. http://www.xinhuanet.com/english/2019-07/26/c_138260582.htm

No author. "Space Vision 2025," Vikram Sarabhai Space Centre, n.d. http://www.vssc.gov.in/VSSC_V4/index.php/space-vision-2025/117-space-vision-2025

No author. "Space: Investing in the Final Frontier," *Morgan Stanley*, November 7, 2018. https://www.morganstanley.com/ideas/investing-in-space

No author. "SpaceX Interplanetary Tranport System (ITS)," *You Tube*, September 27, 2019. https://www.youtube.com/watch?v=0qo78R_yYFA

No author. "Special: China's Air Defense Identification Zone Triggers Mixed Response," *China Daily*, November 27, 2013. http://www.chinadaily.com.cn/china/2013-11/27/content_17135438.htm

No author. "State Media Should Play Due Role in Properly Guiding Public Opinion," *China Daily*, February 22, 2016. http://www.chinadaily.com.cn/opinion/2016-02/22/content_23580181.htm

No author. "The Age of Imperialism (1870-1914)," *Tamaqua K12*, July 15, 2007. https://www.tamaqua.k12.pa.us/cms/lib07/PA01000119/Centricity/Domain/119/TheAgeofImperialism.pdf

No author. "The East India Company and Its Role in ruling India," *Historic UK*, September 2, 2012. https://www.historic-uk.com/HistoryUK/HistoryofEngland/The-East-India-Company/

No author. "The Four Voyages of Christopher Columbus," *Constant Contact*, May 17, 2018. https://myemail.constantcontact.com/The-4-Voyages-of-Columbus.html?soid=1108762609255&aid=c0r8lhGZxC4

No author. "The Legend of Chang'e," n.d. http://www.moonfestival.org/the-legend-of-chang-e.html

No author. "The Ming Voyages," n.d. http://afe.easia.columbia.edu/special/china_1000ce_mingvoyages.htm

No author. "The Northwest Ordinance For Space," *Citizens in Space*, n.d. http://www.citizensinspace.org/2014/04/the-northwest-ordinance-for-space/

No author. "The Opium War," National Army Museum. https://www.nam.ac.uk/explore/opium-war-1839-1842

No author. "The Space Development and Settlement Act of 2019," Alliance for Space Development, n.d. http://allianceforspacedevelopment.org/wp-content/uploads/2019/01/Space-Development-and-Settlement-Act-of-2019.pdf

No author. "The Space Exploration Technologies Corporation is starting to take the lion's share of global launches," *Popular Mechanics*, July 13, 2017. https://www.popularmechanics.com/space/rockets/a27290/one-chart-spacex-dominate-rocket-launches/

No author. "The UAE and the Global Oil Supply," *Embassy of the United Arab Emirates, Washington, DC*, n.d. https://www.uae-embassy.org/about-uae/energy/uae-and-global-oil-supply

No author. "Theoretical Transportation Energy," n.d. https://www.permanent.com/space-transportation-earth-moon.html

No author. "Tiny Bhutan Protests to China Over Roadbuilding in Disputed Territory," *The Strait Times*, June 29, 2017. https://www.straitstimes.com/asia/south-asia/tiny-bhutan-protests-to-china-over-road-building-in-disputed-territory

No author. "Top 10 Fastest Manmade Objects Ever," *The Mysterious World*, March 2, 2014. https://themysteriousworld.com/top-10-fastest-man-made-objects-ever/

No author. "Total Population by Country 2018," *World Population Review*. 2018. http://worldpopulationreview.com/countries/

No author. "Trump: Space is the World's Newest War-Fighting Domain," *BBC*, December 21, 2019. https://www.bbc.com/news/av/world-us-canada-50875940/trump-space-is-the-world-s-newest-war-fighting-domain

No author. "Two Views of Culture: Etic and Emic," *Lumen Learning*, n.d. https://courses.lumenlearning.com/culturalanthropology/chapter/two-views-of-culture-etic-emic/

No author. "UAE Cabinet Approves National Space Strategy 2030," *Khaleej Times*, March 12, 2019. https://www.khaleejtimes.com/news/government/uae-cabinet-approves-national-space-strategy-2030--12

No author. "UAE Space Agency Adopts New Resolution at UN Summit," *Gulf News*, June 23, 2018. https://gulfnews.com/going-out/society/uae-space-agency-adopts-new-resolution-at-un-summit-1.2240898

No author. "UCS Satellite Database," *Union of Concerned Scientists*, November 30, 2018. https://www.ucsusa.org/nuclear-weapons/space-weapons/satellite-database

No author. "Uni.lu hosts Hague Space Resources Group," *University of Luxembourg*, December 12, 2018. https://wwwfr.uni.lu/fdef/actualites/uni_lu_hosts_hague_space_resources_group

No author. "United Arab Emirates GDP," n.d. https://tradingeconomics.com/united-arab-emirates/gdp; Satish Kumar, "UAE Central Bank Says GDP Growth in First Quarter of 2019 was 2.2% Year on Year: WHAM," *Reuters*, June 23, 2019, https://www.reuters.com/article/us-emirates-economy-gdp/uae-central-bank-says-gdp-growth-in-first-quarter-of-2019-was-22-year-on-year-wam-idUSKCN1TO08D

No author. "Which Countries Spend the Most on Space Exploration?," *World Economic Forum*, January 11, 2016. https://www.weforum.org/agenda/2016/01/which-countries-spend-the-most-on-space-exploration/

No author. "Xinhua Headlines: The Return of Long March-5 Inspires China to Reach for the Stars," *Xinhuanet*, December 29, 2019, http://www.xinhuanet.com/english/2019-12/28/c_138663650.htm

NPS Homeland Security Digital Library. "Reopening the American Frontier: Exploring How the Outer Space Treaty Will Impact American Commerce and Settlement in Space," May 23, 2017. https://www.hsdl.org/?view&did=807259

NSS Solar Power Library. n.d. https://space.nss.org/space-solar-power-library/

NSS. "NSS Legislative Blitzes," n.d. https://space.nss.org/nss-legislative-blitzes/

NSS. "NSS Position Paper: Space Debris Removal, Salvage, and Use: Maritime Lessons," October 2019. https://space.nss.org/media/NSS-Position-Paper-Space-Debris-Removal-2019.pdf

NSS. "NSS Statement of Philosophy," *NSS*, August 3, 2017. https://space.nss.org/nss-statement-of-philosophy/

NSS. "Position Paper: A Public/Private COTS-Type Program to Develop Space Solar Power," February 2020. https://space.nss.org/media/NSS-Position-Paper-COTS-Type-Space-Solar-Power-2020.pdf

NSS. "Position Paper: Outer Space Private Investment Corporation (OSPIC)," December 2018. https://space.nss.org/national-space-society-position-papers/

O'Brien, Sean P. "Crisis Early Warning and Decision Support: Contemporary Approaches and Thoughts on Future Research," *International Studies Review* 11, no.1 (2010): 87–104.

O'Connell, Kevin. "Remarks on the Trillion Dollar Space Economy," *Office of Space Commerce*, November 27, 2018. https://www.space.commerce.gov/remarks-on-the-trillion-dollar-space-economy/

O'Hanlon, Michael. *Neither Star Wars nor Sanctuary: Constraining the Military Uses of Space.* Washington, DC: Brookings Institution Press, 2004.

O'Leary, Brain. "Construction of Satellite Solar Power Stations from Non-Terrestrial Material," *Journal of Energy* 1, no. 3 (1977): 155–8.

Olsen, Johan P. and James G. March. "The Logic of Appropriateness," in *The Oxford Handbook of Political Science*. 2004.

O'Neill, Gerald K. *High Frontier: Human Colonies in Space.* New York: William Morrow and Company, 1977.

Obama, Barack. "America Will Take the Giant Leap to Mars," *CNN*, October 11, 2016. https://www.cnn.com/2016/10/11/opinions/america-will-take-giant-leap-to-mars-barack-obama/index.html

Obama-Biden Transition Project: Space Solar Power (SSP) – A Solution for Energy Independence & Climate Change, *SpaceRef*, December 6, 2008. http://www.spaceref.com/news/viewsr.html?pid=30044 original available via archive at: https://web.archive.org/web/20160712094225/http://change.gov/open_government/entry/space_solar_power_ssp_a_solution_for_energy_independence_climate_change/

Ochab, Ewelina U. "Is China Conducting a Crackdown on Religion?," *Forbes*, April 20, 2019. https://www.forbes.com/sites/ewelinaochab/2019/04/20/is-china-conducting-a-crackdown-on-religion/#3dea378c19d3;

Oesterle, Aaron. "The Viability of the Space Settlement Narrative in Politics," (IAC-19. E3.2.9x52875), *International Astronautical Congress*, November 1, 2018.

Office of Technology Assessment (OTA). *Solar Power Satellites*, August 1981. https://space.nss.org/media/1981-OTA-Solar-Power-Satellites.pdf

Office of the Historian. "Barbary Wars, 1801-1805, and 1815-1816," https://history.state.gov/milestones/1801-1829/barbary-wars

Office of the Secretary of Defense, Department of Defense Fiscal Year (FY) 2020 Budget Estimates for Research, Development, Test & Evaluation, Defense-Wide. March 2019. https://comptroller.defense.gov/Portals/45/Documents/defbudget/fy2020/budget_justification/pdfs/03_RDT_and_E/RDTE_Vol3_OSD_RDTE_PB20_Justification_Book.pdf

Office of the Secretary of Defense. "Annual Report to Congress Military and Security Developments Involving the People's Republic of China." https://dod.defense.gov/Portals/1/Documents/pubs/2015_China_Military_Power_Report.pdf

Offworld Website vision, September 9, 2016. https://www.offworld.ai/

Orbit Beyond. n.d. https://www.orbitbeyond.com/

OrbitFab, "Gas Stations in Space," n.d. https://www.orbitfab.space/

Organski, A. F. K. *World Politics* 2d ed. New York: Knopf, 1968.

Origins of NASA Names, "Manned Spaceflight Section IV," n.d. https://history.nasa.gov/SP-4402/ch4.htm

Osinga, Frans. *Science Strategy and War: The Strategic Theory of John Boyd*. The Netherlands: Eburon, 2005.

Osnos, Evan. "The Two Lives of Qian Xuesen," *The New Yorker*, November 3, 2009. https://www.newyorker.com/news/evan-osnos/the-two-lives-of-qian-xuesen

Oxford Dictionary. "Salience," https://en.oxforddictionaries.com/definition/salience

Pace, Scott. Executive Secretary of the National Space Council "Space Development, Law, and Values," IISL Galloway Space Law Symposium Cosmos Club. Washington, DC. December 13, 2017. https://spacepolicyonline.com/wp-content/uploads/2017/12/Scott-Pace-to-Galloway-FINAL.pdf

Pace, Scott. Executive Secretary, National Space Council, "Space 2.0: U.S. Competitiveness and Policy in the New Space Era," *Hudson Institute*, Keynote Address, May 17, 2018. https://www.hudson.org/research/14341-full-transcript-space-2-0-u-s-competitiveness-and-policy-in-the-new-space-era

Pandit, Rajat. "India's First Military Satellite Successfully Launched," *Times of India*, August 30, 2013. https://timesofindia.indiatimes.com/india/Indias-first-military-satellite-successfully-launched/articleshow/22168268.cms

Paris Marx. "We Don't Need Space Colonies, and We Definitely Don't Need Jeff Bezos," *Jacobin*, July 21, 2019. https://jacobinmag.com/2019/07/space-colonies-jeff-bezos-blue-origin.

Pasztor, Andy. "Trump in Space: Transition Focuses on Private-Public Initiatives: Regarding NASA's Future, Steve Bannon Meets with SpaceX's Elon Musk and

Others," *Wall Street Journal*, January 18, 2017. https://www.wsj.com/articles/trump-in-space-transition-focuses-on-private-public-initiatives-1484781119

Paylyk, Oriana. "It's Official: President Trump Has Signed Space Force into Existence," Military.com, December 20, 2019. https://www.military.com/daily-news/2019/12/20/its-official-president-trump-has-signed-space-force-existence.html

Peck, Michael. "Japan is Getting Its Own Space Force (With Better Uniforms?)," *National Interest*, January 21, 2020. https://nationalinterest.org/blog/buzz/japan-getting-its-own-space-force-better-uniforms-115816

Pelton, Joseph N. *The New Gold Rush: The Riches of Space Beckon*. Göttingen, Germany: Copernicus, 2017.

People's Republic of China. "State Administration for Science, Technology, and Industry for National Defense," *The State Council. People's Republic of China*. October 6, 2014. http://english.gov.cn/state_council/2014/10/06/content_281474992893468.htm

Perla, Peter P. *The Art of Wargaming: A Guide for Professionals and Hobbyists*. Annapolis: Naval Institute Press, 1990.

Permanent Mission of the People's Republic of China to the United Nations and Other International Organizations in Vienna. "Statements by Chinese Delegation at the 55th Session of the Scientific and Technical Subcommittee of the Committee on the Peaceful Uses of Outer Space," February 14, 2018. https://www.fmprc.gov.cn/ce/cgvienna/eng/hyyfy/t1535268.htm

Peter J. Carrington. John Scott, Stanley Wasserman, eds. *Models and Methods in Social Network Analysis*. Vol. 28. Cambridge: Cambridge University Press, 2005.

Peter Sage's 2009 TED talk "Space Based Solar Power," March 17, 2016. https://www.youtube.com/watch?v=cT_xNON1_bA

Pettit, Harry. "NASA Headed Towards Giant Golden Asteroid that Could Make Everyone on Earth a Billionaire," *FoxNews*, June 27, 2019. https://www.foxnews.com/science/nasa-headed-towards-giant-golden-asteroid-that-could-make-everyone-on-earth-a-billionaire

Phillips, Scott. "The New Frontier Playbook: The Political Economic Plan to Win the Frontier of Space, 2017. https://newfrontierplaybook.com/

Picardo, Elvis. "Eight of the World's Top Companies Are American," *Investopedia*, October 20, 2018. https://www.investopedia.com/articles/active-trading/111115/why-all-worlds-top-10-companies-are-american.asp

Pillsbury, Michael. *The Hundred-Year Marathon: China's Secret Strategy to Replace America as the Global Superpower*. New York: Henry Holt and Company, 2015.

Pixxel Website. n.d. https://pixxel.co.in/

Planetary Resources. "Asteroid Mining Mission Revealed by Planetary Resources, Inc.," *Youtube*. April 24, 2012. https://www.youtube.com/watch?v=zXXJtSZffVg

Planetary Resources. "Media Coverage," n.d. https://www.planetaryresources.com/category/media/page/25/

Planetary Resources. "Planetary Resources: Our Vision for Humanity's Future in Space," *Youtube*. December 3, 2017. https://www.youtube.com/watch?v=hqW9HtPCnR0&t=9s

Planetary Resources. “President Obama Signs Bill Recognizing Asteroid Resource Property Rights Into Law,” November 25, 2015. https://www.planetaryresources.com/2015/11/president-obama-signs-bill-recognizing-asteroid-resource-property-rights-into-law/

Planetary Resources. “Timeline,” n.d. https://www.planetaryresources.com/company/timeline/

Pletcher, Kenneth. *The Age of Exploration: From Christopher Columbus to Ferdinand Magellan*. New York: Britannica Publications, 2013.

Pontin, Mark Williams. “China’s Anti-Satellite Missile Test: Why?,” *MIT Technology Review*, March 8, 2007. https://www.technologyreview.com/s/407454/chinas-antisatellite-missile-test-why/

Poon, Shuk Waah. “Embodying Maoism: The Swimming Craze, the Mao Cult, and Body Politics in Communist China, 1950s-1970s,” *Modern Asian Studies* 53, no. 5 (September 2019): 1450–85. https://www.cambridge.org/core/journals/modern-asian-studies/article/embodying-maoism-the-swimming-craze-the-mao-cult-and-body-politics-in-communist-china-1950s1970s/955A833A02B8053585F2B1D8631698B0

Porter, Patrick “Why America’s Grand Strategy Has Not Changed,” *International Security* 42, no. 4 (Spring 2018), 9–46. doi: 10.1162/ISEC_a_00311

Posen, Barry R. “Nationalism, the Mass Army, and Military Power,” *International Security* 18, no. 2 (1993): 80–124.

Powell, R. “Stability and the Distribution of Power,” *World Politics 48*, no. 2 (1996): 239–67.

Powers, John. “John Wayne: Icon of America's Booming Confidence,” *NPR*, October 7, 2011. https://www.npr.org/2011/10/07/140870373/john-wayne-icon-of-americas-booming-confidence

PowerSat. “Energy Market Drivers Behind Space Solar Power (SSP),” October 24, 2009. http://www.powersat.com

Pradhan, S. D. “Chinese Grand Strategy of Salami Slicing Continues in its Periphery,” *The Times of India*, April 1, 2018.https://timesofindia.indiatimes.com/blogs/ChanakyaCode/chinese-grand-strategy-of-salami-slicing-continues-in-its-periphery/

Prakash, Rahul and Rajeswari Pillai Rajagopalan. “China Walks the US-India Space Solar Power Dream,” *ORF*, September 21, 2011. https://www.orfonline.org/research/china-walks-the-us-india-space-solar-power-dream/

Prasad, Narayan. “As the Race for Moon Heats Up, Chandrayaan-2 Could Give ISRO an Edge, and Spoils,” *ThePrint*, June 28, 2019. https://theprint.in/science/as-the-race-for-moon-heats-up-chandrayaan-2-could-give-isro-an-edge-and-spoils/254145/

Prasad, Narayan. “India is a Maturing Space Power, but Can Rival the US With an Independent Regulator,” *The Print*, August 8 2019. https://theprint.in/opinion/india-is-a-maturing-space-power-but-can-rival-the-us-with-an-independent-regulator/273886/

Prasad, Narayan. *A Geopolitical and Technological Survey of Potential Use of Space Weapons*. Bangalore: NIAS, 2018.

Press Information Bureau. Government of India. Department of Space. "Space Activities Bill." December 12, 2018. http://pib.nic.in/newsite/printrelease.aspx?relid=186315

Press Trust of India. "Centre Finalises Broad Contours of Defence Space Agency," *NDTV*, June 11, 2019. https://www.ndtv.com/india-news/centre-finalises-broad-contours-of-defence-space-agency-2051699?pfrom=home-topstories

Press Trust of India. "Gaganyaan, India's First Manned Mission, Set to be a Reality by 2022," *Live Mint*, August 15, 2018. https://www.livemint.com/Politics/glFe56EP0V2EkF25mQshXO/Narendra-Modi-announces-Gaganyaan-in-Independence-Day-speech.html

Press Trust of India. "India Successfully Launches EMISAT Satellite," *The Hindu Business Line*, April 1, 2019. https://www.thehindubusinessline.com/news/science/india-successfully-launches-emisat-satellite/article26699384.ece

Press Trust of India. "India to Explore Moon's Surface for Helium 3," *The Hindustan Times*, September 1, 2006. https://web.archive.org/web/20170528033036/https://www.hindustantimes.com/india/india-to-explore-moon-s-surface-for-helium-3/story-K4HAfpqIhRCJm6zeKVsGxM.html

Price WaterhouseCooper (PwC). *The World in 2050*. December 3, 2015. https://www.pwc.com/gx/en/issues/economy/the-world-in-2050.html

PUBLIC LAW 100-685, November 17, 1988. *GovInfo*. https://www.govinfo.gov/content/pkg/STATUTE-102/pdf/STATUTE-102-Pg4083.pdf

Pyle, Rod. *Space 2.0: How Private Spaceflight, a Resurgent NASA, and International Partners are Creating a New Space Age*. Dallas: BenBella Books, 2019.

QUESS (Quantum Experiments at Space Scale), E-Portal Directory. n.d. https://directory.eoportal.org/web/eoportal/satellite-missions/q/quess

Raghuvanshi, Vivek. "Indian Aerospace Command To Operate Military Satellites," *Space News*, April 4, 2005. https://spacenews.com/indian-aerospace-command-operate-military-satellites/

Raine, John. "War or Peace? Understanding the Grey Zone," *IISS*, April 3, 2019. https://www.iiss.org/blogs/analysis/2019/04/understanding-the-grey-zone

Rajagopalan, Megha. "China Trains 'fishing militia' to Sail into Disputed Waters," *Reuters*, April 30, 2016. https://www.reuters.com/article/us-southchinasea-china-fishingboats-idUSKCN0XS0RS

Rajagopalan, Megha. "Chinese Military Force to Take Lead on Cyber, Space Defense," *Reuters*, January 29, 2016. http://www.reuters.com/article/us-china-military-idUSKCN0V714B

Rajagopalan, Rajeswari Pillai. "A First: India to Launch First Simulated Space Warfare Exercise," *The Diplomat*, June 12, 2019. https://thediplomat.com/2019/06/a-first-india-to-launch-first-simulated-space-warfare-exercise/

Rajagopalan, Rajeswari Pillai. "Having Tested Its ASAT Capability, India Should Help Shape Global Space Norms," *The Wire*, March 28, 2019. https://thewire.in/space/having-tested-its-asat-capability-india-should-help-shape-global-space-norms

Rajagopalan, Rajeswari Pillai. "Space Based Solar Power: Time to Put it on the New US-India S&T Endowment Fund?," *ORF Online*, April 2, 2011. https://www.orf

online.org/research/space-based-solar-power-time-to-put-it-on-the-new-us-india-st-endowment-fund/

Rajan, M. S. *Nonalignment and Non Aligned Movement: Retrospect and Prospect.* New Delhi: Vikas Publications, 1990.

Ralf, Eric. "SpaceX to Mature Starship Moon Landing and Orbital Refueling Tech with NASA's Help," *Teslarati*, July 31, 2019. https://www.teslarati.com/spacex-starship-moon-landing-orbital-refueling-nasa/

Ramesh, M. "India's Start-ups Find Opening in Space," *The Hindu Busines Line*, July 14, 2019. https://www.thehindubusinessline.com/news/science/indias-start-ups-find-opening-in-space/article28427597.ece

RAND. "The U.S.-China Military Scorecard Forces, Geography, and the Evolving Balance of Power, 1996-2017," *RAND, Santa Monica*, 2015. https://www.rand.org/pubs/research_reports/RR392.html

Raslan, Karim. "Ignore Silk Road Hubris, China Will Grow Old Before it Gets Rich," *South China Morning Post*, May 23, 2017. https://www.scmp.com/week-asia/opinion/article/2095326/opinion-ignore-silk-road-hubris-china-will-grow-old-it-gets-rich

Rawlings III, Hunter R. "Thycididian Epistemology: Between Philosophy and History," *Rheinisches Museum für Philologie* Neue Folge, 153, no. 4 (2010): 247–90.

ReBeam Website. n.d. https://nopo.in/web/rebeam/

Reddy, Gadhadar, Narayan Prasad, Divyashree Iyer. "Carbon-Nanotube Based Solar Power (CASSP)," 4th Space Solar Power International Student and Young Professional Design Competition, ESA, October 2015. https://www.esa.int/gsp/ACT/doc/nrg/SPS%202015_winner%20paper.pdf

Rhodes, Christopher J. "Solar Energy: Principles and Possibilities," Science Progress 93, no. 1 (2010): 37–112. https://www.theworldcounts.com/stories/current_world_energy_consumption

Riccucci, Norma M. *Public Administration: Traditions of Inquiry and Philosophies of Knowledge.* Washington, DC: Georgetown Press, 2010.

Rocketeers Website. n.d. https://www.rocketeers.in/

Rodham, Nayaf R F AL. "U.S. Space Policy and Strategic Culture," *Journal of International Studies*, April 16, 2018. https://jia.sipa.columbia.edu/online-articles/us-space-policy-and-strategic-culture

Rogers, Everett M. *Diffusion of Innovations.* New York: Free Press, 2003.

Rogin, Josh. "Wilbur Ross Wants the United States to Dominate Space," *The Washington Post*, May 2, 2018. https://www.washingtonpost.com/news/josh-rogin/wp/2018/05/02/wilbur-ross-wants-the-united-states-to-dominate-space/

Rohit, T. K. "Second Lunar Mission Chandrayaan-2 to be Launched in July," *The Hindu*, May 2, 2019. https://www.thehindu.com/sci-tech/science/second-lunar-mission-chandrayaan-2-to-be-launched-in-july/article27003022.ece

Rosen, Stephen Peter. "Military effectiveness: Why society matters," *International Security* 19, no. 4 (1995): 5–31.

Ross, R. S. and Z. Feng. *China's Ascent: Power, Security, and the Future of International Politics.* New York: Cornell University Press, 1998.

Ross, Wilbur. "A New Space Race: Getting to the Trillion-Dollar Space Economy World Economic Forum, Davos, Switzerland," *Commerce Department*, January 24, 2020. https://www.commerce.gov/news/speeches/2020/01/remarks-secretary-commerce-wilbur-ross-new-space-race-getting-trillion-dollar

Ross, Wilbur. "Launching Toward a $1 Trillion Space Economy," *Orlando Sentinel*, February 28, 2019. https://www.orlandosentinel.com/opinion/os-op-wilbur-ross-space-1-trillion-20190228-story.html, also available at: https://www.commerce.gov/news/op-eds/2019/02/op-ed-commerce-secretary-wilbur-l-ross-launching-toward-1-trillion-space

Rousseau, Jean Jacques. "Discourse on Inequality," translated by G.D.H. Cole. https://www.aub.edu.lb/fas/cvsp/Documents/DiscourseonInequality.pdf879500092.pdf

RT Staff. "Star City adventure: Take a sneak peek into India's first space crew training in Russia," *Russia Today*, February 17, 2020. https://www.rt.com/news/480892-india-first-space-crew-russia/

Rugh, Andrea B. *The Political Culture of Leadership in the United Arab Emirates.* New York: Palgrave, MacMillan, 2007.

Russell, Kendall. "What Will It Take to Unleash NewSpace in India?," *Satellite Today*, February 12, 2018. http://interactive.satellitetoday.com/what-will-it-take-to-unleash-newspace-in-india/

Russett, Bruce and John R. Oneal. *Triangulating Peace: Democracy, Interdependence, and International Organizations.* New York: Norton, 2001.

Russett, Bruce. *Grasping the Democratic Peace: Principles for a Post–Cold War World* Princeton, NJ: Princeton University Press, 1993.

Rynning, Sten. The European Union: Towards a Strategic Culture?," *Security Dialogue* 34, no.4 (December 2003): 479–96.

Sabatier, Paul A. and Christopher M. Weible, eds. "The Advocacy Coalition Framework," in *Theories of the Policy Process.* Edited by Paul A. Sabatier and Christopher M. Weible, 135–71. Boulder: Westview Press, 2019.

Sabatier, Paul A. and Christopher M. Weible, eds. *Theories of the Policy Process.* Boulder: Westview Press, 2019.

Sabatier, Paul A. "An Advocacy Coalition Framework of Policy Change and the Role of Policy-Oriented Learning Therein," in *The Oxford Handbook of Classics in Public Policy and Administration.* Edited by Paul Cairney. Oxford: Oxford University Press, 2016. https://www.oxfordhandbooks.com/view/10.1093/oxfordhb/9780199646135.001.0001/oxfordhb-9780199646135-e-24

Sabine, Peter. "Belt and Road is 'Marshall Plan without a War' Analysts Say, as Beijing and Banks Woo Private Sector Investors," *South China Morning Post*, March 28, 2017. https://www.scmp.com/special-reports/business/topics/one-belt-one-road/article/2082733/belt-and-road-marshall-plan

Sabine, Philip. *Simulating War: Studying Conflict Through Simulation Games.* New York and London: A&C Black, 2012.

Safi, Michael. "India Launches Record Breaking 104 Satellites from Single Rocket," *The Guardian*, February 15, 2017. https://www.theguardian.com/science/2017/feb/15/india-launches-record-breaking-104-satellites-from-single-rocket

Sage, Peter. "Space Energy: An Open Letter From Space Energy Co-Founder Peter Sage Covering," n.d. https://petersage.com/spaceenergy/

Sailors, Cara Leigh. "The Function of Mythology and Religion in Ancient Greek Society," n.d. https://dc.etsu.edu/cgi/viewcontent.cgi?article=3471&context=etd

Saldaña, Johnny. *The Coding Manual for Qualitative Researchers*. Thousand Oaks, CA: SAGE Publications, 2016.

Satsure Website. n.d. http://www.satsure.in/

Savin, Leonid. *US Strategic Culture and Foreign Policy*, Katehon, 2016. http://katehon.com/article/us-strategic-culture-and-foreign-policy

Savitsky, Sasha. "NASA's Juno Spacecraft to Become Fastest Man-Made Object as it Slingshots Around Earth Towards Jupiter," *Fox News*, October 9, 2013, https://www.foxnews.com/science/nasas-juno-spacecraft-to-become-fastest-man-made-object-as-it-slingshots-around-earth-towards-jupiter

Schaler, E. "An Assessment of the Institutional Analysis and Development Framework and Introduction of The Social-Ecological Systems Framework" in *Theories of the Policy Process*. Edited by Paul A. Sabatier and Christopher M. Weible, 215–52. Boulder: Westview Press.

Schell, Orville and John Delury. *Wealth and Power: China's Long March to the Twenty-first Century*. New York: Random House, 2013.

Schneider, Christina J. "Weak States and Institutionalized Bargaining Power in International Organizations," *International Studies Quarterly* 55, no. 2 (June 2011): 331–55.

Schon, Donald A. and Martin Rein. *Frame Reflection: Toward the Resolution of Intractable Policy Controversies*. New York: Basic Books, 1994.

Schwartz, Peter. *The Art of the Long View: Planning for the Future in an Uncertain World*. New York: Penguin Random House, 1991.

Schweller, Randall L. "Unanswered Threats: A Neoclassical Realist Theory of Understanding." *International Security* 29, no. 2 (Fall 2004):159–201.

Schweller, Randall. "Bandwagoning for Profit: Bringing the Revisionist State In," *International Security* 19, no. 1 (Summer 1994): 72–107.

Scoles, Sarah. "China Wants to Make a Mark in Space-But It will Need a Little Help," *Wired*, February 14, 2018. https://www.wired.com/story/china-wants-to-make-a-mark-in-spacebut-itll-need-a-little-help/

Scrivener, A. E. "Private Company Rule in the Congo," *King Leopold's Rule in Africa*, 181–6 London: William Heinemann, 1904. http://www.classzone.com/books/wh_05_shared/pdf/WHS05_027_774a_PS.pdf

Sercel, Joel. "Stepping Stones: Economic Analysis of Space Transportation Supplied from NEO Resources," *NASA*, August 16, 2017. fiso.spiritastro.net/telecon/Sercel_8-16-17/Sercel_8-16-17.pdf; final report here: https://www.nasa.gov/sites/default/files/atoms/files/eso_final_report.pdf

Seven Military Classics of Ancient China, translated by Ralph D. Sawyer. New York: Basic Books, 2007.

SF Encyclopedia. "Clarke's Laws," August 2, 2016. http://www.sf-encyclopedia.com/entry/clarkes_laws

Shackleton Energy Website. http://www.shackletonenergy.com/overview

Shah, Amrita. "Nehru's Belief in Science Fuelled India's Audacious Space Program," *The Print*, November 14, 2017. https://theprint.in/opinion/nehrus-belief-fuelled-indias-space-atomic-programmes/16224/

Sharma, Vishnu. *The Panchantantra*. New Delhi: Penguin, 2007.

Shaurya Karanbir Gurung. "All About the Names of Six Places in Arunachal that China Gave," *The Economic Times*, July 12, 2018, https://economictimes.indiatimes.com/news/defence/all-about-the-names-of-six-places-in-arunachal-that-china-gave/articleshow/58259491.cms

Sheehan, Michael. *The International Politics of Space*. London and New York: Routledge, 2007.

Sheetz, Michael. "Elon Musk's SpaceX Becomes the First to Launch Reused Rocket," *CNBC*, December 15, 2018. https://www.cnbc.com/2017/12/15/elon-musks-spacex-becomes-first-to-launch-reused-rocket-on-a-nasa-mission.html

Sheetz, Michael. "The Space Industry Will be Worth Nearly $3 Trillion in 30 years, Bank of America Predicts," *CNBC*, October 31, 2017. https://www.cnbc.com/2017/10/31/the-space-industry-will-be-worth-nearly-3-trillion-in-30-years-bank-of-america-predicts.html

Shida, Tomio. "History Offers Lessons in Escaping China's Rare-Earth Dominance," *Nikkei Asian Review*, June 25, 2019. https://asia.nikkei.com/Economy/Trade-war/History-offers-lessons-in-escaping-China-s-rare-earth-dominance2

Shiga, David. "Will Obama Pursue Space-based Solar Power?," *New Scientist*, December 22, 2008. https://web.archive.org/web/20160731181015/https://www.newscientist.com/blogs/shortsharpscience/2008/12/will-obama-pursue-space-based.html

Siddhartha V. "Military Dimensions in The Future of The Indian Presence in Space." *USI of India* (2000): 243–58.

Siddiqi, Asif A. *The Red Rockets' Glare Cambridge Centennial of Flight*. Cambridge, UK: Cambridge University Press, 2014.

Sidhartha, V. "Military Dimensions in the Future of the Indian Presence in Space," *Presentation to United Services Institution*, New Delhi, 2000.

Sidhartha, V. "Military Dimensions in the Future of the Indian Presence in Space," Presentation to Centre for Airpower Studies (CAPS), September 7, 2010.

Sieren, Frank. "Sieren's China: Speaking Mandarin on Mars," *DW*, October 19, 2018. https://www.dw.com/en/sierens-china-speaking-mandarin-on-mars/a-45964091

Simon, Herbert A. "Rational Choice and the Structure of the Environment," *Psychological Review* 63, no. 2 (1956):129–38.

Simon, Herbert, A. "A Behavioral Model of Rational Choice," in *Models of Man, Social and Rational: Mathematical Essays on Rational Human Behavior in a Social Setting*. New York: Wiley, 1957.

Sims, Henry P. and Dennis A. Gioia. *The Thinking Organization: Dynamics of Organizational Social Cognition*. San Francisco: Jossey-Bass, 1986.

Singer, J. David. "The Level of Analysis Problem in International Relations," *World Politics* 14, no.1 (October 1961): 77–92.

Singh, Gurbir. *India's Space Programme: India's Incredible Journey from the Third World Towards the First.* Manchester: Astrotalkuk Publications, 2017.

Singh, Jatin. "India, the Next 50 Years in Space; Will She Grab her Manifest Destiny?," *LinkedIn*, June 26, 2019. https://www.linkedin.com/pulse/india-next-50-years-space-she-grab-her-manifest-destiny-jatin-singh/

Singh, Surendra and Srinivas Laxman. "India's First Private Satellite Launched," *Times of India*, December 4, 2018. https://timesofindia.indiatimes.com/india/space-x-launches-indias-first-privately-built-satellite-exceedsat-1/articleshow/66937238.cms

Singh, Surendra. "Chandrayaan-2 Mission Deferred Again, Scheduled for May Launch," *Times of India*, April 1, 2019. https://timesofindia.indiatimes.com/india/chandrayaan-2-mission-deferred-again-scheduled-for-may-launch/articleshow/68675904.cms

Siqi, Cao. "Heroes to Receive Highest Honor in China." *The Global Times*, August 08, 2019. http://www.globaltimes.cn/content/1162994.shtml

Sivan, K. "Presentation to Parliamentary Standing Committee on Science & Technology, Environment and Forests," *Government of India, Department of Space*, September 25, 2019. slide 13.

Skymet Weather Services Pvt Ltd Website. n.d. https://www.skymetweather.com/

Skyrms, Brian and U.C. Irvine. "The Stag Hunt," Presidential Address to Pacific Division of the *American Philosophical Association*, March, 2001. http://www.socsci.uci.edu/~bskyrms/bio/papers/StagHunt.pdf

Skyroot Aerospace Website. n.d. https://skyroot.in/

Slevison, Alice. "Considering China's Strategic Interests in Antarctica," *The Strategist*, February 5, 2016. https://www.aspistrategist.org.au/considering-chinas-strategic-interests-in-antarctica/

Smart Enovations Website. http://www.smartenovations.com/

Smith, Rich. "The $1.1 Trillion Space Industry Prediction You Can't Afford to Miss," *The Motley Fool*, February 16, 2018. https://www.fool.com/investing/2018/02/16/the-11-trillion-space-industry-prediction-you-cant.aspx

Snead, James M. "Energy, SSP, and Jumpstarting America's Spacefaring Future: Why America's Pro-Human Spacefaring Community Should Strongly Advocate Starting Commercial Developmen of Space Solar Power," Presentation at the 2009 International Space Development Conference (ISDC), June 10, 2009. https://www.mikesnead.com/resources/spacefaring/presentation_energy_ssp_and_jumpstarting_america's_spacefaring_future_version_2_for_distribution.pdf

Snyder, Jack L. *The Soviet Strategic Culture: Implications for Limited Nuclear Operations*. Santa Monica, CA: RAND Corporation, 1977.

Snyder, Jack. "The Broken Bargain How Nationalism Came Back," *Foreign Affairs*, March/April 2019. https://www.foreignaffairs.com/articles/world/2019-02-12/broken-bargain

Solar Power Satellites – Selected References. n.d. https://www.spaceacademy.net.au/library/refs/spsrefs.htm

Solaren. "The Solaren Mission: Cost Competitive Space Electricity," n.d. https://www.solarenspace.com/

Soldatkin, Vladimir. "Russia Wants to Join Luxembourg in Space Mining," *Reuters*, March 6, 2019. https://www.reuters.com/article/us-luxembourg-russia-space/russia-wants-to-join-luxembourg-in-space-mining-idUSKCN1QN1OQ

Solomone, Stacey. *China's Strategy in Space*. Switzerland and New York: Springer, 2013.

Somarin, Ali. "Unconventional Oil Exploration, Part 3: Ultra-deepwater Oil," *Thermofisher*, October 21, 2014. https://www.thermofisher.com/blog/mining/unconventional-oil-exploration-part-3-ultra-deepwater-oil/

Sonkavday, N. R. *Space Based Solar Power: An Eternal Source of Power from Space Volume* 1. Pune: First Impression Print Technology, 2013.

Soper, Taylor. "Planetary Power Raises $3.4M to Provide Efficient Energy for Off-Grid Telecom Towers," *GeekWire*, June 1, 2016. https://www.geekwire.com/2016/planetary-power-raises-3-4m/

Space Commodities Exchange (SCX). "Space Commodities Exchange," n.d. http://www.spacecommoditiesexchange.com/

Space Development Steering Committee. "About the Space Development Steering Committee," n.d. http://spacedevelopmentsteeringcommittee.org/about-the-space-development-steering-committee/

Space Frontier Foundation. "About the Space Frontier Foundation: Who We Are," n.d. http://newspace.spacefrontier.org/about/

Space Frontier Foundation. "Obama Space Team Seeks Public Comment On Space Solar Power White Paper," via archive, n.d. https://web.archive.org/web/20150929070923/https://spacefrontier.org/2008/12/obama-space-team-seeks-public-comment-on-space-solar-power-white-paper-submitted-by-sff/

Space Island Group, Clean Energy, Cheap Hydrogen and Weather Control From Space. n.d. https://www.spaceislandgroup.com/solarsat.html

Space Policy Directive-1. Reinvigorating America's Human Space Exploration Program, The White House, December 11, 2017. http://www.whitehouse.gov/presidential-actions/presidential-memorandum-reinvigorating-americas-human-space-exploration-program/

Space Policy Directive-2. Streamlining Regulations on Commercial Use of Space. The White House. May 24, 2018. https://www.whitehouse.gov/presidential-actions/space-policy-directive-2-streamlining-regulations-commercial-use-space/

Space Policy Directive-3. National Space Traffic Management Policy. The White House. June 18, 2018. https://www.whitehouse.gov/presidential-actions/space-policy-directive-3-national-space-traffic-management-policy/

Space Policy Directive-4. Establishment of the United States Space Force. The White House. February 19, 2019. https://media.defense.gov/2019/Mar/01/2002095015/-1/-1/1/SPACE-POLICY-DIRECTIVE-4-FINAL.PDF also available at https://spacepolicyonline.com/news/text-of-space-policy-directive-4-spd-4-establishing-a-u-s-space-force/

SpaceX. "Interplanetary Tranport System (ITS)." *You Tube*. September 27, 2019. https://www.youtube.com/watch?v=0qo78R_yYFA

SpaceX. "Making Humans a Multiplanetary Species" Presentation to IAC 2016, *YouTube*, https://www.youtube.com/watch?v=H7Uyfqi_TE8&t=10s

SpaceX. "Making Life Multiplanetary" presentation to IAC 2017. *YouTube*. September 29, 2017. https://www.youtube.com/watch?v=tdUX3ypDVwI&t=8s

Spero, Joshua B. "Great Power Security Dilemmas for Pivotal Middle Power Bridging," *Contemporary Security Policy* 30, no. 1 (2009): 147–71. https://doi.org/10.1080/13523260902760355

Spires, David. *Beyond Horizons: A Half Century of Air Force Space Leadership*, Revised Edition. Montgomery AL: Air University Press, 1998.

Spudis, Paul D., *The Value of the Moon: How to Explore, Life, and Prosper in Space Using the Moon's Resources*. Washington, DC: Smithsonian Books, 2016.

Staff Report. "Space Agency Launches NewSpace Innovation Programme," *Gulf News*, January 11, 2020. https://gulfnews.com/uae/space-agency-launches-new space-innovation-programme-1.68942607

Staff. "Chandrayaan-2 to Reach Moon by August 20: ISRO," *Times of India*, July 24, 2019. http://timesofindia.indiatimes.com/articleshow/70366062.cms

Staff. "ISRO 'Makes Announcement of Opportunity' for Its Venus Mission," *The Hindustan Times*, April 24, 2017, http://www.hindustantimes.com/india-news/isro-makes-announcement-of-opportunity-for-its-venus-mission/story-radr7VgPoTM40Li7yoVaZL.html

Staff. "ISRO to Increase Frequency of Launches to 12 per year," *The Times of India*, April 23, 2017. http://timesofindia.indiatimes.com/india/isro-to-increase-frequency-of-launches-to-12-per-year/articleshow/58326033.cms

Staff. "UAE Leaders Sign Final Piece of Hope Probe Destined for Mars," *The National*, January 5, 2020. https://www.thenational.ae/uae/science/uae-leaders-sign-final-piece-of-hope-probe-destined-for-mars-1.960118

State Council, People's Republic of China. "China's White Paper on Space Activities 2016." http://english.gov.cn/archive/white_paper/2016/12/28/content_281475527159496.htm

State Council, People's Republic of China. "Chinese Space Activities in 2006." October 13, 2006, http://www.china.org.cn/english/features/book/183672.htm

Stephen, David, et al. *The New Politics of Strategic Resources*. Washington, DC: The Brookings Institution Press, 2015.

Stewart, Terry. "The Second Opium War," Historic UK. n.d. https://www.historic-uk.com/HistoryUK/HistoryofBritain/Second-Opium-War/

Stiglitz, J. E. "Trump and Globalization." *Journal of Policy Modeling*, 2018. https://doi.org/10.1016/j.jpolmod.2018.03.006

Stine, G. Harry. *Space Power: The News the Department of Energy Tried to Bury!* New York: Ace Science, 1981. https://space.nss.org/space-power-by-harry-stine/

Stone, Christopher M. *Reversing the Tao: A Framework for Credible Space Deterrence* 1st ed. Scotts Valley, CA: CreateSpace Independent Publishing Platform, 2016.

Storey, David. "Territory and Territoriality," *Oxford Bibliographies*, July 26, 2017, 2019. https://www.oxfordbibliographies.com/view/document/obo-9780199874002/obo-9780199874002-0076.xml

Strout, Nathan. "The Pentagon Wants to Solve a Deep Space Problem With Three Vehicles," *C4ISRNET*, August 27, 2019. https://www.c4isrnet.com/battlefield-tech/space/2019/08/27/the-pentagon-wants-to-solve-a-deep-space-problem-with-three-vehicles/

Subaihi, Thamer Al. "UAE Space Programme a Conduit for Cooperation," *The National*, November 28, 2015. https://www.thenational.ae/uae/uae-space-programme-a-conduit-for-cooperation-1.67069

Subramanian, Samanth. "India's Frugal Mission to Mars," *The New Yorker*, November 7, 2013. https://www.newyorker.com/tech/elements/indias-frugal-mission-to-mars

Sved, J., G. L. Kulcinski, G. H. Miley. "A Commercial Lunar Helium 3 Fusion Power Infrastructure," *Journal of the British Interplanetary Society* 48 (1995): 55–61.

Tammen, R. "The Impact of Asia on World Politics: China and India Options for the United States," *International Studies Review* 8, no. 4 (2006): 563–580.

Tanham, George K. *Indian Strategic Thought: An Interpretive Essay*, RAND, 1992. https://www.rand.org/pubs/reports/R4207.html

Tansey, Oisin. "Process Tracing and Elite Interviewing: A Case for Non-Probability Sampling," *PS: Political Science and Politics* 40, no. 4 (October 2007): 765–72.

Tate, Karl. "How Golden Spike's Moon Landing Plan Works (Infographic)," *Space.com*, December 6, 2012. https://www.space.com/18805-golden-spike-private-moon-landing-graphic.html

Taylor, Stephen J. "Go West, Young Man": The Mystery Behind The Famous Phrase, *Hoosier State Chronicles*, July 9, 2015. https://blog.newspapers.library.in.gov/go-west-young-man-the-mystery-behind-the-famous-phrase/

Team Indus Blog. "When India set a Date with the Moon," November 13, 2017. https://medium.com/teamindus/when-india-set-a-date-with-the-moon-8c095a7b5f37

TeamIndus Website. n.d. https://medium.com/teamindus

Telzrow, Michael E. "John Wayne: American Icon," *The New America*, August 5, 2016. https://www.thenewamerican.com/culture/history/item/23714-john-wayne-american-icon

Temby, Owen. "What are Levels of Analysis and What do They Contribute to International Relations Theory," *Cambridge Review of International Affairs* 28, no. 4 (2015): 721–42.

Tharoor, Ishaan. "The Dark Side of Winston Churchill's Legacy No One Should Forget," *The Washington Post*, February 3, 2015. https://www.washingtonpost.com/news/worldviews/wp/2015/02/03/the-dark-side-of-winston-churchills-legacy-no-one-should-forget/?fbclid=IwAR2s_r7l3-YblgSJrLd5CkfzZsStknhVffvUquQIeJJev4vmHeZ6Lnlpf04

The 70th Anniversary of the Founding of the People's Republic of China," *Xinhua*, October 1, 2019, http://www.xinhuanet.com/english/special/2019-10/01/c_138438666_181.htm

The Draft Bill can be seen at: https://www.prsindia.org/billtrack/draft-space-activities-bill-2017 and its progress tracked at: https://www.insightsonindia.com/2019/01/01/space-activities-bill-2017/

The Earthlight Foundation, "About," n.d., http://earthlightfoundation.org/about/

The Hague Space Resources Governance Working Group, *Final Report of The Hague Space Resources,* Building Blocks for the Development of an International Framework on Space Resources Activities, November 2019. https://www.

universiteitleiden.nl/binaries/content/assets/rechtsgeleerdheid/instituut-voor-publiekrecht/lucht–en-ruimterecht/space-resources/bb-thissrwg–cover.pdf

The Indian Space Draft bill can be seen at: https://www.prsindia.org/billtrack/draft-space-activities-bill-2017 and its progress tracked at: https://www.insightsonindia.com/2019/01/01/space-activities-bill-2017/

The Mars Society, "Founding Declaration," August 1998. https://www.marssociety.org/founding-declaration/

The Moon Society. "About Us," n.d. https://www.moonsociety.org/about

The National Bureau of Asian Research, "The Legacy of Nehruvianism and the Implications for India's Strategic Culture," *National Bureau of Asian Research*, November 21, 2016. https://www.nbr.org/publication/the-legacy-of-nehruvianism-and-the-implications-for-indias-strategic-culture/

The National Bureau of Asian Research. "Behind the Official Narrative China's Strategic Culture in Perspective," *National Bureau of Asian Research*, November 1, 2016. https://www.nbr.org/publication/behind-the-official-narrative-chinas-strategic-culture-in-perspective/

The National Security Strategy of the United States of America. The White House. December 18, 2017. https://www.whitehouse.gov/wp-content/uploads/2017/12/NSS-Final-12-18-2017-0905.pdf

The Planetary Society, "About Us," n.d. http://www.planetary.org/about/

The Space Show, SolarHigh.org, June 14, 2013. https://www.thespaceshow.com/solarhigh.org

The Space Show. "Heliosat," April 27, 2009. https://thespaceshow.com/heliosat

The State Council, People's Republic of China. "China's White Paper on Space Activities 2016," December 28, 2016. http://english.gov.cn/archive/white_paper/2016/12/28/content_281475527159496.htm

The State Council, The People's Republic of China. "The Belt and Road Initiative," July 18, 2019. http://english.www.gov.cn/beltAndRoad/

The Sun's Energy, University of Tennessee Institute of Agriculture. n.d. https://ag.tennessee.edu/solar/Pages/What%20Is%20Solar%20Energy/Sun's%20Energy.aspx

The U.S. Embassy in Luxembourg. "The United States and Luxembourg Sign Memorandum on Space Cooperation on May 10, 2019," May 10, 2019. https://lu.usembassy.gov/the-united-states-and-luxembourg-sign-memorandum-on-space-co-operation-on-may-10-2019/

The White House. "President Donald J. Trump is Unveiling an America First National Space Strategy," March 23, 2018. https://www.whitehouse.gov/briefings-statements/president-donald-j-trump-unveiling-america-first-national-space-strategy/

The White House. "Presidential Memorandum on Reinvigorating America's Human Space Exploration Program." *The White House*. December 11, 2017. https://www.whitehouse.gov/presidential-actions/presidential-memorandum-reinvigorating-americas-human-space-exploration-program/

The White House. "Recommendations Approved by the National Space Council to President Trump." March 26, 2019. http://www.spaceref.com/news/viewsr.html?pid=52358

The White House. "Remarks by President Trump at a Meeting with the National Space Council and Signing of Space Policy Directive-3," June 18, 2018. https://www.whitehouse.gov/briefings-statements/remarks-president-trump-meeting-national-space-council-signing-space-policy-directive-3/

The White House. "Remarks by Vice President Pence at Fifth Meeting of the National Space Council Huntsville, AL." March 26, 2019. https://www.whitehouse.gov/briefings-statements/remarks-vice-president-pence-fifth-meeting-national-space-council-huntsville-al/

The White House. "Remarks by Vice President Pence on the Future of the U.S. Military in Space," August 9, 2018. https://www.whitehouse.gov/briefings-statements/remarks-vice-president-pence-future-u-s-military-space/

The White House, The National Space Council. A New Era for Deep Space Exploration and Development, July 23, 2020. https://www.whitehouse.gov/wp-content/uploads/2020/07/A-New-Era-for-Space-Exploration-and-Development-07-23-2020.pdf

The World Bank. "Belt and Road Initiative," March 29, 2018. https://www.worldbank.org/en/topic/regional-integration/brief/belt-and-road-initiative;

The World Bank. "GDP Per Capita (Current US$)," n.d. https://data.worldbank.org/indicator/ny.gdp.pcap.cd

The World Bank. "GDP Per Capita UAE," n.d. https://data.worldbank.org/indicator/NY.GDP.PCAP.CD?locations=AE.

The World Bank. "GDP per capita, PPP (current international $) – Luxembourg," 2018. https://data.worldbank.org/indicator/NY.GDP.PCAP.PP.CD?locations=LU

The Wright Brothers, Inventing a Flying Machine, Smithsonian National Air and Space Museum. https://airandspace.si.edu/exhibitions/wright-brothers/online/fly/1903/

Thomas Theorem, *Oxford Index*, n.d. https://oxfordindex.oup.com/view/10.1093/oi/authority.20110803104247382

Thomas, John P. "20,000 Satellites for 5G to be Launched Sending Focused Beams of Intense Microwave Radiation Over Entire Earth," *Health Impact News*, March 9, 2019. http://healthimpactnews.com/2019/20000-satellites-for-5g-to-be-launched-sending-focused-beams-of-intense-microwave-radiation-over-entire-earth/

Thompson, Alexander. "Coercion Through IOs: The Security Council and the Logic of Information Transmission," *International Organization* 60, no.1 (Winter 2006): 1–34.

Thompson, Amy. "SpaceX Launches 60 Starlink Satellites for New Megaconstellation, Misses Rocket Landing," *Space.com*, February 16, 2020. https://www.space.com/spacex-starlink-4-launch-success-misses-rocket-landing.html

Thucydides. *The History of The Peloponnesian War*, 431 BC translated by Crawley, R., 1852. *History of the Peloponnesian War* (No. 455). New York: Penguin Books, 1964. https://www.gutenberg.org/files/7142/7142-h/7142-h.htm

Tian, Nan, Aude Fleurant, Alexandra Kuimova, Pieter D. Wezeman, Siemon T. Wezeman. *Trends in world military expenditure, 2017*. Stockholm International Peace Research Institute, 2018. "Trends in World Military Expenditure 2017,"

SIPRI, 2017. https://www.sipri.org/publications/2018/sipri-fact-sheets/trends-world-military-expenditure-2017

Tiruvalluvar's *Kural,* translated by P.S. Sundaram. New Delhi: Penguin, 2005.

Toll, Ian W. *Six Frigates: The Epic History of the Founding of the U.S. Navy.* New York: W.W. Norton & Company, 2008

Toms, Manu P. "Team Indus ends pursuit of Google Lunar XPrize," *VC Circle*, January 24, 2018. https://www.vccircle.com/team-indus-ends-its-pursuit-of-google-lunar-xprize/

Trans Astranautica Corporation. https://www.transastracorp.com/

True, James L. Bryan D. Jones, Frank R. Baumgartner. "Punctuated equilibrium theory," in *Theories of the Policy Process*. Edited by Paul A. Sabatier and Christopher M. Weible, 55–101. Boulder: Westview Press, 2019.

True, James L., Bryan D. Jones, Frank R. Baumgartner. "Punctuated-Equilibrium Theory: Explaining Stability and Change in Public Policymaking," in *Theories of the Policy Process.* Edited by Paul Sabatier, 2nd Edition. Boulder, Colorado: Westview Press, 2006. https://fbaum.unc.edu/teaching/articles/True_Jones_Baumgartner_2006_chapter.pdf

Trump, Donald J. "Remarks by President Trump in State of the Union Address," *The White House*, February 4, 2020. https://www.whitehouse.gov/briefings-statements/remarks-president-trump-state-union-address-3/

Trump, Donald J. "Amending Executive Order 13803 – Reviving the National Space Council," *The White House*, February 13, 2020. https://www.whitehouse.gov/presidential-actions/amending-executive-order-13803-reviving-national-space-council/

Trump, Donald J. "Presidential Executive Order on Reviving the National Space Council," *The White House*, June 30, 2017. https://www.whitehouse.gov/presidential-actions/presidential-executive-order-reviving-national-space-council/

Tucker, Patrick. "China's Moon Missions Could Threaten US Satellites: Pentagon," *DefenseOne*, October 16, 2018. https://www.defenseone.com/technology/2018/10/chinas-moon-missions-could-threaten-us-satellites-pentagon/152084/

Tumlinson, Rick. "The Summit, the Solution, the Possibility," *SpaceNews*, May 27, 2015. https://spacenews.com/op-ed-the-summit-the-solution-the-possibility/

Turtle, Michael. "The World's Oldest Sovereign State," *Time Travel Turtle*, n.d. https://www.timetravelturtle.com/visiting-san-marino-oldest-country/

Tversk, Amos and Daniel Kahneman. "Advances in Prospect Theory: Cumulative Representation of Uncertainty," *Journal of Risk and Uncertainty* 5, no. 4 (1992): 297–323.

Tzu, Sun. *Art of War*, translated by Lionel Giles. http://classics.mit.edu/Tzu/artwar.html

UAE Space Agency. "Mars 2117," *You Tube*, November 27, 2017. https://www.youtube.com/watch?v=JboqeUqQiv4#action=share

UAE Voice. "National Space Strategy 2030 Gets the Approval of the UAE Cabinet," *UAE Voice*, March 12, 2019. https://www.uae-voice.com/national-space-strategy-2030-gets-approval-uae-cabinet/

UAE. "2030-2117," n.d., https://government.ae/en/more/uae-future/2030-2117

UK Government. *Resource Nationalism: A Horizon Scanning Research Paper By The Resources Demand And Supply Resource Nationalism Community Of Interest.* December 2014. https://assets.publishing.service.gov.uk/government/uploads/system/uploads/attachment_data/file/389085/Horizon_Scanning_-_Resource_Nationalism_report.pdf

Union of Concerned Scientists. "USC Satellite Databasem," March 31, 2019. https://www.ucsusa.org/nuclear-weapons/space-weapons/satellite-database

United Arab Emirates. "National Innovation Strategy," 2015, updated May 30, 2019. https://government.ae/en/about-the-uae/strategies-initiatives-and-awards/federal-governments-strategies-and-plans/national-innovation-strategy

United Arab Emirates. "National Space Strategy 2030," https://www.government.ae/en/about-the-uae/strategies-initiatives-and-awards/federal-governments-strategies-and-plans/national-space-strategy-2030

United Launch Alliance. "Cis-Lunar 1000," *Youtube*, January 11, 2016. https://www.youtube.com/watch?v=uxftPmpt7aA

United Nations Development Program. "Human Development Index (HDI)," *UNDP*, October 12, 2019. http://hdr.undp.org/en/data

United Nations Office for Outer Space Affairs and China National Space Administration, Changsha, China. "United Nations/China Forum on Space Solutions," April 24–25, 2019. http://www.unoosa.org/oosa/en/ourwork/psa/schedule/2019/2019-un-china-forum-on-spacesolutions.html

United Nations Office for Outer Space Affairs. "United Nations and China Invite Applications to Conduct Experiments on-board China's Space Station," May 28, 2018. http://www.unoosa.org/oosa/en/informationfor/media/2018-unis-os-496.html

United Nations Office for Outer Space Affairs. *Capacity-Building in Space Science and Technology*. New York: United Nations, 2008. http://www.unoosa.org/pdf/publications/st_space_41E.pdf

United Nations Office for Outer Space Affairs. "Treaty on Principles Governing the Activities of States in the Exploration and Use of Outer Space, Including the Moon and Other Celestial Bodies," http://www.unoosa.org/oosa/en/ourwork/spacelaw/treaties/introouterspacetreaty.html

United Nations Office for Outer Space Affairs. "UNOOSA and UAE Space Agency Sign MoU for Increased Cooperation." November 08, 2017. http://www.unoosa.org/oosa/en/informationfor/media/2017-unis-os-487.html

United Nations Office of Disarmament Affairs. "Treaty on the Non-Proliferation of Nuclear Weapons (NPT)," https://www.un.org/disarmament/wmd/nuclear/npt/

Unnithan, Sandeep. "India Attains the Capability to Target, Destroy Space Satellites in Orbit," *India Today*, April 28, 2012. https://www.indiatoday.in/magazine/nation/story/20120507-agni-v-launch-india-takes-on-china-drdo-vijay-saraswat-758208-2012-04-28

Unnithan, Sandeep. "India has all the Building Blocks for an Anti-Satellite Capability," *India Today,* April 27, 2012. https://www.indiatoday.in/india/story/agni-v-drdo-chief-dr-vijay-kumar-saraswat-interview-100405-2012-04-27

Urban, Tim. "The Cook and the Chef: Musk's Secret Sauce," *Wait but Why*, November 6, 2015. https://waitbutwhy.com/2015/11/the-cook-and-the-chef-musks-secret-sauce.html

U.S. Army. *Project Horizon: Volume I: Summary And Supporting Considerations*, 1959. https://nsarchive2.gwu.edu/NSAEBB/NSAEBB479/docs/EBB-Moon01_sm.pdf

U.S. Congress Senate Committee on Aeronautical and Space Sciences Investigation of Governmental Organization for Space Activities Hearings 86th Congress 1st Session 1959.

U.S.-China Economic and Security Review Commission (USCC). "Statement of Dr. Namrata Goswami Independent Senior Analyst and Author 2016-2017 Minerva Grantee Before the U.S.-China Economic and Security Review Commission Hearing on 'China in Space: A Strategic Competition?'," April 25, 2019. https://www.uscc.gov/sites/default/files/Namrata%20Goswami%20USCC%2025%20April.pdf

U.S. Department of Defense. "Department of Defense Establishes U.S. Space Command," August 2019. https://www.defense.gov/Newsroom/Releases/Release/Article/1948288/department-of-defense-establishes-us-space-command/

U.S. National Commission on Space. *Pioneering the Space Frontier; The Report of the National Commission on Space*. New York: Bantam, 1986.

USSPACECOM. "United States Space Command Fact Sheet," August 29, 2019. https://www.spacecom.mil/About/Fact-Sheets-Editor/Article/1948216/united-states-space-command-fact-sheet/

Valente, Peter. "The Tiger and the Dragon: India as a Counterbalance to China in the Indo-Pacific," *Modern War Institute, West Point*, September 4, 2018. https://mwi.usma.edu/tiger-dragon-india-counterbalance-china-indo-pacific/

Valmiki. *Ramayana*. New Delhi: Penguin, 2010.

Vance, Ashlee. *Elon Musk: Tesla, SpaceX, and the Quest for a Fantastic Future*. New York: HarperCollins, 2017.

Vaz, Mark Cotta and John H. Hill. *Pan Am at War: How the Airline Secretly Helped America Fight World War II*. New York: Skyhorse, 2019.

Verma, S. S. "Space Solar Power," *Scientific India*, November 19, 2014. http://www.scind.org/41/Space/space-solar-power.html

Verstappen, Stefan H. S. *The Thirty-Six Strategies of Ancient China*. San Francisco: China Books and Periodicals, 1999.

Video. "Trump to Establish Space Force as Sixth Branch of Military," *CNBC*, June 18, 2018. https://www.cnbc.com/video/2018/06/18/trump-space-force-sixth-military-branch.html

Video. "Vice President Mike Pence Calls Space a War Fighting Domain," *The Washington Post*, October 23, 2018. https://www.washingtonpost.com/video/politics/vice-president-mike-pence-calls-space-a-war-fighting-domain/2018/10/23/2f7e5f06-d6d1-11e8-8384-bcc5492fef49_video.html

Video. "When Will China Get to Mars," BBC Interview with Wu Weiren, Chief Designer, Moon Missions, April 20, 2017. http://www.bbc.com/news/av/world-asia-36085659/when-will-china-get-to-mars

Video: "UAE's Hope Probe Bound for Mars is 85 per cent complete," *The Khaleej Times*, April 24, 2019. https://www.khaleejtimes.com/news/general/video-uaes-hope-probe-bound-for-mars-is-85-complete-

Vijay, Tarun. *Saffron Surge: India's Re-emergence on the Global Scene and Hindu Ethos*. New Delhi: Har Anand, 2008.

Vikaspedia. "Space Programmes of India," n.d. http://vikaspedia.in/education/childrens-corner/science-section/isro-created-history-by-launching-104-satellites-together;

Vyasa, Krishna Dwaipayana. *The Mahabharat,* Translated by Bibek Debroy. New Delhi: Penguin, 2010.

Walker, Robert S. and Peter Navarro. "Op-ed | Donald Trump's 'peace through strength' space vision," Space News, October 24, 2016. https://spacenews.com/op-ed-donald-trumps-peace-through-strength-space-doctrine/

Walker, Robert. "Asteroid Resources Could Create Space Habs For Trillions; Land Area of A Thousand Earths," *Science 2.0*, September 17, 2018. https://www.science20.com/robert_walker/asteroid_resources_could_create_space_habs_for_trillions_land_area_of_a_thousand_earths-116541

Wall, Mike. "About 17,000 Big Near-Earth Asteroids Remain Undetected: How NASA Could Spot Them," *Space.com*, April 10, 2018. https://www.space.com/40239-near-earth-asteroid-detection-space-telescope.html

Wall, Mike. "JFK's 'Moon Speech' Still Resonates 50 Years Later," *Space.com*, September 12, 2019, https://www.space.com/17547-jfk-moon-speech-50years-aniversary.html.

Wall, Mike. "New Space Mining Legislation Is 'History in the Making'," *Space.com*, November 20, 2015. https://www.space.com/31177-space-mining-commercial-spaceflight-congress.html

Wall, Mike. "Private Company Orbit Beyond Drops Out of 2020 NASA Moon-Landing Deal," *Space.Com*, July 30, 2019. https://www.space.com/nasa-drops-orbit-beyond-moon-landing-contract.html

Wall, Mike. "Trump Teases 'Space Force' Idea Again for US Military," *Space.com*, May 02, 2018. https://www.space.com/40471-trump-teases-military-space-force-again.html

Wallerstein, Immanuel. *The Modern World-System I: Capitalist Agriculture and the Origins of the European World-Economy in the Sixteenth Century*. New York: Academic Press, 1974.

Walt, Stephen M. "Alliance Formation and the Balance of World Power," *International Security* 9, no. 4 (Spring 1985): 3–34.

Walt, Stephen M. *The Origin of Alliances* (Cornell Studies in Security Affairs). Ithaca, NY: Cornell University Press, 1990.

Waltz, Kenneth N. "The Emerging Structure of International Politics," *International Security* 18, no. 2 (1993): 44–79.

Waltz, Kenneth. *Theory of International Politics*. New York: Random House, 1979.

Wan, Mengjia. "Behind the Official Narrative: China's Strategic Culture in Perspective," an interview with Christopher A. Ford," *National Bureau of Asian Research*, November 1, 2016. http://www.nbr.org/research/activity.aspx?id=718

Wang, Brian. "EIU GDP forecasts to 2050," *Next Big Future*, June 24, 2015. https://www.nextbigfuture.com/2015/06/eiu-gdp-forecasts-to-2050.html

Wang, Brian. "Exponential Industrialization of Space is More Important than Combat Lasers and Hypersonic Fighters," *Next Big Future*, October 26, 2017. https://www.nextbigfuture.com/2017/10/exponential-industrialization-of-space-is-more-important-than-combat-lasers-and-hypersonic-fighters.html

Wang, Kong. "Xinhua Insight: China's Long March into Space," *Xinhua*, April 23, 2016. http://news.xinhuanet.com/english/2016-04/23/c_135306099.htm

Wang, Zheng. *Never Forget National Humiliation: Historical Memory in Chinese Politics and Foreign Relations*. New York: Columbia University Press, 2014.

Wasserman, Stanley and Katherine Faust. *Social Network Analysis: Methods and Applications*. Vol. 8. Cambridge: Cambridge University Press, 1994.

Watkins, Derek. "What China has been Building in the South China Sea," *The New York Times*, February 29, 2016. https://www.nytimes.com/interactive/2015/07/30/world/asia/what-china-has-been-building-in-the-south-china-sea-2016.html

Wattles, Jackie. "SpaceX, One of the World's Most Valuable Private Companies, Just Got More Valuable," *CNN Business*, January 3, 2019, https://www.cnn.com/2019/01/03/business/spacex-valuation-500-million-fundraising-round/index.html

Webb, Michael C. and Steven D. Krasner. "Hegemonic Stability Theory: An Empirical Assessment," *Review of International Studies* 15, no. 2 Special Issue on the Balance of Power (April, 1989): 183–98.

Weber, Mark J., Shirli Kopelman, David M. Messick. "A Conceptual Review of Decision Making in Social Dilemmas: Applying a Logic of Appropriateness," *Personality and Social Psychology Review* 8, no. 3 (2004): 281–307.

Weber, Max. *Wirtschaft und Gesellschaft (Economy and Society)*, 2 vols. (Tuebingen, J. C. B. Mohr, 1925), *The Theory of Social and Economic Organization*. New York, Oxford University Press, 1947.

Wegener, Wolfgang. *The Naval Strategy of the World War*. Anapolis: US Naval Institute Press, 1989.

Weick, Karl E. "Cognitive Processes in Organizations," *Research in Organizational Behavior*, 1, 1979: 41–74.

Weiterin, Hanneke. "France Is Launching a 'Space Force' with Weaponized Satellites," *Space.com*, August 02, 2019. https://www.space.com/france-military-space-force.html

Weitering, Hanneke. "Barack Obama Declares U.S. on Track to Mars," *Scientific American*, October 11, 2016. https://www.scientificamerican.com/article/barack-obama-declares-u-s-on-track-to-mars/

Weitering, Hanneke. "How SpaceX's Starship will Help Establish a Mars base," *Space.com*, April 15, 2019. https://www.space.com/spacex-starship-mars-transportation-plans.html

Weitering, Hanneke. "NASA's New Human Spaceflight Chief Optimistic for Moon 2024 Goals," *Space.com*, December 04, 2019. https://www.space.com/nasa-human-spaceflight-chief-douglas-loverro-moon-2024.html

Wendt, Alexander. "Anarchy is What States Make of It: The Social Construction of Power Politics," *International Organization* 46, no. 2 (Spring 1992): 391-425.

Wenyu, Sun. "China Holds Leading Position in Research of Space Based Solar Power," *People's Daily Online*, November 2, 2017. http://en.people.cn/n3/2017/1102/c90000-9288036.html

Wezeman, Pieter D. and Alexandra Kuimova. "Military Spending and Arms Imports by Iran: Saudi Arabia, Qatar, and the UAE," *SIPRI Fact Sheet*, May 2019.

Whittaker, Richard. "'Cowboys in Space' Exhibit Finds the Frontier in the Stars: Exploring the infinite plains at the Bullock Museum," *Austin Chronicle*, October 11, 2019. https://www.austinchronicle.com/screens/2019-10-11/cowboys-in-space-exhibit-finds-the-frontier-in-the-stars/

Whittington, Mark R. "The Space Force has Gone From Joke to Reality," *The Hill*, December 16, 2018. https://thehill.com/opinion/technology/474713-the-space-force-has-gone-from-joke-to-reality

Wignaraja, Ganeshan. "India could Overtake China – One Day," *Nikkei Asian Review*, January 19, 2018. https://asia.nikkei.com/Economy/India-could-overtake-China-one-day

Williams, Matt. "How Long Does it Take to get to the Asteroid Belt?," *Universe Today*, August 10, 2016. https://www.universetoday.com/130231/long-take-get-asteroid-belt/

Wingo, Dennis. "Creating a Bold and Courageous 21st Century American Space Policy," International Institute of Space Commerce, August 2, 2017. https://iisc.im/wp-content/uploads/2017/08/Creating-a-Courageous-21st-Century-American-Space-Policy-_v4c.pdf

Wingo, Dennis. *MoonRush: Improving Life on Earth with the Moon's Resources*. Ontario, Canada: Apogee Books, 2004.

Winter, Laura. "NATO Declares Space an 'operational domain'," *Al Jazeera*, December 4, 2019. https://www.aljazeera.com/ajimpact/nato-declares-space-operational-domain-191204155031893.html

Wood, A. J. R Russel. *The Portuguese Empire, 1415-1808: A World on the Move*. Washington, DC: John Hopkins University Press, 1998.

Worden, Simon P. and John E. Shaw. *Whither Space Power? Forging a Strategy for the New Century*. Maxwell Air Force Base: Air University Press, 2002.

WorldAtlas. "What is Heartland Theory?," September 26, 2017. https://www.worldatlas.com/articles/what-is-the-heartland-theory.html

Wright, Colin. *Agents, Structures and International Relations: Politics as Ontology*. Cambridge: Cambridge University Press, 2006.

Xiangqing, Meng. "Chinese Dream Include Strong PLA," *China Daily*, October 10, 2013. http://www.chinadaily.com.cn/opinion/2013-10/08/content_17012886.htm

Xiangwei, Wang. "Xi Jinping Endorses the Promotion of Confucius," *South China Morning Post*, September 29, 2014. https://www.scmp.com/news/china/article/1603487/xi-jinping-endorses-promotion-confucius

Xiaohong, Wei and Li Qingyuan. "The Confucian Value of Harmony and its Influence on Chinese Social Interactions," *Cross-Cultural Communication* 9, no. 1 (2013): 60–66,. http://citeseerx.ist.psu.edu/viewdoc/download?doi=10.1.1.958.6384&rep=rep1&type=pdf

Xinhua. "Backgrounder: China's Journey to the Moon," *Xinhua*, December 8, 2012. http://www.xinhuanet.com/english/2018-12/08/c_137658274.htm

Xinhua. "Backgrounder: Xi Jinping's Vision for China's Space Development," April 24, 2017. http://www.xinhuanet.com/english/2017-04/24/c_136232642.htm

Xinhua. "China to Build Deep Space Lab in Luxemborug," *Xinhua*, January 16, 2018. http://www.xinhuanet.com/english/2018-01/16/c_136900477.htm

Xinhua. "China's Space Activities 2000," July 27, 2005. http://www.gov.cn/english/official/2005-07/27/content_17656.htm.

Xinhua. "Details of China's Long Range, Quantum Secured "Unhackable" Messaging Revealed," *Xinhua*, January 20, 2018. http://www.xinhuanet.com/english/2018-01/20/c_136909246.htm

Xinhua. "Full Text of Resolution on Amendment to CPC Constitution," *Xinhua*, October 24, 2017. https://www.chinadaily.com.cn/china/19thcpcnationalcongress/2017-10/24/content_33656521.htm

Xinhua. "Xi Focus: Xi Jinping and China's New Era," *Xinhua*, September 30, 2019, 2020. http://www.xinhuanet.com/english/2019-09/30/c_138437362.htm

Xinhua. "Xi Stresses Cultivating New type of Military Personnel," *Xinhuanet*, November 27, 2019. http://www.xinhuanet.com/english/2019-11/27/c_138587576.htm

Xinhua. China Focus: China Sets Record for Longest Stay in Self-Contained Moon Lab," *Xinhua*, May 15, 2018. http://www.xinhuanet.com/english/2018-05/15/c_137180976.htm

XOVIAN Website. n.d. http://www.xovian.co.in/

X-Prize Org. "Google Lunar X-Prize: The New Space Race," n.d. https://www.xprize.org/prizes/google-lunar

XPrize. Competing Teams, n.d., https://www.xprize.org/prizes/google-lunar/competing-teams

Yingqi, Cheng. "Embryos Growing in Space a "Giant Leap," *China Daily*, April 18, 2014. http://www.chinadaily.com.cn/china/2016-04/18/content_24616857.htm.

Youyi, Huang. "Context, not history, matters for Deng's famous phrase," *The Global Times*, June 15, 2011. http://www.globaltimes.cn/content/661734.shtml

Yu, Miles Maochun. "Understanding China's Strategic Culture through its South China Sea Gambit," *Hoover Institution*, May 9, 2016. https://www.hoover.org/research/understanding-chinas-strategic-culture-through-its-south-china-sea-gambit

Zapata, Edgar. "An Assessment of Cost Improvements in the NASA COTS/CRS Program and Implications for Future NASA Missions," *NASA*, 2017. https://ntrs.nasa.gov/archive/nasa/casi.ntrs.nasa.gov/20170008895.pdf

Zeno's Paradoxes, Internet Encyclopedia of Philosophy, n.d. https://www.iep.utm.edu/zeno-par/#SH1b

Zhaoyi, Pan. "China Starts to Build World's First Space Based Solar Power Plant," *CGTN*, February 18, 2019. https://news.cgtn.com/news/3d3d514f34597a4e32457a6333566d54/index.html.

Ziarnick, Brent and Peter Garretson. "Starfleet Was Closer than You Think," *The Space Review*, March 16, 2015. https://www.thespacereview.com/article/2714/1

Ziarnick, Brent, Peter Garretson, Everett Dolman and Coyote Smith. "Dear President Trump: Here's How to make Space Great Again," *The Space Review*, December 14, 2016. https://www.wired.com/2016/12/dear-president-trump-heres-make-space-great/

Ziarnick, Brent. "The Space Corp Needs Naval Rank," *Politico,* July 26, 2019. https://www.politico.com/story/2019/07/26/space-corps-naval-rank-1433541

Ziarnick, Brent. "The Space Force Strategic Vision Emerges," *The Hill*, September 09, 2019. https://thehill.com/opinion/national-security/460469-the-space-force-strategic-vision-emerges

Ziarnick, Brent. *Developing National Power in Space: A Theoretical Model.* Jefferson, NC: McFarland & Company, 2015.

Ziarnick, Brent. *Tommy's Space Force: General Thomas S. Power and the Air Force Space Program*, Drew Paper No. 34, Maxwell AFB, AL: Air University Press, April 2019.

Zubrin, Robert and Richard Wagner. *The Case for Mars: The Plan to Settle the Red Planet and Why We Must.* Rev. and updated. 1st Free Press trade pbk. Ed. New York: Free Press, 2011.

Zubrin, Robert. *The Case for Space How the Revolution in Spaceflight Opens up a Future of Limitless Possibilities.* New York: Prometheus Books, 2019.

Zuniga, Allison F., Daniel Rasky, Bruce Pittman, Edgar Zapata, Robert Lepsch. "Lunar COTS: An Economical and Sustainable Approach to Reaching Mars," *AIAA*, 2015, NASA. https://www.nasa.gov/sites/default/files/atoms/files/aiaa2015-4408zunigalunarcotspaper_0.pdf

Index

About the Authors

Namrata Goswami is an independent analyst and author on space policy, great power politics, and ethnic conflicts. After earning her PhD in international relations from Jawaharlal Nehru University, she served for nearly a decade as Research Fellow at the MP-Institute for Defence Studies and Analyses (IDSA), New Delhi, working on ethnic conflicts in India's Northeast and China-India border conflict. She has been a visiting fellow at the Peace Research Institute, Oslo, Norway; the La Trobe University, Melbourne, Australia; and the University of Heidelberg, Germany. In 2012, she served as Jennings-Randolph Senior Fellow at the United States Institute of Peace (USIP), Washington, D.C., and was awarded a Fulbright-Nehru Senior Fellowship that same year. Shortly after relocating to the United States, she established her own strategy and policy consultancy in 2016 and won the prestigious MINERVA grant awarded by the Office of the U.S. Secretary of Defense (OSD) to study great power competition in the grey zone of outer space. In 2017, she was awarded a contract with Joint Special Forces University (JSOU) to write a monograph on ISIS in Asia. In April 2019, Dr. Goswami testified before the U.S-China Economic and Security Review Commission on China's space program. Dr. Goswami is the first representative from South Asia chosen to participate in the George C. Marshall European Center for Security Studies NATO Partnership for Peace Consortium (PfPC) "Emerging Security Challenges Working Group." She received the Executive Leadership Certificate sponsored by the Harvard Kennedy School of Government, National Defense University (NDU), and the Asia Pacific Center for Security Studies (APCSS). Her latest book *The Naga Ethnic Movement for a Separate Homeland Stories from the Field* was published by Oxford University Press in March 2020. Goswami's work on space has been published in *The Washington Post, The Diplomat, The Space Review, Gulf News, Asia Policy,* and *The Texas National Security Review.*

Peter A. Garretson is an independent strategy consultant who focuses on space and defense, and a Senior Fellow in Defense Studies and Co-Director of the Space Policy Initiative with the American Foreign Policy Council. He was previously the Director of Air University's Space Horizons Task Force, America's think tank for space, and was Deputy Director of America's premier space strategy program, the Schriever Scholars. Garretson's work on space has been published widely including in *The Hill, Newsweek, The National Interest, The Space Review, War on the Rocks,* and *Politico.*

www.ingramcontent.com/pod-product-compliance
Lightning Source LLC
Chambersburg PA
CBHW050355210425
25406CB00004B/64
9781498583114